Library of
Davidson College

 **Benchmark Papers in Genetics**

Series Editor: David L. Jameson
University of Houston

Published Volumes and Volumes in Preparation

GENETICS AND SOCIAL STRUCTURE / Paul Ballonoff
GENES AND PROTEINS / R. P. Wagner
EUGENICS / Carl Jay Bajema
POPULATION GENETICS / James F. Crow and Carter Denniston
MEDICAL GENETICS / William Jack Schull
ECOLOGICAL GENETICS / W. W. Anderson
QUANTITATIVE GENETICS / R. E. Comstock
GENETIC RECOMBINATION / Rollin Hotchkiss
REGULATION GENETICS / Werner K. Maas
ANIMAL BREEDING / Robert C. Carter
PLANT BREEDING / D. F. Matzinger
DEVELOPMENTAL GENETICS / Antonie W. Blackler and Richard Hallberg
CYTOGENETICS / Ronald L. Phillips and Charles H. Burnham

Benchmark Papers
in Genetics / 2

───── A *BENCHMARK* ® Books Series ─────

GENES AND PROTEINS

Edited by
ROBERT P. WAGNER
The University of Texas at Austin

Dowden, Hutchinson & Ross, Inc.
Stroudsburg, Pennsylvania

Distributed by
HALSTED PRESS *A Division of John Wiley & Sons, Inc.*

Copyright © 1975 by **Dowden, Hutchinson & Ross, Inc.**
Benchmark Papers in Genetics, Volume 2
Library of Congress Catalog Card Number: 75-8851
ISBN: 0-470-91357-6

All rights reserved. No part of this book covered by the copyrights hereon may be reproduced or transmitted in any form or by any means—graphic, electronic, or mechanical, including photocopying, recording, taping, or information storage and retrieval systems—without written permission of the publisher.

77 76 75   1 2 3 4 5

Manufactured in the United States of America.

575.1
G327

LIBRARY OF CONGRESS CATALOGING IN PUBLICATION DATA
Main entry under title:

Genes and proteins.

    (Benchmark papers in genetics ; v. 2)
    Includes indexes.
    1. Biochemical genetics--Addresses, essays, lectures
2. Proteins--Addresses, essays, lectures. I. Wagner, Robert P. [DNLM: 1. Genes. 2. Proteins--Metabolism. W1 BE516 v. 2 / QH447 G327]
QH447.G45   575.2'1   75-8851
ISBN 0-470-91357-6

Exclusive Distributor: **Halsted Press**
A Division of John Wiley & Sons, Inc.

Acknowledgments and Permissions

ACKNOWLEDGMENTS

AMERICAN ASSOCIATION FOR THE ADVANCEMENT OF SCIENCE—*Science*
    Immunogenetic Consequences of Vascular Anastomoses Between Bovine Twins
GENETICS SOCIETY OF AMERICA—*Genetics*
    The Anatomy and Function of a Segment of the X Chromosome of *Drosophila melanogaster*
NATIONAL ACADEMY OF SCIENCES—*Proceedings of the National Academy of Sciences*
    Allelic Strains of *Neurospora* Lacking Tryptophan Synthetase: A Preliminary Immunochemical Characterization
    An Alteration in the Primary Structure of a Protein Predicted on the Basis of Genetic Recombination Data
    Atropinesterase, A Genetically Determined Enzyme in the Rabbit
    Correspondence Between Genetic Data and the Position of Amino Acid Alteration in a Protein
    Enzyme Complementation *in Vitro* Between Adenylosuccinaseless Mutants of *Neurospora crassa*
    On the Evolution of Biochemical Syntheses
    The Fusion of Two Peptide Chains in Hemoglobin Lepore and Its Interpretation as a Genetic Deletion
    Galactosemia, A Congenital Defect in a Nucleotide Transferase: A Preliminary Report
    A Gene Cluster in *Neurospora crassa* Coding for an Aggregate of Five Aromatic Synthetic Enzymes
    The Genetic Control of Adenylosuccinase in *Neurospora crassa*
    Genetic Control of Biochemical Reactions in *Neurospora*
    Genetic Control of the α- and β-Chains of Hemoglobin
    Hydroxyanthranilic Acid as a Precursor of Nicotinic Acid in *Neurospora*
    On a Mechanism of Suppressor Gene Regulation of Tryptophan Synthetase Activity in *Neurospora crassa*
    Suppression of Mutations in the Alkaline Phosphatase Structural Cistron of *E. coli*

PERMISSIONS

The following papers have been reprinted with the permission of the authors and copyright holders.
AMERICAN ASSOCIATION FOR THE ADVANCEMENT OF SCIENCE—*Science*
    Suppression *in Vitro*: Identification of a Serine-sRNA as a "Nonsense" Suppressor
AMERICAN SOCIETY OF BIOLOGICAL CHEMISTS, INC.—*The Journal of Biological Chemistry*
    A Genetic and Biochemical Analysis of Second Site Reversion
THE BIOLOGICAL LABORATORY, LONG ISLAND BIOLOGICAL ASSOCIATION, INC.—*Cold Spring Harbor Symposia on Quantitative Biology*
    Complementation Between Alleles in Heterocaryons
    The Effects of Mutation on the Composition and Properties of the A Protein of *Escherichia coli* Tryptophan Synthetase
    Some Recent Studies Bearing on the One Gene–One Enzyme Hypothesis
CAMBRIDGE UNIVERSITY PRESS—*Biological Reviews*
    The Genetics and Chemistry of Flower Colour Variation
COLD SPRING HARBOR LABORATORY—*Cold Spring Harbor Symposia on Quantitative Biology*
    The Mechanism of Complementation Between *am* Mutants of *Neurospora crassa*

FEDERATION OF AMERICAN SOCIETIES FOR EXPERIMENTAL BIOLOGY—*Federation Proceedings*
    Genetic Fine Structure and Enzyme Formation
MACMILLAN JOURNALS LTD—*Nature*
    Chromosomal Rearrangements and the Evolution of Haptoglobin Genes
    Gene Evolution and the Haemoglobins
    Gene Mutations in Human Haemoglobin: The Chemical Difference Between Normal and Sickle Cell Haemoglobin
PLENUM PUBLISHING CORP.—*Biochemical Genetics*
    Regulatory Interactions Involving Two Hemoglobin Loci of *Chironomus*
SPRINGER-VERLAG, BERLIN—*Die Naturwissenschaften*
    Kynurenin als Augenpigmentbildung auslösendes Agens bei Insekten
THE STONY BROOK FOUNDATION FOR THE QUARTERLY REVIEW OF BIOLOGY—*The Quarterly Review of Biology*
    Chemistry of the "Eye Color Hormones" of Drosophila
UNIVERSITY OF CHICAGO PRESS—*The American Naturalist*
    A Further Analysis of the Pantothenicless Mutants of *Neurospora*
THE WISTAR INSTITUTE PRESS—*The Journal of Experimental Zoölogy*
    Immunogenetic Studies of Species and of Species Hybrids in Doves, and the Separation of Species-Specific Substances in the Backcross

Series Editor's Preface

The study of any discipline assumes mastery of the literature of the subject. In many branches of science, even one as new as genetics, the expansion of knowledge has been so rapid that there is little hope of learning of the development of all phases of the subject. The student has difficulty mastering the textbook, the young scholar must tend to the literature near his own research, the young instructor barely finds time to expand his horizons to meet his class preparation requirements, the monographer copes with a wider literature but usually from a specialized viewpoint, and the textbook author is forced to cover much the same materials as previous and competing texts to respond to the user's needs and abilities.

Few publishers have the dedication to scholarship to serve primarily the limited market of advanced studies. The opportunity to assist professionals at all stages of their careers has been recognized by the publishers and by a distinguished group of editors knowledgeable in specific aspects of the literature of genetics. Some have contributed heavily to the development of that literature, some have studied with the early scholars, and some have developed and are in the process of developing entirely new fields of genetic knowledge. In many cases, the judgments of the editors become a historical document recording their opinion of the important steps in the development of the subject. These editors have selected papers and portions of papers that demonstrate both the development of knowledge and the atmosphere in which that knowledge was developed. There is no substitute for reading great papers. Here you can learn how questions are asked, how they are approached, and how difficult and essential it is to obtain definitive answers and clear writing. My own pleasure in working with this distinguished panel is exceeded only by the considerable pleasure of reading their remarks and their selections. Their dedication and wisdom are impressive.

This volume, *Genes and Proteins*, presents a selection of literature constituting "benchmark" papers in one of the most exciting fields of modern science. Few subjects have excited the scholar as much as the development of an understanding of the relation between genetics and metabolism. Robert P. Wagner, the volume editor, was an author of a text that constituted a "benchmark" and has been a major contributor to the literature in this field. His selections are instructive. Perusal of the material renews the excitement of the times and clearly demonstrates the current problems and challenging new directions.

David L. Jameson

Contents

Acknowledgments and Permissions	v
Series Editor's Preface	vii
Contents by Author	xv
Introduction	1

I. THE FIRST GLIMMERS

Editor's Comments on Paper 1	12
1 Cuénot, L.: L'hérédité de la pigmentation chez les souris *Arch. Zool. Exp. Gén.*, 4e sér., **1**(3): *Notes Rev.*, 33–41 (1903) *English translation:* The Heredity of Pigmentation in Mice	13

II. PROGRESS IN THINKING TO 1930

Editor's Comments on Paper 2	28
2 Muller, H. J.: Further Studies on the Nature and Causes of Gene Mutations *Proc. Sixth Intern. Congr. Genetics*, **1**, 231–255 (1932)	29

III. GENES AND ANTIGENS

Editor's Comments on Papers 3 and 4	56
3 IRWIN, M. R., and L. J. COLE: Immunogenetic Studies of Species and of Species Hybrids in Doves, and the Separation of Species-Specific Substances in the Backcross *J. Exp. Zoöl.*, **73**(1), 85–108 (1936)	58
4 OWEN, R. D.: Immunogenetic Consequences of Vascular Anastomoses Between Bovine Twins *Science*, **102**(2651), 400–401 (1945)	82

IV. GENES AND CHEMICAL REACTIONS

Editor's Comments on Papers 5 and 6	86

5	**WRIGHT, S.:** A Quantitative Study of the Interactions of the Major Colour Factors of the Guinea-Pig *Proc. 7th Intern. Congr. Genetics,* 319–329 (1939)	88
6	**LAWRENCE, W. J. C., and J. R. PRICE:** The Genetics and Chemistry of Flower Colour Variation *Biol. Rev. Cambridge Phil. Soc.,* **15**(2), 35–58 (1940)	99

V. THE RECOGNITION OF GENETIC BLOCKS

	Editor's Comments on Papers 7 and 8	124
7	**BUTENANDT, A., W. WEIDEL, and E. BECKER:** Kynurenin als Augenpigmentbildung auslösendes Agens bei Insekten *Naturwiss.,* **28**(4), 63–64 (1940) *English translation:* Kynurenine as an Agent Which Causes the Formation of Eye Pigment in Insects	125
8	**EPHRUSSI, B.:** Chemistry of the "Eye Color Hormones" of *Drosophila* *Quart. Rev. Biol.,* **17**(4), 327–338 (1942)	128

VI. ONE GENE–ONE ENZYME

	Editor's Comments on Papers 9 Through 15	142
9	**BEADLE, G. W., and E. L. TATUM:** Genetic Control of Biochemical Reactions in *Neurospora* *Proc. Natl. Acad. Sci.,* **27**(11), 499–506 (1941)	144
10	**SAWIN, P. B., and D. GLICK:** Atropinesterase, A Genetically Determined Enzyme in the Rabbit *Proc. Natl. Acad. Sci.,* **29**(1), 55–59 (1943)	152
11	**HOROWITZ, N. H.:** On the Evolution of Biochemical Syntheses *Proc. Natl. Acad. Sci.,* **31**(6), 153–157 (1945)	157
12	**MITCHELL, H. K., and J. F. NYC:** Hydroxyanthranilic Acid as a Precursor of Nicotinic Acid in *Neurospora* *Proc. Natl. Acad. Sci.,* **34**(1), 1–5 (1948)	162
13	**HOROWITZ, N. H., and U. LEUPOLD:** Some Recent Studies Bearing on the One Gene–One Enzyme Hypothesis *Cold Spring Harbor Symp. Quant. Biol.,* **16**, 65–74 (1951)	167
14	**WAGNER, R. P., and C. H. HADDOX:** A Further Analysis of the Pantothenicless Mutants of *Neurospora* *Amer. Naturalist,* **85**(824), 319–330 (1951)	177
15	**KALCKAR, H. M., E. P. ANDERSON, and K. J. ISSELBACHER:** Galactosemia, A Congenital Defect in a Nucleotide Transferase: A Preliminary Report *Proc. Natl. Acad. Sci.,* **42**(2), 49–51 (1956)	189

VII. MUTANT GENES AND MUTANT PROTEINS

Editor's Comments on Papers 16 Through 18 — 194

16 SUSKIND, S. R., C. YANOFSKY, and D. M. BONNER: Allelic Strains of *Neurospora* Lacking Tryptophan Synthetase: A Preliminary Immunochemical Characterization — 196
Proc. Natl. Acad. Sci., **41**(8), 577–582 (1955)

17 BONNER, D. M., Y. SUYAMA, and J. A. DeMOSS: Genetic Fine Structure and Enzyme Formation — 202
Fed. Proc., **19**(4), 926–930 (1960)

18 INGRAM, V. M.: Gene Mutations in Human Haemoglobin: The Chemical Difference Between Normal and Sickle Cell Haemoglobin — 207
Nature, **180**(4581), 326–328 (1957)

VIII. COMPLEMENTATION AND THE NATURE OF PROTEINS

Editor's Comments on Papers 19 Through 22 — 212

19 GILES, N. H., C. W. H. PARTRIDGE, and N. J. NELSON: The Genetic Control of Adenylosuccinase in *Neurospora crassa* — 214
Proc. Natl. Acad. Sci., **43**(4), 305–317 (1957)

20 CATCHESIDE, D. G., and A. OVERTON: Complementation Between Alleles in Heterocaryons — 227
Cold Spring Harbor Symp. Quant. Biol., **23**, 137–140 (1958)

21 WOODWARD, D. O.: Enzyme Complementation *in Vitro* Between Adenylosuccinaseless Mutants of *Neurospora crassa* — 231
Proc. Natl. Acad. Sci., **45**(6), 846–850(1959)

22 FINCHAM, J. R. S., and A. CODDINGTON: The Mechanism of Complementation Between *am* Mutants of *Neurospora crassa* — 236
Cold Spring Harbor Symp. Quant. Biol., **28**, 517–527 (1963)

IX. ONE GENE–ONE POLYPEPTIDE

Editor's Comments on Paper 23 — 248

23 ITANO, H. A., and E. A. ROBINSON: Genetic Control of the α- and β-Chains of Hemoglobin — 249
Proc. Natl. Acad. Sci., **46**(11), 1492–1501 (1960)

X. ORIGIN OF NEW GENES AND PROTEINS

Editor's Comments on Papers 24 Through 26 — 260

24 INGRAM, V. M.: Gene Evolution and the Haemoglobins — 261
Nature, **189**(4766), 704–708 (1961)

25	SMITHIES, O., G. E. CONNELL, and G. H. DIXON: Chromosomal Rearrangements and the Evolution of Haptoglobin Genes *Nature,* **196**(4851), 232–236 (1962)	266
26	BAGLIONI, C.: The Fusion of Two Peptide Chains in Hemoglobin Lepore and Its Interpretation as a Genetic Deletion *Proc. Natl. Acad. Sci.,* **48**(11), 1880–1886 (1962)	271

XI. GENE ORGANIZATION AND PROTEIN ORGANIZATION

Editor's Comments on Papers 27 Through 29		280
27	YANOFSKY, C., D. R. HELINSKI, and B. D. MALING: The Effects of Mutation on the Composition and Properties of the A Protein of *Escherichia coli* Tryptophan Synthetase *Cold Spring Harbor Symp. Quant. Biol.,* **26**, 11–23 (1961)	282
28	HELINSKI, D. R., and C. YANOFSKY: Correspondence Between Genetic Data and the Position of Amino Acid Alteration in a Protein *Proc. Natl. Acad. Sci.,* **48**(2), 173–183 (1962)	295
29	HENNING, U., and C. YANOFSKY: An Alteration in the Primary Structure of a Protein Predicted on the Basis of Genetic Recombination Data *Proc. Natl. Acad. Sci.,* **48**(2), 183–190 (1962)	306

XII. SUPPRESSORS AND PROTEINS

Editor's Comments on Papers 30 Through 33		316
30	SUSKIND, S. R., and L. I. KUREK: On a Mechanism of Suppressor Gene Regulation of Tryptophan Synthetase Activity in *Neurospora crassa* *Proc. Natl. Acad. Sci.,* **45**(1), 193–196 (1959)	318
31	GAREN, A., and O. SIDDIQI: Suppression of Mutations in the Alkaline Phosphatase Structural Cistron of *E. coli* *Proc. Natl. Acad. Sci.,* **48**(7), 1121–1127 (1962)	322
32	CAPECCHI, M. R., and G. N. GUSSIN: Suppression *in Vitro*: Identification of a Serine-sRNA as a "Nonsense" Suppressor *Science,* **149**(3682), 417–422 (1965)	329
33	HELINSKI, D. R., and C. YANOFSKY: A Genetic and Biochemical Analysis of Second Site Reversion *J. Biol. Chem.,* **238**(3), 1043–1048 (1963)	335

XIII. SUPRAMOLECULAR ORGANIZATION

Editor's Comments on Paper 34   342

34   GILES, N. H., M. E. CASE, C. W. H. PARTRIDGE, and S. I. AHMED:
A Gene Cluster in *Neurospora crassa* Coding for an Aggregate
of Five Aromatic Synthetic Enzymes   343
Proc. Natl. Acad. Sci., **58**(4), 1453–1460 (1967)

XIV. ONE GENE–ONE CHROMOMERE

Editor's Comments on Paper 35   352

35   JUDD, B. H., M. W., SHEN, and T. C. KAUFMAN: The Anatomy
and Function of a Segment of the X Chromosome of
Drosophila melanogaster   353
Genetics, **71,** 139–156 (May 1972)

XV. REGULATION OF PROTEIN SYNTHESIS IN EUKARYOTES

Editor's Comments on Paper 36   372

36   THOMPSON, P., and M. J. HORNING: Regulatory Interactions
Involving Two Hemoglobin Loci of *Chironomus*   373
Biochem. Genetics, **8**(3), 309–319 (1973)

Author Citation Index   385
Subject Index   395

Contents by Author

Ahmed, S. I., 343
Anderson, E. P., 189
Baglioni, C., 271
Beadle, G. W., 144
Becker, E., 125
Bonner, D. M., 196, 202
Butenandt, A., 125
Capecchi, M. R., 329
Case, M. E., 343
Catcheside, D. G., 227
Coddington, A., 236
Cole, L. J., 58
Connell, G. E., 266
Cuénot, L., 13
DeMoss, J. A., 202
Dixon, G. H., 266
Ephrussi, B., 128
Fincham, J. R. S., 236
Garen, A., 322
Giles, N. H., 214, 343
Glick, D., 152
Gussin, G. N., 329
Haddox, C. H., 177
Helinski, D. R., 282, 295, 325
Henning, U., 306
Horning, M. J., 373
Horowitz, N. H., 157, 167
Ingram, V. M., 207, 261
Irwin, M. R., 58
Isselbacher, K. J., 189

Itano, H. A., 249
Judd, B. H., 353
Kalckar, H. M., 189
Kaufman, T. C., 353
Kurek, L. I., 318
Lawrence, W. J. C., 99
Leupold, U., 167
Maling, B. D., 282
Mitchell, H. K., 162
Muller, H. J., 29
Nelson, N. J., 214
Nyc, J. F., 162
Overton, A., 227
Owen, R. D., 82
Partridge, C. W. H., 214, 343
Price, J. R., 99
Robinson, E. A., 249
Sawin, P. B., 152
Shen, M. W., 353
Siddiqi, O., 322
Smithies, O., 266
Suskind, S. R., 196, 318
Suyama, Y., 202
Tatum, E. L., 144
Thompson, P., 373
Wagner, R. P., 177
Weidel, W., 125
Woodward, D. O., 231
Wright, S., 88
Yanofsky, C., 196, 282, 295, 306, 325

Introduction

This collection of papers deals with one of the two most fundamental questions in biology. What does the genetic material do? We now have an answer to that question, or, to put it more accurately, we have a partial answer well supported by a great amount of direct experimental evidence. Starting in the 1950s, it became certain that the genetic material is DNA, except in some viruses, and that DNA determines the amino acid sequence and, therefore, the specificity of proteins. In this process, which may be called gene action, RNAs of various types act as intermediaries. There are, of course, other activities of DNA and RNA that we do not understand, among them regulation in the eukaryotes. However, the gene–protein relationship is certainly descriptive of the cardinal gene function. Other functions are subordinate.

The selection of papers for this volume was necessarily under stringent constraints. Of the many thousands of papers published on this topic since 1902, perhaps several hundred are worthy of being called classics in the general area of gene action. Thirty-six are reproduced here. I have been extremely eclectic and confined myself for the most part to papers involving eukaryotes. This may cause some molecular biologists to wonder, but the problem of gene action was fairly well delineated long before phage and bacteria were first used in genetic research. There was a molecular biology of sorts in the latter part of the last century, and the outlines of the general problem of gene action were in fact reasonably clearly stated at that time. That is, there is a hereditary material in the cell which has a physical basis, and this material determines something which controls the things that go on in the cell.

The idea that the genetic material may have control over the synthesis of proteins in the cell has its roots in the nineteenth century. The activity at that time in cell biology, principally in continental Europe, was intense, and many of the ideas developed from about 1850 to 1900 form the principal substance of modern biological thought. Indeed it is not an exaggeration to state that very few really new and original ideas have been generated in biology since this period in the last century.

To such men as Claude Bernard (1813–1878) and Albrecht Kossel (1853–1927) we owe the first published speculations about the role of the nucleus versus that of the cytoplasm in the functioning of cells. In the period around 1875 to 1890 they advanced the idea that the nucleus of the cell controls the formation of organic matter in the cell, and Kossel (1891) proposed that the nuclein (nucleoprotein) of the nucleus plays a leading role in this control. From this it was but a short step to the conclusion of E. B. Wilson (1856–1939) made in the last days of the nineteenth century. In the first edition of his *The Cell in Development and Inheritance,* published in 1896, we find the following prescient statement:

> In its physiological aspect, therefore, inheritance is the recurrence, in successive generations, of like forms of metabolism and this is effected through the transmission from generation to generation of a specific substance or idioplasm which we have seen reason to identify with chromatin (pp. 326–327).

The term idioplasm was coined by Karl von Nägeli (1817–1891). Idioplasm was hypothesized by him in 1884 to be an extremely complex substance that formed the physical basis of heredity. Later observers, chief among them being August Weismann (1834–1914), as well as Wilson, concluded that the idioplasm of Nägeli is contained in the nucleus of the cell and is in fact the chromatin of the chromosomes.

By the end of the nineteenth century, biologists looked upon the cytoplasm as "*itself a product of nuclear activity;* and it is just here that the general *role* of the nucleus in metabolism is of such vital importance to the theory of the inheritance" (Wilson, 1886). They recognized that there had to be a hereditary substance with a physical existence which in some way controlled and guided the metabolism of the cell. But how?

There was no end of speculation. Yves Delage (1854–1920) stated one of the more interesting hypotheses in 1895:

> The only way of explaining this control of the nucleus over the cytoplasm is to grant that the nucleus sends physical particles into the cytoplasm which shape it and give each cell its specific kind of activity; this is the intracellular migration of pangenes.[1]

The existence of enzymes was recognized by the 1890s (they were then called "diastases" or "ferments"), but their chemical nature was not understood. Indeed it was not until 1926 that, as a result of J. B. Sumner's (1887–1955) crystallization of urease, enzymes were recognized to be proteins. But some biologists were thinking about enzymes in connection with inheritance as early as 1890. For example, Hans Driesch (1867–1941) suggested that the nuclear idioplasm was a storehouse of ferments.

At the close of the nineteenth century the connection between the genetic material and proteins had not been made, but the stage was set for the work of the next century, which opened in 1900 with the rediscovery of Mendelism. The sudden realization of the significance of Mendel's 1865 paper, initially by Hugo De Vries (1848–1935), Karl

---

[1] A rough translation of the French text. This idea is really from Hugo De Vries who in 1889 developed a theory of inheritance in his book *Intracelluläre pangenesis.*

Correns (1864–1933), and Erich von Tschermak (1871–1962),[2] led to a great deal of activity almost instantaneously. The Mendelian laws were verified again and again by such pioneers as William Bateson (1861–1926)[3] and Lucien Cuénot (1866–1951) among a number of others. It was also quickly recognized that the Mendelian ratios could be explained by the meiotic process. W. L. Sutton (1877–1916) described in 1903 how the random orientation of homologous chromosomes on the meiotic spindle could account for the independent segregation of allelic pairs of genes. This union of Mendelian genetics and cytology was the beginning of genetics as we know it today.

The work of the early geneticists was primarily directed toward a better understanding of the mechanics of inheritance or what we now call transmission genetics. T. H. Morgan (1866–1938) with his students A. H. Sturtevant (1891–1970) and C. B. Bridges (1889–1938) put the genes on the chromosomes and worked out the basic aspects of linkage and crossing over. Little attention was paid to the question of gene action, primarily because there were many other problems to solve with accessible tools.

However, three men stand out in the first decade of this century as the forerunners of biochemical genetics. Cuénot, A. E. Garrod (1857–1936), and Bateson early recognized that the hereditary material must have a role in guiding the chemical transformations that occur in organisms.

As far as I have been able to determine, Garrod was the first observer to recognize that certain inherited human conditions have a chemical basis. In 1902 he pointed out that alkaptonuria in man appears to be inherited in the Mendelian fashion, and that the condition is the result of an "alternative mode of metabolism." He also suggested that two other inherited abnormalities, albinism and cystinuria, have a chemical basis. Nowhere in this paper, however, does he suggest a connection between these alterations and enzymes. According to Sturtevant (1965), Garrod consulted with Bateson about the time this 1902 paper was written, and Bateson suggested the probability that the failure of the body to carry out certain metabolic reactions (as would be the case in the three abnormal conditions mentioned) was due to the absence of certain ferments (enzymes). I have not been able to find Bateson's printed discussion of this, although Sturtevant apparently did. The idea is entirely in keeping with Bateson's known views about the nature of recessive mutant alleles; that is, their effect is due to the absence of something. This and related topics were treated at length by Bateson in 1909 in his book *Mendel's Principles of Heredity*, in which it is clear that he believes that certain inherited conditions in both plants and animals are due to the absence of ferments.

In searching through the literature of the first decade of this century, I came upon Cuénot's (1903) paper, which is the first in this collection. It will be obvious to the reader that Cuénot was certainly among the first to recognize the connection between the hereditary material and enzymes. The question of who actually was the first to make this connection is too academic to pursue further. I think that the trio of Cuénot, Garrod, and Bateson should be recognized as the founders of the gene → enzyme hypothesis.

---

[2] Although von Tschermak is usually regarded as part of this triumvirate, it is not clear that he really understood Mendel until after he read De Vries's and Correns's papers in 1900.
[3] See, for example, Reports 1–5 to the Evolution Committee of the Royal Society (1902–1909).

Introduction

After this initial foray into biochemistry little of note was done in this area for the next 30 years with the exception of the flower pigment work by a group of workers in England, which is described in Paper 6 of this series, and the work of Sewall Wright with coat color in the guinea pig.

In 1917 Wright discussed the genetic and biochemical data that he and others had collected to explain the inherited differences in coat color in mammals. It was recognized that coat colors were derived from various forms of melanin, which in turn were derived from tyrosine by the activity of oxidases in the hair follicles. Wright stated in this paper that his and the investigations of others indicated

> first that melanin is produced by the oxidation of certain products of protein metabolism by the action of specific enzymes; second, that this reaction takes place in the cytoplasm of cells probably by enzymes secreted by the nucleus; third, that various chromogens are used, the particular ones oxidized depending on the character of the enzyme present; and finally that hereditary differences in color are due to heredity differences in the enzyme element of the reaction.

This statement clearly shows that Wright was relating genes and enzymes. His later work and analyses culminated in an attempt to explain the effects of various different alleles of coat color genes on the phenotype with enzyme kinetics. This analysis is reported in Paper 5 published in 1939. Little of note occurred in the 1920s, a low point in interest in gene function and structure. The general malaise is well documented by the contents of the 3rd edition of E. B. Wilson's book *The Cell in Development and Heredity*, published in 1925. Although many advances in cytology and embryology were reported as having occurred since 1900, Wilson deleted much of the speculation he made in the 1896 and 1900 editions regarding the relation of chromatin and its nuclein to the metabolic processes. This was due in part to the fact that during this period biologists were coming to the conclusion that the nucleic acids could not be the genetic material, but that the important hereditary component of the chromosomes was protein. This point of view led to an interesting paper by Max Delbrück (1906–), one of the prime figures in the founding of molecular genetics. In 1941 Delbrück speculated on the possibility of the autocatolytic synthesis of protein to explain chromosome replication, and presented a chemical mechanism to explain how this might occur. That he was barking up the wrong tree became evident about 10 years later.

Other important developments did occur from 1920 to 1940, however, which were necessary for further advances for gene action studies. First, a great deal more was learned about enzymes and proteins, and it was established that enzymes were proteins. Second, the structure of nucleic acids as polynucleotides containing the bases, adenine, thymine, cytosine, and guanine in DNA, and adenine, uracil, cytosine, and guanine in RNA.

Also, starting about 1930 there was much speculation about allelic genes and their phenotypic effects, as illustrated by Muller in 1932 (Paper 2). In addition, in this period immunogenetics got its start, first mainly through the efforts of Karl Landsteiner (1868–1943) and others working with blood groups in man. This development was enhanced considerably by its application to animal forms that could be bred in the laboratory, as exemplified by the studies of M. R. Irwin (1897–   ) and Leon Cole (1877–1948) on dove hybrids (Paper 3). Their findings have since been expanded manyfold and constitute an important aspect of gene–protein relations as studied today.

The plant pigment studies reviewed in Paper 6 indicated that there was a direct relation between specific genes and specific biochemical reactions. That is, there seemed to be a one gene–one reaction relation. However, plant material, valuable as it was, had serious experimental shortcomings. One could not manipulate the mutant "phenotypes" by adding a substance from the outside and cause the plant to produce a pigment it normally did not produce.

Experiments were begun in the 1930s, however, which involved animals that could be manipulated. The meal moth, *Ephestia kühniella*, and the fruit fly, *Drosophila melanogaster*, were used to analyze the formation of eye pigments. In these two insect species a brown pigment is normally formed in the eyes and testes, but mutants occur which have a block that prevents its formation. Ernst Caspari (1909–   ) showed in 1933 that a mutant of Ephestia (a) with pale eyes and testes developed darker eyes and testes when testes from a normal larva (a^+) were implanted into the mutant larva. The reciprocal implantation of an a testis into an a^+ larva resulted in an expected darkening of the mutant testis. This experiment demonstrated that a diffusible substance went from the a^+ to the a tissue and allowed the formation of the brown pigment. The substance was originally thought to be a hormone, but in 1940 was shown by Adolph Butenandt (1903–   ) and his coworkers to be kynurenine (Paper 7). Kynurenine is now known to be an intermediate in the synthesis of the brown pigment from tryptophan.

In 1935 George Beadle (1903–   ) and Boris Ephrussi (1901–   ) used the transplantation technique to implant eye imaginal discs from mutant larvae of *Drosophila melanogaster* into normal larvae, and vice versa, and noted the effects of the host tissue on the implant as well as the implant on the host. The results of this very important work are summarized in Paper 8. One consequence of Beadle's and Ephrussi's results was the discovery of two diffusible precursors in the synthesis of the brown eye pigment of *D. melanogaster*, the v^+ substance (shown to be kynurenine) and the cn^+ substance (now known to be hydroxykynurenine). Their transplantation experiments demonstrated conclusively that the v^+ substance was a precursor to the cn^+ substance. Hence they were on the way to demonstrating a metabolic pathway by using mutants blocked at different steps in the pathway. This finding set the stage for the decade starting in 1940. Biochemical genetics had been reborn, and this time it was not to submerge as it did in the early 1900s.

Despite findings of this nature, the concepts of genes and gene action had not changed radically between 1917 and 1940, as evidenced by this statement made by J. B. S. Haldane (1892–1964) in 1942:

> It is not unreasonable to expect that enzymes will be found among the immediate products of gene action. A still further speculation is that the process by which genes produce their immediate products is the same as that by which they reproduce themselves, and that the antigen produced by a

Introduction

gene presumably during its "resting" period, differs from the parent gene essentially in not being anchored to a chromosome (probably by union with nucleic acid).

What is remarkable about this statement is that Haldane anticipated that the mechanism of replication of the genetic material is the same as the mechanism involved in producing its product. Although the replication of DNA is not identical to the transcription of DNA to form RNA, the immediate gene product, they certainly are very similar processes.

Because of his findings with *Drosophila,* Beadle became convinced that a further study of genetic mutants with specific blocks in the metabolic pathway would be highly desirable. *Drosophila,* however, had definite drawbacks as experimental material. An organism was needed that could be grown on a chemically defined simple medium, and that had a life cycle with the haploid phase dominant so that recessive mutations could be recognized soon after they occurred. Teamed with Edward Tatum (1909–   ), who was a biochemist, Beadle turned to *Neurospora,* a fungus that grows on a simple medium with only sucrose and biotin needed as organic supplements, and that has a haploid vegetation phase and an easily manipulated sexual phase. The discovery of biochemical mutants in *Neurospora,* described in Paper 9, acted as a catalyst to bring many more workers into the area of biochemical and molecular genetics and to ensure continued growth in this branch of genetics.

By isolating a large number of different kinds of mutants with different nutritional requirements, Beadle and Tatum were able to show that it was probable that a single mutation could cause the loss of activity for a single enzyme. From this the "one gene–one enzyme" hypothesis was derived. Cuénot (1903) and Garrod (1908) now received substantial support for their hypothesis. Various workers began to demonstrate that the presence or absence of a specific enzyme activity could indeed be inherited (for example, Paper 10). Others occupied themselves with obtaining more evidence for the one gene–one enzyme aphorism (Paper 13), evolutionary considerations (Paper 11), or the working out of biochemical pathways using *Neurospora* mutants (Paper 12).

At first there was a considerable reluctance on the part of many geneticists to accept the one gene–one enzyme hypothesis, particularly on the part of agriculturally oriented breeders. They could not visualize how enzymes could have anything to do with milk production, crop yield, and so forth. But the overwhelming mass of evidence that accumulated through the 1950s, both with *Neurospora* and with *Escherichia coli,* convinced everyone of its general validity.

Investigations confined to *Neurospora* and *E. coli* mutants alone and using only the techniques of isolating biochemical mutants and working out metabolic pathways would not have led to further advances of any great consequences. As important as the work was, it was not contributing to a definitive understanding of genes or even proteins and enzymes.

Almost coincident with the time the *Neurospora* work was first conceived, bacteriophage were beginning to be investigated with respect to their life cycle and with emphasis on the events that occur during replication of the phage in the bacterial cell.

Introduction

This work, which attracted the participation of many physicists and chemists, led to the demonstration by Hershey and Chase (1952) that the DNA of the T_2 phage is probably the part that carries and transmits the genetic information. This observation reawakened interest in the earlier finding of Avery, MacLeod, and McCarty (1944) that *Pneumococcus* could be transformed with DNA. Many biologists were unwilling to accept the transformation work as direct proof of the significance of DNA, and at first transformation received little attention. However, the Hershey and Chase experiments served to bring again to the attention of the biological community the probablity of DNA being the genetic material. This development was shortly followed by the announcement by Watson and Crick (1953) of their proposed structure for DNA, which has since been fully confirmed and accepted as the genetic material. The papers relating to the demonstration that the genetic material is DNA are reproduced in the volume *Microbial Genetics,* which is part of this series.

Granted that DNA is the genetic material, the question now was, what does DNA do? The obvious answer was that it had something to do with the synthesis of proteins, because of the postulated gene–enzyme hypothesis put forward earlier. The volume *Microbial Genetics* also reproduces many of the papers important to an understanding of how it was shown that DNA was transcribed, and the RNA so formed translated with the formation of protein polypeptides with specific amino acid sequences. An important part of this work was the working out of the genetic code, which was essentially completed by 1965.

In the interim period between 1950 and 1965, however, many other important developments occurred, which were to fit into a matrix that was quickly being formed.

The gene–enzyme hypothesis of the 1940s could not be very specific about what the gene–enzyme relationship was for lack of experimental data. For example, when a gene mutates and causes the formation of a genetic block, is this the result of the complete absence of a protein otherwise made in the presence of the wild-type allele or is an altered protein made with little or no enzymatic activity?

It soon became evident that the latter possibility was the case, as Papers 14 and 16 show, although this could have been predicted from Muller's (1932) and Wright's (1939) papers in this series. This was a point of immense importance. The discovery of CRM proteins in the tryptophanless mutants of *Neurospora* must be considered one of the most important developments in working out our present concept of gene and protein. In brief, it gave workers in this area the feeling that they were working with something rather than with the absence of something. Something present, but altered, was going to be much more easily explained than the complete absence of something. This may sound excessively vague to the modern reader, but I believe that it described the kind of speculation that was going on about 1950 to 1955. By 1960 a rather close tie had been made between genes and altered proteins through the work of David Bonner (1916–1965) and his many students and coworkers. Some of this work, which was certainly among the most important carried out in the 1950s, is described in Paper 17.

Even though it was evident by 1957 that genes and proteins had to be closely related, the essential data were not available to prove it. This was provided by turning to what is usually considered poor material for basic biochemical-genetic investigations,

Introduction

human material. In 1957, Ingram demonstrated that the human inherited disease, sickle-cell anemia, was the result of the production of a hemoglobin that differed from normal hemoglobin in a single amino acid out of about 300 (Paper 18). Thus what appeared to be a point mutation in man resulted in a substitution of a single amino acid, in this case a valine for the glutamate ordinarily present at a specific site in the amino acid sequence of normal hemoglobin. This was a breakthrough at the right time, and led subsequently to a great deal of activity relating to the nature of the genetic code.

The 1950s will probably be remembered as the most important decade in the history of genetics since the first decade of the 1900s. The realization of the significance of the strucutre of DNA, RNA, and proteins and their interrelationships can only be considered as important as the discovery of the Mendelian laws and their explanation through cell biology.

The 1960s formed a period of consolidation and marshaling of evidence. Furthermore, just prior to and during this period, a great deal was learned about proteins themselves through the application of genetic analysis. Examples are the studies on complementation in *Neurospora,* as reported in Papers 19 through 22, and on the nature of hemoglobin in Paper 23. The demonstration that allelic mutants of *Neurospora* complement to form heterocaryons which grow on minimal medium without the addition of the growth factors required by the homocaryotic mutants led to the realization that functional proteins are not necessarily all single polypeptides (monomers), but some are in fact dimers, tetramers, and so on, of the monomeric polypeptides. This was an important contribution of genetics to biochemistry.

A further contribution to this same general area came from the study of the human hemoglobin variants, which culminated in the demonstration that human hemoglobin is a tetramer made up of two different polypeptides, the α and β chains, each under the control of a different gene (Paper 23). At this time it became necessary to change the one gene–one enzyme hypothesis to the one gene–one polypeptide hypothesis.

The discovery of a series of different hemoglobin chains in addition to the α and β chains led to considerable speculation about the origin of new genes as well as new polypeptides (Papers 24, 25, and 26). This contributed a new dimension to our understanding of proteins. It now became evident that many functional proteins existed as a number of types distinguishable from one another by their polypeptides, but all having essentially the same function. These are now called isozymes as applied to proteins with enzymatic functions.

For many years a group of mutant genes had been recognized that had the peculiar effect of suppressing the mutant phenotypic effects of seemingly unrelated genes. Nothing much could be done about analyzing these until new biochemical techniques became available. In 1959 it was shown that the apparent alteration of the internal environment by the mutation of a second gene could allow the mutant protein produced in the presence of the first mutant gene to become functionally active (Paper 30). This was followed by a series of investigations dealing with suppressors in *E. coli.* Some suppressor effects turned out to involve tRNA genes, as suggested in Paper 28 and more directly demonstrated in Papers 31 and 32. These findings added a further dimension to the understanding of the gene–protein relationships. Genes not only

Introduction

determine the amino acid sequences of polypeptides, but determine the activity of other proteins indirectly and control the synthesis of proteins via their control of transcription of the vital tRNAs needed in translation. It was also shown in *E. coli* that a mutation at a second site in an already mutant gene controlling the production of an inactive protein could cause a reversion to the wild-type activity (Paper 33). This type of "intragenic" suppression points to the importance of the tertiary structure of a protein. Whereas a single amino acid substitution may render a protein inactive as an enzyme by changing its tertiary structure, a second, different substitution could bring the tertiary structure back to a semblance of the wild-type condition.

It appeared certain by 1960 that genes determine the sequence of amino acids in polypeptides, but it still remained necessary to show experimentally that the sequence of codons in the DNA controlling the synthesis of a specific polypeptide corresponded to the sequence of amino acids determined by the codons. The pioneer work proving this very important point was done by Yanofsky and his coworkers, as shown in Papers 27, 28, and 29. Later the matter was definitely settled in *E. coli* by a report from Yanofsky's group (see *Microbial Genetics* in this series) and by Sarrabhai et al. (1964) in T_4 phage.

With this done it could be said that the first phase of working out the gene–protein relation was finished. It took 60 years of hard work, minus a few years of quiet interlude especially during the 1920s. But it is no exaggeration to say that thought about the matter was continuously sustained from about 1900 on.

Like everything in science the work is not finished. Genetics, whether molecular or developmental or population, is not dead, although some areas are not flourishing because of a lack of new ideas. The last three papers in this volume are included to give the reader some idea of what is currently topical and important in the area of molecular genetics.

After finding that many functional proteins are made up of more than a single polypeptide, it was discovered that some enzymes are also aggregated to form complexes. An example of this work is given in Paper 34. This kind of investigation, combined with genetic techniques, will eventually give us a better understanding of structure and its morphogenesis. Rather than merely thinking of genes→proteins only, we must also think about genes→proteins→supramolecular structure. Also, we must consider the problem of explaining the formation of the complex polysaccharides in the cell and their association with proteins. So far no one has explained how polysaccharides are formed. Is it via a code?

In the few bacteria that have been worked with, the correlation between the length of DNA in their cells and the number of different polypeptides contained in their protein is fairly close (if we assume three nucleotides are needed for each amino acid coded). Such a correlation is not true in the eukaryotes. They have 10- to 100,000-fold more DNA per cell nucleus than is found in a bacterial chromosome, yet the number of different kinds of polypeptides that they synthesize for their needs is probably not significantly different from the bacteria. The question of what all this extra DNA is doing is an important one, because for one thing we would like to be able to define the eukaryote gene more precisely. This is made difficult by the presence of so much

Introduction

reiterated DNA in eukaryote chromosomes. What about the regions called genes on chromosomes? Does each of them have a single function; that is, does each control the formation of a single kind of polypeptide as in the bacteria? Questions such as these can only be answered by direct experimental work; Paper 35 is an example of one approach. It also signals the renaissance of *Drosophila melanogaster* as an important organism in genetics research.

Part of the function of the excess DNA in eukaryotes is thought to be regulation of gene activity. Paper 36 gives an example of one way this problem is being approached indirectly by the fingerprinting of isoenzymes.

Selected Bibliography

Caspari, Ernst. 1933. *Arch. Entw. Mech. 130:* 352–381.
Delage, Yves. 1895. *La Structure du protoplasma et les théories sur l'hérédité et les grands problèmes de la biologie générale.* Paris.
Delbrück, Max. 1941. *Cold Spring Harbor Symp. Quant. Biol. 9:* 122–126.
Garrod, A. E. 1902. *Lancet 2:* 1616–1620.
Haldane, J. B. S. 1942. *New Paths in Genetics.* Harper & Row, New York.
Sarabhai, A. S., et al. 1964. *Nature 201:* 13.
Sturtevant, A. H. 1965. *A History of Genetics.* Harper & Row, New York.
Wilson, E. B. 1896. *The Cell in Development and Inheritance.*
Wright, Sewall. 1917. *J. Heredity 8:* 224–235.

I
The First Glimmers

Editor's Comments on Paper 1

1 **Cuénot:** *L'hérédité de la pigmentation chez les souris*
English translation: *The Heredity of Pigmentation in Mice*

Generally, in writings dealing with the origins of biochemical genetics, the original paper of Garrod in 1902 is cited as a major step forward in the understanding of the relation between genes and metabolic reactions (see *Medical Genetics* in this series). It cannot be denied that this is so, because Garrod does point out that alkaptonuria, albinism, and cystinuria in man are almost certainly inherited biochemical differences. However, nowhere in the paper does Garrod state that the inherited abnormal conditions he describes are the result of a deficiency in an enzyme activity.

Cuénot, however, in the paper reproduced here, makes the reasonably clear statement, based on his analysis of results from crosses with different color strains of mice, that there must be a relation between what he calls *mnemons* and diastases (enzymes). The word "mnemon," which he borrows from the French biologist Coutagne, comes from a Greek root, *mneme*, meaning memory. The implication is that for each enzyme there is a mnemon inherited through the gametes which dictates the presence of a specific enzyme. Here we have, therefore, the first glimmer of the one gene–one enzyme hypothesis later to be boldly and clearly stated by Beadle, Tatum, and their colleagues in the 1940s.

If more attention had been paid to Cuénot's theoretical considerations, genes might today be called mnemes. Although perhaps more difficult to pronounce than gene, mneme is more descriptive of what we now know to be the function of genes. Also, geneticists would now be called mnemeticists, which is hardly an easily remembered name.

Selected Bibliography

Bateson, W. 1909. *Mendel's Principles of Heredity.* Cambridge University Press, New York.
Dunn, L. C. 1965. *A Short History of Genetics.* McGraw-Hill, New York.
Garrod, A. E. 1902. The incidence of alkaptonuria: a study in chemical individuality. *Lancet 2:* 1616–1620.
———. 1909. *Inborn Errors in Metabolism.* Oxford University Press, Oxford.
Haldane, J. B. S. 1954. *The Biochemistry of Genetics.* Macmillan, New York.
Sturtevant, A. H. 1965. *A History of Genetics.* Harper & Row, New York.

L'HÉRÉDITÉ DE LA PIGMENTATION CHEZ LES SOURIS*

(2ᵐᵉ NOTE)

par L. Cuénot

Professeur à la Faculté des Sciences de Nancy

I. Hérédité de la pigmentation chez les Souris noires

Dans une note précédente, j'ai montré que les croisements entre Souris grise sauvage et Souris albinos suivaient rigoureusement la règle de Mendel (type *Pisum*), quant au caractère différentiel présence de pigment et absence de pigment. Les hybrides de première génération sont toujours, sans exception, identiques à la Souris grise, comme l'avaient déjà constaté Crampe (**1885**) et Haacke (**1897**), c'est-à-dire que le caractère pigment est dominant par rapport au caractère absence de pigment. Ces hybrides, croisés entre eux, fournissent 3 gris pour 1 albinos, soit une forme pure revenue au type gris, 2 gris hybrides, et une forme pure revenue au type albinos, conformément au schéma suivant :

Editor's Note: An English translation follows this article.

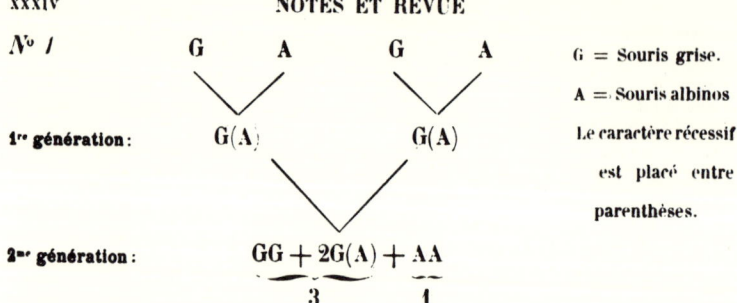

G = Souris grise.
A = Souris albinos
Le caractère récessif est placé entre parenthèses.

D'autre part, des Souris grises hybrides (hétérozygotes suivant l'expression de Bateson), croisées avec des albinos, donnent autant d'individus gris (hybrides) que d'albinos, conformément au schéma suivant :

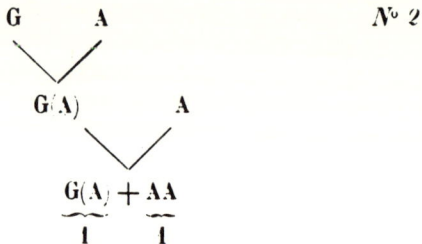

Parmi les produits de croisement entre des Souris grises hybrides de troisième génération et des Souris albinos, j'ai trouvé plusieurs fois des individus à pelage noir, constituant une variété nouvelle, une *mutation régressive*, comme dirait de Vries ; cette variété n'est du reste pas nouvelle, et plusieurs auteurs, notamment Castle (**1903**), rapportent l'avoir obtenue dans des élevages analogues. Elle se distingue du type gris sauvage par sa belle couleur noire veloutée, un peu moins foncée sous le ventre, et l'absence totale de teinte fauve ; les poils, examinés au microscope, présentent une grande quantité de granules pigmentaires noirs, mais on n'y trouve jamais la moindre trace des granules jaunes qui colorent l'extrémité des poils de la Souris grise. La démarcation entre les deux formes est si nette que je n'ai jamais été embarrassé pour ranger dans l'une ou l'autre catégorie les centaines de Souris grises et noires de mes élevages.

Les croisements entre la mutation noire et les albinos suivent exactement la règle de Mendel, comme on pouvait s'y attendre. Les hybrides de 1re génération sont toujours pigmentés, gris ou noirs (je montrerai plus loin qu'on peut obtenir à volonté l'une ou l'autre teinte, suivant l'albinos qu'on emploie); la descendance des hybrides noirs, croisés entre eux, comprend 3 individus noirs pour 1 albinos (schéma n° 1). Enfin, si l'on croise les hybrides noirs avec des albinos, on obtient autant d'individus pigmentés que d'albinos (schéma n° 2). La règle de MENDEL s'applique donc strictement.

Il est du reste très probable que, chez les Mammifères tout au moins, les vrais albinos (yeux rouges, pelage pouvant être blanc ou partiellement coloré) sont toujours récessifs par rapport aux variétés pigmentées à yeux noirs : les produits de croisement ont toujours les yeux noirs et un pelage pigmenté de couleur variable, mais les albinos réapparaissent dans leur descendance suivant la règle de MENDEL. Cette généralisation est autorisée par les expériences et observations plus ou moins complètes de divers auteurs, notamment celles de HAACKE et de VON GUAITA (Souris valseuses du Japon, grises ou noires tachetées de blanc), de CRAMPE (*Mus decumanus* gris, noirs ou tachetés), de CASTLE (Cobayes et Lapins), de RASPAIL (Lapin gris de garenne croisé avec Lapin russe à yeux rouges) de RÖRIG (Chevreuil albinos croisé avec Chèvre normale ou noire), de FARABEE (Nègre albinos croisé avec une Négresse ordinaire).

Après avoir obtenu des Souris noires de race pure, c'est-à-dire de parfaits homozygotes, suivant l'expression de BATESON, j'ai recherché quel était le résultat du croisement entre la mutation noire et le type gris sauvage. Là encore, la règle de MENDEL s'applique rigoureusement : les hybrides de 1re génération sont toujours gris; donc le caractère gris est dominant par rapport au caractère noir. Ces hybrides gris, croisés entre eux, fournissent 3 individus gris pour 1 noir (schéma n° 1); d'autre part, si on croise ces hybrides gris par des noirs de race pure, on obtient autant d'individus noirs que de gris (schéma n° 2).

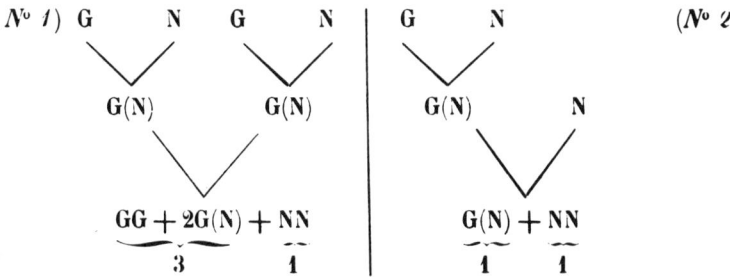

G = Souris grise ; N = Souris noire.

Enfin, pour achever de démontrer que le noir est bien récessif par rapport au gris, il suffit de croiser entre eux des individus noirs comptant dans leur lignée ancestrale un nombre quelconque de parents gris : on obtient uniquement des produits noirs et le gris ne réapparaît jamais.

Autant qu'on peut en juger en dépouillant les généalogies très embrouillées de Crampe, il est très probable que la dominance du gris sur le noir se présente aussi chez *Mus decumanus* : en effet, Crampe a observé : 1° que les variétés noires croisées avec le type gris donnaient exclusivement des produits gris ; 2° que ces produits gris, croisés entre eux, fournissaient un mélange de gris et de noirs ; 3° que les variétés noires, croisées entre elles, avaient des descendants uniquement noirs, le gris qui pouvait exister dans leur lignée ancestrale ne réapparaissant jamais.

II. Transmission héréditaire de pigmentation par les Souris albinos

Les biologistes qui poursuivent actuellement des études expérimentales sur l'hérédité, par des croisements entre animaux de couleur différente, admettent implicitement que la variété albinos à yeux rouges est une forme pure, toujours identique à elle-même, et récessive par rapport aux variétés pigmentées avec lesquelles on la croise [1].

L'albinos ne semble pas contenir de pigment en puissance, ni dans son soma, ni dans son plasma germinatif, puisque des albinos croisés entre eux donnent indéfiniment des albinos, sans que réapparaisse jamais le caractère pigmenté ; il semble donc que lorsqu'on opère des croisements avec des albinos, il n'y a aucun intérêt à connaître la couleur de leur ancêtres pigmentés plus ou moins proches. L'expérience que je vais rapporter montre au contraire que dans certaines conditions, les Souris albinos sont parfaitement aptes à transmettre la couleur des individus pigmentés qu'elles comptent parmi leurs ascendants.

Je possède des Souris albinos *identiques d'aspect* (pelage blanc pur, yeux rouges) qui ont trois origines ancestrales différentes : 1° dans l'ascendance des unes, depuis six générations au moins, les parents pigmentés ont tous été gris ; 2° d'autres proviennent du

---

[1] Voir les travaux de Crampe (**1885**), Haacke (**1895-1897**), von Guaita (**1898-1900**) qui ne connaissaient pas la loi de Mendel, puis ceux de Cuénot (**1902**), Darbishire (**1902**), Castle (**1903**).

croisement de deux Souris noires, dont l'ascendance est variable ; 3º d'autres encore proviennent du croisement de deux Souris jaunes, dont l'ascendance est plus ou moins compliquée.

Des Souris *noires* (pelage d'un noir de velours, yeux noirs) sont réparties en trois lots :

Le 1er est croisé par des albinos à parenté *grise*.

Le 2me est croisé par des albinos à parenté *noire*.

Le 3me est croisé par des albinos à parenté *jaune*.

On peut faire l'expérience d'une façon différente, en donnant à un même mâle noir, successivement, trois femelles albinos, appartenant aux trois catégories précitées.

On obtient :

Dans le 1er lot, toujours des Souris *grises*.

Dans le 2me lot, toujours des Souris *noires*.

Dans le 3me lot, un mélange de Souris *jaunes* et *grises*, ou bien de Souris *jaunes* et *noires*.

On voit donc que l'ascendance des albinos a une influence bien nette sur la teinte du pelage de leurs descendants ; je n'ai cité que cette expérience, mais elle est corroborée par beaucoup d'autres, tout aussi démonstratives, mais plus compliquées, que je n'expose pas, faute de place. Cette constatation, qui n'avait pas encore été faite jusqu'ici, donne la clé des résultats contradictoires obtenus par les auteurs qui ont fait des croisements entre albinos et individus pigmentés (Lapins, Souris, Rats) ; sans s'en douter, ils ont opéré avec des albinos de valeur différente, et par suite la couleur des produits a paru échapper à toute règle. Certainement, les Souris albinos que vendent les marchands ont des origines ancestrales variées, par conséquent une influence héréditaire variable, malgré l'identité de leur aspect extérieur.

Mais comment interpréter ce résultat dans les idées actuelles sur la constitution du plasma germinatif ? Les travaux anciens et récents sur l'hybridation expérimentale, bien plus que les raisonnements théoriques, ont amplement démontré que l'existence de plasmas ancestraux est tout à fait inadmissible, et notre explication devra avant tout s'interdire d'y recourir.

Je rappellerai tout d'abord que le pelage des Souris grises est formé de poils colorés par deux pigments différents, un brun noirâtre et un jaune, tandis que chez les Souris noires, il n'existe que le pigment noirâtre, le jaune manquant d'une façon totale ;

chez les Souris jaunes, le pigment jaune prédomine de beaucoup, le pigment noirâtre pouvant être présent en petite quantité ou tout à fait absent. D'autre part, on sait que les auteurs qui ont récemment étudié la genèse des pigments mélaniques, BIEDERMANN, VON FÜRTH et H. SCHNEIDER, GESSARD, admettent que ces pigments résultent de l'action d'une diastase oxydante (tyrosinase) sur une substance chromogène ; il y a de bonnes raisons pour supposer que les choses se passent de même pour les pigments des poils ; il y aurait donc dans ceux-ci soit deux chromogènes différents et une seule diastase, soit un seul chromogène et deux diastases, l'une pour le pigment noirâtre, l'autre pour le pigment jaune. Adoptons provisoirement, pour la commodité du langage, cette dernière hypothèse.

Le plasma germinatif d'une Souris grise doit contenir en puissance les trois substances qui, par leurs réactions réciproques, produiront plus tard les dépôts pigmentaires des poils ; et sans doute ces trois subtances sont contenues à l'état potentiel dans autant de particules matérielles du plasma germinatif (particules représentatives ou substances qualitatives de l'œuf = *mnémons*[1]). Chez une Souris grise, il y a trois mnémons, un pour le chromogène et deux pour les deux diastases ; chez une Souris noire, il y a seulement deux mnémons, l'un pour le chromogène et l'autre pour la diastase formatrice du pigment noirâtre.

Quant aux albinos, tout s'explique si l'on admet que leur plasma germinatif renferme seulement les mnémons des diastases, celui du chromogène manquant totalement. Dans ces conditions, il ne peut se former de poils colorés chez l'albinos, puisqu'il manque une des substances indispensables à la réaction, mais on comprend facilement que l'albinos transmettra à sa progéniture soit les mnémons pour les deux diastases, soit un seul mnémon, s'il n'en possède qu'un.

L'expérience que j'ai rapportée plus haut peut maintenant s'interpréter (je me bornerai aux deux premiers lots, le troisième étant un peu plus compliqué) : quand on croise un gamète de Souris noire par un gamète de Souris albinos à ascendance grise, on additionne le chromogène du premier gamète avec les deux diastases du second, et l'hybride a nécessairement un pelage gris.

Quand on croise le même gamète de Souris noire par un gamète

[1] Ce terme de *mnémons* est emprunté à COUTAGNE (**1902**).

d'albinos à ascendance noire, on n'introduit que la diastase formatrice du pigment noirâtre, l'autre faisant défaut, et il est tout naturel que les hybrides aient toujours, sans exception, un pelage noir.

Cette explication, toute hypothétique et provisoire qu'elle soit, rend parfaitement compte de ce que produisent les croisements les plus variés entre Souris grises, noires et albinos, quelles que soient les complications de leurs généalogies ; j'ai même pu prévoir, avec son aide, les couleurs que devaient donner certains croisements non encore essayés, et l'expérience a vérifié constamment la prévision. Elle a encore l'avantage de substituer à la notion de dominance (du gris et du noir sur l'albinos, et du gris sur le noir), notion qui n'est que l'expression du fait constaté, une explication d'ordre chimique, susceptible de vérification expérimentale. Mais il reste à prouver qu'elle s'applique aux autres variétés colorées de la Souris ; c'est ce que montreront mes expériences en cours.

Nancy, 3 Mars 1903.

*
* *

A titre d'exemple, j'ai donné dans le tableau ci-contre le schéma de deux croisements assez compliqués ; les combinaisons entre les gamètes, suivant les règles mendéliennes, permettent de prévoir : 1º la couleur des produits de la 2ᵐᵉ génération ; 2º le nombre relatif des différentes variétés obtenues ; 3º la constitution intime et par suite la valeur héréditaire des hétérozygotes et homozygotes de 2ᵐᵉ génération.

Iᵉʳ Exemple

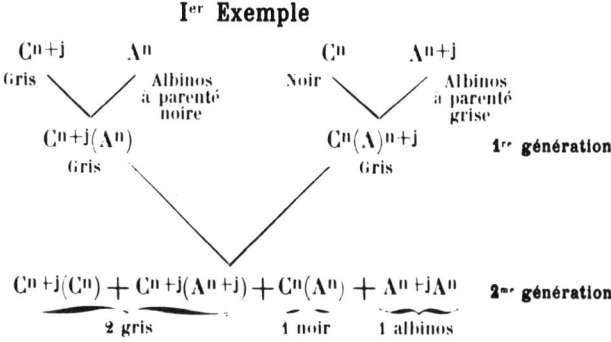

EXPLICATION DES LETTRES

C = chromogène.
A = absence de chromogène ou albinos.
n = diastase pour le pigment noirâtre.
j = diastase pour le pigment jaune.
Dans la formule des zygotes, le caractère récessif est placé entre parenthèses.

IIme Exemple

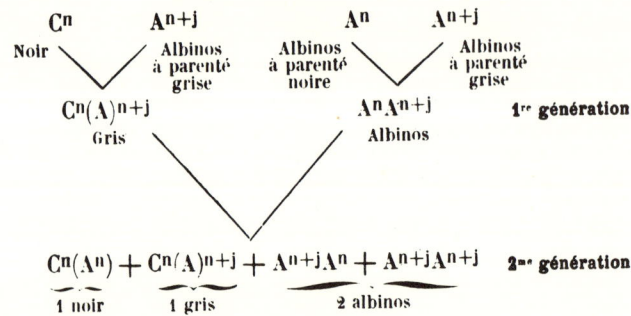

EXPLICATION DES LETTRES

C = chromogène.
A = absence de chromogène ou albinos.
n = diastase pour le pigment noirâtre.
j = diastase pour le pigment jaune.
Dans la formule des zygotes, le caractère récessif est placé entre parenthèses.

Index Bibliographique

1902. BATESON. Mendel's principles of heredity. (*Cambridge*).
1903. CASTLE. Mendel's law of heredity. (*Proc. American Acad. of Arts and Sciences*, vol. 38, p. 535).
1902. COUTAGNE. Recherches expérimentales sur l'hérédité chez les vers à soie. (*Bull. scient. France et Belgique*, t. 37).
1885. CRAMPE. Die Gesetze der Vererbung der Farbe. I. Die Eigenschaften der Spezies und der Varietäten. (*Landwirth. Jahrbücher*, Bd 14, p. 379).
1885. —— II. Die Veränderung der Varietäten bei Fortpflanzung in Farben-Inzucht. (*même recueil*, p. 539).
1902. CUÉNOT. La loi de Mendel et l'hérédité de la pigmentation chez les Souris. (*Arch. Zool. exp.* [3], t. 10, Notes et Revue, p. XXVII).
1902. DARBISHIRE. Note on the results of crossing Japanese waltzing Mice with European Albino Races. (*Biometrika*, vol. 2, p. 101).
1903. FARABEE. Notes on Negro albinism. (*Science*, vol. 17, p. 75).
1901. FÜRTH (von) et H. SCHNEIDER. Über tierische Tyrosinasen und ihre Beziehungen zur Pigmentbildung. (*Beiträge zur chem. Physiol. und Pathol.*, Bd 1, p. 229).
1902. GESSARD. Tyrosinase animale. (*C. R. Soc. Biologie*, t. 54, p. 1304).

1898. GUAITA (von). Versuche mit Kreuzungen von verschiedenen Rassen der Hausmaus. (*Berichte der naturforsch. Gesells. zu Freiburg*, Bd 10, p. 317).

1900. —— Zweite Mittheilung über Versuche mit Kreuzungen von verschiedenen Hausmausrassen. (*même recueil*, Bd 11, p. 131).

1895. HAACKE. Ueber Wesen, Ursachen und Vererbung von Albinismus und Scheckung und über deren Bedeutung für vererbungstheoretische und entwicklungsmechanische Fragen. (*Biol. Centralblatt*, Bd 15, p. 45).

1897. —— Grundriss der Entwickelungsmechanik. (*Leipzig*).

1902. RASPAIL. Note sur une race de Lapins albinos issue du croisement d'une femelle de Lapin russe et d'un mâle garenne (Lepus cuniculus). (*Bull. Soc. Acclimatation*, 49me année, p. 170).

1897. RÖRIG. Ueber Haltung und Fortpflanzung von Rehen in zoologischen Garten und Kreuzung abnorm gefärbter Rehe im Francfurter Garten. (*Zool. Garten*, t. 38, p. 171).

1902. VRIES (de). Die Mutationstheorie, Bd II. Die Bastardirung. (*Leipzig*).

1903. —— La loi de Mendel et les caractères constants des hybrides. (*Comptes-rendus Acad. Sc. Paris*, t. 136, p. 321).

1

The Heredity of Pigmentation in Mice

L. CUÉNOT

This article was translated expressly for this Benchmark volume by R. P. Wagner, The University of Texas at Austin, from Arch. Zool. Exp. Gén., 4ₑ sér., 1(3): Notes Rev., 33–41 (1903).

I. Heredity of Pigmentation in Black Mice

In a previous note, I have shown that crosses between wild gray mice and albino mice obey the law of Mendel (type Pisum) with respect to the contrasting characteristics: presence and absence of pigment. The hybrids of the first generation are always, without exception, identical to the gray parents, as had already been demonstrated by Crampe (1885) and Haacke (1897); that is, the pigmented condition is dominant to the absence of pigment. These hybrids, crossed among themselves, give 3 gray to 1 albino, comprised by a pure gray type, 2 gray hybrids, and a pure albino type, conforming to the following schema:

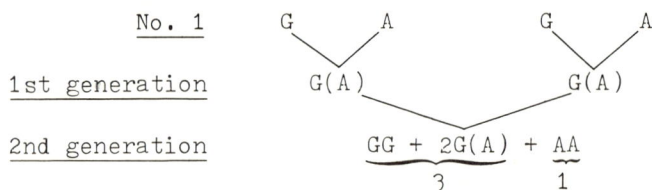

Here G indicates a gray mouse, A an albino mouse, and the recessive characteristic is in parentheses.

The gray hybrid mice (heterozygotes, using the terminology of Bateson) when crossed with albinos, give as many gray individuals (hybrids) as albinos, conforming to the following schema:

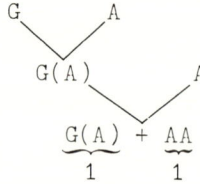

Among the products of crosses between the gray hybrid mice of the third generation and the albino mice, I found several times individuals with black pelage, constituting a new variety, or a *regressive mutation*, according to de Vries. This variety is not, however, new, for several authors, notably Castle (1903), report having obtained them by an analogous type of breeding. It is distinguished from the wild gray type by its beautiful velvety black color, a little less dark in the underbelly, and the total absence of a fawn color tint. Examined under the microscope, the hairs show a great quantity of black pigmented granules, and one never finds the least trace of the yellow granules that color the extremities of the hairs of the gray mouse. The difference between the two forms is so distinct that I am never uncertain as to how to classify gray and black offspring from my breeding experiments.

Crosses between the black mutant and the albinos follow Mendel's Law as exactly as one might expect. Hybrids of the first generation are always pigmented gray or black. (I will show later that one can obtain one or the other color depending on the albino strain used.) The gray hybrids when crossed among themselves produce 3 black to 1 albino offspring (schema no. 2). Thus Mendel's Law applies strictly.

It is also very probable, at least among the mammals, that true albinos (red eyes, coat color white to partial color) are always recessive to varieties with pigmented eyes. That is, the offspring of crosses between the two always have colored eyes and variably colored fur, but the albinos reappear among their descendents following Mendel's Law. This generalization is more or less supported by the experimental results and observations from a variety of authors, notably those of Haacke and von Guaita (Japanese waltzing mice with gray or black–white spotted fur), of Crampe (*Mus decumanus*, gray, black, or spotted), of Castle (guinea pigs and rabbits), of Raspail (gray domestic rabbits crossed with Russian red-eyed rabbits), Rörig (albino goats crossed to normal or black), and of Farabee (albino Negros mated with normally pigmented Negresses).

After having obtained pure breeding black mice, that is, perfect homozygotes using Bateson's terminology, I tested the result of crosses between the homozygous mutant blacks and the wild gray type. Mendel's Law was again found to apply: the first-generation hybrids are always gray. Thus the gray characteristic is dominant to the black. These hybrids, when crossed among themselves, gave progeny in a 3 gray to 1 black ratio (schema no. 1). On the other hand, if one crosses the gray hybrids to the homozygous blacks, one obtains as many black as gray offspring (schema no. 2):

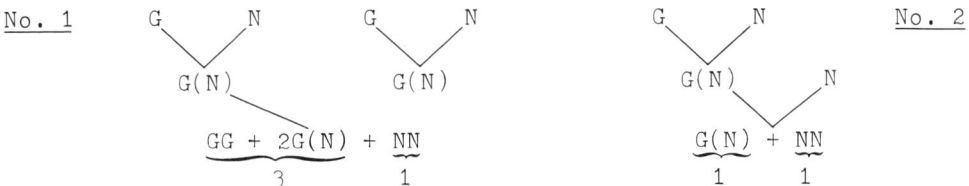

Here G indicates gray; N, black.

Finally, to demonstrate that the black is definitely recessive to gray, it is necessary only to cross among themselves black individuals who have in their lineage ancestors with gray pelage. One obtains only black offspring and the gray never reappears.

II. Hereditary Transmission of Pigmentation in Albino Mice

The biologists who pursue experimental studies on heredity by crossing animals of different colors definitely state that the albinos with red eyes are a pure breeding form, always identical and recessive with respect to pigmented varieties with which they are crossed.

Albinos do not have the capability of making pigment either in their soma or in their germ plasma, since albinos crossed among themselves give albinos only, with no reappearance of the pigmented characteristic. It seems as though there is no importance to know the color of their distant or close ancestors when making crosses with albinos. The experiments that I am going to report show on the contrary that under certain conditions albino mice are quite capable of transmitting the color of the pigmented individuals that they had among their ancestors.

I have identical albino mice (white pelage and red eyes) which have three different ancestries: (1) those which had gray ancestors at least six generations before; (2) those arising from crosses between two black mice whose pedigree is variable; (3) those arising from crosses between yellow mice whose pedigree is complicated.

The black mice (black pelage, black eyes) are divided into three groups:
1. The first group is crossed to albinos of gray parentage.
2. The second group is crossed to albinos of black parentage.
3. The third group is crossed to albinos of yellow parentage.

One can also do the experiment differently by mating each black male successively to three albino females belonging to the three categories given above.

From these experiments the following types of progeny are obtained:
1. All gray mice.
2. All black mice.
3. A mixture of yellow and gray mice, or yellow and black.

Thus it is evident that the ancestors of albinos have a definite influence on the color of the pelage of their descendants. I have cited only one of the many experiments that corroborate this finding. This demonstration, which has never been made before until now, gives the clue to the contradictory results obtained by authors who have made crosses between albinos and pigmented individuals (rabbits, mice, rats); without suspecting it, they have used albinos from different backgrounds, and as a consequence the colors of the offspring appeared not to follow all rules. Certainly, the mice sold by merchants have different ancestral origins, and consequently, a different hereditary background, in spite of the identity of their external appearance (as albinos).

But how to interpret these results with the present ideas about the consititution of the germinal material? The older and more recent work on experimental hybridization, as well as the theoretical thought, has amply demonstrated that the existence of ancestral plasma* is entirely without basis, and our explanation should above all avoid resorting to it.

---

*[*Translator's Note:* This phrase appears as "plasma ancestraux" in the French text. It was invented by August Weismann, and its definition is too complicated to be given here.]

I recall to the reader first that the pelage of the gray mice is formed of hairs colored by two different pigments, a dark brown and a yellow, whereas in the black mice, there exists only the black pigment. The yellow pigment is totally absent in the blacks. In the yellow mice the yellow pigment predominates and the blackish pigment is present either in small quantity or completely absent. On the other hand, the authors who have recently studied the genesis of the melanic pigments (Biedermann, von Fürth, and H. Schneider, Gessard) agree that these pigments result from the action of an oxidizing enzyme (tryosinase) on a chromogen. There are good reasons to believe that the same thing occurs in hairs. There should be in them either two different chromogens and a single enzyme, or a single chromogen and two enzymes, one for the blackish pigment and the other for the yellow. For purposes of discussion we adopt provisionally the last hypothesis.

The germinal material of a gray mouse should contain the three substances which by their reciprocal reactions will later produce the deposits of pigment in the hairs, and without doubt these three substances are contained in a state of potentiality in as many material particles of germinal material (particles represented by qualitative substances of the egg = mnemons[1]). In the gray mouse there are three mnemons, one for the chromogen and two for the two enzymes. In the black mouse there are only two mnemons, one for the chromogen and the other for the enzyme forming the blackish pigment.

As for albinos, all is explained if one admits that their germinal material includes solely the mnemons for enzymes and completely lack those for the chromogen. Under these conditions they are not able to form colored hairs because they lack one of the substances indispensible for the reaction, but one can understand easily why the albinos will transmit to their offspring either the mnemons for the two enzymes or a single mnemon if the albino only possesses one.

The experiments that I have reported above may now be interpreted. (I will limit myself to two primary interpretations, the third being a little too complicated.) When one crosses a gamete from black mice with a gamete from albino mice with gray ancestors, one combines the chromogen of the first gamete with the two enzymes of the second, and the hybrid necessarily has a gray pelage.

When one crosses the same gamete from a black mouse with a gamete from albinos with black ancestors, one only introduces the enzyme forming the blackish pigment, the other being lacking, and it is expected that the hybrids will always have, without exception, a black pelage.

This explanation is completely hypothetical and provisional but explains perfectly the variable products of the crosses between gray, black, and albino mice whatever may be the complexity of their geneologies. I have even been able to predict with its help, the colors that should result from crosses still not tried, and the experiment has uniformly verified the prediction. The hypothesis also has the advantage of substituting the notion of dominance (of gray and black to albinos, and of gray to black), a notion that is only an observation with an explanation of a chemical nature, susceptible to experimental

---

[1]This term is borrowed from Coutagne (1902). [*Translator's Note:* Literally it refers to particles with a memory.]

The Heredity of Pigmentation in Mice

verification. But it remains to be proved that it applies to other color varieties among mice; that is what I will show in my future experiments.

Nancy, March 3, 1903

As an example, I give, in the figure below, the scheme for two rather complex crosses. The combinations between the gametes, following Mendelian rules, allow these predictions: (1) the color of the progeny of the second generation; (2) the relative number of different forms obtained; (3) the constitution and therefore the hereditary value of the heterozygotes and homozygotes of the second generation.

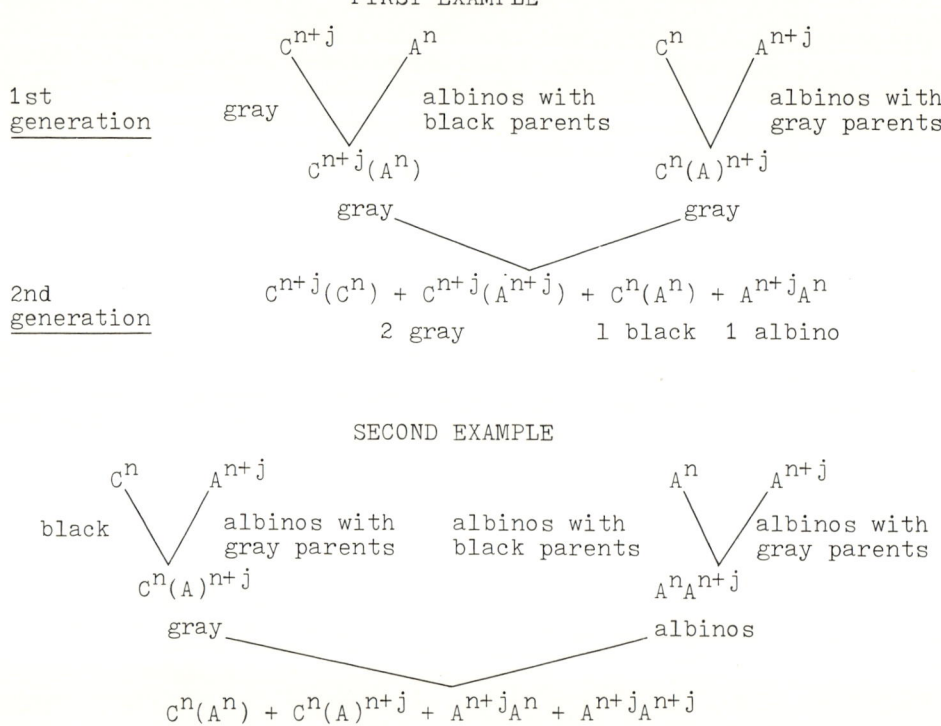

Here C indicates chromogen; A, absence of chromogen or albino; n, enzyme for blackish pigment; j, enzyme for yellow pigment; and the recessive characteristic is in parentheses.

[*Editor's Note:* References will be found at the end of the preceding, original article.]

II
Progress in Thinking to 1930

Editor's Comments on Paper 2

2 **Muller:** *Further Studies on the Nature and Causes of Gene Mutations*

The second part of Muller's speculative paper, reproduced here, represents a significant contribution to thought about gene action. It was written at a time when relatively little experimental work was being done in this area, and had considerable influence on thought about gene action for a considerable number of years after its appearance.

By using X rays, Muller and his colleagues at the University of Texas obtained a vast number of mutant *Drosophila melanogaster.* Upon considering the many accumulated different allelic forms, Muller came to the conclusion that they may be grouped into a number of different categories, such as hypomorphs, amorphs, antimorphs, and neomorphs. He based his conclusions on the effect on the phenotype and considered both quantitative and nonquantitative effects.

Although the terms themselves are not in very wide usage today, the concepts behind them are—the most important being that a mutant allele (besides having zero effect, e.g., no pigment produced) may have a similar but lesser effect than the wild type. Although in this paper Muller does not make an explicit statement about the nature of the gene product, he does imply that the various alleles in a series may differ in the activity or dosage of their products.

Early in this century William Bateson had proposed the presence–absence hypothesis to explain the phenotypic effects of alleles. He posited that an allele of the inherited normal condition in which something is produced is in effect the absence of something or a change to something entirely new. A. H. Sturtevant took strong exception to this in 1913, proposing that genes may undergo changes to different allelic conditions by mutation. Muller carried this point of view several steps further to give a much clearer conception of allelic differences.

Selected Bibliography

Haldane, J. B. S. 1927. The comparative genetics of colour in rodents and carnivora. *Biol. Rev. 2:* 199–212.

Muller, H. J. 1922. Variations due to change in the individual gene. *Amer. Naturalist 56:* 32–50.

Plunkett, C. R. 1926. The interaction of genetic and environmental factors in development. *J. Exp. Zool. 46:* 181–244.

Reprinted from *Proc. Sixth Intern. Congr. Genetics*, **1**, 231–255 (1932)

FURTHER STUDIES ON THE NATURE AND CAUSES OF GENE MUTATIONS

H. J. Muller, University of Texas, Austin, Texas

[*Editor's Note:* In the original, material precedes this excerpt.]

ON THE CHARACTER OF MUTATIONS

Methods of attacking the problem of whether mutations are merely quantitative changes

Probably some geneticists would welcome the problematical connection between induced gene mutations and rearrangements, and between the latter and chromosome contacts, as evidence for the view that gene mutations, or at any rate those produced by irradiation, are merely due to losses or transfers—the latter in some cases perhaps involving additions—of chromosome material of a type previously present. They would take it as evidence for a presence-and-absence, or at any rate for a quantitative, interpretation of mutational changes. Perhaps they might now extend the interpretation to parts of genes, or sub-genes, in order to account for cases like the scute or truncate series, but, so far as any given kind of gene material was concerned, they would see in the mutation process only a mechanical loss or diminution of the gene, by subtraction of material from the chromosome, or—as they would have to say in the case of some reverse mutations, for example—an increase of the gene, such as might be caused by its overgrowth or by the attachment to the chromosome of homologous material from a sister or homologous chromatid. Further plausibility is lent such a view by the fact that many allelomorphic series do give the phaenotypic appearance of being quantitative in their basis.

Fortunately X-rays provide us with a new tool which helps to shed light on these questions concerning the character of the mutations produced by them and by other influences. That is, we can induce gene rearrangements and so get fragments of chromosomes containing normal or mutant genes at given loci. We can then add or subtract such fragments, creating hyperploidy or hypoploidy, and can thus determine what the effects of changing

232 *Further Studies on the Nature and Causes of Gene Mutations*

the quantity of a given gene material really are. These known effects of purely quantitative changes may then be compared with the effects that were produced by the mutations themselves.

It has sometimes been assumed that one can judge the phaenotypic effect of different quantities of a gene simply by comparison of the appearances of heterozygotes and of homozygotes of the two opposite types, or, as a greater refinement, by comparison of the different grades of heterozygotes in polyploids. However, the situation in these cases is hopelessly complicated by the fact that in the comparison of such types we deal not merely with a difference in the dosage of one allelomorph, but always with a simultaneous and opposite difference in the dosage of the other allelomorph, since we must always reckon with a *substitution* of one allelomorph for the other, when chromosome fragments are not added or subtracted. We cannot legitimately assume in advance of the evidence that either the one or the other allelomorph is a mere absence, and so we cannot tell to what extent the observed effects may be due to the changed dosage of the one, to what extent to that of the other allelomorph, or to an interaction process. For example, in a comparison of the homozygous eosin-eyed Drosophila, the intermediate colored eosin-white compound, and the homozygous white, it need not be assumed, *a priori,* that the eosin gene has the effect of producing color, and produces more in double dose. It might be assumed instead (or in addition) that the white gene inhibited color, and inhibited more strongly in double dose. It might even be conceived that both allelomorphs inhibited the pigmentation which genes in other loci tended to produce, but that white was a more effective inhibitor than eosin.

STERN (1929) used actual dosage differences of a given allelomorph in his determination that each additional dose of mutants of the bobbed series adds to bristle length, up to a certain limit. In his work, instead of a small chromosome fragment, the practically inert Y chromosome served to furnish the extra doses. MOHR and BRIDGES, in their studies on deficiencies, realized that they might be dealing with real dosage differences, but at that time other interpretations, such as a peculiar sort of chain mutation, were not excluded. In an attempt to answer this question, however, I have examined cases in which there were known to be actual losses of the same region as was involved in the above cases of deficiencies, and find the effects to be the same.

Thus, for comparison with the Notch-8 deficiency of MOHR (1919, 1923), in which a piece near the left end of the X chromosome, extending from the left of white (1.7) nearly to echinus (5.5), is "deficient," we have certain cases of PATTERSON's (1932c) produced by X-raying. In these, a relatively large piece was removed from the left end of the X chromosome, though at the same time the very left end, which he has found (1932b) to be necessary for the life of the fly, was provided in advance, in the form of a fragment (called duplication X1 or "theta") attached to the right end. These known losses of the w-c_c region result in Notch wings, and allow recessives of the w, f_a and c_c loci, present in the homologous chromosome, to manifest themselves just as they would in a compound having them in one chromosome and the most extreme possible allelomorph of that sort in the other. I find females having apricot in one X chromosome and either white, MOHR's Notch-8 deficiency, or one of these known losses in the other, all to be indistinguishable from one another in shade. Again, to parallel BRIDGES' forked deficiency (1917), I have obtained, by X-raying special stocks, known losses in the region of forked, which allow forked in the other chromosome to show to an exaggerated degree. And OFFERMANN and I, studying BURKART's (1931, 1932) Blond translocation, have been able to show that flies can be obtained from it which lack the right end of the second chromosome (this having been transferred to the X); in such flies the recessive speck, if present in the other second chromosome, manifests itself, and there is a plexus-like venation, as in BRIDGES' "Plexate deficiency."

There is now some evidence from Drosophila, but more especially from maize (McCLINTOCK 1931), that the two breaks in cases of double breakage within a chromosome may be at any distance apart, not being limited in their proximity by any principle of interference as rigorous as that which applies to crossings over. In view of this, and the above parallelisms, there can now be no reasonable doubt that the original proved "deficiencies" were small deletions, that is, actual removals of small regions, and so the studies involving them may now take their place definitely with the dosage studies. Later, I shall again refer to the results from this source. In the meantime, before the status of these deficiencies was established, I undertook, with the assistance of MISS LEAGUE, purposely to produce fragments containing known genes, and to use these for studying the effects of dosage changes.

Hypomorphic mutations

The first locus which we undertook to study was that of white eye. We chose first flies containing the moderately pigmented mutant allelomorph of white called eosin, in which the color is considerably lighter than the normal red, and is distinctly sexually dimorphic, being much lighter in the male than in the female. By irradiation we produced a deleted X chromosome containing this gene. It was then found that the addition of this fragment to a male or female which was otherwise an ordinary eosin caused the eye color to become darker, more nearly like the normal red. This shows that the actual effect of the eosin gene is not to inhibit color, as might have been thought by comparison of it with red, but to produce color, since the addition of more of it results in more color,—*only it does not produce as much color as the normal "red" allelomorph does*. In the male, the addition of the fragment raises the dosage to two, and results in a color like that of the ordinary eosin female, which of course has two doses, while adding the fragment to the female, and so raising the dosage to three, results in a still darker color. This shows that the sexual dimorphism of eosin is due to the difference in dosage normally existing between the two sexes, and not to a difference in the action of the gene in male and female.[7] That the above observed results were not to be explained as effects of the excess dosage of other genes than eosin in the extra fragment was shown by producing a slightly smaller deleted X chromosome, not containing the locus of eosin, and repeating the same tests with it. It was found to have no effect upon the eye color.

The allelomorph of eosin known as apricot, which has a similar coloration except that male and female are alike, was then tried in the same way as eosin. It was thought that it might not show a phaenotypic effect of dosage changes, since the female with two doses looks like the male with one dose, but it responded similarly to eosin, additional doses darkening the color. Two doses of apricot in the male, therefore, give a considerably darker color than two doses in the female. Evidently it is the difference in dosage of other genes in the X chromosome of male and female which, interacting with the effect of apricot, causes the color, for a given dosage of apricot, to be darker in male than in female, in fact, just enough darker so that one dose in the male gives about the same phaenotype as two doses in the female. The same is presumably true of most of the other members of the white series of allelomorphs, which, except for eosin and ivory, look nearly the same in the two sexes.[8] The important thing for us now, however,

---

[7] For this reason, eosin cannot legitimately be used as an indicator of sex in such experiments as those of BRIDGES, in which he sought to demonstrate the female character of haploid tissue. That the haploid tissue was dark eosin, as in a female, was doubtless due to the fact that one dose of eosin, with one dose of all other genes, involves the same ratio as two eosins in a diploid, and was not due to the tissue being female. In the present author's opinion haploid tissue of Drosophila containing but one X should in fact be female, but the matter cannot be demonstrated by the use of eosin as a sex marker.

[8] In a recent publication, MORGAN, BRIDGES and SCHULTZ (1931) include cherry among the strongly sexually dimorphic members of the white series. This was certainly not true of the original cherry (see SAFIR 1913). The present sexually dimorphic stock, labelled "cherry Abnormal," contains neither cherry nor Abnormal abdomen, but is doubtless an ordinary eosin that either displaced the cherry by contamination or was mislabelled.

is that apricot, like eosin, is a mutant gene which produces an effect similar to that of the normal allelomorph, but a lesser effect. That is, it works in the same direction (towards the same *superficial* end result) as the normal allelomorph, but not so strongly. It is, *in this sense,* like a lesser-normal. I therefore call it a "hypomorphic" mutant.

The above results agree perfectly with the findings of Mohr that if either apricot or eosin is in one X chromosome of a female, and the other X has Notch-8 deficiency, which includes a deficiency for this locus, the color is lighter than in the homozygous female. As mentioned above, the same result was obtained when this part of one X was known to have been removed by X-rays. Thus, one dose of this gene produces an effect less like normal than two, and two doses less than three.

Similar tests involving known additions or losses of fragments, or both, were then applied to genes in a number of other loci. A deleted fragment containing the gene scute-1 was first produced and was used to study the effect of increased dosages of scute-1, a gene which is said to "remove" certain bristles. (See, for example, Sturtevant in these *Proceedings.*) As with apricot, eosin, and bobbed, so here, the addition of an extra dose of scute in male or female made the individual more nearly normal, in this case almost completely normal, while the presence of two extra doses tended to result in slightly more of certain bristles than are present in the normal. Scute-1 is therefore a hypomorph. It does not "remove" bristles, except by comparison with normal. It produces them, though not as efficaciously.

In line with this conclusion derived from hyperploids, Agol (1932) found, by the use of a chromosome (from scute-19) from which we knew the extreme left end, containing the scute locus, had been removed, that a female with just one dose of scute-1 has fewer bristles than one with two. The test of the effect of underdoses, as seen in hypoploids, is obviously as valid and informative regarding these problems as the test involving overdoses in hyperploids. What Mohr has named the "exaggeration phenomenon" shown by deficiencies is, then, in our terminology, the lesser effect of one dose of a hypomorphic gene than of two doses. By this test the other mutant allelomorphs of scute, in which other groups of bristles tend to be absent, are also hypomorphic, as Agol (1932) found; facet is hypomorphic, as shown by Mohr's deficiencies and Patterson's cases of known losses; and forked is hypomorphic, as shown by my experiment previously cited. In elucidation of the test for forked, it may be explained that in this experiment females were made up which possessed one entire X bearing forked and having attached to its right end an extra piece consisting of the region from Bar to the right end; these females also possessed another X that had contained the scute-8 inversion but that had had the distal ("left") end of this chromosome removed up to a point between forked and scalloped. Hence all regions were present in double dose except a small region between scalloped and Bar, containing the forked locus. These haplo-forked hypoploids were markedly forked, phaenotypically.

Tests thus far indicate that most mutant genes (both spontaneous and induced) are hypomorphs, inasmuch as they show "exaggeration" with deficiencies, as MOHR has pointed out, or at least give a form having about the same degree of abnormality as the homozygous mutant. The latter relation would be expected in cases like white eye, where the mutant gene had nearly reached the bottom of the scale of effectiveness and hence itself had almost as little normal effect as the deficiency had. This latter type of mutant may, descriptively, be called "amorphic."

These hypomorphs and amorphs are just the kind of mutants which the few remaining advocates of the presence-and-absence hypothesis, and the advocates of purely quantitative mutation, require as evidence for their views. It should be noted, however, that their having a lesser effectiveness than the normal allelomorph by no means proves that they themselves involve material losses. They may consist of partial inactivations, or they may give rise to processes that lead in a somewhat different direction, and hence do not work so effectively in the observed direction, or they may involve conflicting tendencies. Moreover, a given mutant allelomorph (whether spontaneous or induced) may be very hypomorphic, or practically amorphic, in regard to one kind of activity of the normal gene, and normal or nearly normal in regard to another kind of activity. This is well exemplified in the scute series, in which each different allelomorph acts hypomorphically only in respect to its own peculiar combination of bristles, and is normal or nearly so in its action on other bristles. Since, in a comparison of different allelomorphs, the amount or intensity of effectiveness may vary separately from the types of effect, and both of these in turn may vary separately from the number or extensity of the effects, advocates of the quantitative view would here be driven to admit the existence of various parts of the gene, and to assume that these parts could vary quantitatively more or less independently of one another. This would be a distinct retreat from the simple hypothesis of quantitative variation of the gene as a whole.

Whatever the explanation of hypomorphism may be, it is of interest to observe that the finding that most mutant genes are of this type conforms to WRIGHT's contention (1929; see also MULLER 1928b, pp. 259-260) that gene mutations should in the majority of cases involve more or less inactivation of the processes governed by the normal gene, and that these less active genes should more often act as recessives to the normal than as dominants. This implies that one dose of the normal gene usually has an effect more nearly like that of two doses than of no dose. Whether the latter principle is a primary one, however, or is due to the past selection of modifiers, is another question.

On the compensation of the effects of dosage differences between the sexes, and on dominance

In the above connection, it will be worth while to make somewhat of a digression, to consider a curious fact that has emerged from the results concerning hypomorphs. That is, it appears that in the great majority of the cases of hypomorphic sex linked genes, one dose in the male produces about as strong or at times even a slightly stronger effect in the direction of normality than do two doses in the female. This must of course be due to the interaction of other genes in the X chromosome, whose simultaneous change in dosage affects the reaction.[9] In some cases at least it has been possible to show, by studies of the effects of different chromosome pieces, (a) that genes other than the genes for sex are acting as the "modifiers" in question, (b) that the modifiers responsible for the dosage compensating effect on different loci are to some extent different from one another, and (c) that more than one modifier may be concerned for a specific locus.[10] I base these conclusions on various results obtained in work of OFFERMANN, who has been especially active in the study, of PATTERSON, and of myself.

We may for convenience call these genes "modifiers," but with the reser-

---

[9] We arrived at our main results and conclusions regarding this phenomenon of dosage compensation in the spring of 1930. Although we communicated our results to Doctor STERN at that time (prior to the remarks made by STERN and OGURA 1931, upon this topic), we withheld our preliminary report (MULLER, LEAGUE and OFFERMANN 1931) until after certain checks had been carried through.

[10] Judging by certain results recently reported by MORGAN, BRIDGES and SCHULTZ (1931), the second-chromosome mutation Pale (associated with BRIDGES' original translocation) has, in addition to a "diluting" effect, an effect on the different eye colors of the white series similar to that produced by lessening from two doses to one the gene or genes in the X chromosome that are responsible for the dosage-compensation of most members of this series (thus, those allelomorphs of white that are lighter in the male are lightened by Pale, but the others are darkened somewhat). This means that the chemical process affected by Pale is the same as, or in its effect similar to, that affected by the dosage compensator(s) of the X; but, since we have seen that there is no reason to identify the latter with the gene or genes in the X that decide sex, we have no reason to agree with the suggestion of the above authors that "the translocation (Pale) may be closely connected with the sex-determining reaction."

vation that they may sometimes be as important in the causation of the phaenotypic effect as the "primary" gene whose mutations we have available for study. The essential relation is that, in so far as the amount of phaenotypic effect produced by this so-called "primary" gene depends on its dosage, it does not depend at all on the mere "concentration" of this gene in the cell, nor on the relation of its dosage to that of the other genes in general, still less to that of the autosomal genes, but solely on the ratio of its dosage to that of another specific gene or genes which lie in the same chromosome (the X). That a relatively high amount of intra-chromosomal interdependence in regard to dosage expression existed among sex linked genes was realized some time ago (MULLER 1930b) and denoted as "intra-chromosomal genic balance." In that work, however, we were dealing with those relatively rare normal genes, or gene-combinations, which have a quite different effect, visibly, in one dose than in two. The present findings go much further, in showing the existence of a far stronger interdependence, and one which applies not just to a relatively few scattered genes but to the great majority of the individual genes in the X which can be sampled.

Now this great system of "modifiers," all acting to give a similar sort of effect, and probably affecting most of the genes of the X chromosome, must have a function. It cannot be that of giving the male *mutant* as strong, that is, as nearly normal, an expression of its mutant gene as the homozygous female mutant has. It must therefore be a system which acts on the normal allelomorph similarly to the mutant, but the action of which is more readily apparent to our eye in the mutant type. In most cases the normal gene gives, so far as our eye can perceive, practically the same effect in one as in two doses. Nevertheless, there must be some difference which, though imperceptible, is important for survival; otherwise this system of genic interaction would not be thus maintained to keep the same optimum degree of effect in both sexes, despite the different doses. It follows that the dominance of the normal gene over its "absence" is really far from perfect, physiologically (that is, that one dose is not really as effective as two), though it may seem so to the casual genetic observer, and that by selection a system of interacting genes has become established such that the expression of the one dose in the haplo X type is like that of the double dose in the two X type. Bobbed, being present in double dose in male as well as in female, is, *as expected,* an exception to this rule of dosage compensation, in *Drosophila melanogaster.* In *Drosophila simulans,* on the contrary, bobbed does show dosage compensation, and here it is found, correspondingly, that the male carries only one dose of the "normal" allelomorph (the Y being sometimes neutral and sometimes actually "antimorphic" in effect—see page 245—as shown by results of STURTEVANT [1929]).

The question may here be raised: Why were the normal allelomorphs of most of the sex linked genes other than the bobbed of *D. melanogaster* ever "lost" from the Y chromosome if their absence was so deleterious as to require the subsequent evolution of this complicated compensation system? The case of the Y of *simulans* shows that they can be thus "lost," or, better to say, changed in expression like a loss, and this would seem to point to the importance of *accidental* multiplication, not guided by selection, as an occasional evolutionary process. It may be, however, that most of the genes in the X, unlike bobbed, never were present in the Y, in anything like their present form, at least; that is, that the male has had but one dose of them from the beginning of their existence as such. In that case, they must have arisen either as duplications, as "neomorphs" (see page 246), or both, after the present sex-determining system had already become established. This question might be answered definitely by genetic analysis in a species in which we knew that a part of the X had been derived from an autosome (for example, *D. hydei* or *"obscura"*?).

The existence in the X of "modifiers" of such a specific kind that, by their change in dosage, they modify the amount of effect of other sex linked genes to the extent required to make the male and female alike, indicates that specific modifiers of gene action are plentifully available. In the case of some of the sex linked genes arising in the manner last suggested (so as to have existed in different dosages in the two sexes from the start) it is possible that the dosage compensation did not result from the selection of mutations in these modifiers but that the "primary" genes themselves were so selected at the time of their origination as to be, *ab initio*, adapted in their action to the other, preëxisting genes in the X which we now call "modifiers." And even when the compensation did not thus exist from the start, it is likely that the inter-adaptation of primary gene and modifier did not always occur through changes in the modifier alone but also through changes in the primary gene that made the latter sensitive to the modifier (such changes in the primary gene as would be involved in the mutation of eosin to apricot, for example). Thus, where there is only one modifier causing the dosage compensation of a gene that did not have this property to begin with, the chances seem *a priori* to be equal that the dosage compensation arose (if by one step) by a further change in the primary gene itself, or by a mutation in the modifier; the greater the number of modifiers, the larger the rôle that their mutations have probably played in the process, as compared with mutations in the primary gene. We should also remember that mutations could also take place in other genes, for example, autosomal genes, which would serve to bring the primary gene and the modifier into the reciprocal relation with one another which they now have. But, however all that may be, the results do give evidence of the availability of "modifiers," or, to put it more precisely, of mutations which cause certain specific types of nicely adjusted genic interaction, favorable for survival, and not having this survival value too much obscured by pleiotropic effects.

The above conclusion would appear to lend support to FISHER's theory of the origin of dominance, inasmuch as on that theory, too, specific modifiers (albeit of a somewhat different kind), without important other effects of their differences from their own parent genes, are called for. It would also allow us to adopt to a certain extent the suggestions of HALDANE, concomitantly. There is, however, an important difference between the mechanism of selection for dosage compensation here studied and that postulated either by FISHER or by HALDANE for the modification of dominance. For in the former the selective moment, if I may call it so, exists throughout the population, while in the latter it is supposed to be limited to a comparatively small minority. Thus the difficulty is encountered that the pressure of the selection in question may be too small, as compared with that of mutation, or of the selection for even very weak pleiotropic effects.

I believe that the above difficulty can be avoided and a better case made out for the origin of dominance by selection if we assume that this selection has had a somewhat different mechanism from that previously postulated. I prefer rather to postulate that the mutations favoring dominance—the genes or genetic conditions which tend to make the heterozygote like the homozygote—have been selected and are maintained not so much for their specific protection against heterozygosis at the locus in question as to provide a margin of stability and security, to insure the organism against weakening or excessive variability of the character by other and more common influences—environic and probably also genetic. These modifiers must so affect the reaction set going by the primary gene in question as to cause this gene, when in two doses, to be near an upper limit of its curve of effectiveness,[11] that is, in a nearly horizontal part of the curve, not so readily subject to variation by influences in general, *including* reduction in the dosage

[11] That is, the curve expressing the relation of amount of phaenotypic effect (the ordinate) to the amount or concentration of gene material (the abscissa)—a curve which must usually, in its right-hand portion, rise with ever decreasing slope, approaching a horizontal limit, as seen, for instance, in STERN's studies on bobbed and in ours on scute and apricot.

of the primary gene. This does not mean that the phaenotype is necessarily made any more extreme, for counter-checks can be set up. That is, the level of the curve as a whole and its shape, as well as the region wherein it approaches a horizontal limit, are also adjustable, by means of modifying mutations that reframe the conditions under which the reaction takes place. Such modification (see FORD 1930), and not merely an increase in potency of the "primary" gene, will be necessary in the numerous cases in which the curve of effectiveness did not, originally, approach the horizontal within physiologically acceptable limits (or did not do so at all).

It should be distinctly understood that the crux of the above view of the origin of dominance lies in the proposition that, where a change in gene dosage causes a perceptible change in its phaenotypic expression (that is, when it is in a noticeably sloping part of its "curve of effectiveness"), it is likely that the degree of expression of the character will be modifiable to an unfavorable extent by environic and by other genetic changes. This seems reasonable, *a priori,* inasmuch as some of the disturbing influences would be expected to act by altering the reaction in a way similar to that whereby the change in gene dosage would alter it, and hence would tend to be similarly effective.

But we need not rely on *a priori* reasoning alone. There is a significant amount of experimental evidence already existing to show that there is considerably more phaenotypic variability in the expression of hypomorphic mutant genes than of their normal allelomorphs. Now, these hypomorphs evidently cause a reaction of a type similar to that of their normal allelomorphs, but a weaker or lesser reaction, one which, unlike that of the normal allelomorphs, is much affected by dosage changes. This variability is true of all the known hypomorphs yet studied: namely, all the hypomorphs of the scute series, the white series, the forked series, and the bobbed series (excluding the amorphs, which afford a converse test of the same proposition). The same variability applies also, as we should expect, to the effect of normal allelomorphs in single dose in those relatively rare cases in which the single dose has a perceptibly different (that is, lesser) effect than two: these cases comprise Notch wings, Plexate venation, and several Minute bristle conditions. All told, the evidence given above may be sufficient to show the truth of our proposition as a usual rule. It is not necessary to claim for it, nor do we believe that it has, the validity of a universal law.

In conclusion, we may call attention to the bearing of a further fact, derived from our study of dosage compensation, on the problem of dominance. We have seen that in all probability many of the normal genes in the X chromosome have this dosage compensation, despite the fact that, even in the female, there is hardly a perceptible difference between the effect of one dose and of two. This indicates that, even though the normal gene produces its effect in what appears to us to be a nearly horizontal region of its curve of effectiveness (where changes in dosage produce little discernible effect), nevertheless there is a distinct influence, unfavorable to the organism and perceptible in its survival rate, if the effect is made either slightly stronger or slightly weaker. The disadvantage of a stronger effect is shown by the fact that, in the female, the strength of effect has become fixed at so low a level as to call for dosage compensation. For in a sense the dosage compensation may as rightly be regarded as a means of keeping the female from having too strong an action of the gene as a means of giving the male a strong enough action. If twice as high potencies in the female were biologically acceptable, this relatively simple change should often have been utilized (that is, have survived) whereby the male would automatically have been provided with a sufficiently high potency to obviate the need for dosage compensation. We must conclude, then, that in the fixing of the conditions determining dominance too, it was not feasible merely to increase the potency of the "primary" gene; instead, the characteristics of its curve of effectiveness had somehow to be altered.

Experimental evidence of a different nature, indicating that dominance is not a primary property of genes but must have become developed by selection, is given in the section on neomorphs (see page 248).[12]

Hypermorphic mutations

We must now return from our digression, which has perhaps helped us to understand why hypomorphic mutant genes usually show dosage changes better than do the normal genes from which they were derived, and are recessive to the latter. The question next arises: are all mutant genes hypomorphic? This can be answered categorically in the negative.

Since it has been found that there are reverse mutations of hypomorphic mutant genes, such as scute, apricot, and forked, both spontaneously and as a result of irradiation, we must regard the allelomorphs thereby resulting not as hypomorphic but as hypermorphic to their immediate progenitor genes. Whether or not such a change involves a real increase of material is a doubtful question, subject to the same considerations as applied, con-

[12] I am indebted to Doctor C. R. PLUNKETT for calling my attention to the fact that in a paper (1932) presented independently to the same congress he has espoused what is essentially the same viewpoint regarding dominance as that given in the above section.

versely, to the hypomorphic mutations. TIMOFÉEFF-RESSOVSKY (1929, 1931a), as well as PATTERSON and myself (1930), has discussed in some detail the varying frequency of such changes for different loci and allelomorphs.

Now if hypermorphic changes of already mutant allelomorphs may occur, resulting in partial or complete reverse mutations, there might well be hypermorphic mutations of normal genes also, resulting in changes of a type opposite to that of our ordinary mutations. Usually these would be difficult or impossible to detect, on account of the fact previously referred to that two doses of the normal gene are already at nearly the maximum point in the curve of effectiveness. Thus such changes would be apt to escape observation. Very likely, however, NASARENKO's (1930) mutant "abrupt" is a hypermorphic mutation of the normal allelomorph of Notch, for it is at or near the Notch locus, and it and Notch deficiency counteract each other instead of showing an exaggeration effect.

Antimorphic mutations

What evidence have we for other mutational changes than such as could be explained as mere diminutions and increases? The dominant (somewhat variegated) allelomorphs of brown eye in chromosome II are a case in point. When there is one dose of the recessive brown and one of the normal gene, the latter dominates and the phaenotype is red. But, as GLASS and I have found (see GLASS 1932), when to the above complex a dose of the dominant allelomorph of brown is added, the result is a brownish (somewhat variegated) color. It may be explained that this combination is produced by making up a fly that is a compound of recessive brown and dominant brown, and carries as excess a fragment of the second chromosome derived from BRIDGES' "Pale" translocation; this fragment contains the normal allelomorph of brown. The resulting brownish color shows us that the addition of dominant brown to a heterozygote of normal and recessive brown has a real effect and involves the addition of some kind of gene material different in its effect from the material in the normal gene. This effect, the color change, lies in the same direction from normal as does that of the recessive brown, as comparison of the colors indicates. This is shown more conclusively by the fact that while a hyper-diploid containing one dose of dominant brown and two of normal has practically normal red eyes, a hyper-diploid otherwise similar to the above but with a dose of recessive brown substituted for one of the normals has brownish (somewhat variegated) eyes— that is, the substitution of recessive brown in place of normal results in a better manifestation of dominant brown. But the recessive brown itself acts practically as an amorph, since the addition of a dose of it, as an extra, has practically no effect either on the incompletely brown color of the heterozygote of dominant brown and normal or on the red color of the heterozygote of recessive brown and normal. Hence the dominant brown represents something that differs from normal, in its effect, in the same direction as a loss does, but more strongly. In a sense, it has an actively negative value. More accurately, it has an opposite action to that of the normal allelomorph, competing with the latter when both are present.

A similar conclusion may be drawn with regard to the mutant gene ebony, of chromosome III. For, starting with a hyperploid containing two ebony genes and one normal (derived from translocation II-III26—PAINTER and MULLER 1929), as a basis of reference, we find that the subtraction of one ebony makes the color lighter, while the subtraction of the normal makes it darker. I would term such antagonistic mutant genes, having an effect actually contrary to that of the gene from which they were derived by mutation, *antimorphic*.

Abnormal abdomen may now be interpreted to be a member of this class, as shown by results in MOHR's experiments with Notch-8 deficiency. In the first place, it is to be observed that the gene for Abnormal produces a change in the same direction as a loss of the normal gene. This is shown by the fact that if we start with a heterozygous fly having one Abnormal and one normal gene (this is somewhat Abnormal in appearance), the substitution of a real loss (Notch-8 deficiency) for the normal gene in it intensifies the Abnormal abdomen character. But the Abnormal gene, though thus producing a change in the same direction as a loss of the normal gene, acts more strongly in this same direction than a mere loss does. This in turn is shown by the fact that homozygous Abnormal flies are still more Abnormal in appearance than are the compounds of Abnormal and deficiency. That is, the degrees of phaenotypic Abnormality, as found by MOHR, were as follows:

$$\frac{Ab.}{Ab.} > \frac{def.}{Ab.} > \frac{norm.}{Ab.} > \frac{norm.}{def.} = \frac{norm.}{norm.}$$

Since in the first three terms of the series the gene represented below was always the same, the observed differences prove the degree of abnormal effects to be in the order $Ab > def > norm$.

It may be mentioned that a recessive allelomorph of Abnormal has been produced by X-rays. It will be seen that in such cases the recessive mutant, though classifiable as an amorph or possibly a weak hypomorph, probably involves no mere loss of material, since what is apparently a still greater change in the same direction gives a gene which again has a demonstrably active influence.

Unless we make the very improbable assumption that the Y may contain other active genes than bobbed influencing the same character, we may also include among antimorphs the gene existing in the Y of most races of *Drosophila simulans* (see STURTEVANT 1929) which (unlike the bobbed allelomorphs reported upon by STERN 1929) actually decreases the bristle length of males containing bobbed in their X. It is also possible that the genetic conditions designated as Minute 1^2 and Plexate include antimorphs—in fact, such is the conclusion which we should ordinarily draw from a recent report; on the other hand, an earlier report interprets these conditions as deficiencies (see MORGAN, STURTEVANT and BRIDGES 1927, and MORGAN, BRIDGES and SCHULTZ 1931). Possibly the apparent contradiction is due to the effect of dosage changes of other genes in the added fragments rather than to the genes in question (that is, an intraregional dosage interdependence). Fortunately, this possibility can rather easily be put to the test in these cases (in part at least), since a smaller fragment involving the region in question is available in the Blond translocation, and others can rather readily be manufactured. In the meantime, the "position effect" interpretation is not excluded here, nor is that of gene mutation accompanying breakage.

Neomorphic mutations

Somewhat different from the negatively acting, competing mutant genes, or antimorphs, is the class which I am provisionally terming "neomorphs." A good example is the dominant mutant, Hairy wing, near the left end of the X chromosome. The homozygous Hairy wing female is about twice as hairy as the heterozygous Hairy wing female or the Hairy wing male (this constituting an exception to the dosage compensation rule for sex linked genes). The relatively low grade hairiness of the heterozygous as compared with the homozygous female, in this case, is due solely to the single dose condition of the gene for Hairy wing and not at all to a possible influence of the normal allelomorph in the heterozygote. For if a small piece containing this region be broken off of a normal X chromosome, and added either to the heterozygous or homozygous Hairy wing female or to the Hairy wing male, there is no diminution of the hairiness. On the other hand, if a small piece containing a Hairy wing gene be added to an individual otherwise normal, Hairy wing will show. The normal allelomorph thus fails to compete. It itself acts like an *amorph,* so far as its detectable effect on the character under consideration is concerned. Yet it is no mere absence; it has a material existence, for Hairy wing has arisen at the same locus several times (including at least twice by irradiation).

We must conclude from the above results that the mutation to Hairy wing does not result from an addition of material transferred from another locus (since the mutation always reappears at the same locus). It must rather be a change in the nature of the gene at the original locus, giving an effect not produced, or at least not produced to an appreciable extent, by the original normal gene. If the effect had been produced to some appreciable extent by the normal gene also, then the addition of a dose of the normal to the Hairy wing individual should have actually increased hairiness.

The fact that normal genes may thus act as amorphs with regard to a particular character affected by their mutations should serve as another warning against regarding mutant genes that seem to be amorphic or hypomorphic as really involving a mere absence or loss of material. The obtaining of reverse mutations from near-amorphs, such as eosin from white, gives further evidence for this conclusion.

The same kind of finding as above noted for Hairy wing—namely, lack of effect on the character when extra doses of the normal allelomorph are added—was observed by OFFERMANN in studying the spontaneously arisen dominant, Blond, of BURKART. This interpretation holds only if we regard Blond as having its locus in the X chromosome. This is uncertain as Blond lies near the break of a mutual translocation involving X and II (see BURKART 1932), but as Blond follows the sex linked rule of dosage compensation it is in all probability in the X. We are, however, making sure of its neomorphism by testing also the effect of adding an extra dose of the suspected region of chromosome II.

Bar eye is a third neomorph. It is well known that STURTEVANT has considered Bar as having no normal allelomorph, at least none at the same locus as itself. However, the recently reported finding, by DOBZHANSKY (1932), of a second Bar-like mutation ("baroid"), induced by X-rays at the same locus as the old, indicates to me that this locus normally contains a gene that is subject to this particular type of mutation, although DOBZHANSKY still believes that the normal allelomorph was somehow transported there from another locus, at the time of the mutation. BRIDGES' original Bar-deficiency of 1915 (published upon in 1917), which we may now interpret definitely as a loss, shows that the absence of the Bar-locus in the non-Bar chromosome of a heterozygous Bar female has the same effect on the Bar eye character as the presence of the normal allelomorph itself, and STURTEvant's work on chromosomes which have lost the Bar locus by unequal crossing over is an indication in the same direction. (There is a possibility that in the origination of Bar a gene became duplicated *in situ,* and that one of the resulting twins mutated at the same time. On this rather special hypothesis the mutation would have been of the neomorphic type. But in that case the normals formed from Bar by unequal crossing over would not represent complete "absence.") On the other hand, increased doses of Bar give the abnormal effect more strongly, just as we find for Hairy wing and Blond, and unlike the situation in the case of hypomorphs.

While THOMPSON (1929) has raised some objection that we may here be adding and subtracting only a part of the gene, in getting these effects, this possibility is ruled out in some recent studies of OFFERMANN using a strong allelomorph of Bar ("Super-Bar," B^s, found by STONE) that exists in a chromosome fragment. The addition of fragments containing the whole Bar gene had the expected effect of increasing the bar-like character of the eye in a clear-cut fashion. OFFERMANN likewise proved that this result could not be due to the excess dosage of other genes in the piece. Bar, then, is a mutation of a normal gene, giving a gene that produces a new effect, foreign to the original gene, and not competing with the latter. It is very probable, however, that the new effect is in some way related to that of the normal allelomorph. For it is evident that Bar obeys the usual rule of sex linked genes, having the male, with his one dose, much more nearly like the homozygous female, with her two doses, than like the heterozygous female (see also the case of Blond, and note the contrast with that of Hairy wing).

A recently published mention by MORGAN, BRIDGES and SCHULTZ (1931) of the lack of effect of changes in dosage of a fragment containing the normal allelomorph of Bristle on the degree of expression of this second chromosomal dominant leads to the conclusion that it also must belong in the class of neomorphs.

It might yet be possible to evade the obvious conclusion that gene mutations, including those produced by X-rays, involve qualitative changes, changes in the kind of structure and not merely in the quantity of the gene or its parts. For it might be postulated that in all cases of neomorphs there was an imperceptible rudiment of the part which produced the effect in question, already present in the normal gene, and that this part merely became vastly increased in amount by the "mutation." Or it might be postulated that all such changes were "position effects," caused by gene rearrangements. While there are an exceptionally large number of rearrangements both among known neomorphs and antimorphs, there are cases—Hairy wing, Bristle, Dominant eyeless, Abnormal abdomen—which do not involve such changes, unless we suppose the rearrangement to be on such a minute scale as to escape detection. Both these paths of escape into the ultra-small would, however, be pure speculations, the burden of proof for which would rest upon the advocate thereof.

It does not seem to be a coincidence that more loci have yielded hypomorphs than neomorphs, and that even loci which have yielded neomorphs have done so with relative infrequency. These results, if corroborated by more extensive work, would speak for the correctness of the principle put forward by WRIGHT (1929; see also MULLER 1928b, pp. 259-260) that mutations having an effect in the direction of losses (that is, those that tend to be disorganizing and inactivating) should in general be more frequent than those causing increased or new effects. But while this principle is necessary as one basis for WRIGHT's theory of dominance, it is not, alone, sufficient for a derivation of the latter; neither is it contradictory to the general viewpoint put forward by FISHER that the usual dominance of normal genes has been developed through natural selection. It is to be noted, further, that the hypomorphs tend to be recessive, and the neomorphs "dominant." This again is in line with WRIGHT's view, but it is also in line with FISHER's (since any given neomorph originates so infrequently that there has been much less chance for selection to have affected its mode of expression), and it is still more in line with the idea previously offered (p. 240), that selection has worked primarily towards the stabilization of the reactions of the normal, homozygous genes. (In the latter case, even rather frequently recurring neomorphs would tend to be dominant.)

When, however, we examine into the type of dominance found, we obtain a result of greater apparent significance. For while the recessiveness of the hypomorphs is usually fairly complete, as generally expected, *the "dominance" of the neomorphs is in most cases far from complete, being of the "intermediate" type.* Now this result is exactly what we should expect if dominance of the nearly complete type has been developed by selection (especially, if by the type of selection advocated on page 240), but it is a considerable surprise, in fact, it seems contradictory to the idea that such dominance is usually a primary property of the gene. It will therefore be important to examine further cases with reference to this question.

While we have spoken above of the general trends of the results, it should be emphasized that no absolute rules can be made with regard to the dominance of the different classes of mutants. A known loss like Notch-8, Plexate, and at least three known Minute bristle conditions, may be dominant or semi-dominant in its effect, and therefore an amorph or a hypomorph may be likewise. In these cases one dose of the normal gene has distinctly less effect than two. On the other hand, neomorphic genes may be so "weak" in their effect that two doses are required before they rise to the level of visible manifestation. This was very nearly true in the case of a certain Hairy wing mutant, and in the case of baroid in the female; under certain genetic conditions (for example, in the presence of ZELENY's modifier, called "emarginate") it was true of Bar itself, and under certain environmental conditions it was true of Abnormal abdomen. For the same reason, we cannot make absolute rules regarding the exaggeration of recessives and dominants by deficiencies. If the recessive or near-recessive should be a neomorph, like baroid, it will not show exaggeration by a deficiency; if the dominant should be hypomorphic, as in the case of the absence of coxal bristles in some scutes, it will be exaggerated by a deficiency. But the more usual case is the recessive hypomorph (for example, eosin, facet), which shows exaggeration, the amorph (like white) which shows no effect, and the semi-dominant neomorph (for example, Bar) and antimorph (for example, Abnormal), which show instead an apparent inhibition by a deficiency.

On our interpretation of most gene mutations as qualitative structural changes, even the distinction into classes above outlined is not an absolute one, and reflects rather the gene's final behavior than its real structure. So we may expect to find genes, for example, that are hypomorphic in one respect and neomorphic in another. Possible examples of this are scute-8, scute-12, and scute-M-4 (in deleted X 24); the two latter show certain semi-dominant Hairy wing effects, as well as hypomorphic scute characters, but it is as yet uncertain whether these effects are really referable to the same locus or represent group mutation or possibly effects of changed position.

Multiple allelomorphs forming non-quantitative series

There are already numerous cases known in which it can be shown that a given mutation has markedly changed a gene only in regard to certain of the effects which the original gene produced, while another mutation in the same gene changed it more pronouncedly in some other respects. This has been shown *par excellence* with regard to the various hypomorphic changes possible in the scute locus in the studies on scute allelomorphs carried on by the Moscow geneticists. One of their most important contributions lies in showing the richness of the different patterns of change possible in a given gene, since thus far very few of the numerous allelomorphs are indistinguishable from one another. That the tendency to certain kinds of groupings of effects on the different bristles is partly an expression of certain real features of gene structure, and will help us to understand the arrangement of gene parts, is also a reasonable conclusion.

Attempts to explain the matter in a simple quantitative way, as in GOLDSCHMIDT's criticisms, or by means of developmental relations, as in the Plunkett-Sturtevant-Schultz hypothesis of diffusion of influences from a center, fall in the face of the facts. We do not have time to mention the various logical difficulties which the latter hypothesis encounters in its actual working out. Suffice it here to say that a study of numerous gynandromorphs involving various scute allelomorphs has been carried out in our laboratory, chiefly by PATTERSON, and that the results show clearly that the development of bristles, in so far as it is under the influence of the scute gene, is not governed by one or a few centers, but is in its major features autonomous at the site of each bristle. On the other hand, later work throws grave doubt on the possibility of grouping all the effects into one exact line (this is equally against both the unmodified sub-gene hypothesis and the theories of GOLDSCHMIDT, STURTEVANT, et cetera). And the evidence that such a line, if it represents gene parts in a one-to-one correspondence, may be cut without destruction of either piece, is still to be found (see page 222).

This still leaves the locus of scute the most suitable yet found for the study of multiple allelomorphism and gene structure, and it leaves the sub-gene hypothesis, or some modification of it, as a possible interpretation, although the way is not as clear and easy as before. It will, I think, be profitable to follow the method there used, that of concentrating on intensive studies of the different kinds of mutations possible in individual genes, as induced by irradiation and otherwise.

Such studies as we have carried out on other loci than scute have shown somewhat similar phenomena, and in some respects amplify our view. For example, the cases now known are fairly numerous in which different recessive mutant allelomorphs of the same locus have effects which are to some extent, or almost wholly, different in their character or in their location on the organism. Thus, mutant allelomorph 1 may affect character A very much and B very little or not at all, while allelomorph 2 affects A little and B much. Such allelomorphs, when crossed, usually form a compound that is more normal than either. For, in respect to each character effect or body region, the more normal effect is usually the more dominant; that is, the compound is usually in each respect more like that allelomorph which has a more nearly normal effect on that character or region. This was evident, for example, in EMERSON's (1911) allelomorphs giving different combinations (= patterns) of red *versus* white silk, cob, grain, et cetera, in corn. In Drosophila, the first case was that of the truncate series (MULLER 1919,

1922b), which concerns not only different regions but different characters, and obeys the same rule throughout. Thus, in this case, the cross of vortex bristles by oblique wings was found to give a compound that was sensibly normal. To explain those members of this series which showed two or more of the effects at once, the interpretation of group mutation of neighboring but physiologically entirely distinct genes was early considered but it was rejected, chiefly because studies on the action of modifying genes as well as of "chief" genes at other loci showed the different developmental effects in question to be physiologically related. In this case, it was also observed that the groupings of effects of different allelomorphs fitted in with no linear series rule. The normal-appearing compound of achaete and scute-1 (found by DUBININ to be allelomorphs) falls under the same category as the vortex-oblique cross. So too may the normal compound of split bristles and recessive notch wings (GLASS and MULLER unpublished), and also certain effects observed by DOBZHANSKY (1930a) in the Stubble series of allelomorphs. The list could be considerably extended.

There are, however, exceptional cases, in which the compound is not more like the normal in respects in which the two allelomorphs differ. The best case of this is the appearance of leg-like antennae in the compound between aristopedia, which has such an effect, and its allelomorph spineless, which does not, as found by STURTEVANT (1929). A few of the missing bristle effects in scute crosses show a similar tendency; so too does the extra bristle effect in crosses of split bristle and facet-eye (see below).

We now have to report exceptions of the opposite type also, namely, those in which the compound is more like normal in respect to effects in which both allelomorphs are similarly abnormal. One such case is that of lozenge-eye in combination with a particular spectacled-eye allelomorph of it. The compound has a practically normal eye but has the female infertility common to both, and their mutual allelomorphism is further shown by the fact each gives a distinctly mutant eye type when crossed with still other members of the series (see PATTERSON and MULLER 1930, AGOL 1930). Another case is that of the ommatidial disarrangement in split bristles and facet-eye. Both of these mutants cause ommatidial disarrangement, yet (as with spectacled and lozenge) the compound has a normal eye (MULLER unpublished). Their allelomorphism is shown not only by their linkage but by their behavior with other mutual allelomorphs (notches) and by the appearance of extra bristles in the compound, as in split bristles by itself (see above). In such cases as these, we must draw the conclusion that **the two allelomorphs, although acting on the very same body region, and having** superficially similar effects on that region, nevertheless attain these effects through the intermediation of qualitatively different developmental processes. Further studies of the relations in such series are needed.

Ultimately, too, we must undertake the still more difficult study of the effects of successive mutations in the same gene, to discover, if possible, principles governing their continued evolution. Such evolution, as I see it, implies the possibility of qualitative change in the gene as a necessary condition. The foregoing illustrations, if taken together, afford, I believe, considerable experimental evidence for the existence of such a phenomenon, both as a natural occurrence and as a result of irradiation. And this conclusion remains likely no matter whether the mutational effects of irradiation are of a direct or an indirect nature.

For the rest, I fear that the present paper has raised far more questions than it has solved. But if some of these questions may thus have been opened to attack, our time may not have been wasted.

The author wishes to acknowledge with thanks the assistance of the Committee on the Effects of Radiation on Living Organisms, of the NATIONAL RESEARCH COUNCIL of the United States, in the prosecution of experiments referred to in the foregoing.

LITERATURE CITED

AGOL, I. J., 1930 Evidence of the divisibility of the gene. (Abst.) Anat. Rec. **47**:387.
   1932 Das Sichtbarmachen der verborgenen allelomorphen scute-Teile mit Hilfe von Faktorenausfällen (deficiencies). Biol. Zbl. **52**:349-367.
BOLEN, H. R., 1931 A mutual translocation involving the fourth and the X chromosomes of Drosophila. Amer. Nat. **65**:417-422.
BRIDGES, C. B., 1917 Deficiency. Genetics **2**:445-465.
   1919 Duplications. Abst. Anat. Rec. **15**:357.
   1923 The translocation of a section of chromosome II upon chromosome III in Drosophila. Abst. Anat. Rec. **24**:426.
BURKART, A., 1931 Investigaciones genéticas sobre una nueva mutación de *Drosophila melanogaster* determinante de excepciones hereditarias. Rev. de l. Fac. de Agron. y Veter. Univ. de Buenos Aires, Ent. II. Tomo **7**:393-491.
   1932 (in press in Biol. Zbl.)
DOBZHANSKY, T., 1929 Genetical and cytological proof of translocations involving the third and the fourth chromosomes of *Drosophila melanogaster*. Biol. Zbl. **49**:408-419.
   1930a The manifold effects of the genes Stubble and stubbloid in *Drosophila melanogaster*. Z. indukt. Abstamm.-u VererbLehre. **54**:427-457.
   1930b Translocations involving the third and the fourth chromosomes of *Drosophila melanogaster*. Genetics **15**:347-399.
   1931 Translocations involving the second and the fourth chromosomes of *Drosophila melanogaster*. Genetics **16**:629-658.
   1932 The baroid mutation in *Drosophila melanogaster*. Genetics **17**:369-392.
DOBZHANSKY T., and STURTEVANT, A. H., 1931 Translocations between the second and third chromosomes of Drosophila and their bearing on Oenothera problems. Pub. Carnegie Instn. **421**:29-59.

1932 Changes in dominance of genes lying in duplicating fragments of chromosomes. Proc. Sixth Int. Congress Genetics **2**:45-46.

DUBININ, N. P., 1930a Issled. stup. allelomorphisma. Tsentrovaya teoria gena achaete-scute. J. Exp. Biol. (Russ.) **6**.

1930b On the nature of artificially deleted X chromosomes. J. Exp. Biol. (Russ.) **7**:365-368.

EFROIMSON, W. P., 1931 Die transmutierende Wirkung der X Strahlen und das Problem der genetischen Evolution. Biol. Zbl. **51**:491-506.

EMERSON, R. A., 1911 Genetic correlation and spurious allelomorphism in maize. Ann. Rep. Nebraska Agric. Exp. Sta. **24**:

FERRY, L., N. T. SHAPIRO and B. N. SIDOROFF, 1930 On the influence of temperature on the process of mutation, with reference to Goldschmidt's data. Amer. Nat. **64**::570-574.

FISHER, R. A., 1928 The possible modifications of the responses of the wild type to recurrent mutations. Amer. Nat. **62**:115-126.

1930 The genetical theory of natural selection. Oxford Univ. Press. 272 pp.

FORD, E. B., 1930 The theory of dominance. Amer. Nat. **64**:560-565.

GERSHENSON, S. M., 1930 Phenomenon of reinversion in the sex chromosome of *D. melanogaster*. Reports to Fourth All-Union Congress of Zoologists, etc., p. 7 (Russ.)

GLASS, H. B., 1932 A study of dominant mosaic eye-color mutants in Drosophila. Proc. Sixth Int. Congress Genetics **2**:62-63.

GOLDSCHMIDT, R., 1929 Experimentelle Mutation und das Problem der sogennanten Parallelinduktion. Biol. Zbl. **49**:437-448.

1931 Die entwicklungsphysiologische Erklärung des Falls der sogenannten Treppenallelomorphe des Gens scute von Drosophila. Biol. Zbl. **51**:507-526.

GRAUBARD, M. A., 1932 Inversion in *Drosophila melanogaster*. Genetics **17**:81-105.

HALDANE, J. B. S., 1930 A note on Fisher's theory of the origin of dominance and on a correlation between dominance and linkage. Amer. Nat. **64**:87-90.

HAMLETT, G. W. D., 1926 The linkage disturbance involved in the chromosome translocation I of Drosophila, and its probable significance. Biol. Bull. **51**:435-442.

HANSON, F. B. and F. HEYES, 1929 An analysis of the effects of the different rays of Radium in producing lethal mutations in Drosophila. Amer. Nat. **63**:201-213.

1932 Radium and lethal mutations in Drosophila. Amer. Nat. **64**:335-345.

HANSON, F. B., F. HEYES and E. STANTON, 1931 The effects of increasing X-ray voltages on the production of lethal mutations in *Drosophila melanogaster*. Amer. Nat. **65**:134-142.

JOLLOS, V., 1930 Studien zum Evolutionsproblem. I. Biol. Zbl. **50**:541-554.

MCCLINTOCK, B., 1931 Cytological observations of deficiencies involving known genes, translocations and an inversion in *Zea mays*. Missouri Agric. Exp. Sta. Bull. **163**: 30 pp.

MOHR, O. L., 1919 Character changes caused by mutation of an entire region of a chromosome in Drosophila. Genetics **4**:275-282.

1923 A genetic and cytological analysis of a section deficiency involving four units of the X chromosome in *Drosophila melanogaster*. Z. indukt. Abstamm.-u Vererb-Lehre. **32**:108-232.

1927 Exaggeration and inhibition phenomena. Norsk Videnskaps-Akad. i Oslo. I. no. **6**:19 pp.

1929 Exaggeration and inhibition phenomena encountered in the analysis of an autosomal dominant. Z. indukt. Abstamm.-u. VererbLehre. **50**:113-200.

MORGAN, T. H., A. H. STURTEVANT and C. B. BRIDGES, 1927 The constitution of the germ material in relation to heredity. Carnegie Instn. Year Book **26**:284-288.

MORGAN, T. H., C. B. BRIDGES and J. SCHULTZ, 1930, 1931 The constitution of the germinal material in relation to heredity. Carnegie Instn. Year Book **29**:352-359 and **30**: 408-415.

MULLER, H. J., 1919 A series of allelomorphs in Drosophila with non-quantitative relationships. Address before Amer. Soc. of Nat. at Princeton, December 31, 1919. Title in Science, 1920.

1922a A lethal gene which changes the order of the loci in the chromosome map. Address before Genetics Sections at Toronto, December, 1921; title in Anat. Rec. **23**:83.

1922b Variation due to change in the individual gene. Amer. Nat. **56**:32-50.

1927 Artificial transmutation of the gene. Science **66**:84-87.

1928a The effects of X-radiation on genes and chromosomes. Science **67**:82-83.

1928b The problem of genic modification. Z. indukt. Abstamm.-u Vererblehre. Supl. Bd. **1**:234-260.

1928c The production of mutations by X-rays. Proc. Nat. Acad. Sci. Washington **14**:714-726.

1928d The measurement of gene mutation rate in Drosophila, its high variability, and its dependence upon temperature. Genetics **13**:279-357.

1930a Radiation and genetics. Amer. Nat. **64**:220-251.

1930b Types of visible variations induced by X-rays in Drosophila. J. Genetics **22**:299-334.

MULLER, H. J., and E. ALTENBURG, 1919 The rate of change of hereditary factors in Drosophila. Proc. Soc. Exper. Biol. and Med. **17**:10-14.

1930 The frequency of translocations produced by X-rays in Drosophila. Genetics **15**:283-311.

MULLER, H. J., B. B. LEAGUE and C. A. OFFERMANN, 1931 Effects of dosage changes of sex-linked genes, and the compensatory effects of other gene differences between male and female. Abst. Anat. Rec. **51**:110.

MULLER, H. J., and L. M. MOTT-SMITH, 1930 Evidence that natural radioactivity is inadequate to explain the frequency of "natural" mutations. Proc. Nat. Acad. Sci. Washington **16**:277-285.

MULLER, H. J., and T. S. PAINTER 1929 The cytological expression of changes in gene alignment produced by X-rays in Drosophila. Amer. Nat. **63**:193-200.

1932 The differentiation of the sex chromosomes of Drosophila into genetically active and inert regions. Z. indukt. Abstamm.-u. VererbLehre. **62**:316-365.

MULLER, H. J., and W. S. STONE, 1930 Analysis of several induced gene-rearrangements involving the X chromosome of Drosophila. Abst. Anat. Rec. **47**:.

NASARENKO, I. I., 1930 Ein Fall wahrscheinlicher Verdoppelung eines Chromosomstückes bei *Drosophila melanogaster*. Biol. Zbl. **50**:385-392.

OFFERMANN, C. A., and H. J. MULLER, 1932 Regional differences in crossing over as a function of the chromosome structure. Proc. Sixth Int. Congress Genetics **2**:143-145.

OLIVER, C. P., 1930a The effect of varying the duration of X-ray treatment upon the frequency of mutation. Science **71**:44-46.

1930b Complex gene rearrangements induced by X-rays. Abst. Anat. Rec. **47**:

1932 An analysis of the effect of varying the duration of X-ray treatment upon the frequency of mutations. Z. indukt. Abstamm.-u. VererbLehre. **61**:447-488.

PAINTER, T. S., and H. J. MULLER, 1929 Parallel cytology and genetics of induced translocations and deletions in Drosophila. J. Hered. **20**:287-298.

PATTERSON, J. T., 1931 Continuous versus interrupted radiation and the rate of mutation in Drosophila. Biol. Bull. **61**:133-138.

1932a A new type of mottled-eyed Drosophila due to an unstable translocation. Genetics **17**:38-59.

1932b A gene for viability in the X chromosome of Drosophila. Z. indukt. Abstamm.-u. VererbLehre. **60**:125-136.

1932c Lethal mutations and deficiencies produced in the X chromosome of *Drosophila melanogaster* by X-radiation. Amer. Nat. **66**:193-206.

PATTERSON, J. T., and H. J. MULLER, 1930 Are "progressive" mutations produced by X-rays? Genetics **15**:495-578.

PLOUGH, H. H., and P. T. IVES, 1932 New evidence of the production of mutations by high temperature, with a critique of the concept of directed mutations. Proc. Sixth Int. Congress Genetics **2**:156-158.

PLUNKETT, C. R., 1926 The interaction of genetic and environmental factors in development. J. Exp. Zool. **46**:181-244.

1932 Temperature as a tool of research in phenogenetics: methods and results. Proc. Sixth Int. Congress Genetics **2**:158-160.

ROKITZKY, P. T., 1930 Über das Hervorrufen erblicher Veränderungen bei Drosophila durch Temperatureinwirkung. Biol. Zbl. **50**:554-566.

SAFIR, S. R., 1913 A new eye-color mutation in Drosophila. Biol. Bull. **25**:45-51.

SEREBROVSKY, A. S., 1929 A general scheme for the origin of mutations. Amer. Nat. **53**:374-378.

SEREBROVSKY, A. S., and N. P. DUBININ, 1930 X-ray experiments with Drosophila. J. Hered. **21**:259-265.

STADLER, L. J., 1930 Some genetic effects of X-rays in plants. J. Hered. **21**:2-19.

STERN, C., 1929 Über die additive Wirkung multipler Allele. Biol. Zbl. **49**:231-290.

1931 Zytologisch-genetische Untersuchungen als Beweise für die Morgansche Theorie des Faktorenaustausches. Biol. Zbl. **51**:547-587.

STERN, C., and OGURA, S., 1931 Neue Untersuchungen über Aberrationen des Y chromosoms von *Drosophila melanogaster*. Z. indukt. Abstamm.-u. VererbLehre. **58**:81-121.

STURTEVANT, A. H., 1926 A crossover reducer in *Drosophila melanogaster* due to inversion of a section of the third chromosome. Biol. Zbl. **46**:697-702.

1929 The genetics of *Drosophila simulans*. Pub. Carnegie Instn. **399**:1-62.

1931 Known and probable inverted sections of the autosomes of *Drosophila melanogaster*. Pub. Carnegie Instn. **421**:1-27.

1932 Paper in these Proceedings, Vol. I.

STURTEVANT, A. H., and T. DOBZHANSKY, 1930 Reciprocal translocations in Drosophila and their bearing on Oenothera cytology and genetics. Proc. Nat. Acad. Sci. Washington. **16**:533-536.

STURTEVANT, A. H., and C. R. PLUNKETT, 1926 Sequence of corresponding third-chromosome genes in *Drosophila melanogaster* and *D. simulans*. Biol. Bull. **50**:56-60.

STURTEVANT, A. H., and J. SCHULTZ, 1931 The inadequacy of the sub-gene hypothesis of the nature of the scute allelomorphs of Drosophila. Proc. Nat. Acad. Sci. Washington **17**:265-270.

THOMPSON, D. H., 1925 Evidence on the structure of the gene. Amer. Nat. **59**:91-94.

1929 The side-chain theory of the structure of the gene. Abst. Anat. Rec. **44**:288.

TIMOFÉEFF-RESSOVSKY, N. W., 1929 The effect of X-rays in producing somatic genovariations of a definite locus in different directions in *Drosophila melanogaster*. Amer. Nat. **63**:118-124.

1931a Reverse genovariations and gene mutations in different directions. J. Hered. **21**:67-70.

1931b Die bisherigen Ergebnisse der Strahlengenetik. Ergebn. d. mediz. Strahlenforschung **5**:130-228.

VAN ATTA, E. W., 1932 Dominant eye-color in Drosophila. Amer. Nat. **66**:93-95.

WRIGHT, S., 1929a Fisher's theory of dominance. Amer. Nat. **63**:274-279.

1929b The evolution of dominance. Amer. Nat. **63**:556-561.

ZELENY, C., 1918 Full-eye and emarginate-eye from bar-eye in Drosophila without change of the bar gene. Anat. Rec. **14**:89.

III
Genes and Antigens

Editor's Comments on Papers 3 and 4

3 **Irwin and Cole:** *Immunogenetic Studies of Species and of Species Hybrids in Doves, and the Separation of Species-Specific Substances in the Backcross*

4 **Owen:** *Immunogenetic Consequences of Vascular Anastomoses Between Bovine Twins*

No proper discussion of the background to the gene–protein relationship would be complete without recalling the important contributions from immunology.

The relation between genes and antigens had been firmly established long before Irwin and Cole did their work. The human A and B red blood cell antigens were first recognized by Landsteiner in 1900 and shown to be inherited by the Hirschfelds in 1919. Irwin and Cole's contribution was primarily to introduce a system more amenable to experimentation than man, and then to exploit this system brilliantly.

The ringneck and pearlneck dove are closely related enough to be crossed and produce fertile hybrids. By making the appropriate antisera in rabbits, Irwin and Cole were able to demonstrate that the red blood cells of the two dove species share antigens in common, which they refer to as "biochemical substances." Also, each has a set of unique antigens. Antigens common to both species are present in the hybrid, as well as those specific to the pearlneck parent. The antigens are assumed to be gene products because they segregate according to Mendelian laws and not by cytoplasmic inheritance.

There is clear evidence from the data that at least 10 to 20 different antigens were segregating in their backcross experiments. The authors point out that these serological differences are in reality chemical differences, and imply a rather direct gene–antigen relationship, which even in the 1930s was considered by many to imply a gene–protein difference. Hence it was evident as far back as the 1930s that the methodology of immunology was a powerful tool in the elucidation of the most subtle differences in antigenic structure and therefore possibly in protein structure. Subsequent work based on the dove experiment made it evident that one could in fact distinguish extensive allelic differences in cattle.

Another finding of considerable interest from the results of the dove experiments was that an antigen was produced in the F_1 hybrid which had no counterpart in either parent. This "hybrid substance" aroused considerable interest at the time as an example of gene interaction.

Owen makes one important point in his brief note. The analysis of results from the study of the cellular antigenic components of cattle twins shows that, as a result of vascular anastomosis *in utero,* the twins may exchange embryonic cells which are ancestral to the erythrocytes of the adult animals. The result is that the animals show certain blood types which they cannot pass on to their offspring. In addition, although these antigens are "foreign" to the twins, they do not form antibodies to them. Thus there is a period in embryogeny when a "foreign" antigen is present and the host will not respond with the formation of antibodies, and when it does become immunoresponsive, it will tolerate the antigen as one of its own, even though it does not carry the gene for the antigen in its other cells. This early somatic cell inheritance study also demonstrated that cells operate independently in the formation of antigens.

Selected Bibliography

Fudenberg, H. H., and N. Warner. 1970. Genetics of immunoglobulins. *Adv. Human Genetics 1:* 131–209.

Hildemann, W. H. 1973. Genetics of immune responsiveness. *Ann. Rev. Genetics 7:* 19–36.

Hirschfeld, L., and H. Hirschfeld. 1919. Serological differences between the blood of different races. The results of researches on the Macedonian front. *Lancet 2:* 675–679.

Landsteiner, K. 1900. Zur Kenntnis der antifermentativen, lytischen und agglutinativen Wirkungen des Blutserums und der Lymphe. *Zbl. Bakt. 27:* 357–362.

———, and A. S. Wiener. 1941. Studies on an agglutinogen (Rh) in human blood reacting with anti-Rhesus sera and with human isoantibodies. *J. Exp. Med. 74:* 309–320.

IMMUNOGENETIC STUDIES OF SPECIES AND OF SPECIES HYBRIDS IN DOVES, AND THE SEPARATION OF SPECIES-SPECIFIC SUBSTANCES IN THE BACKCROSS[1]

M. R. IRWIN AND L. J. COLE[2]

ONE FIGURE

INTRODUCTION

Following the work of Nuttall ('04) many investigators have utilized serological technics, principally the precipitin reaction, to distinguish between the serum proteins of species. It is not the purpose of this paper to review the extensive literature on this subject; leading references to papers involving the precipitin technic may be found in reports published by Baier ('33), Boyden ('34), Hicks and Little ('31), Landsteiner and Van der Scheer ('24), and Sasaki ('35). Reichert and Brown (see Reichert, '14) have employed crystallography as a means of differentiating the hemoglobin systems of species. By a similar technic, these workers also showed differences between starches of various species.

It is only recently that anti-erythrocyte serum as a means of differentiating species has been employed. Landsteiner and co-workers ('24, '24 a, '25, '31) have shown that it is possible by the use of hemagglutinating serum to distinguish

[1] Paper from the department of genetics, Agricultural Experiment Station, University of Wisconsin, no. 190. Published with the approval of the director of the station. This investigation was supported in part by grants from the committee for research in problems of sex, National Research Council (grant administered by Prof. L. J. Cole), and from the Wisconsin Alumni Research Foundation. A preliminary paper has been published, and a report given (Irwin, '32, '32 a).

[2] Mr. C. D. Gordon, research assistant in genetics, has given much valuable aid in these experiments.

between the corpuscles of closely related species, a distinction which for the serum proteins of closely allied species is performed with difficulty, if at all, by the precipitin reaction.

MATERIALS AND METHODS

Representatives of the domesticated Ring dove (Streptopelia risoria), an Asiatic species (Spilopelia chinensis) commonly called Pearlneck, and of F_1 (hybrids) from male Pearlnecks × Ring dove females have served as donors of corpuscles for injections and agglutinations. In this paper, the terms F_1 or hybrid will designate the species hybrid. All F_1 birds from which living offspring have been obtained in backcrosses to Ring doves (producing 'first backcross generation,' 'quarter-Pearlnecks,' or '1/4 P.N.') are progeny of one Pearlneck male. The offspring of matings of F_1 individuals to Ring doves are specified in the following tables by numbers, and the members of each family by letters. These doves are a part of experiments designed (L.J.C.) to examine differential sexual mortality and fertility, particularly in the hybrids and backcrosses of the species involved. One of us (M.R.I.) is responsible for the serological work reported.

Antisera were prepared by injecting rabbits with washed erythrocytes from representatives of each species and of the F_1. The agglutinations were performed by adding to 0.1 cc. of the immune serum in varying dilutions (by halves) 1 drop of a 2.5 per cent saline suspension of washed red blood cells. For the absorptions, the immune serum was diluted according to its titre and mixed with a proportionate amount (generally one-half) of washed cells. The mixture was agitated gently at intervals, and allowed to stand at room temperature from 1 to 2 hours. The process was repeated until absorption was complete at the original dilution. Second and third absorptions were sometimes completed at ice box temperature. Readings of agglutinations were generally made after 2 hours at room temperature, and again if possible after storage in the ice box. The last trace of agglutination was determined by microscopic examination.

ABSORPTION OF AGGLUTININS

The differentiation of species, of the F_1's from their parents, and of the individuals of the backcross generation from one another and from the F_1 parents, is based on agglutinations of the corpuscles following absorption of agglutinins from the immune serum. A comprehensive explanation of this useful technic is given by Krumwiede et al. ('25). The principle of this reaction rests upon the presence within the cells of many antigens, which have exact counterparts, presumably, in the immune serum. The absorption technic may be illustrated in a hypothetical case by employing capitals to represent the agglutinogens (antigens) of the cells, and small letters to denote the corresponding agglutinins (antibodies) in the antisera, as follows:

	Agglutinogens		*Agglutinins*
Cell group 1.	A B C D E	Antiserum 1.	a b c d e
Cell group 2.	C D E F G	Antiserum 2.	c d e f g
Cell group 3.	A E F G H	Antiserum 3.	a e f g h

Antiserum 1 would contain the agglutinins *a b c d e* and because of the presence of each and of all of these it would agglutinate the cells of group 1. For these cells it would be the homologous serum. It would also agglutinate group 2 cells in virtue of agglutinins *c, d* and *e,* and group 3 cells in virtue of agglutinins *a* and *e*. For these cells it is termed the heterologous serum.

If antiserum 1 were completely absorbed by cells of group 1, it would not agglutinate these cells, nor those of groups 2 and 3. If, however, exhaustion were performed on this antiserum by cells of group 2, removing agglutinins *c, d* and *e,* but leaving *a* and *b,* the absorbed serum would not bind the cells of this group (2), but would react with those of group 1 because of agglutinins *a* and *b,* also with group 3 by virtue of agglutinin *a*. Similar illustrations would apply for the specificity of the other antisera.

Reciprocal absorptions of antisera 1 and 2, with subsequent agglutinations of the cells of these two groups, constitute a test for the homologous or heterologous antigenic structure of the cells. This double cross-absorption method is often called the 'mirror test.' Its parallelism to reciprocal matings in genetic procedure is obvious.

EXPERIMENTAL

A summary of the reactions by means of which the cells of Pearlneck, Ring dove and their F_1 may be distinguished each from the others, is given in table 1. The direct agglutination tests for each serum fail to differentiate the cells. However,

TABLE 1

Serological relationships between the blood cells of Pearlnecks, Ring doves, and their hybrids

IMMUNE SERUM	ABSORBED BY CELLS OF	FIRST SERUM DILUTION	AGGLUTINATIONS ON CELLS OF		
			Pearlneck	Ring dove	F_1
Pearlneck (A)	—	90	15,360	15,360	15,360
Pearlneck (A)	Ring dove	90	7,680	0	7,680
Pearlneck (A)	Pearlneck	90	0[1]	0	0
Pearlneck (A)	F_1	30	60	0	0
Pearlneck (B)	—	180	46,080	46,080	46,080
Pearlneck (B)	Ring dove	180	23,040	0	23,040
Pearlneck (B)	F_1	90	180	0	0
Pearlneck (C)	—	90	11,520	11,520	11,520
Pearlneck (C)	Ring dove	45	5,760	0	5,760
Pearlneck (C)	F_1	45	90	0	0
Pearlneck (D)	—	90	23,040	23,040	23,040
Pearlneck (D)	Ring dove	90	11,520	0	11,520
Ring dove (1)	—	30	15,360	15,360	15,360
Ring dove (1)	Pearlneck	30	0	1,920+	1,920+
Ring dove (1)	F_1	45	0	180	0
Ring dove (2)	—	90	23,040	23,040	23,040
Ring dove (2)	Pearlneck	45	0	5,760	5,760
Ring dove (2)	F_1	45	0	360	0
F_1	—	60	15,360	15,360	15,360
F_1	Ring dove	45	5,760	0	5,760
F_1	Pearlneck	45	0	2,880	2,880
F_1	Ring dove and Pearlneck	45	0	0	180+

The numbers are the serum titres; i.e., the highest serum dilutions at which the last trace of agglutination was observed microscopically.

A dash (—) in this and all succeeding tables indicates that the absorption or combination was not made.

[1] 0 = no trace of agglutination at first serum dilution.

when anti-Pearlneck serum is exhausted by the cells of Ring doves, with no trace of agglutination for Ring dove cells at the serum dilution used for the absorptions, the agglutinins remaining cause clumping of the cells of both Pearlneck and the F_1 at one-half the dilution for the direct agglutinations. This constant difference occurred without exception for each of several different anti-Pearlneck sera employed, and for all the many tests of each of these absorbed sera.

If anti-Pearlneck serum is absorbed by cells of the F_1, the agglutinins for Ring dove cells are removed at the same time, the exhausted fluid in these tests producing agglutination only with Pearlneck cells at a low titre. Absorption by Pearlneck cells removed all agglutinins for the three types of cells.

Similar absorptions of anti-Ring dove serum and of anti-F_1 serum with the cells as indicated in the table show various interactions which, combined with the results described above, may be interpreted as follows. The erythrocytes of the Pearlnecks have antigenic components in common with Ring doves (the agglutinins for the common substances are those absorbed) as well as components specific for themselves—'species-specific' Pearlneck[3] substances not shared by Ring doves.

In like manner, the cells of Ring doves, in addition to the constituents shared with Pearlnecks, also possess substances specific for themselves, i.e., 'species-specific' Ring dove substances. There is some indication that these specific Ring dove substances are quantitatively less than those of Pearlneck for itself, as suggested by the greater drop in the titre of Ring dove serum absorbed by Pearlneck cells than is found in the reciprocal test.

The F_1 cells lack a small part of the total complement of both Pearlneck and Ring dove cells, since each anti-parental serum absorbed by hybrid cells still retains agglutinins for the parental cells of the anti-serum employed. The results of single absorptions of anti-F_1 serum by Pearlneck and by Ring dove cells with subsequent agglutinations indicate

[3] Throughout this paper the species-specific Pearlneck substances refer only to those not shared with Ring dove.

further that there is a quantitative difference in the species-specific substances of the cells of the two parental species, i.e., Pearlneck cells absorb more agglutinins from F_1 antiserum than do Ring dove cells.

A very significant finding is that some agglutinins remain in anti-F_1 serum for F_1 cells after exhaustion with both Pearlneck and Ring dove cells. Therefore, the cells of this hybrid contain a substance or substances not found in either parent. This is a new, or 'hybrid' substance, as before reported

TABLE 2

Test for individual differences between species hybrids in the content of specific Pearlneck substances in their cells

CELLS OF	ANTI-PEARLNECK SERUM, FIRST ABSORBED BY RING DOVE CELLS, THEN BY THE CELLS OF EACH OF THE SEVERAL F_1 BIRDS BELOW. SERUM DILUTION 1:30									
	241D	241G	241Q	241Y	241Z	241A$_2$	241K$_2$	382C	382D	575D
Pearlneck	+	+	+	+	+	+	+	+	+	+
Ring dove	0	0	0	0	0	0	0	0	0	0
F_1 individuals										
241D	0	0	0	0	0	0	0	0	0	0
241G	0	0	0	0	0	0	0	0	0	0
241Q	0	0	0	0	0	0	0	0	0	0
241Y	0	0	0	0	0	0	0	0	0	0
241Z	0	0	0	0	0	0	0	0	0	0
241A$_2$	0	0	0	0	0	0	0	0	0	0
241K$_2$	0	0	0	0	0	0	0	0	0	0
382C	0	+	0	+	+	±	±	0	±	0
382D	0	0	0	0	0	0	0	0	0	0
575D	0	0	0	0	0	±	0	0	0	0

Symbols: + = agglutination; ± = trace; 0 = no agglutination.

(Irwin, '32). The cells of all the F_1 progeny tested (sixteen) were alike in containing this new substance.

There exists the possibility of differences between F_1 individuals in 'species-specific' Pearlneck and/or Ring dove substances. The inability to obtain sufficient amounts of cells from each hybrid for the many agglutinin absorptions required has prevented a comprehensive search for these differences. A suggestion of dissimilarities between the cells of F_1 birds in their content of specific Pearlneck substances is shown in two cases (382C and 575D, table 2). These reactions are not uniformly consistent. A few trials of the segregation of Ring dove substances in the cells of the hybrids, while suggestive, are not sufficiently inclusive to be given. To be conclusive, it would be necessary to perform similar tests for segregation of the specific substances of each parent at a much lower dilution of the immune sera.

Individual differences within a species are also to be expected, and, if pronounced, would be a source of error in the experiments to follow. Particularly is this true of the Ring dove species, whose cells have been used for the absorptions

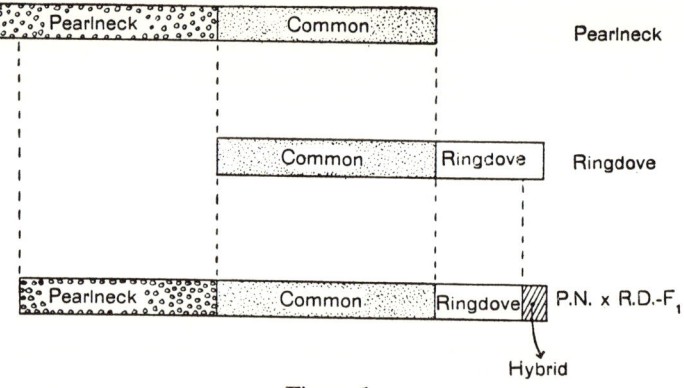

Figure 1

of anti-Pearlneck serum. It may be said that cellular differences have been noted within the Ring doves, but these are detectable only at a much lower dilution of the antiserum than has been used in these tests for the differentiation of species-specific substances. However, it is necessary to pool the cells from many Ring doves for the absorptions, in order to provide the amounts of absorbed serum required for the several tests. It is, therefore, considered unlikely that variations in cellular components within the Ring dove have led to errors in this report, particularly not in the detection of characters specific for Pearlneck.

A diagrammatic representation of the antigenic components of the corpuscles of these species and of their hybrids is shown in figure 1. This is based on the explanation above of the relationships given in table 1.

THE FIRST BACKCROSS PROGENY

With a technic for determining specific antigenic characters of a species and their presence in the F_1, it is possible by an examination of the cells of the offspring of a backcross, in this case to Ring dove, to determine whether or not the species-specific substances (Pearlneck) are divisible and heritable.

The tests as summarized in table 3 were performed to ascertain the degree to which each of these first backcross progeny resembled, and differed from, the F_1 parent. The possibility that each of the quarter-Pearlnecks may have recovered, as a result of the backcross, at least a part of specific Ring dove components lacking in the F_1 is given in the first column,[4] the presence or absence of the 'hybrid' substance in the second, and the quantitative relationship to the F_1 on the basis of specific Pearlneck substances in the third. Second absorptions of anti-Pearlneck serum first exhausted by Ring dove cells were made, using the cells of the F_1 and of thirteen individual backcrossed hybrids, thus providing this number of different specific reagents with which to test the cells of the other 'quarter-Pearlnecks' as given (table 3, columns 4 to 14).

It was not possible to carry out all these tests simultaneously, hence the titre for the hybrid cells was employed as the criterion of the uniformity of the reactions. Only a slight difference in titre for F_1 cells has been noted for each of the various second absorption reagents; seldom, possibly 10 per cent of the time, has a difference exceeding one dilution been observed (see table 4 for variation of such test sera for F_1 cells). Complete exhaustion at the first dilution of the agglutinins for Ring dove cells and for those of each backcrossed hybrid as given allows the general conclusion that the individual differences displayed are due to a separation of different component specific Pearlneck substances.

[4] The numbering of columns in the tables begins with the first column of data.

EXPLANATION OF TABLES

The titres for the various sera represent the highest dilution at which the last trace of agglutination could be observed under the microscope. Two different immune sera were used for the majority of the tests of the second absorptions of table 3, thereby providing duplicate tests for approximately three-fourths of these backcrossed progeny, excepting the 458 family. The titres for these two sera differed, hence a numerical value corresponding to the serial dilution is used, allowing a grouping and a direct comparison of the results of many experiments.

Dilution	Titres of Serum A	Titres of Serum B	Dilution	Titres of Serum A	Titres of Serum B
1	90	180	6	2,880	5,760
2	180	360	7	5,760	11,520
3	360	720	8	11,520	23,040
4	720	1,440	9	23,040	46,080
5	1,440	2,880			

The first dilutions of sera C and D were 45 and 90, respectively, and were used in the tests summarized in table 4. Symbol '0' in the tables indicates no trace of agglutination at the first dilution.

It is obvious that the most exact test of cell individuality between any two birds is that of reciprocal second absorptions with subsequent agglutination reactions, the 'mirror test.' For example, in table 3 birds 292E and 292M are different in that the cells of 292E agglutinate with serum absorbed by 292M, whereas those of 292M show no clumping in serum exhausted by corpuscles of 292E. Hence, all the antigenic substances of 292M are found in 292E, with additional components in 292E. This comparison does not show, however, whether the substances by which 292E differs from 292M are identical with those which distinguish 292G (and others) from 292M.

Equally exact but less comprehensive is the comparison of the cells of any two or more birds on the basis of the presence or absence of clumping when mixed with one of the specific reagents. Thus, those birds whose cells are agglutinated by serum exhausted by the cells of 189L are surely

TABLE 3

Agglutinations of the cells of first backcrossed hybrids ('quarter-Pearlnecks') by various absorbed sera

CELLS OF	TITRES FOLLOWING ABSORPTION OF			TITRES OF ANTI-PEARLNECK SERUM FIRST ABSORBED BY RING DOVE CELLS, THEN BY EACH OF THE FOLLOWING (FIRST DILUTION, SERUM A = 1:90, SERUM B = 1:180)														
	Anti-Ring dove serum by F_1 hybrid cells[1]	Anti-F_1 serum by Pearlneck and Ring dove cells[1]	Anti-Pearl neck serum by Ring dove cells[2]	F_1	189L	189V	213K	281F	281W	281Z	281C$_2$	290Y	292E	292G	292M	330F	330M
Ring dove	360	0	0	0	0	0	0	0	0	0	0	0	0	0	0	0	0
F_1	0	180	8	0	5	7	6	8	6	8	6	5	5	5	6	8	6
189L	90	180	8	0	0	8	6	8	5	7	5	4	4	5	5	4	5
189V	—	—	5	—	0	0	3	0	4	5	4	3	2	0	2	3	3
189H$_2$	—	—	5	—	3	3	5	6	4	5	4	2	3	0	3	6	3
189L$_2$	90	0	6	0	3	1	5	4	4	—	6	2	3	1	3	—	4
189M$_2$	90	0	5	—	—	0	4	±	3	2	5	3	2	0	—	3	4
189M$_3$	120	—	5	—	0	0	0	0	0	5	2	3	3	0	4	3	3
213K	360	90	8	0	3	8	5	7	4	5	5	3	3	3	4	4	6
253J	90	90	6	—	2	1	5	1	5	5	5	2	2	1	2	2	+4
278Z	120	—	8	0	3	4	5	6	6	6	6	5	5	4	5	4	5
281A	90	180	8	0	5	7	5	8	5	—	5	4	2	6	5	+	6
281E	45	180	8	0	2	7	5	8	5	—	5	5	±	6	—	6	6
281F	360	0	6	0	±	3	5	0	+4	5	5	3	2	1	5	4	4
281J	90	180	8	—	5	7	4	8	5	5	5	5	4	6	5	4	6
281K	360	+90	2	0	0	2	0	2	1	2	0	0	0	1	0	+	2
281M	90	180	8	—	5	7	5	8	5	—	5	4	4	6	6	6	4
281T	90	180	8	0	4	7	5	8	5	—	5	5	3	6	6	6	6
281W	90	90	8	—	0	8	3	8	0	5	4	3	0	4	0	4	5
281Z	180	—	4	0	0	1	±	2	0	0	4	3	0	2	0	4	4
281C$_2$	45	90	8	0	3	8	4	8	3	5	0	4	0	4	0	4	5
290J	120	0	5	0	2	2	5	2	5	5	4	3	2	0	5	1	4
290Y	180	180	8	0	5	8	5	8	6	7	5	0	3	5	5	6	5
290F$_2$	360	90	8	0	5	7	4	8	5	5	6	2	3	6	5	4	4
290V$_2$	90	180	5	—	—	—	—	—	—	4	—	—	2	—	5	2	3
290Y$_2$	—	—	8	—	4	7	4	7	5	5	4	4	1	6	—	3	4
292E	45	180	8	0	5	7	5	8	6	7	6	5	0	6	4	8	5

67

	Titre 1	Titre 2																
292G	180	180	8	0	5	7	5	8	6	7	6	4	4	0	5	7	6
292M	45	180	8	0	4	7	3	8	5	6	5	4	0	5	0	8	5
292U	360	90	8	0	3	7	3	8	5	6	5	4	1	5	1	6	5
292V	360	90	8	—	4	7	3	8	5	6	5	4	3	5	3	6	5
330F	120	0	5	0	3	4	4	4	5	5	5	2	1	3	3	5	3
330K	360	0	5	0	1	1	4	2	4	4	4	3	1	3	1	0	3
330M	180	90	8	—	3	6	2	6	4	5	5	2	3	4	3	6	0
355H	+180	180	5	—	4	—	5	1	5	5	5	3	2	3	5	3	5
355J	+180	180	8	—	5	—	4	8	4	5	5	3	0	4	0	3	4
355T	90	0	5	—	4	—	5	1	5	4	5	2	2	3	4	1	3
355V	+180	90	8	0	5	—	3	8	4	5	5	5	2	3	2	6	4
355Y	180	0	5	0	4	—	5	1	4	5	4	4	1	3	5	3	4
355Z	180	45	8	0	5	—	4	8	4	6	5	4	1	4	—	5	5
458A	90	—	8	0	—	—	3	8	4	—	3	5	++	4	—	8	—
458F	180	—	8	0	—	—	3	8	4	—	3	5	++	5	—	8	—
458L	90	—	6	0	—	—	—	—	6	—	—	—	+++	—	—	—	—
458M	360	—	6	0	—	—	—	—	6	—	—	0	—	±±	—	—	—
458P	90	—	8	0	—	—	—	—	2	—	—	++	+++	++	—	—	—
458U	45	—	8	0	—	—	—	—	5	—	—	90	+++	++	—	—	—
458V	180	—	8	0	—	—	—	—	2	—	—	++	—	++	—	—	—
458W	180	—	8	0	—	—	—	—	6	—	—	++	++	±	—	—	—
458Y	90	—	4	0	—	—	—	—	2	—	—	—	±	±±	—	—	—
458Z	90	—	8	0	—	—	—	—	—	—	—	++	++	+	—	—	—
458B₂	90	—	8	0	—	—	—	—	2	—	—	90	±	+++	—	—	—
458C₂	90	—	8	0	2	—	—	—	4	—	—	++	++	+++	—	—	—

The titres as given in the first two columns represent the highest serum dilution at which agglutination was visible microscopically. In the other columns, the digits are the dilutions for the agglutinations; see text for explanation. Symbols: $+++$ = complete agglutination; $++$ = marked agglutination; $+$ = agglutination; $\pm$ = trace; 0 = no agglutination—all at first serum dilution. A dash (—) indicates that the combination was not made.

[1] First serum dilutions = 1:45.
[2] First serum dilutions = 1:180.

TABLE 4

Agglutinations of cells of quarter-Pearlnecks by various absorbed sera

CELLS	TITRES FOLLOWING ABSORPTIONS OF		TITRES OF ANTI-PEARLNECK SERUM FIRST ABSORBED BY RING DOVE CELLS, THEN BY EACH OF THE FOLLOWING BIRDS (FIRST DILUTION, SERUM C = 1:45, SERUM D = 1:90																	
	Anti-Ring dove serum by F_1 hybrid cells[1]	Anti-Pearlneck serum by Ring dove cells	F_1	458A	458F	458G	458L	458M	458P	458U	458V	458W	458Y	458Z	458A_2	458B_2	458C_2	458D_2			
Ring dove	360	0	0	0	0	0	0	0	0	0	0	0	0	0	0	0	0	0			
F_1	0	8(9)	0	5,7	5,6	8	7,8	7,8	6,7	1,2	6,7	3,5	8	6,7	8	6,7	6,7	8			
458A	90	8(9)	0	0	2	7,8	8	7,8	3,6	0	4,5	2,4	7,8	2,5	8	4	2	7			
458F	180	8(9)	0	1	0	8	7,8	7,8	2,5	1	4,5	3,4	7,8	3,5	8	4	1	7			
458G	180	1(2)	0	0	0	0	0	0	0	0	0	0	0	0	0	0	0	0			
458L	90	6(7)	0	6	6	6	0	2,3	6	0	6	±	5,6	6	5,6	6	6	6			
458M	360	6(7)	0	6	6	6	1	7,8	6	0	6	1	5,6	3,5	5,6	6	6	6			
458P	90	9	0	2+	2,3	7	8	7,8	0	0	3,4	2,4	7,8	6	7,8	4	1,2	7			
458U	45	8	0	4,6	4,6	7	6,8	7,8	5	0	5,6	2,4	7,8	3,5	7,8	4	4	—			
458V	180	9	0	0	2	—	8	8	1,2	0	0	2,4	8	6	7,8	2	3	8			
458W	180	8(9)	0	5,7	5,6	—	7	7,8	6	0	6	0	8	2	8	6	6	8			
458Y	90	5	0	1,2	1,2	4	2	3	2,4	0	4	0	8	5,6	3,4	2	0	4			
458Z	90	9(7)	0	±	0	—	7	7	2,3	0	2,4	0	8	2,3	8	3	±	8			
458A_2	180	2(3)	0	0	1	2	2	1,3	±	0	0	0	0	0	0	1	0	3			
458B_2	90	8(7)	0	0	0	—	8	8	±	0	1	2,5	8	1	8	0	0	8			
458C_2	90	9(7)	0	3,4	2,3	—	8	8	2,3	±	4,5	2,5	8	3	8	5	0	8			
458D_2	90	1	0	0	0	1	±	±	0	0	0	0	0	4,6	0	0	0	0			
189L	90	8	0	6	6	—	—	—	+6	1	6	—	—	6	—	6	6	—			
281W	90	8	0	0	0	—	—	—	2	0	3	—	—	6	—	3	0	—			
290Y	180	8	0	6	6	—	—	—	6+	2	6	1	—	6	—	6	6	—			
292E	45	8	0	6	6	—	—	—	6+	1	6	2	—	6	—	6	6	—			
292G	180	8	0	7	7	—	—	—	6+	1	6+	2	—	6	—	6	6	—			
281K	360	2	0	0	0	—	±	±	0	0	±	0	±	0	1	0	0	—			

Digits represent serial dilution at which the last trace of agglutination was visible microscopically. If more than one number is given, they refer to differences between or within sera C and/or D for the cells involved.

[1] First serum dilution = 1:45.

69

different in their antigenic complex from 189L, and from any other bird whose cell substances are the same as, or less than but completely contained in, those of 189L.

The comparison of any two birds by virtue of the titres for their cells of any one absorbed serum is less exact. A difference of one dilution, unless confirmed by repeated tests, is probably not even suggestive of significance. Although it cannot be said with certainty that a contrast in titre of two or three dilutions signifies individuality of the cells under test, such a difference would at least be indicative of the presence of unlike substances.

DISTRIBUTION AND SEGREGATION OF SPECIFIC PEARLNECK SUBSTANCES

It is obvious that many (thirty-one of fifty-three) of the 'quarter-Pearlnecks' are indistinguishable from the F_1 when their cells are tested only quantitatively for their content of specific Pearlneck substances, i.e., with anti-Pearlneck serum absorbed by Ring dove cells (third column, table 3, and second column, table 4). For convenience, cells agglutinating in this specific serum at the first dilution (as titre of 180, serum B) are assigned to class I; those agglutinating at the second, third, fourth, fifth, sixth, seventh, eighth and ninth dilutions, respectively, to classes II to IX. Thus there are two birds in class I, two in class II, thirteen in class IV and/or V, five in class VI, and the remaining thirty-one in class VIII and/or IX, in which also are found the F_1 cells. A differentiation between individuals within each of these various classes is not possible until after second absorptions are made.

The agglutinations of F_1 cells after second absorptions with the several birds of class VIII show that none of these back-cross hybrids contains Pearlneck substances equal to those of the hybrid. While not given in either table, similar absorptions and many subsequent agglutinations have been performed with the cells of all but five of the thirty-one birds of class VIII; agglutinins for F_1 cells invariably remained. Hybrid cells usually showed agglutination at serum dilutions of five or six following second absorptions by the cells of class VIII, a reduction of two to three dilutions from that of the first absorbed serum. Only the cells of 458U (table 4) reduced the titre further for F_1 cells. Absorption by F_1 cells, however, completely removes the agglutinins for the cells of all the quarter-Pearlnecks. The slight difference between the corpuscles of the Pearlneck and those of the hybrids is, therefore, not restored in the first backcross generation.

Cells of individuals of classes I, II, IV, V and VI (281F, 281Z and 330F in table 3, and 458G, L, M, Y, A_2 and D_2 in table 4) do not by absorption reduce appreciably, if at all, the titre for cells of F_1 nor for those of members of class VIII. Hence the ratio between the serum titre and the quantity of substances in the cells in agglutination and absorption is surely not linear.

In this connection one should examine the reciprocal tests between such cells as those of 189L and 281W; cells of 189L contain all of and more than the substances of 281W, but no significant difference in titre is displayed against the 281W reagent by the cells of the F_1 and of 189L (6 and 5 respectively). A parallel comparison is shown by the interactions of 292E, 292M and 281W. However, the serum (B) used in these tests was very powerful; subsequent trials involving other sera and the same birds displayed differences, but not so marked. This suggests that immune sera from different rabbits may not be entirely uniform, a fact entirely in keeping with many immunological findings. Pertinent evidence to illustrate this point will be reported elsewhere.

A specific means of comparing any two birds whose corpuscles provided reagents for testing against the cells of others is a contrast of their reciprocal reactions, and, in addition, the agglutinations against each absorbed serum of all the other cells. Thus, the test sera provided by the cells of 189L and 292E show that these two quarter-Pearlnecks have different specific Pearlneck substances. The cells of both remove all agglutinins for 281W (class VIII); therefore, the antigenic complex of the cells of 281W is possessed by each. Other substances provide the differences between these two birds; the cell components of $189M_2$ and $189M_3$, themselves dissimilar, are a part of the biochemical make-up of the cells of 189L, but differ somewhat from those of 292E, while the cells of 292E possess all the substances of $281C_2$ which are not completely contained in 189L.

The reactions of all other cells in table 3 to the test sera after absorption by the cells of 292E and of 292M, provide a very critical analysis of the delicacy of the test. Since all the antigenic components of 292M are found in 292E, with additional substances in 292E, it is to be expected that the absorbed serum titre of 292E for the other cells would be less than for the reagent of 292M. The table shows that for a considerable number of cells it was less, and in no case more; however, the agglutination titers for the cells of these two birds against the other test sera were very similar.

On the basis of the tests as summarized in table 3, there are no two birds with identical cell reactions among the thirteen whose cells provided the specific reagents. In order to establish all the differences and/or similarities in cell composition, reciprocal absorption tests for each bird against all the others would be necessary. It was not possible to perform all these tests, but second absorption fluids from fourteen other birds, selected because of similarities in cell reactions for the reagents of table 3, were tested with cells in ninety-four combinations. Specifically, these tests showed distinct differences between all members of families 281 and 292 and for many others as well. It is improbable that, within the relatively small number of quarter-Pearlnecks in the experiments, there were two with exact duplication of these substances. Aside from the unquestionable individuality of the cells of family 458, table 4, no two being identical, a point of much interest is that all the birds of class VIII except U and W possess few if any of the antigenic components of L and M (class VI).

If one attempts to calculate the number of different substances necessary to account for the differences displayed only within family 458 in table 4, much speculation is involved. For example, a minimum of two substances is required to account for the difference between the cells of G and D_2; these are both found in A_2, with others, which in turn are present in Y, requiring the presence of not less than four components in the cells of Y. Presumably some of these four substances are found in L and in M, plus two more to account for the difference between these two birds, necessitating at least six substances in all. After this assured minimum number, it is a matter of conjecture only as to the number of different

TABLE 5

The absorption by two 'quarter Pearlnecks' of varying amounts of the specific Pearlneck substances

CELLS	TITRES OF ANTI-PEARLNECK SERUM FIRST ABSORBED BY RING DOVE CELLS, THEN BY THE FOLLOWING COMBINATIONS (SERUM B, FIRST DILUTION = 180)				
	281A and 292E	281A and 292M	189L and 292G	213K and 290Y	281W and 292M
Ring dove	0	0	0	0	0
F_1	280	720	0	360	+1440
189L	—	—	0	+180	+1440
281A	0	0	—	—	—
281W	—	—	0	180	0
213K	—	—	0	0	—
290Y	—	—	0	0	+1440
292E	0	720	—	—	—
292G	—	—	0	0	+1440
292M	—	0	—	360	0

substances necessary to explain the results given in tables 3 and 4. A conservative estimate would place the number as more than ten, and possibly less than twenty, separate cell components specific for Pearlneck, as revealed in the backcross offspring.

Although no single backcross bird has yet been obtained which has the full complement of specific Pearlneck substances found in the F_1, there probably are combinations of two birds the sum of whose substances equals those of the hybrid. To test this possibility, the cells of a few pairs of class VIII individuals were combined in second absorptions, with results as shown in table 5.

It is seen that the sum of the Pearlneck substances of only one of the pairs (189L and 292G) combined in this manner equals the amount found in the cells of the F_1. These exists no means of determining in advance which cell combinations equal the substances of the F_1.

One of us (Cole, '30) has found that sex-linked plumage characters (blond and white) in Ring doves have a dominant allele (dark) in the Pearlneck. The backcross birds were therefore divided into their respective antigenic classes according to sex and plumage color as given in table 6, to

TABLE 6

A serological classification of the cells of quarter-Pearlnecks according to sex and a sex-linked character

SEX	ANTIGENIC CLASS OF CELLS						
	I	II	IV	V	VI	VIII	Totals
♂♂	1	2	0	4	3	16	26
♀♀	1	0	1	8	2	15	27
Totals	2	2	1	12	5	31	53
COLOR	$X^2 = 3.587$, $P = 0.61$						
Dark	1	2	1	5	2	20	31
Blond or white	1	0	0	7	3	11	22
Totals	2	2	1	12	5	31	53

$$X^2 = 4.549, P = 0.47$$

determine a possible difference in distribution associated with sex, or with a sex-linked character.

It is seen that the values of X^2 are no greater than expected in random assortment, hence it can be concluded that major portions of the biochemical characters are not sex-linked to a detectable extent.

SPECIFIC RING DOVE SUBSTANCES IN THE BACKCROSS (QUARTER-PEARLNECKS) GENERATION

Since the first generation hybrid lacks a fraction of the substances found in the Ring dove, and not in the Pearlneck, the possible recovery of these substances, in whole or in part, in the offspring of matings of these hybrids to Ring doves becomes a matter of interest. Possible individual differences in this respect among these progeny are shown in the first column of tables 3 and 4, based on cell reactions with anti-Ringdove serum absorbed by F_1 cells. A more critical comparison of the variation of several backcross birds is given in table 7.

These few comparisons show clearly that the substances of Ring doves, not present in the cells of the F_1, are divisible into distinct parts, the number of which is uncertain. No

TABLE 7
Segregation of specific Ring dove substances in 'quarter-Pearlnecks'

CELLS	TITRE OF ANTI-RING DOVE SERUM ABSORBED FIRST BY PEARLNECK CELLS THEN BY EACH OF THE FOLLOWING (FIRST SERUM DILUTION 1: 120)						
	189L	189V	213K	281F	281W	290Y	292G
Ring dove	240	240	+120	120	240	150	120
Pearlneck[1]	0	0	0	0	0	0	0
189L	0	0	0	0	0	0	0
189V	240	0	0+	±120	240	0	0
213K	240	0	0	0	240	0	0
281F	240	±	0	0	120	0	0
281W	240	0	0	0	0	0	0
290Y	120	0	0+	0	120	0	0
292G	+240	0	0+	0	120	0	0

[1] No agglutinations at a serum dilution of 1: 60.

bird in these tests has recovered all the specific Ring dove substances. Probably only a few, if any, of the quarter-Pearlnecks possess all the Ring dove substances normally found in the cells of that species.

THE 'HYBRID' SUBSTANCES IN THE QUARTER-PEARLNECKS

The production in the cells of the hybrid of a substance, or substances, not found in the cells of either parental species is the first instance recorded of its particular kind. Hence the distribution of this character in the cells of the backcross generation is of special importance, since thereby its unity or complexity, and, possibly, the specific genetic factors involved from at least one species parent may be determined.

There is some indication from the distribution (second column, table 3) that quantitative differences of the substance exist; certainly its absence in the cells of several of the birds tested is beyond question. It is found in the cells of all quarter-Pearlnecks of class VIII, not in all those of class VI or lower. While these tests are not complete, particularly for family 458, a prediction as to the presence or absence of the 'hybrid' substance in the cells of any bird can be made on the basis of whether or not all the cell components of 281K, in which at least a part of the hybrid character is found, are a part of its biochemical complex (see table 4). Further studies on this subject will be reported later.

DISCUSSION

Evidence has been presented which shows that the erythrocytes of two dove species share biochemical substances and are distinguishable by virtue of other cell components which are species-specific. The technic employed allows the assay of these species-specific components in the F_1; in this cross a slight but definite difference from each parent is displayed in this respect. The characters common to both species presumably are present undiminished in the hybrid.

If these biochemical components of the cells are the result of the action of genes, a segregation of the characters specific for the Pearlneck is to be expected in offspring from backcrosses of the F_1 to Ring doves. Progeny from hybrid males only have been examined, thus eliminating the possible complication of cytoplasmic inheritance. The most critical genetical evidence that these data afford is that absorption of anti-Pearlneck serum by F_1 cells removes completely the agglutinins for the cells of the backcross progeny, a finding easily explained on genetic grounds, but otherwise not at all understandable.

Assuming that these chemical characters are produced by genes, it follows that in the F_1 only the genes producing the substances common in the parental species find homologues; those which give rise either to the specific Pearlneck or to the specific Ring dove substances are simplex.[5] Therefore, in the quarter-Pearlnecks the phenotype and genotype for the specific Pearlneck substances are the same. Heterozygosity in the original species may well account in part for the lack in the F_1 cells of a small fraction of the components specific for each parent.

The distribution of specific Pearlneck characters in the first and subsequent backcrosses would follow that normally expected from such matings, provided a random assortment of genes prevails. The two extremes would be, on the one hand, identity with the F_1, and on the other, with the Ring dove. Neither extreme has been observed in the fifty-three backcross individuals tested. There has, however, been noted a distinct tendency on the part of the majority of these birds to approach quantitatively the F_1, implying that a random assortment of the Pearlneck characters has not occurred. If this be true, and is found to obtain in subsequent backcross progeny, the obvious genetic interpretation implies a grouping of the genes producing these characters. Future reports will present substantiating evidence, and carry further the analysis of specific Pearlneck substances.

Between ten and twenty different biochemical constituents, implying at least as many genes, which in this experiment are not alleles, are required to explain the segregation of the Pearlneck substances in the first backcross generation. The question of their possible linkage relationships must await the isolation of these characters in unit form. It is conceivable that in the Pearlneck some chromosomes may carry a gene or genes producing only specific Pearlneck substances, others may carry genes for specific and for common components, and still others have genes giving rise only to the substances shared by the two species. Presumably an analogous situation exists for any animal species in relation to another.

It is not possible, lacking demonstrable segregation, to estimate the number of different cell components common

---

[5] An alternative explanation would be to assume dominance over alleles in the other species. In our opinion, this assumption is not as reasonable as the above.

to these species. However, unless a different system prevails than has been found for the specific Pearlneck components, they also are multiple. Buchbinder ('34) has reported that a heterophile antigen (single or plural) is shared to some extent in widely related birds. The occurrence in class Mammalia of Forssman's antigen has been investigated by Landsteiner ('33, p. 45), who also has demonstrated the sharing of definite cell components in man and the higher apes ('25).

Our studies show that the antigenic characters specific for Pearlneck, not shared by Ring dove, are multiple, requiring many genes (not alleles) to account for their presence. Assuming that the substances common to the two species are also numerous, it becomes evident that the number of cellular components within the species is indeed very great. Obviously, these substances can be detected within a species only if the causative genes, for any or all, have alleles with different effects.

At present there are only a few definite agglutinogens of cells within all animal species which have been investigated genetically. Many of the immunological findings of individual differences in bloods are cited by Wiener ('35). The various results referred to by him show very strikingly that a wealth of material awaits a genetic analysis.

All previous experimental data dealing with the antigenic effect of allelic genes have shown only an independent action. For example, genes which produce in man agglutinogens M and N when alone (Landsteiner and Levine, '28) produce MN when acting together. The specific 'hybrid' substance of this F_1 offers evidence that a different biochemical character of the cells may be produced by the interaction of two or more genes than is produced in conjunction with their normal alleles. It is logical to assume that this new substance is the result of interaction of genes which within each species produce only species-specific effects.

A matter for speculation is whether or not these particular genes in the F_1 produce singly their species-specific components, as well as the new substance by their interaction.

It seems reasonable to assign a part of the difference in the species-specific substances between the hybrid and each of its parents to the interaction of the genes responsible for the hybrid character. Direct evidence on this point can be obtained only after the particular gene or genes of at least one parent are isolated.

In this connection a dissimilarity between this effect of complementary action and other well-known genetic examples of such interactions should be emphasized. Since the blood cells arise from tissue laid down very early in embryonic development, the production of biochemical components of the erythrocytes may reasonably be considered as a direct action, or very nearly so, of the genes, thereby avoiding the complications of the chain of many reactions presumably involved in the formation of other genetic characters. Still another difference exists in that the counterparts of all the chemical characters, both common and species-specific, are removed from the reagent before the hybrid substance can be detected. Thus there are no other genic effects for either species with which this new character might be confused.

In the light of our present knowledge of the specificity of immunological reactions, it may be stated that the genetic characters studied in this paper are probably due to structural chemical differences within the cells.[6] Arguments to the effect that immunology is a branch of chemistry have been advanced by Wells ('27). The very exact differentiations by serological methods of closely related chemical compounds performed by Landsteiner and Van der Scheer are pertinent evidence for the above statement. The discovery of haptenes by Landsteiner and associates has added greatly to the knowledge of the chemistry of immunologic specificity, particularly with reference to the chemistry of the cell, as well as giving weight

[6] The idea of a multiplicity, or mosaic, of antigens within cells, although generally accepted among bacteriologists for bacterial cells, possibly may not always be strictly true (compare Burnett, '34). However, we propose to show in succeeding papers that the specific Pearlneck substances of these experiments are multiple, and specific in that they are separable by genetic segregation. They are therefore presumably produced by the action of definite genes.

to the thesis that serological differences are in reality chemical differences. Excellent surveys of the specific work in this field are given by Landsteiner ('33) and by Marrack ('34). The biological significance of many of these findings has been discussed by Landsteiner ('28, '31 a).

SUMMARY AND CONCLUSIONS

The technic involved in these tests is essentially that of producing counterparts of a complex phenotype (antigen) in an outside reagent (immune serum), from which a part or parts may be removed at will. By this method the red blood cells of two species (Pearlneck and Ring dove) have been assayed for species-specific and for common substances. A new constituent has been found in the cells of the hybrid due to the interaction of genes which in each parent presumably produce species-specific characters. The distribution of specific Pearlneck components in the offspring of the F_1 mated to Ring doves shows 1) that many different substances are involved and 2) that there is a tendency for many of these to stay together, inferring a similar tendency for the genes responsible. In brief, the biochemical composition of the red blood cells is a quantitative genetic character which in these studies has been separated into many of its possible combinations.

LITERATURE CITED

BAIER, J. C., JR. 1933 Quantitative studies on precipitins. Physiol. Zoöl., vol. 6, pp. 91–125.

BOYDEN, A. 1934 Precipitins and phylogeny in animals. Am. Nat., vol. 68, pp. 516–536.

BUCHBINDER, LEON 1934 A new heterophile antigen common to avian erythrocytes and some varieties of genus Pasturella; its significance in the classification of birds. J. Immunol., vol. 26, pp. 215–231.

BURNETT, F. M. 1934 Antigenic differences between related bacterial strains: a criticism of the mosaic hypothesis. Brit. J. Exp. Path., vol. 15, pp. 354–359.

COLE, L. J. 1930 A triple allelomorph in doves and its interspecific transfer. Abstract. Anat. Rec., vol. 47, p. 389.

HICKS, R. A., AND C. C. LITTLE 1931 The blood relationship of four strains of mice. Gen., vol. 16, pp. 397–421.

IRWIN, M. R. 1932 Dissimilarities between antigenic properties of red blood cells of dove hybrid and parental genera. Proc. Soc. Exp. Biol. and Med., vol. 29, pp. 850–851.

———— 1932 a On heritable individual differences in the biochemical composition of the red blood cells in dove hybrids. Proc. 6th Internat. Cong. Genetics, vol. 2, pp. 103–104.

KRUMWIEDE, C., G. COOPER AND D. J. PROVOST 1925 Agglutinin absorption. J. Immunol., vol. 10, pp. 55–239.

LANDSTEINER, K. 1928 Cell antigens and individual specificity. J. Immunol., vol. 15, pp. 589–600.

———— 1931 Serological tests with the blood of Cavia porcellus and Cavia rufescens. Proc. Soc. Exp. Biol. & Med., vol. 28, pp. 981–982.

———— 1931 a Individual differences in human blood. Science, vol. 73, pp. 403–409.

———— 1933 Die Spezifizität der serologischen Reaktionen, Berlin.

———— 1933 a Immunchemische Spezifizität. Reale Accad. d'Italio, Convegno, Volta.

LANDSTEINER, K., AND J. VAN DER SCHEER 1924 On the specificity of agglutinins and precipitins. J. Exp. Med., vol. 40, pp. 91–107.

———— 1924 a Serological examination of a species-hybrid. I. On the inheritance of species-specific qualities. J. Immunol., vol. 9, pp. 213–219. II. Tests with normal agglutinins. Ibid., vol. 9, pp. 221–226.

LANDSTEINER, K., AND P. MILLER, JR. 1925 Serological studies on the blood of the primates. I. The differentiation of human and anthropoid bloods. J. Exp. Med., vol. 42, pp. 841–852. II. The blood groups in anthropoid apes. Ibid., vol. 42, pp. 853–862. III. Distribution of serological factors related to human isoagglutinogens in the blood of lower monkeys. Ibid., vol. 42, pp. 863–872.

LANDSTEINER, K., AND P. LEVINE 1928 On the inheritance of agglutinogens of human blood demonstrable by immune agglutinins. J. Exp. Med., vol. 43, pp. 731–749.

MARRACK, J. R. 1934 The chemistry of antigens and antibodies. Privy Council, Medical Research Council, Special Report Series, No. 194.

NUTTALL, G. H. F. 1904 Blood immunity and blood relationship. Cambridge, The University Press.

REICHERT, E. T. 1914 The germplasm as a stereochemic system. Science, vol. 40, pp. 649–661.

SASAKI, K. 1935 Precipitation test for a hybrid between Japanese Long-tailed fowl and White Leghorn. Z. f. Züchtung, Bd. 32, S. 95–108.

WELLS, H. G. 1927 Immunology as a branch of chemistry. Columbia University Press, New York.

WIENER, A. S. 1935 Blood groups and blood transfusion. Chas. C. Thomas, Baltimore.

IMMUNOGENETIC CONSEQUENCES OF VASCULAR ANASTOMOSES BETWEEN BOVINE TWINS[1]

RAY D. OWEN

ALMOST thirty years have passed since Lillie[2] used the demonstrated union of the circulatory systems of twin bovine embryos of opposite sex to explain, on an endocrine basis, the frequent reproductive abnormalities of the female twin. Since the appearance of Lillie's paper, the freemartin, as the modified female is called, has become an important example of the effects of hormones on sex-differentiation and sexual development in mammals.[3] Consequences other than endocrinological of nature's experiment in parabiosis have, however, received little attention.

Estimates of the frequency of identical as compared with fraternal twinning indicate that the former is relatively rare in cattle.[4] Tests for inherited cellular antigens in the bloods of more than eighty pairs of bovine twins show, however, that in the majority of these pairs the twins have identical blood types. Identity of blood types between full sibs not twins is infrequent, as might be expected from the large number

[1] From the Departments of Genetics (No. 346) and Veterinary Science, University of Wisconsin, in cooperation with the Bureau of Animal Industry, U. S. Department of Agriculture. This is part of a program aided by grants from the American Guernsey Cattle Club, the Holstein-Friesian Association of America, the Rockefeller Foundation and the Wisconsin Alumni Research Foundation. Appreciated contributions to various phases of the investigation have been made by Professor M. R. Irwin, C. J. Stormont and Mary W. Yeas. This study has been possible only through the generous cooperation of workers at many state experiment stations and of numerous private breeders of cattle.

[2] F. R. Lillie, SCIENCE, 43: 611–613, 1916.

[3] See ''Sex and Internal Secretions,'' edited by Edgar Allen (Williams and Wilkins, Baltimore, 1939), for general discussions of and references to the literature on the freemartin.

[4] D. Sanders, *Zeit. für Züchtung B*, 32: 223–268, 1935.

of different, genetically controlled antigens[5, 6] (now approximately 40) identified in the tests. If, therefore, the frequent identity of blood types in twin pairs can be explained neither as the result of monozygotic twinning nor as chance identity between fraternal twins, nor as the sum of these two factors, it is evident that some mechanism is operating to produce frequent phenotypic identity of blood types in genetically dissimilar twins. The vascular anastomosis between bovine twins, known to be a common occurrence,[2] provides an explanation.

Three additional, independent sources of evidence help to define the action of this mechanism. (1) One twin sire failed to transmit to any of his twenty progeny certain of the antigens found in his blood. In other words, the genotype of this bull as determined from his progeny appeared to lack factors responsible for some of the antigens found in his phenotype. Tests showed that cells containing these antigens could have been derived from his twin, whose genotype did contain the necessary factors. (2) In a case of superfecundation in cattle, involving twins of opposite sex and by different sires,[7] the twins had identical blood types, each possessing two antigens the genetic factors for which could not have come from his own sire or from the dam. Cells containing these critical antigens could in each case have been derived from the co-twin. (3) It has been demonstrated, by a simple immunological technique developed for this purpose, that there is a mixture of two distinct types of erythrocytes in certain twins.

These facts are consistent with the conclusion that an interchange of cells between bovine twin embryos

[5] L. C. Ferguson, *Jour. Immunol.*, 40: 213–242, 1940.
[6] L. C. Ferguson, C. Stormont and M. R. Irwin, *Jour. Immunol.*, 44: 147–164, 1942.
[7] A description and discussion of this case will be published elsewhere by Mr. B. H. Roche, who called it to our attention and provided us with blood samples from the animals involved.

occurs as a result of vascular anastomoses. Since many of the twins in this study were adults when they were tested, and since the interchange of formed erythrocytes alone between embryos could be expected to result in only a transient modification of the variety of circulating cells, it is further indicated that the critical interchange is of embryonal cells ancestral to the erythrocytes of the adult animal.[8] These cells are apparently capable of becoming established in the hemapoietic tissues of their co-twin hosts and continuing to provide a source of blood cells distinct from those of the host, presumably throughout his life.

Several interesting problems in the fields of genetics, immunology and development are suggested by these observations. Most of them are still largely speculative and will not be considered here. An application that may be mentioned is the tool now provided by the blood tests for selecting, with a high degree of reliability, those heifers, born twin with bulls, that are potentially not freemartins but normal, fertile individuals. A heifer whose blood type is the same as her twin brother's will very probably be a freemartin, while a difference in even a single antigen between twins of opposite sex may indicate that vascular anastomosis did not occur, and therefore that the heifer will be normal. Thus clinical observations on the heifer alone, probably not always reliable when the heifer is young, can be supplemented by an objective laboratory test applicable as soon as the twins are born. Possible limitations of this application, as well as a more complete presentation of the data and further discussion of the implications of the present study, will be included in another paper.

[8] *Cf.* H. E. Jordan, *Physiol. Rev.*, 22: 375–384, 1942.

IV
Genes and Chemical Reactions

Editor's Comments on Papers 5 and 6

5 Wright: *A Quantitative Study of the Interactions of the Major Colour Factors of the Guinea-Pig*

6 Lawrence and Price: *The Genetics and Chemistry of Flower Colour Variation*

Wright carried out a series of investigations on inheritance of coat color in guinea pigs for a period of more than 40 years. His first definitive paper, analyzing his early observations, appeared in 1917. In it he follows the line of thought originally stated by Cuénot (Paper 1). He suggested that ". . . heredity differences in color are due to hereditary differences in the enzyme element of the reaction." By this he meant the conversion of a chromogen to a pigment by an enzyme.

Following a period of more than 20 years of further work after the 1917 paper, Wright wrote the short paper presented here in which he makes some penetrating observations on gene action. First, he obviously believed that the most probable immediate product of genes are enzymes. He then proceeds to develop a mathematical theory in an attempt to explain the quantitative data given in Table 1. This is an early example of the use of model making in genetics. It leads to a kinetic explanation of the relation between enzyme and substrate concentration, and a physiological explanation of dominance, i.e., a situation in which the enzyme by virtue of active "dominant" alleles is in excess relative to its substrate.

Wright also points out the bewildering complexity of gene interaction involved in producing a phenotype—the kind and amount of pigment. He shows that by careful analysis one can begin to explain how enzymes act in concert and in progression.

Work on the color pigments in flowers started early in this century soon after the rediscovery of Mendel's work. Wheldale (later, Mrs. Onslow) began publishing papers about the anthocyanins in flower petals in 1909. Her results led J. B. S. Haldane to interest others in these studies, which culminated in a fascinating picture of the actions and interactions of genes segregating in the various species of cultivated plants.

The review article reproduced here describes the progress of the work done by a large number of investigators, both geneticists and chemists, up to the beginning of World War II. It brings out several important points: (1) The production of specific pigments is controlled by specific genes. (2) Genes act in a sequence to produce end products derived by a series of reactions from precursors. In other words, a one gene–one reaction hypothesis is advanced. This was a natural predecessor to the one gene–one enzyme hypothesis and supported it with sound evidence. (3) In metabolism there appears to be competition for substrates, which are present in limited concentration. Thus a limited substrate, potentially capable of being converted into a number of different end products, would actually be converted into only so many products as the competing genes are able to effectively catalyze by their control of the various reaction rates.

Nowhere in this paper is a gene–enzyme relation explicitly stated, but the idea is implied. This body of work was an important step forward in our understanding of the roles of genes in metabolism.

After the discovery of paper chromatography in the 1940s, some of the data reported here were modified, but no new significant conclusions were arrived at.

Selected Bibliography

Geissman, J. A., and G. A. L. Mehlquist. 1947. Inheritance in the carnation, *Dianthus* caryophylbis: IV. The chemistry of flower color variation, I. *Genetics 32:* 410–433.

Jorgensen, E. C., and T. A. Geissman. 1955. The chemistry of flower pigmentation in *Antirrbinum majus* color genotypes: III. Relative anthocyanin and aurone concentrations. *Arch. Biochem. Biophys. 55:* 389–402.

Wright, S. 1917. Color inheritance in mammals. *J. Hered. 8:* 224–235.

―――. 1941. The physiology of the gene. *Physiol. Rev. 21:* 487–527.

―――. 1960. Postnatal changes in the intensity of coat color in diverse genotypes of the guinea pig. *Genetics 45:* 1503.

A Quantitative Study of the Interactions of the Major Colour Factors of the Guinea-Pig

S. WRIGHT

INTRODUCTION

Animal or plant breeding appears at first sight to be a very crude type of physiological investigation. Yet from one viewpoint it is a rather penetrating method. There is the opportunity of making exactly one or two, or in some cases more replacements of one of the ultimate physiological agents of living matter without any experimental disturbance. A given measurable character may be affected by many such sets of alternative agents, and the quantitative effect of each gene replacement can be studied in a great diversity of different genetic backgrounds, as well as under different environmental conditions. There are, of course, obvious limitations. It is impossible to attain such certainty of interpretation as in reactions carried out with isolated materials in the test-tube. Advance in physiological genetics will require maximum utilization of both methods.

The colours of animals and plants have long been recognized as among the most favourable characters for study of gene physiology, both because of the multiplicity of gene combinations which can be distinguished and because of the probable relative directness of the relations between primary gene action and observed character. I wish to review the results of studies of coat colour in the guinea-pig which have been conducted from this viewpoint.

MEASUREMENT OF PIGMENTATION

The coat colours of the guinea-pig, as of other mammals, fall into two main series, dark (or melanic) colours and yellow (or xanthic) colours. The actual pigments of these series are undoubtedly closely related and all belong to the group known as melanins. In our laboratory, more than 10,000 animals distributed among several hundred different genotypes have been graded at birth, and in many cases later, by means of squares of skin in which the colour of the hair progresses by scarcely perceptible changes from white (grade 0) to intense red (grade 13) in the xanthic series and from white (grade 0) to intense black (grade 21) in the melanic series. There are minor differences in quality within the melanic series, but for the most part these are not great enough to interfere with grading. Our grades were found by Kröning to agree very well with ones which he established.

The set of average grades of the various genotypes establishes the *order* of intensity but does not, of course, measure the relative *quantities* of pigment. If, however, the average quantity in representatives of each grade can be established, it becomes possible to

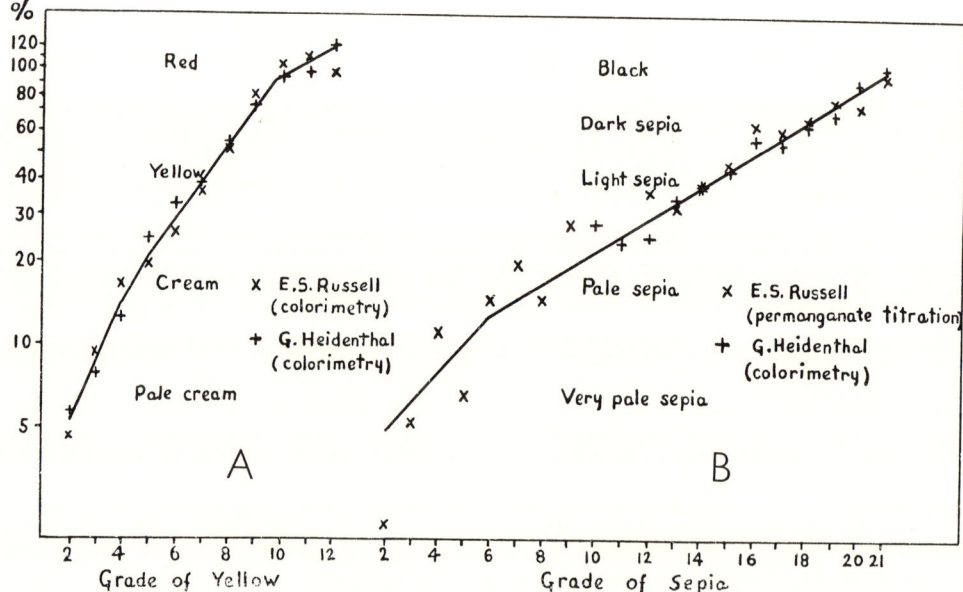

Fig. 1. Amount of pigment by colorimetry or titration with KMnO₄ plotted against grade. Intense red (A, left) and intense black (B, right) taken as 100% (E. S. Russell and G. Heidenthal).

transform the grade averages into quantitative averages.

The measurement of quantity of melanin presents various theoretical and practical difficulties. Probably there is no method available at present that can be relied upon completely. In our laboratory we have found excellent agreement between the results of two widely different methods of measuring melanic pigmentation: titration with $KMnO_4$ of pigments extracted from weighed samples of hair by the method developed by Einsele (Dr Elizabeth Russell) and colorimetric comparison of alkaline solutions of similarly extracted pigment (Dr Gertrude Heidenthal). There was fair agreement with weight of pigment in one comparison, although, as in Einsele's results, there was some indication that pigment from lower alleles of the albino series may be slightly lighter in colour per unit weight. Colorimetric measurement of grades of yellow by both Mrs Russell and Miss Heidenthal have given consistent results. Titration with $KMnO_4$ was also tried with yellow. While the results were rather variable they indicated that a typical intense yellow (grade 10) has only about 20% as much reducing capacity per unit weight of hair as an intense black (grade 21).

Some of the results are shown in Fig. 1 A, B, where the logarithms of the relative colorimetric values are plotted against the grades. A succession of straight lines was fitted by least squares to the weighted averages in each case and used as a basis for transformation of the grade averages.

PHYSIOLOGY OF PIGMENTATION

It is probable from somatic mutations (Wright and Eaton, 1926; Dunn, 1934) that the major colour genes of mammals act locally. We are thus dealing with intracellular rather than organismic physiology.

It has been well established that the melanins come from tyrosin or related substances (such as dopa) under the influence of oxidative enzymes. Danneel and Schaumann seem to have cleared up difficulties with earlier studies of skin extracts. They have devised means of separating dopa oxidase from inhibiting substances present in extracts from black rabbits. Danneel has given evidence for a chain of three processes: first, one that is suppressed by X-rays and second the anoxidative formation of dopa oxidase, occurring under continuous high temperatures (over 37°) in intense rabbits (C) but only after a period of exposure to temperatures below 33° in Himalayans ($c^H c^H$) and failing completely in albinos (cc). His third reaction, resulting in pigment formation, requires oxygen and is inhibited by HCN. The above temperature effect is in harmony with the previous observations of Schultz and others that pigment develops in Himalayan rabbits only in regions of the skin that have been exposed to cold.

Another method of demonstrating the occurrence of dopa oxidase has been from the reaction of frozen sections of skin to buffered solutions of dopa. Schultz and Kröning have found differences among a number of genotypes in rabbits and guinea-pigs by this method. In our laboratory, Dr W. L. Russell has used this method extensively.

MATHEMATICAL THEORY

To make full use of the quantitative determinations of the effects of gene substitution, it is necessary to apply a mathematical theory. This is a hazardous venture in connexion with complex processes of the living cell. Yet it seems better to attempt to use some simple quantitative theory consistently than to restrict oneself to merely verbal reasoning. It is usually easy to suggest a plausible interpretation of any particular interaction, but difficult to appreciate the implications throughout the entire network of reactions unless one introduces numbers.

Pigment is being produced throughout life. While certain processes are irrevocably determined early in development, it appears that the action of most of the genes in which we are interested here, or at least of active products of these, must take place more or less continuously. It will accordingly be assumed that the observed quantities reflect the rates of certain reactions during a phase of flux equilibrium rather than differences in duration or timing of the reactions (Wright, 1929, 1934).

An observed character difference may be at any number of removes in a chain reaction from the causal difference. Moreover, as illustrated in Fig. 2A, it need not be connected by a direct chain at all. A change at one point in the network of interacting processes may decrease or increase production in collateral processes through competition effects. We will assume as short and direct a chain of processes in each case as will suffice, but there is nothing to prevent elaboration if further data make it necessary.

One type of elementary reaction that needs consideration is that of Fig. 2B. An agent, A, whose rate of production $\dot{A}$ is affected by differences in a given gene, interacts with a substrate S to produce the observed product, P, or a precursor. The rate of production, $\dot{P}$, depends jointly on the concentrations (A) and (S) which, under flux equilibrium, depend on the balance between the rates of production $\dot{A}$ and $\dot{S}$ and the rates of dissipation whether in formation of the product $c(A)(S)$, or otherwise, $b(A)$ and $d(S)$. The rates of production are related by the equation

$$\dot{P} = \frac{c}{bd}(\dot{S} - \dot{P})(\dot{A} - \dot{P}).$$

Fig. 2C represents the case in which the agent is a catalyst, produced in a certain quantity E, a portion of which is free (F) and the rest (B) bound in an intermediate product with the substrate. The equation is the same as in the preceding case on substituting $b(E)$ for $\dot{A}$.

zygotes with an amorph, but as an antimorph in heterozygotes with an allele which produces a more efficient product. Such an allele might be called a mixomorph.

The greatest weakness of the attempt to make physiological deductions from the study of factor

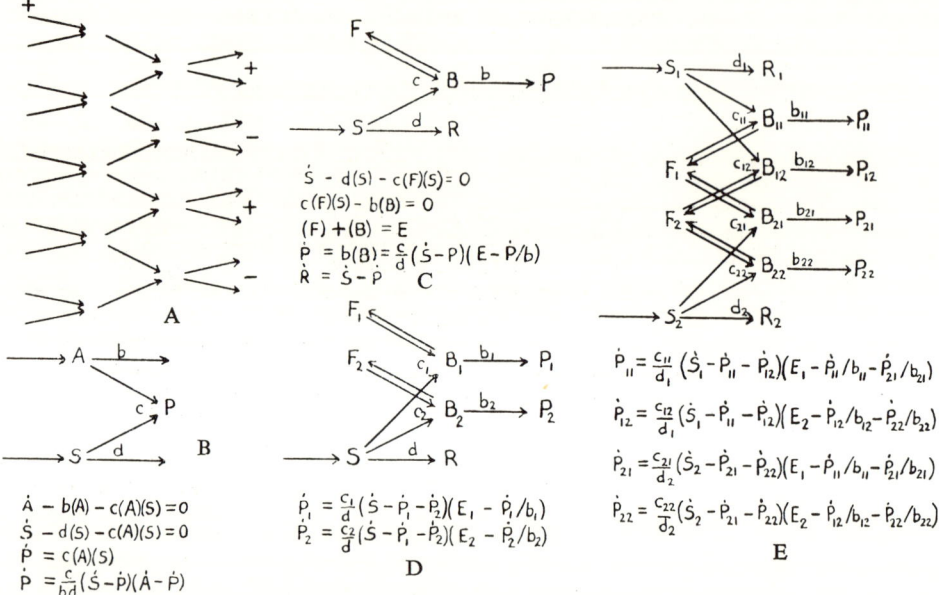

Fig. 2. Reactions under flux equilibrium: A, upper left, illustration of possible indirectness of relation between variation in cause and effect; B, lower left, case of non-catalytic interaction of agent (A) and substrate (S); C, upper middle, case of catalytic action; D, lower middle, case of two catalytic agents; E, right, case of two catalytic agents and two substrates.

If E is the gene itself, the amount of E is the same for all alleles (with the exception of deficiencies and duplications), but the different alleles may differ in the rate coefficients c and b. Moreover, in this case the products may well differ in specificity in the same sense as the alleles, if as previously suggested (Wright, 1927) production is related to the process by which genes duplicate themselves between cell divisions. Thus products formed equally freely from the substrate (same c and b) may differ in efficiency in relation to the characters actually observed. There is no difficulty on this basis in understanding how certain alleles may, in Muller's terminology, be amorphs ($c=0$), others hypomorphs, others hypermorphs in relation to type; and others antimorphs (P freely formed but with zero efficiency and thus destroying substrate that might otherwise be utilized). An allele that produces freely a product of low efficiency would behave as a hypomorph in hetero-

interactions arises from this possibility that gene substitution may bring about not only quantitative differences along a chain of reaction but also transmit differences in specificity. It is possible to dismiss any interaction effect as merely an inexplicable manifestation of specificity, an emergent in the sense of Lloyd Morgan. On the other hand, it is obvious that the greater the number of steps in the reaction chain between gene and product, the less the likelihood that any differences in specificity among alleles will be carried through to the latter, and thus the greater the likelihood that the effects of gene replacement at this point will be purely quantitative. We will treat products as differing merely quantitatively as far as possible.

Fig. 3 A–C is intended to bring out the relation between the rates of production of substrate and of a particular product under the influence of a gene that plays an irreplaceable role as a catalyst. Different rate coefficients are assumed in this figure. In Fig. 3 B, where both c/d and b are finite, there is an approach to complete dominance of the active phase of the gene over an inactive phase if the substrate is much in defect, practically no dominance if the substrate is in excess, with intermediate degrees of dominance between. Two extreme cases are shown in Fig. 3 A, C. In Fig. 3 A (c/d finite, $b=\infty$), the effect has a destructive effect. Again the gene has a multiplicative effect (in this case in the direction of decreased amounts of product) if c/d is finite, $b=\infty$, while it has a subtractive or threshold effect if $c/d=\infty$, b is finite. In interpretation it is convenient to use the extreme cases as far as possible because of their simplicity.

If alleles differ in the rate constants c and b rather

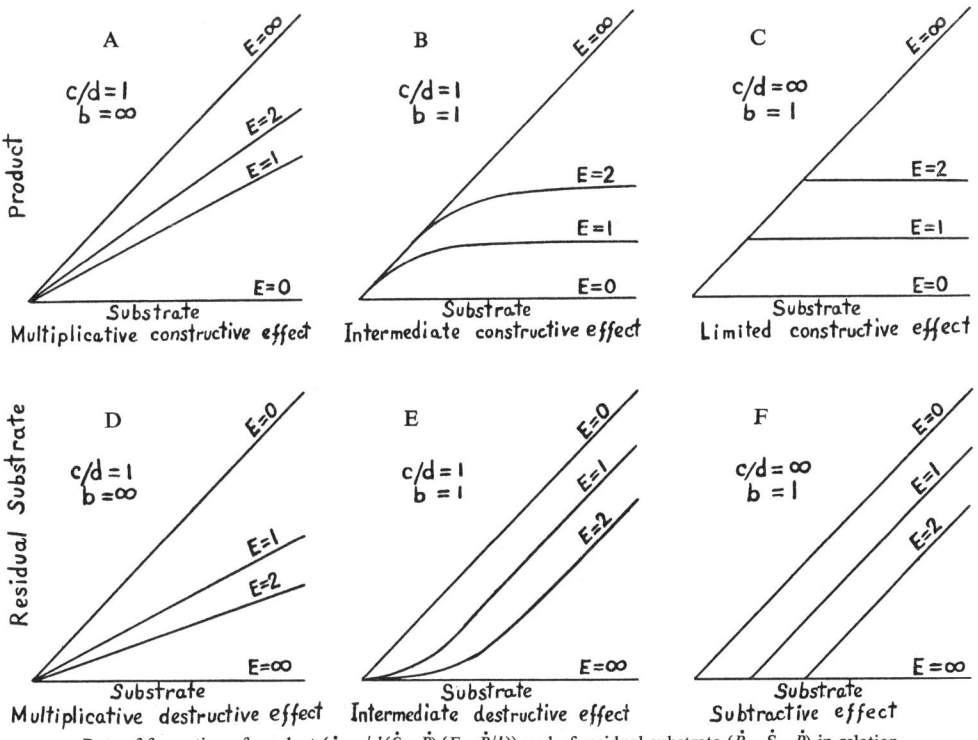

Rate of formation of product ($\dot{P}=c/d(\dot{S}-\dot{P})(E-\dot{P}/b)$) and of residual substrate ($\dot{R}=\dot{S}-\dot{P}$) in relation to rate of formation of substrate ($\dot{S}$), amount of enzyme (E) and rate constants c/d and b.

Fig. 3. Relations between substrate and product in reactions brought about by one or two representatives of catalytic agent with various rate onstants. A, B, C, upper row, constructive action; D, E, F, lower row, destructive action.

of the gene is strictly multiplicative, the rate of formation of the product being always a certain multiple of the rate of formation of the substrate $\left(\dot{P}=\dfrac{cE\dot{S}}{cE+d}\right)$. The degree of dominance $\left(\dfrac{2cE+d}{2cE+2d}\right)$ is the same for all rates of formation of substrate. In Fig. 3 C ($c/d=\infty$, b finite), there is a ceiling effect; the product depends merely on which is the limiting factor, substrate or gene. Fig. 3 D–F shows the relations between substrate and product where the gene

than in quantity (E) the effects of heterozygosis may become somewhat complicated as indicated in Fig. 2 D. There are further complications if two or more substrates compete for combination with the gene as indicated in Fig. 2 E, the system used in interpreting the effects of the albino series.

THE DISTRIBUTION OF COLOUR

The colour factors of the guinea-pig may be divided primarily into those that determine the distribution of colour in the coat irrespective of its kind, and those that determine quality and quantity of colour in the coloured areas.

The factors that determine the common piebald pattern (principal gene S, s) are representatives of the first group. It is generally agreed that there is no dopa reaction whatever in the white areas, but this probably indicates absence of pigment cells rather than mere absence of dopa oxidase.

MELANIC DIFFERENTIATION

Fig. 4 shows the quantity of pigment (melanic above, xanthic below) in some of the more important gene combinations. The effects of the compounds of the albino series, arranged in what seems the most significant order, are shown in combination with genes of other series. A condensed tabulation of the quantities of pigment in the more important combinations is given in Table 1.

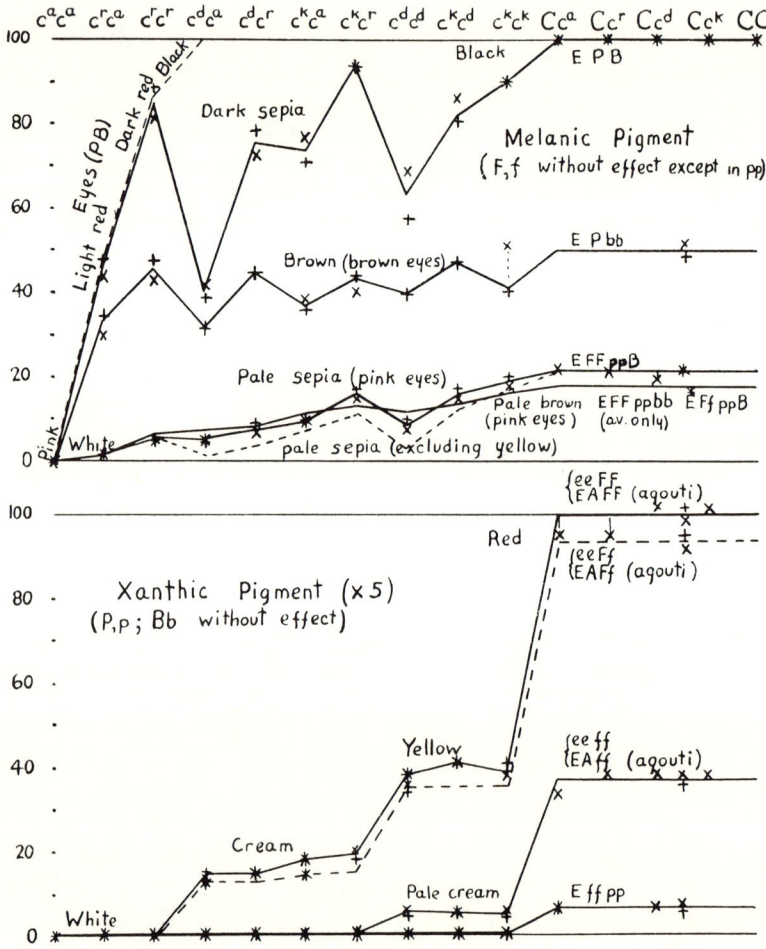

Fig. 4. The average quantities of pigment (by transformation of grades) in various gene combinations. Melanic pigment above. Xanthic pigment below. +, quantities from average grades published in 1927. ×, quantities from average grades during the period 1933 to 1938.

Table 1. *The average concentrations of pigment in the hair of guinea-pigs at birth.*

	Sepia		Brown		Yellow			Eye colour (E, e; F, f without effect)		
	EPFB EPfB	EpFB	EPFb EPfb	EpFb	EpFB Epfb	ePFB ePFB epFB epfB	ePfB ePfb epfB epfb	PB	Pb	pB pb
C-	100	21	50	17	6	100	36	Black	Brown	Pink
$c^k c^k$	90	18	42	15	0	38	5	Black	Brown	Pink
$c^k c^d$	82	15	47	13	0	41	5	Black	Brown	Pink
$c^d c^d$	64	9	40	11	0	38	5	Black	Brown	Pink
$c^k c^r$	94	14	43	13	0	19	0+	Black	Brown	Pink
$c^k c^a$	73	9	37	11	0	18	0+	Black	Brown	Pink
$c^d c^r$	75	7	44	7	0	14	0+	Black	Brown	Pink
$c^d c^a$	40	5	31	6	0	14	0+	Black	Brown	Pink
$c^r c^r$	84	5	45	6	0	0	0	Dark red	Dark brown-red	Pink
$c^r c^a$	46	1	33	1	0	0	0	Light red	Light brown-red	Pink
$c^a c^a$	0	0	0	0	0	0	0	Pink	Pink	Pink

The sepias and browns are given on a scale in which intense black (EPBC) is 100. The yellows are given on a scale in which intense yellow (eeFFC) is 100. The latter actually has only about 20 % as much capacity for reduction of $KMnO_4$ as intense black. Replacement of aa, assumed above, by A replaces the sepia, brown or yellow of E combinations by the yellow of the corresponding e combinations in a subterminal band in each hair in the e combinations. There is no effect. The averages for the b combinations, especially pb, are based on relatively inadequate numbers.

The most important factors in distinguishing the melanic and xanthic colours are those of the series E, e^p, e. With E present, the entire coat, the eye, and skin tend to exhibit melanic pigmentation. With ee the entire coat (at birth) exhibits only xanthic pigment (if any). There seems to be only melanic pigmentation with ee in the skin although a reduced amount compared with E. The eyes (in ee) are melanic and indistinguishable from those of animals with E. With $e^p e^p$ and $e^p e$ the coat shows a tortoise-shell mixture of the colours found in otherwise similar animals of constitutions E and ee. The yellow area is considerably greater on the average in $e^p e$ than in $e^p e^p$, although there is much overlap (Chase). Transplantation experiments indicate that the pattern is irreversibly determined early in development (Seevers and Spencer). This process we will call melanic differentiation. It is affected by minor modifiers in tortoiseshells. The most important modifying condition, however, is white spotting. With no white (SS $e^p e^p$) tortoiseshells characteristically show a mixture of relatively few yellow hairs on a melanic background. In tricolours (ss $e^p e^p$) there is more yellow, and the melanic and xanthic colours tend to be segregated into distinct spots (Wright, 1917; Chase, 1939). It appears that the same factors that affect the capacity of the primordial pigment cells to reach all parts of the skin also affect their tendency to undergo melanic differentiation.

The characterization of the coat with E present as melanic and with ee as xanthic requires qualification. A small amount of xanthic pigmentation seems to be present in certain combinations mixed with the prevailing melanic pigment (E $c^d c^d$ F pp B yellowish sepia in contrast with E $c^r c^r$ F pp B slaty sepia).

Moreover, certain factors bring out pure xanthic pigmentation in spite of E. Thus agoutis, EA, show xanthic colour (in subterminal bands in the hair) of the same intensity as that shown by otherwise similar animals in which E is replaced by ee. With $A^r A^r$ and $A^r a$, involving a gene introduced into the guinea-pig from crosses with another species (*Cavia rufescens*), there is a similar pattern except that the extent of yellow on both back and belly is much reduced. With aa there is no such pattern. Genes A and A^r must act *before* there has been any interference of the melanic processes with the xanthic process to account for the full intensity of the xanthic colour.

Pure xanthic pigmentation is also exhibited in the combination E C ff pp. In this case, however, the intensity is much reduced. There is only about 17 % as much pigment as in ee C ff pp. It must be concluded that in this case the melanic process is blocked (by simultaneous failure of P and F) *after* it has interfered with the xanthic process.

That the postulated melanic differentiation of the embryonic skin is not an absolute prerequisite for melanic pigmentation is shown by the fact that in hair that is purely xanthic at birth (as in the whole coat of ee) considerable melanic sootiness may appear with age, especially after temporary thinning of the hair with exposure to low temperature). This sootiness is affected by modifiers (such as P, p; B, b) in the same way as the pigment in the areas of melanic differentiation. It is also inhibited subterminally by the agouti factor, A.

THE ALBINO SERIES

The albino series plays a fundamental role in the determination of both kinds of pigment. With $c^a c^a$, there is no pigment at birth in any combination with other genes (white with pink eyes). The allele C determines the highest intensity of pigment found with each combination. It is completely or almost completely dominant over all of the other alleles in all combinations. At one time, the averages indicated considerable incompleteness of dominance in the combination E Cc^{a}pp, but more recent data indicate that this was due to inadequately controlled modifying factors.

There are peculiarities in the effects of intermediate alleles. Only three alleles (C, c^{kd}, c^{ra}) are distinguishable in effect on xanthic pigment, resulting in four successive levels with respect to quantity if the compounds are arranged as in Fig. 4. The genotypes ee c^{r}c^r and ee c^{r}c^a are as white as albinos (ee c^{a}c^a), though differing in eye colour. The four compounds in which either c^k or c^d is heterozygous for either c^r or c^a (ee c^{kd}c^{ra}) have from 14 to 19% as much pigment as intense reds (C). A slight difference is indeed indicated between c^{k}c^{ra} and c^{d}c^{ra}, but this may be due to modifiers. The three compounds represented by c^{kd}c^{kd} are practically indistinguishable with 38 to 41% as much pigment as in ee C, and thus more than twice as much as in the heterozygotes c^{kd}c^{ra}.

On considering the sepias (EPB) in the same order one finds a succession of waves in place of the levels in the yellows. At the level of no yellow, sepia rises from c^{a}c^a (white) through c^{r}c^a (46%) to c^{r}c^r (84% of the pigmentation of black, C). With the appearance of yellow in c^{d}c^a, sepia drops to 40%, rising through c^{d}c^r (75%) and c^{k}c^a (73%) to a new high level in c^{k}c^r (94%), while yellow remains unchanged. With the rise in yellow in c^{d}c^d, sepia falls to 64%, to rise at this level of yellow through c^{k}c^d (82%) to c^{k}c^k (90%).

The melanic pigmentation of the eyes gives a third order. There is no pigmentation in the pink eyes of albinos. In c^{r}c^a the eyes are much darker but show a strong red reflexion through iris and pupil. They are still darker in c^{r}c^r, but the pupil still appears red. In c^{d}c^a and all higher compounds the eyes appear black. Thus c^{d}c^a has more black in the eyes than c^{r}c^r but less black in the coat.

In previous papers (Wright, 1916, 1917, 1925, 1927) it has been suggested that the albino series determines a linear series of rates of production or of potencies of a substance necessary for any pigmentation in the order c^a, c^r, c^d, c^k, C, and that the irregularities are due to a difference in threshold depending on the presence or absence of E product, with competitive interference in the hair follicles but not in the eyes of the xanthic with the melanic process above the threshold at which the former can be formed. The hypothesis that a higher threshold for yellow than black is characteristic of the albino series has been borne out by the later discoveries of similar conditions in rabbits, rats, mice, cats, etc. Some evidence from other cases was adduced for a competition between the xanthic and melanic processes, although no other cases involving the albino series seem to have been described. Difficulties were recognized which have become more obvious on attempting a quantitative interpretation. We will return to these after considering certain other combinations.

SPECIFIC MODIFIERS OF MELANIC PIGMENTATION

The replacement of P by pp greatly reduces the amount of melanic pigment in the coat (C pp = 21% of CP), the skin, and the eyes (pink), but has no effect whatever on xanthic pigment. The effect is exaggerated in lower compounds, especially in c^{r}c^r pp with only 5% of c^{r}c^r P; and in c^{r}c^a pp, which is sometimes pure white at birth and has only a trace of colour at best. The reduction of c^{r}c^a pp to less than 50% of c^{r}c^r pp is taken to indicate a slight subtraction effect. The order of the grades is different from that with P (e.g. c^{r}c^r pp < c^{d}c^d pp but c^{r}c^r P > c^{d}c^d P). This may be due to the presence of considerable xanthic pigment in such compounds as c^{d}c^d pp which is wholly lacking in c^{r}c^r pp. The compounds c^{d}c^a pp, c^{d}c^r pp and c^{d}c^d pp have a very yellowish appearance (in contrast with c^{k}c^a pp and c^{r}c^r pp) and occasionally have approached pure light yellow so closely that a breeding test was necessary to establish the constitution.

The replacement of B by b has effects which in several ways resemble those due to replacement of P by p. In both cases melanic pigment is reduced in hair, skin and eyes. In neither case is there any effect on xanthic pigment. There are, however, two important differences. The browns (E P bb) appear to differ qualitatively from the dark-eyed sepias (EPB), while the pink-eyed sepias (E pp B) do not. The higher compounds of the albino series are more alike in browns than in dark-eyed sepias while, as just noted, the reverse is true in the pink-eyed sepias. Thus the ratio c^{r}c^r to C is 84% in dark-eyed sepias, 91% in browns but 25% in pink-eyed sepias. The ratio c^{r}c^a to c^{r}c^r is 54% in dark-eyed sepias, 73% in browns and probably less than 25% in pink-eyed sepias, where the quantity in c^{r}c^a pp is too small to measure. Dunn and Einsele have found a similarly slight effect on brown of compounds of the albino series in mice. The double recessive, pink-eyed browns, E pp bb, differ qualitatively from the pink-eyed sepias but apparently very little in quantity. Where E P bb has about 50% as much pigment as EPB, E pp bb has about 80% of the intensity of E pp B. Among the lower compounds, the grade assigned pink-eyed browns was practically the same as that assigned the corresponding pink-eyed sepias.

These facts can be explained on the hypothesis that the C and P series affect the quantity of an agent which acts on a limited melanic substrate according to the intermediate theory, and that the product is completely transformed into sepia if **B** (completely dominant) is present but is transformed into brown according to the intermediate theory if **B** is absent. With **P** present, the substrate is to a considerable extent the limiting factor, and the sepias show a damping of the effects of the higher compounds of the albino series and the browns exhibit a more pronounced damping of this sort. If, however, the quantity of the agent is reduced to only about 5% of its value by replacing **P** by **pp**, the agent becomes the limiting factor in the main. On this hypothesis the true quantitative relations among the compounds of the albino series are presented more faithfully in the pink-eyed browns and sepias (at least after correcting for a probable late subtractive effect and for the presence of xanthic pigment above $c^r c^r$) than in black-eyed sepias. The results can also be interpreted on the basis of separate sepia and brown substrates.

THE ALBINO SERIES AGAIN

It will be convenient here to return to consideration of the irregularities of the albino series. We may distinguish three alternative interpretations of the threshold difference between xanthic and melanic colours: (1) There may be a subtraction from the primary effects of the *compounds*, in the absence of melanic differentiation, which reduces $c^r c^a$ and $c^r c^r$ to white. (2) There may be a subtraction from the primary effects of the separate genes. (3) The difference in threshold may be in the potency of the genes themselves relative to yellow and sepia substrates. Under (1) $c^d c^r$ and $c^k c^r$ should produce more yellow than $c^d c^a$ and $c^k c^a$, if c^r is higher than c^a in a quantitative series. Since this is not the case, this alternative is eliminated. Under (2) there should be little or no competitive effect of the xanthic c^d reaction with the melanic c^d reaction, since under this hypothesis the genes of the albino series (or their products) combine with E product in preference to the inhibiting substance, and with the latter in preference to the xanthic substrate. As for competition between the reactions for utilization of another substance of limited quantity, this sort of competition should be greatest with C present and should be negligible with c^d if the effect of c^d is as low relative to C as is indicated in the pink-eyed sepias and browns. Since the only evidence for successful competition of yellow with sepia comes in the c^d compounds (the melanic process almost suppressing yellow in the presence of C), this alternative must either be ruled out or the reversed effects of c^d and c^r in eye and hair be attributed to specificity rather than to differences in competition. This brings us to the third alternative, viz. that c^r and c^a are specifically unable to act on the xanthic substrate although able, in widely differing degrees, to act on the melanic substrate. In the mathematical theory, it is necessary to postulate increasing values of both rate constants, **c** and **b**, in relation to both substrates, apparently implying four dimensions of variability among alleles. However, the ratio of **c** to **b** may be constant in each case, reducing the number of dimensions to two, which in turn may not be independent. There is no difficulty in assigning values to the constants that will account for the observed differences in order of effect of c^d and c^r in coat and eyes on the hypothesis that there is a yellow process only in the former.

The close similarity if not identity of c^d and c^k in effect on yellow is another difficulty in applying the original hypothesis of a single quantitative order. We will here merely accept this identity with respect to yellow, assigning higher rate constants to c^k in effect on sepia.

The complete or nearly complete dominance of **C** over the other alleles in all combinations, including ones in which the intensities of yellow and sepia are greatly reduced (e.g. ee Cc^a ff, E Cc^a ff pp, E Cc^a FF pp), implies that **C** (or a product) is in excess relative to its substrate in a reaction *preceding* action of F or P. We assume that **C** is in excess relative to its immediate substrate while its lower alleles are in defect.

THE COMBINATIONS WITH F, f

The intensity of yellow is markedly reduced by replacing FF by ff. There is relatively (but not absolutely) more reduction in $c^{kd} c^{kd}$ than in C. In $c^{kd} c^{ra}$, xanthic pigment is almost or wholly lacking. The fact that F is not completely dominant over f suggests an approach to a multiplicative effect. The data are in fair agreement with such an effect if supplemented by a later small subtractive effect.

This replacement has no effect on black (ECffP) or dark-eyed sepias (E $c^k c^a$ ff PB, E $c^d c^d$ ff PB, E $c^d c^a$ ff PB). There is also no effect on brown-eyed browns (EC ff P bb, E $c^d c^d$ ff P bb, etc.). As we have attributed the relatively low intensity of sepia in the c^d combination to a competitive effect of yellow, it must be supposed that the yellow process is affected by F, f only *after* this competitive action is completed.

The data above suggest that F, f are specific modifiers of xanthic pigment. This idea is, however, upset by consideration of the combinations with pp. In the combinations EC ff pp B and EC ff pp bb, black and brown are replaced by a pale yellow: E $c^k c^k$ ff pp and lower compounds of the albino series are pure white. This peculiar relation of F, f to P, p can be explained most easily by the hypothesis that F can substitute weakly for P in its reaction and that there is no other substitute. The fact that P is completely dominant

327 *Interactions of the Major Colour Factors of the Guinea-Pig*

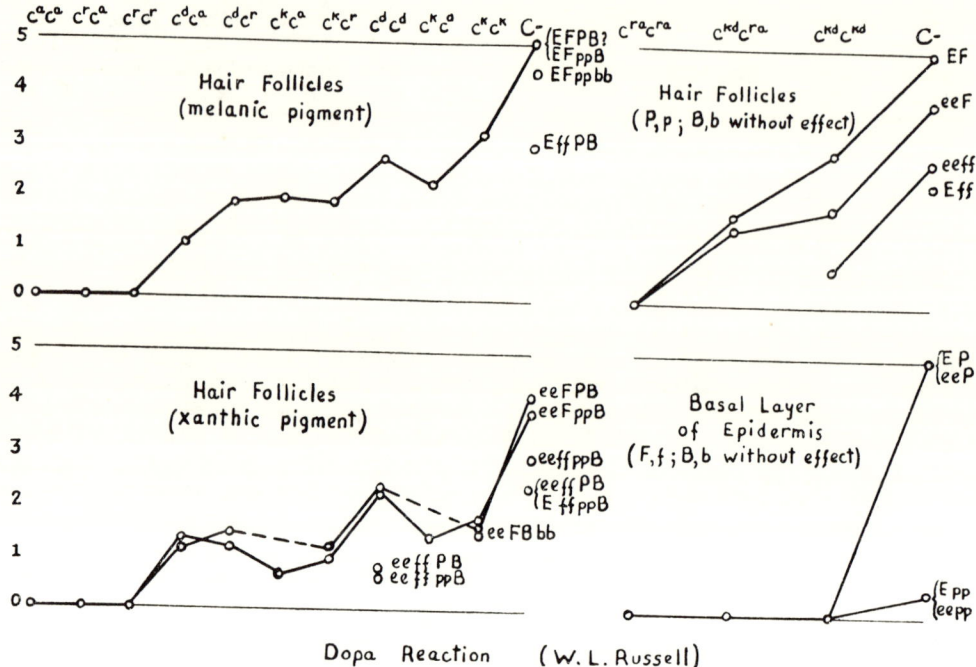

Fig. 5. Grade of dopa reaction in various genotypes from W. L. Russell.

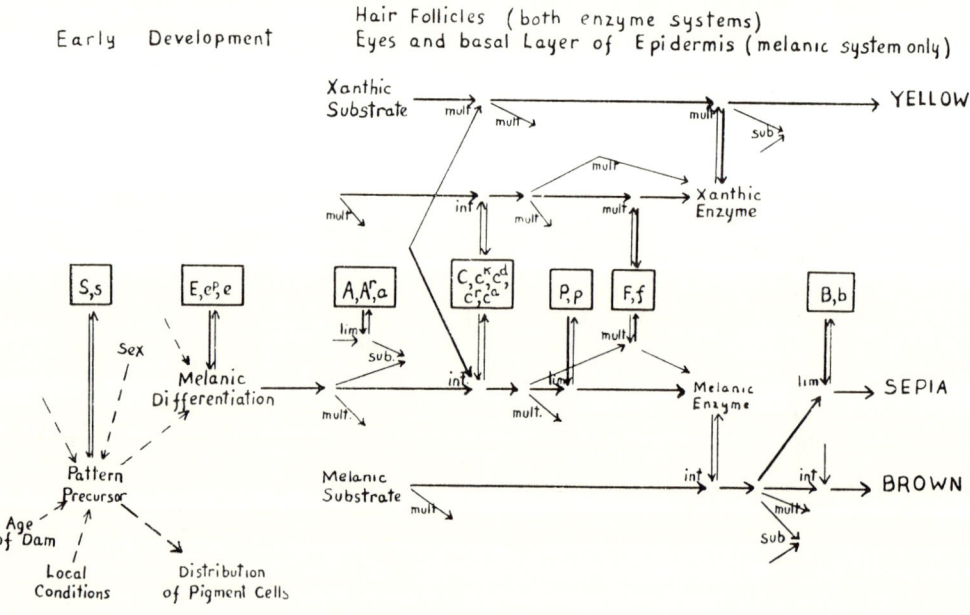

Fig. 6. System of reactions leading to production of yellow, sepia and brown pigment as deduced from gene interactions.

over **p** implies that it is in excess relative to its substrate, transforming all that there is. This explains the absence of any effect on melanic pigment of replacing **F** by **f** where **P** is present. If **P** fails, **F**

The great reduction of the yellow revealed in **E ff pp** (17%) in comparison with **ee ff pp** implies a severe inhibition of yellow by the melanic process at some stage *before* the action of **P** and **F**, but *later*

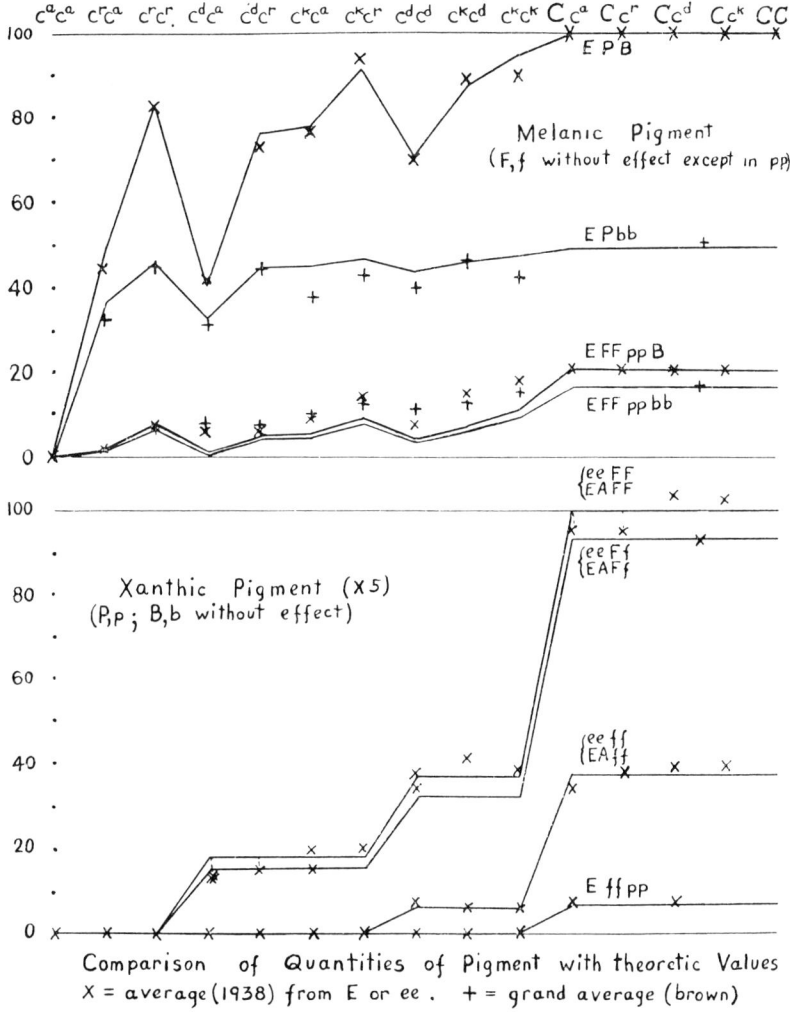

Fig. 7. Average quantities of melanic pigment (above) and xanthic pigment below, compared with theoretical values obtained by substituting appropriate numbers in the system of Fig. 6. Averages based on data for period 1933-8, except in case of brown where grand averages for this and earlier data are used.

carries on the melanic process but at only about 5% of its normal rate. It must be supposed to be approximately multiplicative in effect. Its imperfect dominance (**E Ff pp** being lighter and yellower than **E FF pp**) is in harmony.

than the agouti reaction. This is also indicated by the small amount of yellow associated with melanic pigment in blacks **E C F P B** and pale sepias **E C F pp B**. The latter often fade almost to white, when adult, without revealing any appreciable yellow coloration.

On the other hand, as noted, combinations of $c^d c^d$, $c^d c^r$ and $c^d c^a$ with E F pp (especially E Ff pp) are markedly yellowish in appearance. This seems to imply that the competitive effect of sepia on yellow is less in these cases, which means that it occurs either at the same time or later than the melanic C reaction. Thus the P and F reactions must follow the C reaction if this reasoning holds.

THE DOPA REACTION

Two very important additional indicators of the activities of these genes are furnished by Dr W. L. Russell's studies of the dopa reaction. His results can apparently be interpreted on the hypothesis that the dopa reaction of hair follicles (affected by the C series and by F, f but not by P, p or B, b irrespective of the actual colour of the animal) is an indicator of a xanthic enzyme, while the dopa reaction of the basal layer of the epidermis (strong only in CP, weak or absent in C pp, absent in $c^k c^k$ P, etc., unaffected by F, f; B, b) is a weak indicator of a melanic enzyme (Fig. 5). It is postulated that B acts on either the substrate or product of action of the melanic enzyme but not on the enzyme itself. Finally, since replacement of e by E slightly increases rather than decreases the strength of the dopa reaction in the hair follicle, it must be supposed that the inhibitory action of the melanic process on yellow occurs on the substrate or product of action of the xanthic enzyme but not on the enzyme itself (of which the dopa reaction is assumed to be an indicator).

There are diverse effects of temperature which indicate that in some cases the positive reactions are increased by higher temperatures (e.g. ee ff), while in othes cases inhibitory reactions are increased (e.g. E $c^r c^a$ PB, E $c^a c^a$ PB, E $c^a c^a$ P bb, etc.), in some cases there seems to be little or no effect (e.g. ee FF), while in other cases there seems to be a conflict of opposing tendencies (e.g. EF pp B). It does not appear practicable at present to attempt to locate these effects more precisely.

INTERPRETATION

The simplest interpretation of the sequence and nature of the reactions (multiplicative, intermediate, limiting, subtractive) to which we are led by the foregoing considerations is represented in Fig. 6. The consistency of the scheme has been tested by assigning numbers to the reaction constants throughout. The extent of agreement is indicated in Fig. 7 and is undoubtedly capable of some improvement.

REFERENCES

CHASE, H.B. (1939). *Genetics*, 24, 67.
DANNEEL, R. and SCHAUMANN, K. (1938). *Biol. Zbl.* 58, 242–60.
DUNN, L.C. (1934). *J. Genet.* 29, 317–26.
DUNN, L.C. and EINSELE, W. (1938). *J. Genet.* 36, 145–52.
EINSELE, W. (1937). *J. Genet.* 34, 1–18.
HEIDENTHAL, G. (1938). Thesis. The University of Chicago. (In the Press.)
KRÖNING, F. (1930). *Z. indukt. Abstamm.- u. VererbLehre*, 53, 355–67.
—— (1930). *Roux Arch. Entw. Mech. Organ.* 121, 470–84.
MULLER, H.J. (1932). *Proc. 6th Int. Congress Genetics*, 1, 213–55.
RUSSELL, ELIZABETH S. (1939). *Genetics*, 24, 332–55.
RUSSELL, W.L. (1939). *Genetics*, 24, 645–67.
SCHULTZ, W. (1918). *Z. indukt. Abstamm.- u. VererbLehre*, 20, 27–40.
—— (1925). *Arch. EntwMech. Org.* 105, 677–710.
SEEVERS, C.H. and SPENCER, D.A. (1932). *Amer. Nat.* 66, 183–9.
WRIGHT, S. (1916). *Publ. Carneg. Instn*, no. 241, pp. 59–121.
—— (1917). *J. Hered.* 8, 224–35, 476–80.
—— (1925). *Genetics*, 10, 223–60.
—— (1927). *Genetics*, 12, 530–69.
—— (1934). *Amer. Nat.* 68, 24–53.
WRIGHT, S. and EATON, O.N. (1926). *Genetics*, 11, 335–51.

THE GENETICS AND CHEMISTRY OF FLOWER COLOUR VARIATION

By W. J. C. LAWRENCE AND J. R. PRICE

(John Innes Horticultural Institution, Merton)

(*Received 16 March 1939*)

CONTENTS

		PAGE
I.	The chemistry of flower colour variation	35
	(1) The anthocyanins	36
	(2) The anthoxanthins	38
	(3) The carotinoids	39
II.	The inheritance of colour variations	40
	(1) The role of genes in pigment production: qualitative effects	40
	(2) The role of genes in pigment production: quantitative effects	42
	(3) Gene control of pigment modification	43
III.	The biogenesis of the anthocyanins and anthoxanthins	46
IV.	Dominance relationships and competition	50
IV.	Summary	55
VI.	References	55
	Addenda	57

I. THE CHEMISTRY OF FLOWER COLOUR VARIATION

COLOUR is one of the characters of plants and animals most frequently used in genetical investigations. Separation of colour types depended until recently on visual comparisons alone. These, since they represent only a first analysis, are always inadequate and may sometimes be misleading. For a further understanding of the developmental processes involved we need a knowledge of the chemical structure and properties of the pigments responsible.

It is with the flower pigments, especially the anthocyanins and anthoxanthins, that chemical analysis has gone farthest. In other cases, such as eye colours in *Drosophila* (Beadle & Ephrussi, 1936, 1937) and chlorophyll deficiencies in *Zea Mays* (Phipps, 1929), *Trifolium pratense* (Williams, 1937) and *Datura* (Inman & Blakeslee, 1938), something is known of the pigments which are affected, but nothing of the nature of the changes they undergo. It is in the flower pigments that gene action can be examined for the first time in its fundamental sense, namely as governing simple chemical changes: oxidation, reduction, methylation and glycoside formation.

The great majority of flower pigments belong to three main classes, the anthocyanins and the anthoxanthins, both of which are sap-soluble, and the carotinoids, which are generally found in the plastids and are not sap-soluble. The term anthocyanin, introduced by Marquart (1835) and used by other early workers, denoted simply the red and blue sap-soluble pigments of plants. It is still used in this broad sense by botanists, but in chemical usage the term is generally reserved for what are known as the hydroxy-flavylium salts, which include most, but not all, of the red and blue plant pigments. Full details of the chemistry of the hydroxy-flavylium salts can be found in the publications of Willstätter (1913, 1914, 1915, 1916), Robinson[1] and Karrer[2] and their collaborators; as much as is relevant to this review is outlined below.

(1) *The anthocyanins*

With two exceptions the anthocyanins (in the limited sense) are derived from three hydroxy-flavylium salts, pelargonidin (I), cyanidin (II) and delphinidin (III), which differ only in the number of hydroxyl groups in the 2-phenyl nucleus:

These substances, known as anthocyanidins, are the colour-producing part of the anthocyanin molecule. They do not occur free in nature, but in combination with one or two molecules of a sugar, this compound being an anthocyanin.

As shown, cyanidin has one more hydroxyl group in the molecule than pelargonidin, and delphinidin two more. This is one of the principal factors upon which variation in flower colour depends, as an increase in the number of oxygen atoms (in the form of —OH groups) causes a marked increase in blueness of tone. The scarlet pelargonium, deep red rose and purple delphinium are good examples of colours due to pelargonidin, cyanidin and delphinidin derivatives respectively.

A second variable involving structural difference in the anthocyanidin molecule concerns the 3'- and 5'-hydroxyl groups,[3] which may or may not be methylated, that is, have the hydrogen atom of the hydroxyl group replaced by a methyl group, —CH_3. Other conditions being the same, methylated anthocyanins or anthocyanidins are redder than the corresponding unmethylated types, e.g. peonidin (IV) is redder than cyanidin and malvidin (V) is redder than delphinidin.

[1] Two series of papers in the *J. chem. Soc.* "A synthesis of pyrylium salts of anthocyanidin type", parts I–XXII, 1922–34; "Experiments on the synthesis of anthocyanins", parts I–XXVI, 1926–34.
[2] A series of papers in *Helv. Chim. Acta*, "Über Pflanzenfarbstoffe", 1927–32.
[3] It affects three groups in the anthocyanin hirsutin, found in *Primula hirsuta*, which is a 3':5':7-*O*-trimethyl delphinidin derivative. This, however, only occurs in some four species.

Reference to formulae I, II and III shows that there is one methylated derivative of cyanidin—3′-O-methyl cyanidin (peonidin), and two of delphinidin, the 3′-O-methyl derivative (petunidin) and the 3′:5′: O-dimethyl derivative (malvidin).

As mentioned above, the anthocyanins are compounds involving one or two molecules of a monose. Of these sugar molecules, one is always attached at the 3-position. If a second sugar molecule is present, it may either be attached directly to the first one or be linked with the anthocyanidin at position 5-. Hence there are two classes of glycosides: (*a*) those with a monose or a biose attached at position 3-, (*b*) and those with monose units at both 3- and 5-. These two classes differ in colour, the 3:5-dimonosides (VI) being bluer than the corresponding 3-type (VII).

For the terminology employed in describing these glycosides see Robinson & Robinson (1932).

So far three factors influencing the colour of anthocyanins have been dealt with, namely:

(i) the number of substituent hydroxyl groups in the anthocyanidin molecule,
(ii) the methylation of hydroxyl groups,
(iii) the position of attachment of sugar molecules.

Combinations of these three give rise to twelve anthocyanins each slightly different in colour, but together covering a wide range from scarlet to purple. The three factors are all dependent upon structural changes in the anthocyanin molecule, that is, the differences are internal. Conditions external to the molecule may also affect the colour of the anthocyanins. The most important of these conditions is a phenomenon known as copigmentation which will be referred to in connexion with the anthoxanthins.

Modification of flower colour can also be brought about by variation of a second external factor: the pH of the cell sap. The colour of pure natural or synthetic anthocyanins, all of which are indicators, varies from red to blue according to the pH of the solution, being blue at a high pH and red at a low one (see Robinson & Robinson, 1933). Comparisons of the colour of different anthocyanins should therefore be made in solutions of the same pH (Robertson & Robinson, 1929). The suggestion that many flower colour variations are due to changes in the pH of the cell sap was made by Willstätter & Everest (1913), Buxton (1932) and others. This was proved correct by Philip-Smith (1933) making use of the indicator nature of the anthocyanins themselves and by Scott-Moncrieff (1936) using a glass electrode. It should be emphasized that the values given by the latter direct method are not those of the uncrushed petal, and are nearly always on the acid side of the neutral point; the recorded pH of blue flowers is in itself insufficient to account for the blueness of tone (Robinson & Robinson, 1933). Nevertheless, significant pH differences are found between red and blue flowers of genetically related plants containing the same anthocyanin. These differences are usually of the order of 0·5–1·0 pH (Scott-Moncrieff, 1936).

It has been suggested by Robinson & Robinson (1933) that the anomalous behaviour of many blue flowers in showing a cell sap pH of less than 7·0 is due to the existence of colloidal solutions of the anthocyanin, which are bluer than a true solution at the same pH.

(2) *The anthoxanthins*

The anthoxanthins are sap-soluble pigments, usually glycosides, that range in colour from ivory to yellow. They are closely related chemically to the anthocyanins, though more varied with respect to the numbers and positions of substituent groups, and they fall into two classes, the flavones and flavonols. These classes differ in that the flavones have no substituent hydroxyl group at position 3-.

The anthoxanthins most commonly found in flowers are the flavonols quercetin (VIII, e.g. in *Bougainvillea*, Price & Robinson, 1937) and kaempferol (IX, e.g. in *Crocus*, Price et al. 1938) and the flavones apigenin (X, e.g. in *Antirrhinum*, Wheldale & Bassett, 1913) and luteolin.

In general, increase in the number of hydroxyl groups slightly intensifies the colour.

There are four ways in which the anthoxanthins may be concerned in flower colour:

(a) In flowers which have no anthocyanin they may be directly responsible for some or all of the colour.

(b) When a yellow anthoxanthin occurs together with an anthocyanin, the resultant colour is a blend of the two.

(c) In the presence of anthocyanins, ivory anthoxanthins do not contribute independently to the colour, but they may do so indirectly by their "copigmenting" action. Copigments are substances which when present in the same solution as an anthocyanin form weak additive complexes that are much bluer than the anthocyanin alone. Willstätter & Zollinger (1916) observed that the addition of tannin to a solution of malvidin 3-glucoside produced this effect. Robinson & Robinson (1931) and later Lawrence (1932) pointed out that such complexes play an important part in producing variations of flower colour. The anthoxanthins are not the only natural copigments; tannins and other unidentified substances also act in this way. Copigmentation varies in degree according to the nature of the anthocyanin; delphinidin derivatives are most readily copigmented and pelargonidin derivatives least so. It is also dependent upon the nature of the anthoxanthin, ivory anthoxanthins usually being the most effective.

(*d*) The structural similarity of the anthocyanins and anthoxanthins suggests that their syntheses in the plant may be correlated. Evidence from *Dahlia* (Lawrence & Scott-Moncrieff, 1935) shows that the two classes of substances are formed from the same starting materials which are probably limited in quantity. This results in competition between the two, and if most of the precursor is utilized in the synthesis of one pigment, then of necessity less of the other is produced. Thus the presence of much anthoxanthin may lead to almost complete suppression of anthocyanin, producing delicately flushed flowers. Further, if the anthoxanthin concerned is a copigment for the anthocyanin, then there is a modification of colour as well as of intensity.

(3) *The carotinoids*

Although there are no combined chemical and genetical data on the inheritance of the non-sap-soluble flower pigments, some mention of them must be made. They comprise a number of yellow or orange substances, xanthophylls and carotins, which are carried by plastids or are dissolved in oils (Möbius, 1885). In the absence of anthocyanins they are either solely responsible for flower colour, or are supplemented by yellow anthoxanthins. In the presence of anthocyanins the colour is a blend of the two.

To summarize the position, variation in flower colour may be brought about in the ways shown in Table I. Changes are shown in one direction only; the reverse may be inferred.

Table I

Anthocyanins:	
i. Increase in number of hydroxyl groups	Increased blueness
ii. Alteration from 3- to 3:5-sugar types	Increased blueness
iii. Methylation of one or more hydroxyl groups	Increased redness
iv. Increase in pH of the cell sap	Increased blueness
v. Copigmentation (anthoxanthins, tannins, etc.)	Increased blueness
Anthoxanthins:	
vi. Increase in number of hydroxyl groups	Increased yellowness. Alteration of background. Change in copigment effect
vii. Interaction of anthocyanins and anthoxanthins	Partial suppression of one or both types
Plastid pigments:	
viii. Alteration in nature of plastid pigment	Yellow ⇌ orange. Alteration of background

II. THE INHERITANCE OF COLOUR VARIATIONS

It is clear from the preceding account that the visual method of classifying flower colour variations is analytically inadequate. A change from red to blue, for example, can be brought about in five different ways. Should the plant under examination make use of two or more of these ways, as is often found, the inheritance may be confusing or even completely misleading. In addition to facilitating genetical classification, a knowledge of the chemistry of the substances concerned constitutes a first step towards an understanding of the mechanism of gene action. The question of the biogenesis of the anthocyanins, anthoxanthins and related substances necessarily involves the recognition of the successive stages of their synthesis, and one method of attack is by the separation of the effects of the genes controlling these stages. Genetics and biochemistry must go hand in hand, since any information bearing on synthesis *in vivo* means a step forward in the interpretation of genetical data, and conversely.

The first to realize this was Wheldale (1914) who attempted to analyse the pigments in various genotypes of *Antirrhinum majus*. Unfortunately little was known at that time of the chemistry of the anthocyanins, and perhaps the main outcome of Wheldale's work was to draw attention to the potentialities in this field. Later Scott-Moncrieff, at the suggestion of Haldane, began a series of combined chemical and genetical studies of flower colour variation, and it is from these studies that the greater part of our knowledge of the subject has been derived (see Scott-Moncrieff, 1936; Lawrence & Scott-Moncrieff, 1935; Beale *et al.* 1939; Lawrence *et al.* 1939; Haldane, 1935, 1937).

For convenience of presentation, we shall consider the action of genes involved in pigment production under two heads—"qualitative" and "quantitative". The former refers to the production or non-production of a pigment, and the latter to difference in the amount of pigment. The distinction is obviously arbitrary, and breaks down in extreme cases. Modification of chemical structure is, of course, regarded as a qualitative change.

(1) *The role of genes in pigment production: qualitative effects*

In many garden plants there are coloured and white-flowered varieties. The white forms usually originate as mutants from the coloured, and differ from them in respect of a single gene. Thus the hybrid from crossing the coloured and white-flowered forms of *Phaseolus multiflorus* resembles the coloured parent (Tschermak, 1904). On self-fertilization of this hybrid, coloured and white-flowered forms segregate in the ratio 3 : 1. The presence or absence of anthocyanin is here determined by a single gene, which in the recessive condition is apparently inactive. In like manner, the production of anthoxanthins may be controlled by a single gene. Magenta flowers of *Primula acaulis* differ from the red in that the former contain large amounts of an anthoxanthin copigment not present in the latter. This difference has been shown to be due to a single gene (Buxton, 1932; Scott-Moncrieff, 1932, 1936).

Flower colours

The formation of the carotinoid or so-called "plastid" pigments is controlled in the same way. In *Cheiranthus Cheiri* (Scott-Moncrieff, 1936) the presence or absence of a deep yellow non-sap-soluble pigment is governed by a gene **Y**. In flowers containing anthocyanin the yellow pigment has a strong background effect, and the flower colour is brown or orange as compared with the purple or pink of the plants recessive for **Y**.

These examples deal with dominant genes which give rise to the formation of some pigment. In certain cases, however, it is the recessive form which is pigmented, while the dominant allelomorph is apparently negative. For example, in *Primula sinensis* (de Winton & Haldane, 1933) the white-flowered type is dominant to the coloured, and in *Papaver Rhoeas*, the production of an anthoxanthin copigment is a recessive character (Scott-Moncrieff, 1936).

In an account of the genetics of *Zea Mays*, Eyster (1934) lists thirteen groups of complementary genes which affect chlorophyll production. One of these groups includes eleven recessive genes giving rise to albino seedlings containing at most only traces of chlorophyll. That is, eleven dominant genes are necessary for the normal production of chlorophyll and the carotinoid pigments. There is also a second group of seven dominant genes which are necessary for the production of chlorophyll alone. In these two groups alone, then, eighteen dominant genes had been recognized in 1934 (and others have since been found) which are necessary for the production of chlorophyll. These genes probably act in series and control successive stages in the synthesis of chlorophyll (see Haldane, 1937). The identification of dominant genes is of course only possible when mutant recessives arise. In the case of the flower pigments no instance is known in which so many genes have been identified, but there are several plants where more than one gene is necessary for the production of anthocyanins or anthoxanthins.

In *Cheiranthus Cheiri* (Scott-Moncrieff, 1936) and *Lathyrus odoratus* (Bateson *et al.* 1905) two complementary genes are necessary for anthocyanin formation. Similarly, in *L. odoratus*, an anthoxanthin copigment is produced only when the genes **K** and **M** are present (Beale *et al.* 1939). The position in *Pharbitis nil* (Hagiwara, 1932) is more interesting. Four complementary genes **Ca**, **C**, **A** and **R** are necessary for anthocyanin formation. The flowers of all plants recessive for **Ca** are dead white and contain no anthoxanthin. The dominant gene **Ca** produces some anthoxanthin; **C** in the presence of **Ca** produces much anthoxanthin and **R** and **A** together with **Ca** and **C** produce anthocyanin. In addition, dominant whites are known (Hagiwara, 1929).

Matthiola incana (Saunders, 1928) and *Linum usitatissimum* (Tammes, 1922; Searle, 1926) carry respectively two and three complementary genes for anthocyanin formation. In both cases the effect of these genes alone is sufficient to produce only an extremely light pink, hardly distinguishable to the eye from white. An additional gene (or genes) is necessary for the development of coloured phenotypes of average intensity.

(2) *The role of genes in pigment production: quantitative effects*

A gene is said to be dominant when it produces the same phenotypic effect in both the homozygous and the heterozygous conditions. Such a gene has been mentioned in the case of *Phaseolus multiflorus*, where the heterozygote from coloured × white resembles the homozygous coloured parent. There are numerous other examples. However, in *Mirabilis jalapa* the F_1 from crossing crimson × yellow is intermediate, i.e. it has orange-red flowers and segregates 1 crimson : 2 orange-red : 1 yellow in F_2 (Marryat, 1909). This quantitative difference between the homozygous and heterozygous forms indicates that one gene of the kind segregating does not have the same effect as two. Such characters are said to be incompletely dominant. A theory to account for them, advanced by Goldschmidt (1927, 1938), is that differences in rates of reaction can determine whether dominance is complete or incomplete. If this is so, the phenotypic expression of the gene concerned should be dependent upon the temperature, since temperature is one of the factors controlling reaction velocity. The amount of an end-product, as well as its rate of formation, should be greater at a higher temperature provided the temperature coefficient of competing processes is not higher than that of the reactions leading to the end-product under consideration. No quantitative work has yet been done on flower pigments, but cases are known where modification follows temperature variation.

Klebs (1906) observed a variation in the flower colour of *Campanula Trachelium* from white, in heated greenhouses, through pale blue to deep blue in the cold. In *Primula sinensis* (de Winton, private communication) the effect of a dominant gene inhibiting anthocyanin formation in the petals is incomplete in phenotypes containing anthocyanin in the stigma. Such plants show an increase in the amount of anthocyanin in the petals when kept at a lower temperature. The possibility of a rise in temperature causing an increase in the amount of anthocyanin is not realized in these two examples. Klebs considered this to be due to the fact that growth at high temperatures is so rapid that anthocyanin intermediates are not available in sufficient quantity. However, Kuilman (1930, quoted by Karstens, 1938) found that at 5° C. seedlings of *Fagopyrum esculentum* develop an anthocyanin, cyanidin 3-glycoside (Karstens, 1938), slowly and after some time the amount of pigment remains constant. *At higher temperatures* (25°, 30°) *the anthocyanin appears more quickly*, though the quantity is less than at lower temperatures. By analogy with all genetically analysed cases, this process must be gene controlled. It shows, then, that the expression of the genes governing anthocyanin formation is dependent upon reaction velocity. Probably the majority of quantitative differences in gene-controlled processes—incomplete dominance, intensifying and diluting effects—can be explained in this way.

In diploid plants incomplete dominance is found less frequently than in polyploids. In the latter more than two of each gene may be present, and additive effects are common. Johnson & Miller (1938) found that the amounts of total carotinoids and of β-carotene in the endosperm of *Zea Mays* is approximately in direct proportion to the number of "dominant" **Y** genes. The vitamin A activity is also in close agreement (Mangelsdorf & Fraps, 1931). The results are shown in Table II.

Table II

Endosperm genotype	Total carotinoids		β-carotene		Relative vitamin activity
	%	Ratio	%	Ratio	
YYY	0·000465	3·3	0·000131	3·1	7·50
YYy	0·000282	2·0	0·000079	1·9	5·00
Yyy	0·000139	1·0	0·000042	1·0	2·25
yyy	0·000042	0·3	0·000011	0·3	0·05

In flowers of *Dahlia variabilis* the effect of the gene **A** governing anthocyanin formation is additive, the flowers of an **AAAA** genotype containing more anthocyanin than those of **AAAa**, **AAAa** more than **AAaa** and so on (Lawrence & Scott-Moncrieff, 1935).

Many genes have been identified whose end-effects are purely quantitative; they alter the amount of pigment produced. For instance in *Papaver Rhoeas*, the genes **C** and **B** increase the amount of anthocyanin (Scott-Moncrieff, 1936). In *Primula sinensis* the gene **I** has the opposite effect; it reduces the amount of anthocyanin (Scott-Moncrieff, 1936). In addition there are intensifying and diluting genes whose action is not effective over the whole petal, but is restricted to certain areas, giving rise to patterns.

Another factor causing quantitative differences in pigment production is interaction. In *Primula sinensis* (Scott-Moncrieff, 1936) the copigmented magenta flowers have less anthocyanin than the corresponding uncopigmented red flowers, which contain no anthoxanthin. The product of the gene **B** controlling anthoxanthin formation thus interacts with the product of the genes governing anthocyanin formation and the change from the dominant (**B**) to the recessive (**b**) results indirectly in an increase in the amount of anthocyanin.

(3) *Gene control of pigment modification*

The previous two sections deal with chemical processes about which we know little or nothing, except that they terminate in the formation of an anthocyanin or anthoxanthin. We shall now consider examples of gene action in which we do know what the gene is doing, though not yet how it does it.

In the first place, the number of hydroxyl groups in the 2-phenyl ring of the anthocyanidin molecule is determined genetically. Wit (1937) found that in *Callistephus hortensis* the change pelargonidin → cyanidin → delphinidin is controlled by three multiple allelomorphs. Thus if a homozygous plant containing cyanidin 3-glycoside is crossed with one containing pelargonidin 3-glycoside, the progeny all contain cyanidin 3-glycoside and on self-fertilisation give three plants with cyanidin derivatives to one with the pelargonidin derivative. Similarly, a cross between plants with delphinidin and cyanidin derivatives respectively yields progeny containing the delphinidin anthocyanin, and both parental types segregate in F_2.

There are two principal colour classes in *Verbena hybrida* (Beale & Scott-Moncrieff, unpublished). The flowers of these are pigmented with derivatives of pelargonidin and delphinidin respectively, the production of which is determined by three allelomorphs. In *Streptocarpus* (Lawrence *et al.* 1939) two genes **O** and **R** determine the state of oxidation of the anthocyanin. Plants recessive for **O** and **R** produce pelargonidin derivatives, **Ro** produce cyanidin derivatives and **RO** or **rO** produce delphinidin derivatives. The situation in *Lathyrus odoratus* (Beale *et al.* 1939) is the same as in *Streptocarpus*. A further example is *Primula sinensis*. Here a gene **K** is responsible for the production of a delphinidin derivative, malvidin 3-galactoside, while its recessive gives rise to pelargonidin 3-glycoside.[1]

As we shall show, there is good reason to suppose that the production of delphinidin involves one more stage than that of cyanidin, and that this stage is one of oxidation. Similarly the production of pelargonidin requires one more stage than that of cyanidin, in this case a reduction. Hence the genes referred to are bringing about oxidation or reduction processes. As in the more general problem of pigment production, we find that these processes are not always carried to completion. In *Streptocarpus* the delphinidin and the pelargonidin derivatives sometimes contain traces of cyanidin derivatives (Lawrence *et al.* unpublished) and in *Verbena* both the pelargonidin and delphinidin derivatives may contain small amounts of anthocyanin derived from cyanidin.

A second type of chemical change which is genetically controlled is that of glycoside formation. In *Verbena* the anthocyanins are either 3-monosides or 3:5-dimonosides and the difference between these two classes is genetically determined. Again, in *Streptocarpus* there are two main glycosidal types, the 3:5-dimonoside and a second which consists of a mixture of 3:5-dimonoside with 3-pentose-glycoside. A dominant gene **D** gives rise to 3:5-dimonoside and its recessive **d** to the mixtures of 3-pentose-glycoside and 3:5-dimonoside. This situation is more complex than that in *Verbena*, as recent results indicate that several genes, all of which are hypostatic to **D**, are concerned in the production of the mixtures.

In *Verbena* the action of the gene determining the glycosidal type must be to bring about union with a hexose molecule at position 5 or conversely to remove a hexose residue at this point. Here also examples are found where the reaction is not completed and mixtures of 3-monoside and 3:5-dimonoside result.

A third chemical factor causing colour variation in the anthocyanins is methylation of hydroxyl groups. Evidence from *Streptocarpus* (Lawrence *et al.* unpublished) shows that this is genetically controlled, but no simple segregation of methylated and unmethylated types has yet been observed, so more than one gene pair is responsible. Nevertheless, it is worth mentioning that incompleteness of methylation

[1] Probably galactoside, but the sugar has not yet been identified.

is very common; for example, mixtures of malvin with small amounts of petunin have been found in *Lathyrus odoratus* (Beale *et al.* 1939), while in *Streptocarpus* comparable plants from the same family containing malvidin and peonidin derivatives respectively show a wide difference in the extent to which methylation is incomplete. In plants carrying the methylating gene or genes, unmethylated anthocyanins of the delphinidin series occur only in small amounts (*ca.* 1–5 %), but in the cyanidin series it is usual to find as much as 50 % of unmethylated anthocyanin.

The extent of methylation is correlated not only with the degree of oxidation, but also with the glycosidal type. The 3-pentose-glycosides of the delphinidin series usually contain small amounts of anthocyanins derived from petunidin or delphinidin which are not present in the 3:5-dimonosides. The same applies in the cyanidin series where methylation is more complete in the 3:5-dimonosides than in the 3-pentose-glycosides. Perhaps the most interesting interaction of this type was found by Beale *et al.* (1939) in *Lathyrus odoratus*. In the delphinidin series, the anthocyanins are fully methylated when the copigment (quercetin) is present, but when this is absent methylation is incomplete and mixtures of malvin, petunin and delphin are found.

Robinson & Robinson (1931) devised a set of qualitative tests by which rapid and accurate identification of anthocyanins is possible with small amounts of material. These tests were soon applied to the study of the inheritance of flower colour variations. Unfortunately no such tests are available for the identification of anthoxanthins, for which tedious large-scale processes are still necessary. For this reason there are few precise data available concerning the inheritance of the flavones and flavonols.

Wheldale & Bassett (1913) found that a gene **Y** produces luteolin in the lips and apigenin in the tube of *Antirrhinum* flowers and a second gene **I** modifies the luteolin to apigenin. However, recent preliminary work (Price, unpublished) has cast doubt on the validity of the identification of the yellow pigment as luteolin. It is possible that the substance is a chalkone.

All cyanic flowers of the garden *Streptocarpus* contain anthoxanthins, but some are copigmented and others not. The difference between the copigmented and uncopigmented forms is determined by a single gene which apparently modifies the structure of the noncopigmenting anthoxanthin in such a manner that it becomes capable of forming an additive complex with the anthocyanin. A similar situation is found in certain genotypes of *Primula sinensis*, which contain large amounts of an anthoxanthin that does not act as a copigment.

Two further identifications of anthoxanthins in genetic material are noteworthy. First, Beale *et al.* (1939) identified quercetin, accompanied by a small amount of kaempferol, in the flowers of *Lathyrus odoratus*. The nature of the flavonol is unaffected by modification of the anthocyanin. Secondly, Sando & Bartlett (1922) isolated quercetin, as the glucoside iso-quercitrin, and later the closely related cyanidin 3-monoglucoside (Sando *et al.* 1935) from husks of *Zea Mays*.

Gene control of pH differences in the cell sap of flowers has already been referred to. The cases known are *Primula sinensis*, *Primula acaulis*, *Papaver Rhoeas*,

Tropaeolum majus (Scott-Moncrieff, 1936), *Lathyrus odoratus* (Beale *et al.* 1939) and *Trifolium pratense* (Price & Williams, unpublished).

Two examples may be cited to illustrate the independent segregation of the genes whose action we have been considering, and to show how a wide range of flower colours results from the various combinations of only a few genes. Three of the major gene pairs governing flower colour in *Primula sinensis* are **K–k**, **B–b** and **R–r**. **K** gives rise to malvidin 3-galactoside and the recessive to the corresponding pelargonidin anthocyanin. **B** produces an anthoxanthin copigment, and **R** increases the acidity of the cell sap. On self-fertilization of a plant heterozygous for all three genes the following classes and ratios are obtained:

27 **KBR** magenta; copigmented malvidin anthocyanin; more acid cell sap.
9 **KbR** red; uncopigmented malvidin anthocyanin; more acid cell sap.
9 **KBr** blue: copigmented malvidin anthocyanin: less acid cell sap.
3 **Kbr** slaty; uncopigmented malvidin anthocyanin; less acid cell sap.
9 **kBR** almost white; pelargonidin anthocyanin + anthoxanthin; more acid cell sap.
3 **kbR** coral; pelargonidin anthocyanin, no anthoxanthin; more acid cell sap.
3 **kBr** almost white; pelargonidin anthocyanin + anthoxanthin; less acid cell sap.
1 **kbr** pale coral; pelargonidin anthocyanin, no anthoxanthin; less acid cell sap.

Similarly with the five genes, **A** (necessary for anthocyanin production), **R, O, D** and **I** (producing anthoxanthin copigment) in *Streptocarpus*, we obtain eleven different colours, blue, purple, blue-mauve, mauve, bluish magenta, magenta, bluish rose, rose, pink, salmon and white, which segregate in normal Mendelian ratios from the appropriate crosses.

III. THE BIOGENESIS OF THE ANTHOCYANINS AND ANTHOXANTHINS

The biogenesis or mechanism of synthesis of the anthocyanins has inspired numerous publications from 1682 to the present day. A good account of earlier theories is given by Onslow (1925). The possibility that there is a synthetical interrelation with the anthoxanthins was recognized by the same author (Wheldale, 1909) before the structural formulae of the anthocyanins had been worked out. She supposed (1911) that the anthocyanins were oxidation products of flavones or flavonols. When it was shown that the change from flavonol to anthocyanidin must involve reduction, not oxidation, the earlier theory was discarded in favour of a reduction process (for example, see Shibata, 1915).

Other theories, involving tannins, have also been put forward, but the most important contribution is that due to Robinson (1934) who suggests that the anthocyanins, flavones, flavonols and the related catechins and flavanones may all be derived, in different ways, from the same intermediate. In other words, the production of the anthocyanins and anthoxanthins is parallel, not sequential. The hypothetical intermediate (XII) can be built up from two hexose units and one triose unit by a series of aldol condensations, and dehydrations. Oxidation at carbon atom (1), dehydration between (2) and (3) and ring closure would give cyanidin.

Oxidation at carbon atom (3), followed by dehydration and ring closure, would give luteolin, and oxidation at (2) and (3), or at (1) and (3), and then ring closure would give the flavonol quercetin. Support for the parallel syntheses of anthocyanidins and anthoxanthins is furnished by genetical observations of the com-

petition in development between the two classes of substances, particularly in *Dahlia variabilis* (Lawrence & Scott-Moncrieff, 1935; Robinson, 1936). Robinson's theory was put forward on the basis that the structural units are C_6–C_3–C_6 and that these residues have a state of oxidation comparable with that of a carbohydrate. If the aldol condensations take place in the direction $C_6 \to C_3 \to C_6$ and the C_6 groups become phenolic nuclei, the left-hand ring will have two and the right-hand ring three hydroxyl groups. This is found to be so in the majority of anthocyanins, flavonols and catechins, which contain catechol and phloroglucinol nuclei respectively. If this reasoning is justified, it follows that the 2-phenyl ring of the anthocyanins and flavonols represents the hexose unit that first became attached to the three-carbon fragment and it might be expected that this part alone would occur in nature. This is indeed the case, and the orientation of the hydroxyl groups in the benzene ring in such substances is commonly of the catechol type. This theory implies that cyanidin and quercetin are synthetically the simplest members of their classes and that pelargonidin and kaempferol or delphinidin and myricetin require an extra stage of reduction or oxidation respectively. A survey of the frequency of distribution (Lawrence *et al.* 1939) provides strong confirmation of this view.

Robinson & Robinson, by qualitative tests (1931, 1932, 1934) identified the anthocyanins in large numbers of flowers and fruits. These surveys were then extended to young leaves, in which the appearance of anthocyanin is only transient (Price & Sturgess, 1938) and to autumn leaves (Lawrence *et al.* 1938). The colour of flowers and fruits is known to be of importance to the organism in attracting insect pollinators and bird distributors respectively. Since the colour is altered by a change in the state of oxidation of the anthocyanidin, a mutant form containing an anthocyanin derived from a different anthocyanidin will be selected if the new colour is more advantageous to the plant than the old. In flowers and fruits therefore we may expect to find a greater variation in the nature of the anthocyanidin than in those parts of the plant where the anthocyanin has little or no value, and is not subjected to selection for colour. This was found to be so, as shown in Table III.

The selection of one anthocyanin type in preference to another is shown by the fact that over 90% of species whose flowers contain pelargonidin derivatives originate from tropical or other hot countries. In temperate climates, bees are the

Table III

	Number of genera containing cyanidin derivatives expressed as a percentage of the total number of genera examined
	%
Autumn leaves	95
Young leaves	93
Permanently pigmented leaves	76
Fruits	69
Flowers	50

commonest pollinators. Von Frisch (1937) found that bees are sensitive to blue but are red-blind and so a mutation which results in reduction to a pelargonidin derivative would be harmful to bee-pollinated species and would tend to be eliminated. On the other hand, a mutant form containing a delphinidin derivative is likely to be advantageous in cool climates and selection would operate in its favour. Conversely in hot climates the pollinating agents may be red-sensitive, e.g. butterflies. Then a mutation to the redder pelargonidin anthocyanin has a greater chance of survival. It is of interest that in a number of tropical plants whose flowers contain cyanidin derivatives, the anthocyanin is accompanied by a deep yellow carotinoid pigment, the mixture having much the same colour as pelargonin. Thus, when a mutation arises producing this advantageous effect it is selected regardless of the way in which the effect is brought about.

The flowers of the majority of "good" species are pigmented by one anthocyanin only, but there are a few cases in which mixtures are encountered. This is of interest in connexion with the theory of dominance put forward by Muller (1932). Muller supposes that dominance is developed by the selection of mutations which provide a margin of stability to the expression of the gene, insuring the organism against excessive variability of the character either by environmental or genetic influences. On this view, if a certain flower colour has a positive selective value, the production of the pigments concerned should take place under optimum conditions—conditions which not only allow the various stages in the synthesis to be completed, but also provide a "margin of safety" enabling the reactions to proceed normally in different environments. In all probability the decisive factor is the rate at which the reactions take place. Mutations increasing the velocity of the more important stages, such as that which differentiates delphinidin from cyanidin, should therefore be selected and in the majority of species or stable communities we should not expect to find mixed anthocyanins.

The simplest class of anthocyanin mixture is due to incomplete methylation; for example, *Geranium psilostemon* contains malvin together with a small amount of petunin. More important are those in which the anthocyanins are derived from anthocyanidins with a different degree of oxidation. In these species we find mixtures of pelargonidin and cyanidin derivatives, and of delphinidin and cyanidin derivatives, but not mixtures based on pelargonidin and delphinidin. The cyanidin is usually, but not always, present in small quantities only. This is taken to mean that the production of both pelargonidin and delphinidin involves at least one more stage than that of cyanidin, and that this stage is not always carried to completion. It should be emphasized that there is no suggestion that cyanidin glycosides themselves are oxidized or reduced—these processes would take place at a stage prior to the actual formation of the anthocyanins. Phylogenetic classification of the flowers which had been examined gave further evidence in support of the hypothesis.

Recently, Bancroft & Rutzler (1938) have revived the hypothesis that anthocyanins are formed from anthoxanthins. They consider that this is so in some cases, but that in others the leuco-anthocyanins, as suggested by Robinson (1936) may be the precursors. The leuco-anthocyanins are colourless substances which on treatment with acids are converted into anthocyanidins. The conversion of leuco-anthocyanins to anthocyanins may take place in the plant under certain circumstances, such as in the development of autumnal coloration, but it is doubtful whether it is often so. Only two examples are on record where the anthocyanidin from both the anthocyanin and the leuco-anthocyanin have been identified in the same part of a plant, namely the flowers of *Hydrangea hortensis* and the fruits of *Vitis heterophylla* (Lawrence *et al.* 1939). The former contained delphinidin pentoseglycoside and the latter malvidin 3:5-dimonoside, but the leuco-anthocyanin from both yielded cyanidin.

Bancroft & Rutzler assert that the anthocyanins in red autumn leaves of sumach, dogwood and barberry are due to reduction of flavones. This conclusion appears to be based solely upon the fact that the green leaves contain flavones but no leuco-anthocyanins; evidence which is far from convincing.

The isolation of cyanidin 3-glucoside and the corresponding quercetin 3-glucoside from *Zea Mays* led Sando to favour the possibility that anthocyanins are reduction products of flavonols, regardless of the fact that cyanidin and quercetin derivatives are by far the commonest representatives of their classes (Gisvold & Rogers, 1938). On this basis alone it would be surprising if they were not frequently found together. Sando stresses the fact that previously only free flavonols had been studied in connexion with anthocyanins, whereas he and his collaborators found agreement not only between the aglycones but also in the nature of the sugar residue. This agreement obviously does not help to distinguish between the various theories which have been put forward.

Keeble & Armstrong (1912a, 1912b) considered that anthocyanins are formed by the action of an oxidase or peroxidase on a chromogen. That an oxidation process is involved in anthocyanin formation is quite clear from the results of Kuilman (1930), Karstens (1938) and others. Karstens showed that in *Fagopyrum esculentum* formation of anthocyanin depends upon the availability of carbohydrates, both in respect of their quantity and distribution. He further showed that a photochemical reaction involving oxidation and a second oxidation process which can take place in the dark are the essential steps in anthocyanin synthesis. The fact that there appear to be two oxidation reactions need not be an objection to Robinson's hypothesis; as this author himself points out: "only the general direction of the process is suggested; the biochemical detail may be much more complex". The C_3 fragment may be the product of a degradation requiring oxidation.

IV. DOMINANCE RELATIONSHIPS AND COMPETITION

The following is an extract from a paper by Moore (1910): "...if we take the view that life processes are chemical in their nature, we must of necessity accept the consequences which follow the application of chemical laws, and concede that such laws hold just as truly for chemical reactions within the living organism as for reactions *in vitro*...it seems reasonable that a clear explanation of the variations from Mendel's law of dominance lies...in the domain of physical chemistry." It is evident that the action of every gene must be capable of interpretation in chemical terms. Large numbers of chemical reactions, some more or less independent, others parallel or sequential, are proceeding simultaneously in every cell of a living organism. If we consider the possibilities for any single reaction, then a gene can be controlling this reaction in one of the following ways:

(1) In controlling supplies of the necessary reactants (such a gene might control a previous stage in the synthesis, or the permeability of the cells).

(2) In removing the reaction product—this would, for example, be necessary in a balanced reaction with an equilibrium point favouring the starting material.

(3) In supplying an enzyme or other catalyst.

(4) In controlling conditions, such as pH, which if unfavourable may lower the reaction velocity or even completely inhibit the reaction. Some of these points have been more fully discussed by Haldane (1932).

In the previous sections, the effects of genes involved in pigment production were separated for convenience into qualitative and quantitative. This distinction however is clearly an arbitrary one, since some qualitative differences may be straightforward presence and absence, e.g. of an enzyme, while others may be due to interaction. Many qualitative differences also arise from interaction, but in other cases both allelomorphs may act in the same way to a different degree, particularly by influencing reaction rates to different extents.

Dominance or recessiveness of a character is determined by the resemblance of the heterozygote to one or other of the homozygous forms. As Goldschmidt (1938) points out, dominance strictly is not an attribute of the gene, but is a phenotypic result of the action of the gene in relation to its genetic and external environments. For example, the action of the "dominant white" gene in *Primula sinensis* varies according to whether it is associated with the genes for "green stigma" or "red stigma" and also, as we have already seen, it varies with the temperature. It is by the consideration of reaction rates and competition that the various aspects of dominance relationships are best co-ordinated. Strong support for this view is found in the modification and sometimes reversal of dominance by changes in temperature, which we know may profoundly influence reaction velocities (see Goldschmit, 1938). Dominance is thus determined by the reaction rate in the heterozygote relative to the rates in the two homozygotes. We shall now see how this idea is in keeping with the data on pigment inheritance.

Examples of complete and incomplete dominance in the production of plastid pigments, anthoxanthins and anthocyanins have already been mentioned. Complete dominance is the usual state of affairs, and in such cases heterozygotes certainly bear a close resemblance to the dominant forms, though it has not yet been established by quantitative comparisons that dominance is ever 100 % complete. Nevertheless, the chain of reactions must proceed at a rate which does not vary greatly with a change in dosage of the active allelomorph.

As would be expected, it is impossible to lay down definite rules concerning the dominance or recessiveness of reactions leading to structural modification of the pigment molecule, but a higher state of oxidation is usually dominant to a lower. Thus the delphinidin pigment types in *Streptocarpus* are dominant to cyanidin, though the oxidation process is sometimes incomplete (see p. 44). This is best understood in relation to the rate at which oxidation takes place and to the competition between the oxidation process and the normal reactions leading to the formation of cyanidin. The pelargonidin types in *Streptocarpus* are much paler than, and recessive to, the cyanidin types. The rate of formation of the anthocyanin is evidently slowed down by the introduction of the reduction process so that the anthocyanin is produced in smaller amount than the corresponding cyanidin derivative. Moreover, cyanidin derivatives are also present, so the reduction is not carried to completion. It is not surprising then that reduction in the heterozygote cannot compete with the normal reactions leading to cyanidin, with the result that pelargonidin forms are recessive.

In species which have only the pelargonidin and delphinidin series of anthocyanins, e.g. *Verbena* and *Primula sinensis*, the oxidation and reduction processes are in competition with one another. Where two processes, A and B, are in competition for an intermediate a change of conditions affecting reaction velocity may lead to partial or complete suppression of the end-product of A, with consequent increased formation of the end-product of B. In extreme cases the process B may take place so slowly that in normal organisms its effect is unrecognizable. But when one of the genes controlling B mutates to a more efficient condition, an entirely new substance or character—the end-product of B—appears. The same result could be achieved by a mutation lowering the rate of A. In *Primula sinensis* delphinidin derivatives are as a rule completely dominant to pelargonidin. Oxidation evidently proceeds at a much greater rate than reduction, so that the heterozygote contains no (detectable) pelargonidin. As in *Streptocarpus*, the flowers containing pelargonidin are very pale. However, there is a mutant gene known as "Dazzler" (**Dz**) which increases the production of pelargonidin. As a result, the flowers of **k Dz** forms are deep salmon and contain a much larger amount of pelargonidin 3-monoside than those of **k dz**. The **K Dz** forms contain malvidin 3-monoside mixed with a smaller amount of pelargonidin 3-monoside. This is evidently due to an acceleration of the reduction process to a point where it competes on only slightly less than equal terms with the oxidation. That it does not completely "block" the oxidation process is not surprising in view of the fact that **k** plants heterozygous for **Dz** are intermediate in intensity between the two homozygotes and **K** plants heterozygous for **Dz** are usually indistinguishable from **K dz**. Moreover, **Dz** interacts with **B**, the gene governing anthoxanthin production, and the suppressing effect of **B** on **k Dz** forms is greater than in **K dz**, supporting the idea that the reaction rate determined by **Dz** is less than that determined by **K**.

The somewhat similar situation with regard to the anthocyanins in *Papaver Rhoeas* can be explained along the same lines.

At an early stage of this work, when less information was available, Scott-Moncrieff (1936) found it desirable to ascribe the behaviour of the "Dazzler" gene, and its analogues in *Papaver*, to a "specific" pigmentation process, though it is evident that even at that time certain contradictions between dependence and independence were inherent in this view. The idea of competitive processes outlined here does away with the necessity for postulating "specific" pigmentation, i.e. wholly independent processes to account for supposedly exceptional cases.

In *Verbena* there is little or no difference in intensity between the pelargonidin and delphinidin series, and competition between oxidation and reduction processes is presumably more equally balanced. This results in flowers of some plants containing mixtures of pelargonidin and delphinidin derivatives.

The functionally tetraploid *Dahlia variabilis* furnishes an excellent example of competition. The inheritance of flower colour in this plant was analysed semi-quantitatively by Lawrence & Scott-Moncrieff (1935). There are four principal genes governing flower colour: **A** and **B**, both of which produce anthocyanin, **I** which gives rise to the flavone apigenin, and **Y** which produces a deep yellow substance. This yellow pigment has not yet been identified. Schmid and his collaborators (1928, 1932, 1933) consider it is an isomer of apigenin, but work in progress (Price, unpublished) does not support this, though showing it to be closely related. The products of the genes **A**, **B**, **I** and **Y** interact with one another, and these genes have different competitive values. Moreover, they appear to control to different extents the quantity of "raw material" available for pigment production. This suggests that the intermediate (S), which the pigment genes are drawing upon, is preceded by a balanced reaction which is competing with other processes. The amount of "S" which is formed and converted to anthocyanin or anthoxanthin would then be dependent on the rate at which "S" was utilized. With the introduction of more pigment genes, the rate of utilization may be increased and consequently more may become available. This is evidently the case in *Primula sinensis* also, where processes controlled by **K** and **B** use up the intermediate more rapidly than when **K** is acting alone. As a result, the suppression of anthocyanin in **KB** plants, as compared with **Kb** where anthoxanthin is absent, is not in proportion to the increase in anthoxanthin content. In *Dahlia*, however, there is an upper limit to the amount of intermediate available, and accumulation of the pigment genes **A**, **B**, **I** and **Y** cannot increase the total pigment production beyond a certain point. The genes **Y** and **B** appear to be completely dominant in the simplex condition, which of course means that one **B** or **Y** gene can utilize all the available intermediate. That it is the limitation of the quantity of intermediate which causes simplex **Y** and **B** to be dominant is shown by their interaction, which is dependent upon dosage.

A is cumulative up to quadriplex, but **I** appears to reach its maximum of production when duplex. The main features of the competition between the processes controlled by these genes are as follows. **Y** strongly suppresses the effects of **I** and **A**; for example, Y_1I_3 plants[1] contain a little apigenin, but Y_2I_2 contain none. Similarly A_1Y_1 plants contain little or no anthocyanin while A_4Y_1 have a small amount. The competitive value of **B** is greater than that of **I** or **A**, consequently in B_1Y_1 plants there is an appreciable amount of anthocyanin, though so far as one can tell, less of it than of the yellow substance. In B_1Y_2 plants the anthocyanin intensity is reduced. **I** interacts strongly with **A**, but almost complete suppression of anthocyanin formation occurs only in such extreme cases as A_1I_3 and A_1I_4. I_1 is unable to suppress anthocyanin production by B_1, but the intensity in B_1I_4 plants is considerably decreased as compared with B_1. These quantitative variations are, then, the expression of a fairly straightforward competition. But there is one further point of importance. The genes **A** and **B** both produce anthocyanin, they are not complementary, and may have arisen from the same ancestral gene and control the same process. They are not specific for any particular anthocyanin, and no gene modifying the nature of the anthocyanin has yet been identified. Yet two anthocyanins, cyanin and pelargonin, may occur separately or together in *Dahlia* flowers. When the sum of the potential contributions or "activity values" of the pigment genes is below a certain critical value, the anthocyanin is pure cyanin, e.g. in all **A** and **AI** genotypes. But when this sum exceeds the critical value, as in all **AY** or **BY** genotypes, the anthocyanin is pelargonin or a mixture of pelargonin and cyanin. B_1 plants contain cyanin, but B_2 contain the pelargonin-cyanin mixture. Thus the nature of the anthocyanin seems to be determined by the combined competitive effects of all the pigment genes. Perhaps *Dahlia* is homozygous for a gene or genes governing reduction, which is unable to proceed satisfactorily under all conditions. It is hoped that some explanation of this curious state of affairs will be forthcoming when the structure of the yellow pigment is known.

It will be seen from the preceding account that inherited chemical differences originate in two ways: first from gene action, i.e. the action of a gene in relation to that of its allelomorph, and secondly from gene interaction, i.e. the action of a gene in relation to that of non-homologous genes. These relationships are interdependent, and together they are the determinants of dominance, and sometimes of epistasy.

Bateson (1909) called genes which prevent others from manifesting their effects "epistatic" and the concealed genes "hypostatic". For instance, in *Cheiranthus*

---

[1] The abbreviated symbolism of Lawrence & Scott-Moncrieff for tetrasomic inheritance in *Dahlia* is used in this article. Genetic constitutions are denoted by dominant factors only.

Cheiri the presence of a yellow anthoxanthin is masked when accompanied by a deeper yellow carotinoid pigment. In many such cases, inability to recognize the expression of the hypostatic gene is due to the use of methods which lack precision. For example, in *Dahlia* the deep yellow pigment produced by **Y** is said to be epistatic to the apigenin produced by **I**. But the presence of apigenin can be recognized when chemical tests are used; the same applies to the anthoxanthin in *Cheiranthus*.

However, epistasy is often developmental in character; it may arise from gene interaction. The gene **Y** in *Dahlia* has a greater competitive value than **A**, and it may inhibit the production of anthocyanin by the latter: then, **A** does not manifest itself in the presence of **Y** and is said to be hypostatic to **Y**. Similarly in *Streptocarpus*, competition between oxidation and reduction processes is all in favour of the former, so that **O** is epistatic to **r**. In each of these examples, epistasy is due to the simultaneity of action of the genes.

Epistasy also arises from interference in a chain of reactions, and it would seem that the term epistatic could logically be applied to that member of a pair of complementary genes which is known to have priority of action. But this is not always easy to establish. There are eleven genes in *Zea* controlling the production of both chlorophyll and the carotinoids, and a second group of seven genes which controls the synthesis of chlorophyll only and is without direct effect on that of the carotinoids. We might infer that the first group of genes acts before the second group. Such an inference can be treated more rigorously in the case of the anthocyanins and anthoxanthins. We know from the competition between them that the syntheses of the two classes of substances are interrelated; they not only have a common precursor, but are competing for this precursor at the same time. To prove this last point let us suppose that they draw upon the precursor at different times. If the two competing processes are controlled by two non-allelomorphic genes, each of which is completely dominant in the absence of competition from the other dominant (such as **B** and **Y** in *Dahlia*, or **B** and **I** when the dosage of **I** is two or more), then one of these genes (that which acts first) should have the same competitive effect in the homozygous and heterozygous condition. But the result of competition between **B** and **Y** or **B** and **I** is dependent on the dosage of both. Therefore the synthesis of the two classes of substances does include, at some stage, a simultaneous competition for a common precursor.

In *Pharbitis nil* the gene **Ca** is essential for the production of both anthocyanin and anthoxanthin, therefore its effect is manifest at or before the separation of the two chains of syntheses. On the other hand, the genes **A** and **R**, both necessary for anthocyanin formation, do not directly influence the production of anthoxanthin. Therefore they become effective later than **Ca** and may be said to be hypostatic to **Ca**.

We do not know whether **A** precedes **R** or vice versa, nor do we know whether **C** precedes **Ca**, but we can outline roughly the sequence of events as in Fig. 1.

In *Streptocarpus* the extent of methylation is dependent on the state of oxidation and on the glycosidal type. Evidently the genes controlling methylation become operative later than **O, r** and **D**. Similarly in *Lathyrus*, the correlation between incomplete methylation and absence of anthoxanthin suggests that the methylation of the anthocyanin occurs after the divergence of the anthocyanin and anthoxanthin syntheses.

V. SUMMARY

1. The principal flower colouring matters are the anthocyanins, anthoxanthins and carotinoids. Variation in colour depends upon the presence or absence of one or more of these substances, upon structural alterations in their molecules, changes in the pH of the cell sap, or quantitative changes affecting the amounts of pigment produced.

2. Pigment production is genetically controlled, and in a number of cases complementary genes are involved.

3. Variation in the amount of any pigment is also gene controlled. Such quantitative differences can be interpreted in terms of the velocity of the reactions involved in pigment production.

4. Modifications of the chemical structure of anthocyanins, including the state of oxidation, glycosidal type and probably the degree of methylation are each determined by simple gene relationships. In certain cases reactions are incomplete, giving rise to mixtures of anthocyanins.

5. The synthesis of anthocyanins in the plant is correlated with that of anthoxanthins. Some workers consider that the anthocyanins are formed from flavones or flavonols, but the most comprehensive theory, put forward by Robinson, postulates parallel formation of anthocyanin and anthoxanthin from the same intermediate. On this theory cyanidin and quercetin are synthetically the simplest members of their classes, others requiring additional stages of oxidation or reduction. The above view is supported by statistical analysis of the distribution of anthocyanins in flowers, fruits and leaves.

6. Heritable chemical differences result in the first place from gene action. They may be accentuated or minimized by gene interaction, which can modify dominance relationships and is sometimes the causal factor of epistasy. A study of gene interaction can also help in the determination of the sequence of gene action.

VI. REFERENCES

BANCROFT, W. D. & RUTZLER, J. E. (1938). "Colloid chemistry of leaf and flower pigments. I. Precursors of the anthocyanins." *J. Amer. chem. Soc.* **60**, 2738.

BATESON, W. (1909). *Mendel's Principles of Heredity*. Cambridge.

BATESON, W., SAUNDERS, E. R. & PUNNETT, R. C. (1905). "Experimental studies on the physiology of heredity." *Rep. Evolut. Comm. roy. Soc.* **2**, 154.

BEADLE, G. W. & EPHRUSSI, B. (1936). "The differentiation of eye pigments in *Drosophila* as studied by transplantation." *Genetics*, **21**, 225.

—— —— (1937). "Development of eye colours in *Drosophila*: diffusible substances and their interrelations." *Genetics*, **22**, 76.

BEALE, G. H., ROBINSON, G. M., ROBINSON, R. & SCOTT-MONCRIEFF, R. (1939). "Genetics and chemistry of flower colour variation in *Lathyrus odoratus*." *J. Genet.* **37**, 375.
BUXTON, B. H. (1932). "Genetics of the primrose, *Primula acaulis*." *J. Genet.* **25**, 195.
DE WINTON, D. & HALDANE, J. B. S. (1933). "The genetics of *Primula sinensis*. II. Segregation and interaction of factors in the diploid." *J. Genet.* **27**, 1.
EYSTER, W. H. (1934). "The genetics of *Zea Mays*." *Bibliogr. genet.* **11**, 187.
FRISCH, K. V. (1937). "The language of bees." *Sci. Progr. Twent. Cent.* **32**, 29.
GISVOLD, O. & ROGERS, C. H. (1938). *The Chemistry of Plant Constituents*. Minneapolis.
GOLDSCHMIDT, R. (1927). *Physiologische Theorie der Vererbung*. Berlin.
—— (1938). *Physiological Genetics*. London.
HAGIWARA, T. (1929). "On the role of the factors **C** and **R** in the production of flower colours in *Pharbitis nil*." *Bot. Mag., Tokyo*, **43**, 643.
—— (1932). "On the genetico-physiological studies of the colour-development of flowers in *Pharbitis nil*." *Proc. imp. Acad. Japan*, **8**, 54.
HALDANE, J. B. S. (1932). "The time of action of genes, and its bearing on some evolutionary problems." *Amer. Nat.* **66**, 5.
—— (1935). "Contribution de la génétique à la solution de quelques problèmes physiologiques." *Réun. plen. Soc. Biol.* Paris.
—— (1937). "The biochemistry of the individual." Essay in *Perspectives in Biochemistry*. Cambridge.
INMAN, O. L. & BLAKESLEE, A. F. (1938). "New or modified chlorophylls resulting from a recessive pale mutation in *Datura*." *Science*, **87**, 428.
JOHNSON, I. J. & MILLER, E. S. (1938). "Variation in carotinoid pigment concentration among inbred and crossbred strains of corn." *Cereal Chem.* **15**, 345.
KARSTENS, W. K. (1938). "Anthocyanin and anthocyanin formation in seedlings of *Fagopyrum esculentum* Moench." Thesis, Leyden.
KEEBLE, F. & ARMSTRONG, E. F. (1912a). "The distribution of oxidases in plants and their role in the formation of pigments." *Proc. roy. Soc.* B, **85**, 214.
—— —— (1912b). "The role of oxidases in the formation of anthocyanin pigments of plants." *J. Genet.* **2**, 277.
KLEBS, G. (1906). "Über Variationen der Blüten." *J. wiss. Bot.* **42**, 155.
KUILMAN, L. W. (1930). "Physiologische Untersuchungen über die Anthocyane." Thesis, Amsterdam.
LAWRENCE, W. J. C. (1932). "Interaction of flavones and anthocyanins." *Nature, Lond.*, **129**, 834.
LAWRENCE, W. J. C., PRICE, J. R., ROBINSON, G. M. & ROBINSON, R. (1938). "A survey of anthocyanins. V." *Biochem. J.* **32**, 1661.
LAWRENCE, W. J. C., PRICE, J. R., ROBINSON, G. M. & ROBINSON, R. (1939). "The distribution of anthocyanins in flowers, fruits and leaves." *Phil. Trans. roy. Soc.* B, **230**, 149.
LAWRENCE, W. J. C. & SCOTT-MONCRIEFF, R. (1935). "The genetics and chemistry of flower colour in *Dahlia*: a new theory of specific pigmentation." *J. Genet.* **30**, 155.
LAWRENCE, W. J. C., SCOTT-MONCRIEFF, R. & STURGESS, V. C. (1939). "Studies in *Streptocarpus*, I." *J. Genet.* **37**, 299.
MANGELSDORF, P. C. & FRAPS, G. S. (1931). "A direct quantitative relationship between vitamin A in corn and the number of genes for yellow pigmentation." *Science*, **73**, 241.
MARQUART, L. (1835). *Die Farben der Blüthen, eine chemischphysiologische Abhandlung*. Bonn.
MARRYAT, D. C. E. (1909). "Hybridisation experiments with *Mirabilis Jalapa*." *Rep. Evolut. Comm. roy. Soc.* **5**, 32.
MÖBIUS, M. (1885). "Über den Glanz der *Ranunculus*-Blüten." *Bot. Zbl.* **23**, 115.
MOORE, A. R. (1910). "A biochemical conception of dominance." *Univ. Calif. Publ. Physiol.* **4**, 9.
MULLER, H. J. (1932). "Further studies on the nature and causes of gene mutations." *Proc. 6th Int. Cong. Genet.* **1**, 213.
ONSLOW, M. WHELDALE (as WHELDALE) (1909). "On the nature of anthocyanin." *Proc. Camb. phil. Soc.* **15**, 137.
—— (1911). "On the formation of anthocyanin." *J. Genet.* **1**, 133.
—— —— (1914). "Our present knowledge of the chemistry of the Mendelian factors for flower colour." *J. Genet.* **4**, 109.
ONSLOW, M. WHELDALE (1925). *The Anthocyanin Pigments of Plants*. Cambridge.
ONSLOW, M. W. (as WHELDALE, M.) & BASSETT, H. LL. (1913). "The flower pigments of *Antirrhinum majus*." *Biochem. J.* **7**, 441.
PHILIP-SMITH, E. (1933). "The calibration of flower colour indicators." *Protoplasma*, **18**, 112.
PHIPPS, I. F. (1929). "Inheritance and linkage relations of virescent seedlings in maize." *Mem. Cornell agric. Exp. Sta.* no. 125.
PRICE, J. R. & ROBINSON, R. (1937). "Nitrogenous anthocyanins. IV. The colouring matter of *Bougainvillea glabra*." *J. chem. Soc.* p. 449.
PRICE, J. R., ROBINSON, G. M. & ROBINSON, R. (1938). "The occurrence of Kaempferol in *Crocus*." *J. chem. Soc.* p. 281.
PRICE, J. R. & STURGESS, V. C. (1938). "A survey of anthocyanins, VI." *Biochem. J.* **32**, 1658.

ROBERTSON, A. & ROBINSON, R. (1929). "Note on the characterization of the anthocyanins and anthocyanidins by means of their colour reactions in alkaline solutions." *Biochem. J.* **23**, 35.
ROBINSON, G. M. & ROBINSON, R. (1931). "A survey of anthocyanins. I." *Biochem. J.* **25**, 1687.
—— —— (1932). "A survey of anthocyanins. II." *Biochem. J.* **26**, 1647.
—— —— (1933). "Anthocyanins as indicators and the colours of flowers." *Nature, Lond.*, **132**, 626.
—— —— (1934). "A survey of anthocyanins. IV.". *Biochem. J.* **28**, 1712.
ROBINSON, R. (1934). "The molecular architecture of some plant products." *9th Int. Cong. Chem.* Madrid.
—— (1936). "The formation of anthocyanins in plants." *Nature, Lond.*, **137**, 172.
SANDO, C. E. & BARTLETT, H. H. (1922). "Pigments of the Mendelian colour types in maize: isoquercitrin from brown husked maize." *J. biol. Chem.* **54**, 629.
SANDO, C. E., MILNER, R. T. & SHERMAN, M. S. (1935). "Pigments of the Mendelian colour types in maize: chrysanthemin from purple-husked maize." *J. biol. Chem.* **109**, 203.
SAUNDERS, E. R. (1928). "*Matthiola*." *Bibliogr. genet.* **4**, 141.
SCHMID, L. & HASCHEK, L. (1933). "Der gelbe Dahlienfarbstoff." *S.B. Akad. Wiss. Wien*, 11b, **142**, 123.
SCHMID, L. & SEEBALD, A. (1932). "Der Farbstoff der gelben Dahlien." *S.B. Akad. Wiss. Wien*, 11b, **141**, 32.
SCHMID, L. & WASCHKAU, A. (1928). "Über die Konstitution des Anthochlors der gelben *Dahlia*." *S.B. Akad. Wiss. Wien*, 11b, **137**, 83.
SCOTT-MONCRIEFF, R. (1932). "A note on the anthocyanin pigments of the primrose, *Primula acaulis*." *J. Genet.* **25**, 206.
—— (1936). "A biochemical survey of some mendelian factors for flower colour." *J. Genet.* **32**, 117.
SEARLE, G. O. (1926). "A botanical study of the flax plant. VII. A preliminary account of the genetics of flower colour and other related characters." *Linen Indust. Res. Ass. Memoir*, no. 35, 115.
SHIBATA, K. (1915). "Untersuchungen über das Vorkommen und die physiologische Bedeutung der Flavonderivative in den Pflanzen. I." *Bot. Mag., Tokyo*, **29**, 118.
TAMMES, T. (1922). "Genetic analysis, schemes of co-operation and multiple allelomorphs of *Linum usitatissimum*." *J. Genet.* **12**, 19.
TSCHERMAK, E. v. (1904). "Weitere Kreuzungsstudien an Erbsen, Levkojen und Bohnen." *Z. landw. VersWes. Öst.* **7**, 533.
WHELDALE (see ONSLOW).
WILLIAMS, R. D. (1937). "The frequency of chlorophyll deficient mutants in red clover, *Trifolium pratense*." *Rep. Brit. Ass.* p. 433.
WILLSTÄTTER, R. (1914). "Über die Farbstoffe der Blüten und Früchte." *S.B. preuss. Akad. Wiss.* p. 402.
WILLSTÄTTER, R. & EVEREST, A. E. (1913). "Über den Farbstoff der Kornblume." *Liebigs Ann.* **401**, 189.
WILLSTÄTTER, R. & MALLISON, H. (1914). "Über die Verwandtschaft der Anthocyane und Flavone." *S.B. preuss. Akad. Wiss.* p. 769.
WILLSTÄTTER, R. & ZECHMEISTER, L. (1914). "Synthese des Pelargonidins." *S.B. preuss. Akad. Wiss.* p. 886.
WILLSTÄTTER, R. & ZOLLINGER, E. H. (1916). *Liebigs Ann.* **412**, 195.
WILLSTÄTTER, R. and collaborators (1915). *Liebigs Ann.* **408**, 1–162.
—— —— (1916). *Liebigs Ann.* **412**, 113–251.
WIT, F. (1937). "Contributions to the genetics of the china aster." *Genetica*, **19**, 1.

ADDENDA

1. Since this article was written, a similar review has appeared on the "Genetics and Chemistry of Flower Colour Variation", by R. Scott-Moncrieff (*Ergebnisse der Enzymforschung*, 1939, **8**, 277).

2. The yellow colouring matter of *Dahlia variabilis* has been identified as the chalkone butein,

a substance structurally related to the anthocyanins and flavones, and belonging to the C_6—C_3—C_6 group (Price, *J. chem. Soc.* 1939, p. 1017). Thus, in *Dahlia* there are now three types of similarly constituted pigments instead of two, and competition between three genetically controlled lines of synthesis.

3. Hibbert (*J. Amer. chem. Soc.* 1939, **61**, 725) has discussed the formation of tannins and pigments in the light of recent work on lignin. In his opinion the theory that hexoses are precursors of the phenolic components of tannins, lignins and pigments does not provide a satisfactory solution of the problems of plant synthesis, and he puts forward the view that simpler substances, such as methylglyoxal or its dismutation isomers, may occur as intermediates. In effect, Hibbert's theory does not conflict with that of Robinson outlined in the present review, but carries the question a stage further back.

4. In a reply to Bancroft & Rutzler (1938), Robinson & Robinson (*J. Amer. chem. Soc.* 1939, **61**, 1605) again point out that they do not regard the production of anthocyanins from leuco-anthocyanins as the standard mechanism, but merely as an auxiliary process possibly operative in autumnal reddening and other special cases. G. M. Robinson (*J. Amer. chem. Soc.* 1939, **61**, 1606) has discussed those factors causing variations of flower colour which are not due to changes in the nature of the anthocyanin, e.g. the concentration of the anthocyanin relative to that of co-pigments. She also points out that chlorogenic acid may have some significance in relation to flower colour.

5. R. Harder (*Naturwissenschaften*, 1938, **26**, 713) describes the effect of temperature changes on the development of anthocyanins in several plants. In *Viola* and *Calceolaria*, in addition to the cases mentioned earlier in this article, increase in temperature results in a diminution in the intensity of anthocyanin pigmentation. But in *Petunia* the reverse is the case; the flowers of a variety which are self-coloured violet when grown at 35° C. have large white areas when grown at 25° C. and at 15° C. these white areas extend over nearly the whole of the flower. Not only anthocyanins, but also flavones and leuco-anthocyanins are inhibited in these white areas. A certain strain of *Dahlia variabilis* resembled *Petunia*, having yellow flowers when grown out of doors and red when grown in a hot-house at 30° C. This is probably a strain which is not pure recessive for genes controlling anthocyanin production, but one in which anthocyanin is normally suppressed by interaction with **Y** genes. In both *Petunia* and *Dahlia* Harder found that modifications arising from temperature differences are initiated during a sensitive period prior to the opening of the flower. In *Petunia* a decrease of light intensity also brings about increased anthocyanin production.

V
The Recognition of Genetic Blocks

Editor's Comments on Papers 7 and 8

7 Butenandt, Weidel, and Becker: *Kynurenin als Augenpigmentbildung auslösendes Agens bei Insekten*
English translation: *Kynurenine as an Agent Which Causes the Formation of Eye Pigment in Insects*

8 Ephrussi: *Chemistry of the "Eye Color Hormones" of* Drosophila

The full significance of Paper 7 will best be appreciated after reading Paper 8. Paper 7 precedes the Ephrussi paper in time and is actually a prelude to it.

The significant discovery made here is that a specific compound, kynurenine, is probably a precursor in the formation of brown pigment in insect eyes. The *v* mutants of *Drosophila* and the *a* mutants of *Ephestia* cannot make kynurenine from tryptophan. Hence they lack the brown pigment.

This is probably the first example of modifying a mutant phenotype to the wild type by feeding a specific chemical compound. It was an important prelude to the work with *Neurospora* mutants described in Paper 9.

Ephrussi wrote his review article near the end of the active period of work with the insect eye color "hormones." It therefore serves the highly useful function of documenting an important phase in the search for an understanding of gene function. The article is beautifully written and is made the more valuable to those interested in the history of the development of concepts because of its rather detailed references to A. H. Sturtevant's contributions to this area.

Much of the work described here was done by Ephrussi himself in collaboration with G. W. Beadle. The results they obtained with *Drosophila* stimulated Beadle along with E. L. Tatum to search further afield for a more suitable organism, which they found in *Neurospora,* as described in Paper 9.

Selected Bibliography

Fölling, A. 1934. Über Ausscheidung von Phenylbrenztraubensäure in dem Harn als Stoffwechselanomalie in Verbindung mit Imbezillität. *Hoppe–Selyer's Z. Physiol. Chemie* **227:** 169–176.

Jervis, G. A. 1939. The genetics of phenylpyruvic oligophrenia. *J. Mental Sci.* **85:** 719–762.

———. 1953. Phenylpyruvic oligophrenia: a deficiency of phenylalanine oxidizing system. *Proc. Soc. Exp. Biol. Med.* **82:** 514–515.

Kynurenin als Augenpigmentbildung auslösendes Agens bei Insekten*

ADOLF BUTENANDT, WOLFHARD WEIDEL, und ERICH BECKER

Wir haben gefunden, daß das *Kynurenin*[1]), ein im Harn von Kaninchen nach l-Tryptophan-Gaben auftretendes Stoffwechselprodukt der wahrscheinlichen Konstitution (I),

$$\underset{NH_2}{\underset{|}{C_6H_4}}-C(COOH)=CH-CH(NH_2)-COOH$$
I.

imstande ist, an der hellaügigen *vermilion-(v bw)-Drosophila melanogaster* bzw. an der rotäugigen *a*-Rasse der Mehlmotte *Ephestia kühniella* die für den Gen-abhängigen v^+- (bzw. a^+)-Wirkstoff in bezug auf die Augenausfärbung charakteristischen physiologischen Wirkungen[2]) auszulösen. Auf Injektion einer etwa 1 proz. wäßrigen Lösung des Kynurenins in *Ephestia*-Puppen der *a*-Rasse wurden die normalerweise hellorangeroten Augen der Testtiere bis zur tief kaffeebraunen Farbstufe[3]) ausgefärbt; *v bw*-Tiere von *Drosophila melanogaster* zeigen ein grundsätzlich gleiches Verhalten; bei ihnen ist die Ausfärbung der Augen sowohl durch Injektion von Kynurenin in die junge Puppe als auch nach Zugabe von Kynurenin zum Nährboden der Larven zu erzielen. Die Festlegung der Grenzdosen an Kynurenin, mit denen diese Reaktionen auszulösen sind, muß in weiteren Versuchsreihen erfolgen. Zu den vorstehenden Befunden gelangten wir auf folgendem Wege:

1. Systematische Versuche zur Anreicherung des in Extrakten aus Schmeißfliegenpuppen (*Calliphora erythrocephala*) enthaltenen[4]) Augenpigment bildenden Wirkstoffes führten uns zu hochwirksamen Konzentraten, aus deren Verhalten sich folgende Eigenschaften für den gesuchten Stoff ermitteln ließen: Es liegt ein bis 100° völlig hitzestabiler, gegen verdünnte Säuren weitgehend beständiger Stoff vor, der in Wasser, 95proz. Alkohol und verdünntem Aceton leicht, in reinem Aceton schwer löslich und in Lipoidlösungsmitteln unlöslich ist[5]). Er ist dialysabel, zeigt einen isoelektrischen Punkt um p_H 6, läßt sich in schwach saurem Gebiet besonders leicht mit Butanol aus seiner wäßrigen Lösung ausschütteln und ist mit Phosphorwolframsäure fällbar. Aus dieser Fällung läßt er sich regenerieren. Zu gleichartigen Ergebnissen sind vor uns TATUM und BEADLE[6]) bei der Untersuchung des Pigment bildenden Wirkstoffes aus *Drosophila*-Puppen-Extrakten gelangt, und KHOUVINE und EPHRUSSI[4]) ermittelten für den Wirkstoff aus *Calliphora*-Puppen ähnliche Löslichkeits- und Fällungseigenschaften. Von besonderer Bedeutung für die weitere Charakterisierung der vorliegenden Verbindung wurden die Beobachtungen, daß hochgereinigte Konzentrate beim Erwärmen mit verdünntem Alkali unter Wirksamkeitsverlust einen an o-Amino-acetophenon erinnernden Jasmingeruch aufweisen, und daß mit Sauerstoff eine langsame Zerstörung der physiologischen Wirksamkeit unter starker Vertiefung der normalerweise vorhandenen hellgelben Färbung der Lösung eintritt.

2. TATUM[7]) gelang die Feststellung, daß es gewisse, von ihm nicht identifizierte Bakterien gibt, die in Gegenwart von Tryptophan einen Stoff bereiten, der die physiologischen Wirkungen des „v^+-Hormons" in bezug auf die Augenausfärbung bei *Drosophila* besitzt und in seinen Eigenschaften dem Wirkstoff aus *Drosophila*-Extrakten ähnlich ist. Im Anschluß an diese wichtige Beobachtung konnten wir den Befund erheben, daß das seit 2 Jahren im Kaiser Wilhelm-Institut für Biochemie gezüchtete, durch eine spezifische dehydrierende Wirkung auf Steroidhormone gekennzeichnete *Coryne-Bakterium mediolanum*[8]) imstande ist, auf Agarnährböden oder in Hefewasser bei 37° unter aeroben Bedingungen einen die Augenpigmentbildung auslösenden Stoff zu bereiten unter der Voraussetzung, daß Tryptophan zugegen ist. Alle physiologischen und chemischen Eigenschaften des von den Bakterien gebildeten Wirkstoffes sind identisch mit denen des in *Calliphora*-Extrakten nachgewiesenen Inhaltsstoffes.

3. Die sich aus diesen Versuchen ergebende Vorstellung, daß der gesuchte Wirkstoff ein Abwandlungsprodukt des Tryptophans sein könnte, veranlaßte uns zur physiologischen Prüfung einer Tryptophanderivate. Im Verlauf dieser Untersuchungen wurde die Wirksamkeit des Kynurenins entdeckt. Da die Wirkung weitgehend spezifisch zu sein scheint und alle Kennzeichen des Kynurenins sich mit den bisher bekannten Eigenschaften des natürlich vorkommenden Wirkstoffes aus *Calliphora*-, *Drosophila*- und *Ephestia*-Extrakten decken, halten wir es für wahrscheinlich, daß im Kynurenin der natürliche Pigment bildende Stoff vorliegt. Diese Frage soll durch Versuche zur Isolierung des Kynurenins aus *Calliphora*-Konzentraten endgültig beantwortet werden. Im Anschluß daran ist zu untersuchen, welche Beziehungen zwischen dem sich in der natürlichen Entwicklungskette nach dem v^+-Stoff einschiebenden und wahrscheinlich aus ihm entstehenden cn^+-Stoff, der die Augenausfärbung der *Drosophila*-Mutation *cinnabar* (*cn*) bedingt[2]), und dem Kynurenin bestehen.

Die chemische Untersuchung der Faktoren, die die Pigmentbildung der Insektenaugen bewirken, hat entwicklungsphysiologische Bedeutung für die Frage nach der Verknüpfung von Gen und Außenmerkmal. Ob das nur in Gegenwart eines bestimmten Gens im Organismus unserer Versuchstiere auftretende Kynurenin eine Pigmentvorstufe ist und in den Farbstoff eingebaut wird[9]), oder ob ihm die Wirkungen eines Biokatalysators („Genhormons") zukommen, muß in weiteren Untersuchungen geklärt werden.

Berlin-Dahlem, Kaiser Wilhelm-Institut für Biochemie und Kaiser Wilhelm-Institut für Biologie, Abt. KÜHN, den 16. Januar 1940.

[1]) Zusammenfassung: KOTAKE, Erg. Physiol. **37**, 245 (1935).
[2]) Zusammenfassung: BECKER, Naturwiss. **26**, 433 (1938).
[3]) Vgl. KÜHN, Z. indukt. Abstammgslehre **73**, 419 (1937).
[4]) KHOUVINE, EPHRUSSI, C. r. Soc. Biol. Paris **124**, 885 (1937).
[5]) Vgl. BECKER u. PLAGGE, Biol. Zbl. **59**, 326 (1939).
[6]) TATUM u. BEADLE, J. gen. Physiol. **22**, 239 (1938).
[7]) TATUM, Proc. nat. Acad. Sci. U.S.A. **25**, 486 (1939).
[8]) MAMOLI, Ber. dtsch. chem. Ges. **72**, 1863 (1939).
[9]) Vgl. BECKER, Biol. Zbl. **59**, 597 (1939).

*Editor's Note: An English translation follows this article.

7

Kynurenine as an Agent Which Causes the Formation of Eye Pigment in Insects

ADOLF BUTENANDT, WOLFHARD WEIDEL, and ERICH BECKER

This article was translated expressly for this Benchmark volume by Terry W. Wesley, The University of Texas at Austin, from Naturwiss., *28(4), 63–64 (1940).*

We have found that kynurenine[1] (probable structure shown in I), a metabolic product found in the urine of rabbits after ingestion of 1-tryptophan, reverses the effects of the *v* (vermilon) mutant gene in *Drosophila melanogaster* and the *a* (red eyes) mutant gene in the Flour-moth, *Ephestia kühniella*.[2]

Upon injection of an approximately 1 percent aqueous solution of kynurenine into a *Ephestia* chrysalis of genotype *a/a*, the normally orange to red eyes of the mutants were changed in color to a deep coffee-brown shade.[3] The homozygous *v, bw* mutants of *D. melanogaster* showed fundamentally the same effects. The color of the eyes becomes brownish after the injection of kynurenine into the young pupae and after adding it to the larval food medium. Quantitative estimates of the minimal dosages needed to obtain these effects must be determined by a further series of tests.

We arrived at the findings presented here in the following way:

1. We systematically tested the eye-pigment-forming hormones concentrated from extracts of the blow fly (*Calliphora erythrocephala*).[4] We found present in the extracts an agent that is completely stable at temperatures up to 100°C, is quite resistant to dilute acids, and is easily soluble in water, 95 percent alcohol, and diluted acetone but not in pure acetone or lipoid solutions.[5] It is dialyzable, has an isolectric point around pH 6, can be easily extracted from weakly acid aqueous solutions with butanol, and is precipitable with phosphotungstic acid. It can be regenerated from this precipitate. Tatum and Beadle[6] arrived at quite similar results while examining the pigment-forming hormones from *Drosophila* pupae extracts before we did, and Khouvine and Ephrussi[4] found similar solubility characteristics for the hormone from *Calliphora* chrylises. The observations that highly purified concentrates have a jasmine odor reminiscent of *o*-aminoacetophenon when heated in dilute alkaline solutions and a concomitant loss of activity, and that oxygen causes a slow destruction of activity with a simultaneous strong deepening of the normally bright yellow color of the solution, are of particular importance for the characterization of the compound under consideration.

2. Tatum[7] succeeded in showing that there are certain bacteria, not identified by him, that synthesize a substance in the presence of tryptophan which has the physiological effect of the "v^+ hormone" in *Drosophila* vermilion mutants and is similar in its other characteristics to the v^+ substance from *Drosophila* extracts. In connection with this important observation, we can also state that *Cornyene-Bacterium mediolanum*,[8] cultivated for two years at the Kaiser Wilhelm-Institute for Biochemistry, which is characterized by a specific dehydrating effect on steriod hormones, is capable of synthesizing a substance that causes the formation of eye pigment. The substance is formed when the

bacterium is grown either in soil or yeast water at 37°C under aerobic conditions provided tryptophan is present. All physiological and chemical characteristics of the hormone formed by the bacteria are identical with the substance found in *Calliphora* extracts.

3. The idea that the sought-for hormone could be a derivative of tryptophan, indicated by the above-mentioned tests, caused us to test several known tryptophan derivatives. In the course of these tests the effect of kynurenine was discovered. Because the effect of kynurenine seems to be largely specific, and because the characteristics of kynurenine coincide with the previously known characteristics of the naturally occurring hormone in *Calliphora*, *Drosophila*, and *Ephestia* extracts, we consider it highly probable that it is the naturally occurring compound. This question should be answered once and for all by isolating kynurenine from *Calliphora* extracts. Subsequently, the relationship existing between the cn^+ substance[2] (cinnabar plus substance), which appears in the natural chain of development after the v^+ substance and is probably derived from it (the v^+ substance), and kynurenine itself should be determined.

The chemical investigation of the factors causing pigment formation in insect eyes is of importance in developmental physiology with respect to the question of the link between genes and external characteristics. Whether kynurenine, which only occurs in the presence of a particular gene in the test animals, is a precursor built into the pigment[9] or whether it is a biocatalyst (gene–hormone) must be cleared up by further investigations.

[*Editor's Note:* References will be found at the end of the preceding, original article.]

CHEMISTRY OF "EYE COLOR HORMONES" OF DROSOPHILA

By BORIS EPHRUSSI

Department of Biology, Johns Hopkins University, Baltimore, Maryland

INTRODUCTION

ALL biologists are familiar with the statement that hereditary factors or genes control the development of hereditary traits or characters. How, by what mechanism, they achieve this control is one of the great unsolved problems of physiological genetics. In recent years attempts have been made to approach the solution of this problem from different angles. One of these attempts, the study of the mechanisms of the genetic control of eye color differentiation in *Drosophila*, has made particularly rapid advances since 1935. Several reviews have been published on the progress in this field (Beadle, 1937; Ephrussi, 1938; Becker, 1938; Beadle and Tatum, 1941). The purpose of the present article is to review a particular line of this work, namely, that which has led to the knowledge of the chemical nature of some gene-controlled substances intervening in the development of eye colors in *Drosophila* and representing, so to say, the intermediate links between the genes controlling their production and the final character.

In presenting the results of the work along this line, the author has chosen to follow, as closely as possible, the chronological order. It is indeed probable that, in adopting this order, the author, who has to a certain extent contributed to the development of the subject, was guided by a sentimental impulse. Whatever the reason, the author feels that for him this is the easiest way to tell an "eye-witness" story, in which a series of "lucky accidents" played as important a rôle as the logic of the planned experiments. This alternation of logical steps and of the unforeseen will be in many ways suggestive to the unspecialized reader.

From the latter point of view it is, no doubt, worth noting that the relatively rapid advances of the work related in this article has been, in a great measure, due to a frank and wholehearted cooperation between the different workers, particularly between the groups headed, at Stanford, by G. W. Beadle and, in Paris, by the present writer, and that, in this field of Biology, like in many others, the collaboration of biologists and chemists was a major factor in the achieved advances.

Vermilion in mosaics

The original facts which constitute the starting point of the work discussed in this review have been reported in 1920 by Sturtevant in a short paper published in the *Proceedings of the Society for Experimental Biology and Medicine*. In this paper Sturtevant has described the aberrant behavior of the mutant character vermilion in *Drosophila* gynandromorphs.

Gynandromorphs, that is, individuals composed of male and female parts, have been known in *Drosophila melanogaster* ever since this species has been extensively used in breeding experiments. Morgan and Bridges (1919) have made a detailed study of these exceptional individuals and have explained their origin through the elimination of one of the sex-chromosomes in the course of early cleavages. The study of sex-linked characters in these genetically mosaic individuals has shown that, as a general rule, "the male and female parts and their sex-linked characters are strictly self-determining, each part developing according to its own genetic constitution. No matter how large or small the region may be, it is not interfered with by influences coming from neighboring parts, nor is it overruled by the action of the gonads" (Morgan, Bridges and Sturtevant, 1925).

The aberrant case reported by Sturtevant in the above cited paper acknowledges the first exception to this general rule. A female heterozygous for the sex-linked genes eosin, ruby and forked was crossed with a male carrying in its X-chromosome the recessives scute, echinus, cut, garnet, vermilion and forked. Among the offspring of this cross a gynandromorph was found in which the male parts, including the head, showed the characters scute, echinus, cut, garnet and forked. The female parts showed the forked character only. This appearance of the different parts of the fly was consistent with the interpretation that the animal started its development as a female carrying the maternal X-chromosome which was marked by the recessive genes listed above and that, in the course of one of the early cleavages, this chromosome was lost from one of the blastomeres, leaving the latter with the paternal X-chromosome only. However, on this interpretation, the eyes of the gynandromorph, whose head showed the paternal sex-linked characteristics, should show also the vermilion eye-color. This was not the case and from this observation Sturtevant has drawn the conclusion that "the non-vermilion color was then apparently determined not by the genetic constitution of the eye-pigment itself, but by some other portion of the body". Thus this short note contains the first description of a non-autonomous character in Drosophila and, in the quoted sentence, the first tacit assumption of the intervention of a diffusible substance in the process of eye color differentiation.

A similar behavior of vermilion was later observed by Bridges (1925) and L. V. Morgan (1929) in mosaics of different origin.

The importance of non-autonomous mendelian characters for the analysis of gene-controlled processes has been stressed by Sturtevant in his paper "The use of mosaics in the study of developmental effects of genes" read at the Sixth International Congress of Genetics (1932). In concluding this paper Sturtevant writes:

It is clear that in most cases there is a chain of reactions between the direct activity of a gene and the end-product that the geneticist deals with as a character. One may surmise that any valid generalizations about these reactions are more likely to concern the initial links than the terminal ones. However, it is the terminal ones that are usually more open to experimental attack, since the only index to the effectiveness of a given experimental technique is the condition of the end product. Looked at from this point of view, the type of experiment that I have described may be considered as a beginning in the analysis of certain chains of reactions into their individual links.

This paragraph is quoted *in extenso* because it contains, along with a clear and concise definition of the purpose of Sturtevant's experiments (and of later work derived from these experiments), a precise statement of the methodological limitations involved.

Transplantation in Drosophila

Although Sturtevant's interpretation of the behavior of vermilion in mosaics implied a humoral correlation, the direct demonstration of the existence of hormone-like substances in *Drosophila* was provided only in 1935, with the introduction of a transplantation technique adapted to this species (technique described by Ephrussi and Beadle, 1936). One of the first experiments performed with the help of the new technique was, so to say, a duplication of the situation which arose spontaneously in the case observed by Sturtevant. Eye discs of vermilion larvae were transplanted into wild type larvae and it was found that, under these conditions, the vermilion eye discs developed into adult structures showing the wild type color (Ephrussi and Beadle, 1935; Beadle and Ephrussi, 1936). Since the implanted eyes developed in the abdomens of the hosts without any constant or very definite connections with the surrounding tissues, it was concluded from these experiments that the lymph of the wild type hosts contained a substance responsible for the observed change from vermilion to wild type. Vermilion, the first non-autonomous character known in *Drosophila*, was therefore interpreted as a deficiency for a specific substance. This substance, produced under the control of the wild type allelomorph of the vermilion gene, represents then a link connecting this gene with the corresponding character (Fig. 1, A and B).

Essentially similar results have been obtained by a German group of workers with an eye-color mutant in the moth, *Ephestia kuhniella* (Caspari, 1933; Kühn, Caspari and Plagge, 1935). In the following pages, however, the main emphasis will be placed on the work with *Drosophila* because the existence in this species of a considerable number of eye color mutants has made possible a much further analysis.

The confirmation of the non-autonomous nature of vermilion led to the search for other non-autonomous eye colors. Over two dozen *Drosophila* eye-color mutants have been tested by transplantation experiments similar to those used with vermilion. Eye discs of the different mutant types have been transplanted into wild type hosts and their color characteristics were examined after hatching of the flies. Among the various eye colors thus tested, another eye color, cinnabar, characterized by a similar vermilion-like pigmentation, was found to be also non-autonomous.

Cinnabar eye discs transplanted into wild type hosts differentiated into eyes with wild type color. Following the reasoning applied in the case of vermilion, the conclusion was drawn that the wild type lymph contains a substance responsible for the change of pigmentation of implanted cinnabar eyes. And thus the question arose whether the change from cinnabar to wild type was caused by the same substance which controlled the change from vermilion to wild type. The wild type lymph contains both these substances. The lymph of the mutant cinnabar contains only one of them, namely the substance responsible for the change from vermilion to wild type. The mutant vermilion contains none of these substances. These two substances have been called respectively the v^+ and cn^+ substances or hormones, the exponent $+$ indicating their occurrence in the lymph of animals

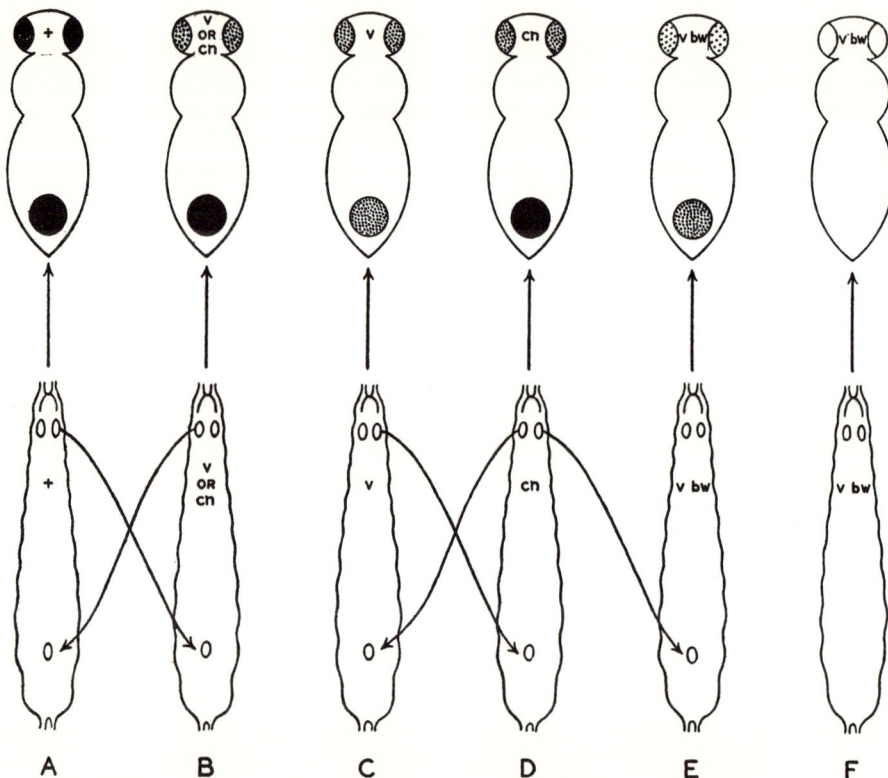

FIG. 1. SUMMARY OF THE RESULTS OF TRANSPLANTATION EXPERIMENTS
Below, host and donor larvae. Above, adult flies containing differentiated implants

from vermilion to wild type or whether the wild type lymph contained two different substances. The answer to this question was given by other transplantation experiments (compare A, B, C, and D of Fig. 1) whose results are summarized in Table I and among which the two reciprocal transplantations (5 and 6) between cinnabar and vermilion are particularly significant.

The results of these reciprocal transplantations make it quite obvious that there are two different substances, one responsible for the change from vermilion to wild type and the other for the change carrying the wild type allelomorphs of the genes vermilion (v) and cinnabar (cn) (Beadle and Ephrussi, 1936).

The above table contains only a few of the transplantations performed. In other transplantation experiments very numerous kinds of combinations have been realized between hosts and implants of different genetic constitutions. These experiments have lead to the important observation that, while a deficiency for the cn^+ substance alone characterizes the flies of certain genotypes, a deficiency or a decrease of concentration of the v^+ substance is always accompanied by a deficiency or corresponding decrease of concentration of the cn^+ substance. On the basis of this observation the assumption was made by Beadle and Ephrussi (1936) that the two substances are formed in the course of a single chain of reactions, of which the v^+ substance represents the first and the cn^+ substance the second link:

$$\rightarrow v^+ \rightarrow cn^+$$

This hypothesis, fundamental for the ensuing experimental developments, naturally did not imply a corresponding simplicity of the chain of reactions when translated in concrete chemical terms.

TABLE 1
Summary of results of transplantation experiments

IMPLANT	HOST	PHENOTYPE OF THE IMPLANT
1. Vermilion	Wild type	Wild type
2. Cinnabar	Wild type	Wild type
3. Wild type	Vermilion	Wild type
4. Wild type	Cinnabar	Wild type
5. Vermilion	Cinnabar	Wild type
6. Cinnabar	Vermilion	Cinnabar

In all transplantation experiments hitherto described, the observed changes in pigmentation were produced in the implants, under the influence of the new environment (host). However, effects of essentially the same sort can be produced as well in the opposite direction, i.e. changes can be induced in the pigmentation of the host eyes through an influence emanating from the implants. If, for example, a cinnabar eye disc is transplanted into a vermilion-brown larva (the reasons for choosing this type of double recessive host will be explained in a moment), the eyes of the host, normally containing but traces of pigment, show a marked darkening (Fig. 1, E and F). This result has been interpreted as a case of release of the v^+ substance produced by the implant.

This type of effect is mentioned here not because it is thought that it represents, in itself, something essentially new, but for the following two reasons. First, because effects of this type and, especially, reciprocal effects of the host on the implant and of the implant on the host, within the same experimental animal, further support the formation of the v^+ and cn^+ substances in the sequence indicated above (see review by Beadle and Ephrussi, 1937); secondly and particularly, because of its technical value. The possibility of producing an effect on the eyes *in situ* has suggested a method for the detection of the diffusible substances in any material, for example in an extract. In other words, the finding of the effects on the hosts has supplied the basis for an assay technique indispensable in the early phases of all hormone work.

In the last mentioned experiment, double recessive vermilion-brown ($v\ bw$) flies were used instead of vermilion (v) flies, because in the almost colorless eyes of the double recessive a very slight change produced by the v^+ substance (i.e. a slight shift from the $v\ bw$ towards the v^+bw phenotype) is much more easily detected than in the eyes of vermilion flies. In other terms, the eyes of the vermilion-brown flies were found to constitute the most sensitive "detectors" of the v^+ substance. Similarly, cinnabar-brown flies can conveniently be used as "reagents" for the detection of the cn^+ substance.

The changes produced in the eyes of these detector flies by implants releasing one of the diffusible substances were found to vary according to the genotype of the implant (Ephrussi and Beadle, 1937). What is emphasized here is not the difference in the implants, but the variability of the effect. This variability was interpreted as indicating that the degree of darkening of the eyes was a measure of the amount of the diffusible substance released into the body fluid of the host. By selecting a series of flies of different genotypes, representing a series of intergrades between the phenotypes of vermilion-brown and brown flies, Tatum and Beadle (1938) have introduced a scale of standards which are comparable to various degrees of modification of the eye-color of vermilion-brown flies under the influence of the v^+ substance and can be evaluated in terms of arbitrary color values. The comparison of an experimental fly with this scale of standards permits the description of the results in terms of these color values, assumed to be related to the amount of v^+ substance by approximate proportionality. One of the first steps of the eye color modification, corresponding to one of the lighter shades in the scale of standards, was taken as indicative of the action of one arbitrary hormone unit. Thus the method of detection of the v^+ substance was transformed into a method of semi-quantitative determination of the hormone.

Transfusion of lymph

As shown above, the existence of the diffusible (v^+ and cn^+) substances has been postulated on the basis of, and offered as an interpretation for, the non-autonomous or dependent differentiation of the color characteristics of certain eye implants. A necessary corollary of this interpretation is that the postulated substances are transmitted from the locus of their production to the reacting organs (host or implanted eyes) through the lymph. The demonstration of their presence in the lymph should therefore be regarded as a supplementary proof of the correctness of the interpretation. This proof has been provided by direct lymph transfusion. The lymph of wild type or cinnabar pupae was injected into "detector" larvae. The flies which developed from the injected larvae showed a clear modification of the eye color, indicating that the injected lymph actually contained the postulated active substance (Beadle, Clancy and Ephrussi, 1937).

While this was a confirmation of the original hypothesis, there still remained one possible objection. The lymph contains, along with its liquid phase, cellular elements which may be held responsible for the modifications of the pigmentation. This objection was met by experiments in which the lymph, prior to injection, was frozen at very low temperatures and thawed several times and centrifuged in order to remove the residues of the destroyed cells. Lymph treated in this manner, when injected into vermilion-brown larvae, produced nevertheless clear modifications of the eye color and thus the experiments have shown that it is really the liquid phase of the lymph which carried the factors responsible for the changes in pigmentation (Harnly and Ephrussi, 1937).

The results just described were the first real indication that it should be possible to isolate the active principles. Before this could be attempted efficiently, however, several purely practical difficulties had to be solved. As in all chemical studies of hormones, the location of the source of active material to be extracted was thought to be particularly important in view of the exceptionally small size of the flies which constitute the natural source of the diffusible substances. Therefore the search was undertaken for another rich and easily available source of the hormones. Extracts of different tissues of higher animals were examined but all gave negative results. Yeast extracts were negative also. Lymph of different insects (*Galleria melonella, Calliphora erythrocephala, Habrobracon juglandis, Ephestia kuhniella, Bombyx mori*, etc.) on the contrary, was found to contain substances which proved to be active in modifying the pigmentation in *Drosophila* (Ephrussi and Harnly, 1936; Beadle, Anderson and Maxwell, 1937). These experiments indicated then, in the first place, that the diffusible substances of *Drosophila* were not species-specific; secondly, they indicated a practical means for obtaining large quantities of extraction material.

Among the different insects rich in the v^+ and cn^+ substances, *Calliphora* was chosen, because the larvae and pupae of this species could be easily purchased in considerable amounts on the Paris market. Thus the first chemical characteristics of the diffusible substances were obtained from the study of extracts of *Calliphora* pupae.

Studies of extracts

That active extracts of the diffusible substances can actually be obtained was very soon shown by extracting *Calliphora* pupae in ethyl alcohol or ethyl alcohol-ether mixtures (Khouvine, Ephrussi and Harnly, 1936). The study of these extracts soon indicated that the active substances were soluble in water and in ethyl alcohol, insoluble in ether and chloroform. They were heat stable and dialysable. Therefore they were neither lipoids, enzymes, nor proteins (Khouvine and Ephrussi, 1937). These results were confirmed by Thimann and Beadle (1937) working with extracts of *Drosophila* pupae. Further study of both *Calliphora* and *Drosophila* extracts has shown that the diffusible substances are very unstable in acids and alkali and are not extractible in most organic solvents. They are precipitated by lead acetate and phosphotungstic acid as well as by Neubergs' reagent considered to be specific for amino-acids (Khouvine, Ephrussi and Chevais, 1938; Tatum and Beadle, 1938). Finally, the study of their rate of diffusion in agar blocks has indicated a molecular weight of 400–500 (Tatum and Beadle, l.c.). It should be pointed out here that throughout all these experiments the behavior of the v^+ and cn^+ substances and their properties appeared strictly identical and no separation of them has been possible.

Taken together, the results of all these studies have suggested the amino-acid like nature of the eye-color hormones and from this point on their further purification and eventual crystallization was attempted. This line of work, however, has not led to the final identification of their chemical structure. While the results of the chemical work described above have played an important rôle in orienting the experiments which finally gave the clue, these experiments, as will be shown, were suggested by a different line of work.

Feeding experiments

In 1937 it was found that the diffusible substances will produce their effects on sensitive hosts not only when injected into the blood stream but also when administered *per os* (Beadle and Law, 1938). Since at that time it was already supposed, on the basis of the chemical investigations summarized above, that the active substances were amino-acids, it was realized that the "feeding technique" offered the possibility of testing various pure amino-acids or their mixtures for their hormone activity on a scale which was not accessible so long as it was necessary to resort to injections. Experiments were consequently started in which *Drosophila* (vermilion brown) larvae were raised on mixtures of agar, glucose and different amino-acids. It was assumed that these media would be sufficiently nutritive to support the completion of the growth and differentiation of larvae which, prior to the experiment (up to a certain rather advanced stage) were grown under standard food conditions. This assumption, however, appeared to be erroneous. In the first experiments only a low per cent of the larvae reached and went through metamorphosis. Only a few flies hatched, in a few of the experimental vials, and all these flies were of a very small size as a result of food insufficiency. However, the experiment led to an interesting observation. In one of the vials some flies developed a modified eye color, similar to that produced by a low dose of the v^+ substance. Was the amino-acid added to the medium in this vial responsible for the change? This was not certain since none of the control flies, raised on the agar glucose mixture without addition of amino acids, reached emergence. The repetition of the experiment on a large scale soon showed that this was not the case. In this new experiment eye color modifications were again observed, but they were observed in several vials containing very different amino acids. This pointed to the conclusion that the added amino acids were not involved at all in the observed effects. As a matter of fact, in this second experiment, some control flies did hatch on a medium containing no added amino-acids, and these flies showed the eye color modification as well. Were then the observed effects due to some impurity in the agar or in the glucose? Control experiments were performed which showed conclusively that the eye color changes could not be ascribed to any one of the constituents of the medium and suggested that they were induced by the general malnutrition of the larvae (Khouvine, Ephrussi and Chevais, 1938). Special experiments performed in order to test this conclusion have shown that partial or total starvation of vermilion-brown larvae will produce the change from v towards v^+, or, in other words, that under the influence of a certain diet, vermilion larvae, normally characterized by the inability to produce the v^+ substance, will be induced to produce it (Khouvine, Ephrussi and Chevais, *l.c.*; Beadle, Tatum and Clancy, 1938). This conclusion has been further confirmed by direct extraction of the v^+ substance from starved vermilion-brown larvae (Beadle, Tatum and Clancy, *l.c.*). Thus a very interesting relationship between the production of the v^+ substance and the general (probably protein) metabolism of the flies has been established. This relationship is an interesting problem of its own, but has no direct connection with the problem discussed here except for the practical conclusion concerning the feeding technique which had to be drawn from the results of these experiments. This conclusion pointed to the necessity of first working out an adequate nutritive medium into which the substances to be tested could be incorporated.

The effect of tryptophane

Among the different substances tested for their nutritive value, a peptone was tested and, while it appeared to be entirely satisfactory in so far as the general growth of the flies was concerned, its assay brought up a new and surprising result. Vermilion-brown flies raised on the medium containing the peptone also showed a modification of the eye-color, similar to that produced by starvation or by the addition of v^+ substance. In this case, however, the effect could hardly be attributed to a deficient diet and therefore seemed rather to indicate really the presence of an active principle in the peptone tested. Consequently experiments of the same type were repeated using several samples of peptones of different origins and various proteoses. These experiments gave a very interesting result, namely that all the peptones tested, except one, gave positive results. The exceptional peptone was gelatin peptone which, as is well known, is characterized by the absence of tyrosin and tryptophane. The next obvious step then was the addition of tyrosin and tryptophane or their mixtures to the medium containing gelatine peptone. These experiments showed conclusively that tryptophane was the substance responsible for the eye color change.

But was tryptophane to be regarded as the v^+ substance? By definition, the v^+ hormone is a substance producing its characteristic effect when injected into the blood-stream. Tryptophane, however, when tested by injection, gave negative results. It became clear therefore that, while tryptophane could probably be regarded as a precursor of the v^+ substance, it could not be considered as being the substance itself. In other words, this finding made it quite probable that the v^+ hormone was a derivative of tryptophane. Consequently experiments were started in order to test the activity of various products of tryptophane transformation, both by feeding them to and injecting them into vermilion-brown larvae. This work has led to the finding of many tryptophane compounds active in the same way as tryptophane but inactive by injection (Khouvine and Ephrussi, unpublished). The work along this line has, however, been interrupted by the war and the decisive developments came from the combined results of two different lines of work, both starting from the discovery of the rôle of tryptophane in the described effects.

Bacterial synthesis of the v^+ substance

Throughout the experiments thus far described no particular attempt was made to control the growth of microorganisms in the media tested. In 1938 Tatum became interested in the nutritional requirements of *Drosophila* larvae and in connection with this work (Tatum, 1939b, 1941) established a technique for growing flies aseptically (technique described in Tatum and Beadle, 1939). One of the results of these studies of larval nutrition was the finding that media such as the agar-peptone mixtures used by Khouvine, Ephrussi and Chevais cannot support, under aseptic conditions, the normal growth and development of *Drosophila* larvae. The normal development of the flies in the experiments of these authors appeared therefore as probably due to the growth of microorganisms in these, by themselves deficient, media. If so, the hormone-like effects of starvation, peptones and tryptophane may have been due also to the intervention of microorganisms. With this idea in mind, Tatum and Beadle (1939) reinvestigated, under aseptic conditions, the effects of starvation peptones and tryptophane. While it was found that the starvation effect can be observed on flies grown under aseptic conditions, it appeared that the effect of tryptophane does not occur in the absence of microorganisms. The latter two effects were thus due to the presence of bacteria or yeast. In the course of these experiments, however, Tatum (1939a) found one vial accidentally contaminated by an aerobic bacillus. The vermilion-brown flies hatched in this vial showed a marked eye color modification. A culture of this unidentified *Bacillus* sp. was isolated from the contaminated vial and it was soon shown that the bacillus, when grown on a tryptophane-containing medium, synthetizes a substance similar in its effects and properties to the v^+ substance. The substance can be added to normal (complete) nutritive media and will produce the same effect. It therefore is clear that starvation plays no rôle in these results. Moreover, the substance extracted from the agar medium on which pure cultures of the bacillus were grown, was found to be active also when injected into vermilion-brown larvae. The method of extraction from the agar medium showed that the solubilities of the active substance were similar to those ascribed to the v^+ substance on the basis of the chemical work related above. Finally, it should be mentioned that this substance produced effects only on vermilion-brown, not on cinnabar-brown flies. But Tatum (*l.c.*) was able to show that, like the v^+ substance, the substance obtained from bacterial synthesis, is converted, within the animal, to cn^+ substance. If vermilion-brown larvae, fed on a medium containing the substance synthetized by *Bacillus sp.* are, after pupation, boiled, crushed and fed to cinnabar-brown larvae, the latter develop into flies with a clearly modified eye color. Thus, the substance synthetized by Tatum's bacillus had all the known attributes of the v^+ hormone. Their identification required, however, their isolation in a pure state. The next step in the direction of this identification was made by Tatum and Beadle in 1940 when they succeeded in crystallizing the product of bacterial synthesis. On the results of this work the authors report in the following terms:

... "This bacterially produced v⁺ hormone has now been obtained in a pure crystalline state. The bacteria were grown on an agar medium containing dead yeast, sugar and l-tryptophane. The agar and yeast were precipitated in 80 per cent alcohol. The hormone was then taken up in a mixture of butyl alcohol, ethyl alcohol and water, and was finally precipitated from absolute butyl alcohol. It was then crystallized from 90 per cent ethyl alcohol. The crystals are very light yellow, elongated plates, usually forming in rosettes. The elementary analysis (made under the direction of Dr. A. J. Haagen Smit, of the California Institute of Technology) supports the empirical formula $C_{21}H_{34}N_2O_{14}$." (Tatum and Beadle, 1940).

It should be added that according to data of Tatum and Beadle one gram of this crystalline material will change from vermilion to wild type approximately one million flies.

The identification of the v⁺substance with kynurenine

The knowledge of the structural formula of the v⁺ substance, however, is connected with still another line of work and due to another group of workers. When the relation of the v⁺ substance with tryptophane was established, tests of the activity of various tryptophane derivatives have been performed by Butenandt and his coworkers in Berlin. Among the substances tested, kynurenine, a substance discovered by Kotake in the urine of rabbits fed on l-tryptophane, gave positive results, both by the feeding and injection tests (Butenandt, Weidel and Becker, 1940a). Thus kynurenine was the first pure chemical of known constitution whose physiological effects were found to be similar to those of the v⁺ substance. The properties of kynurenine solutions are also similar to those of the highly active and highly purified solutions of the v⁺ substance obtained from fly extracts. The results of Butenandt, Weidel and Becker made it very probable that kynurenine should be identified with the v⁺ substance. This impression was further strengthened by the work of Tatum and Haagen Smit (1941) who were able to show that the crystalline product having v⁺ activity and isolated from bacterial synthesis is a sucrose ester of l-kynurenine.

The final proof of the identity of a chemically defined compound with the v⁺ substance, however, should necessarily involve the demonstration of (1) the presence of the substance in flies which, according to transplantation experiments, contain it and (2) its absence in the mutant vermilion. Kikkawa (1941) has undertaken this work using the Otani-Honda method for the demonstration of kynurenine. The results of Kikkawa's work can be summarized as follows. Kynurenine can be detected in the mutant cinnabar and in its equivalent (mutant white-I) in *Bombyx*. From the eggs of the latter kynurenine sulphate has been obtained in crystalline form and shown to be active when injected into vermilion-brown *Drosophila* larvae. Tests for kynurenine in vermilion larvae and pupae are negative. Thus the first two criteria of the identification of kynurenine with the v⁺ substance were fulfilled. However, the tests for kynurenine in wild type larvae and pupae were negative, while the original scheme of Beadle and Ephrussi calls for the presence of the v⁺ substance in wild type lymph. Kikkawa assumes that these negative results are due to the rapid conversion of kynurenine into cn⁺ substance (which is the compound responsible for the eye color change) and thus reconciles the scheme of Beadle and Ephrussi with the observed distribution of kynurenine.

The two remaining questions for which at least a partial answer can be given at present are those concerned with the origin and the fate of kynurenine.

According to Kotake, kynurenine is derived from tryptophane via α-oxy-tryptophane:

(TRYPTOPHANE) → (α-OXYTRYPTOPHANE) → (KYNURENINE)

This substance was only recently isolated by Wieland and Witkop (1940) from hydrolized phalloidin. Butenandt, Weidel and Becker (1940b) have injected solutions of α-oxy-tryptophane into vermilion-brown larvae and obtained positive, although weaker results than by injecting equivalent concentrations of kynurenine. It can be considered therefore that kynurenine arises from tryptophane via α-oxy-tryptophane *in vivo* also and α-oxy-tryptophane can be regarded as a precursor of kynurenine.

The nature of the cn^+ substance

Since the cn^+ substance is known to be derived from the v^+ substance and since the chemical nature of the latter seems to be established, it should be possible now to gain some insight into, if not determine the nature of, the cn^+ substance also. The most natural idea is to look for substances having a cn^+ activity among the derivatives of kynurenine.

stance, both found in human urine and responsible for Ehrlich's diazo-reaction, are derived from kynurenine. Although the solubilities of these substances are known to be different from those of the cn^+ substance, Kikkawa has applied the diazo-reaction to the mutants of *Drosophila* and *Bombyx* in which transplantation experiments have demonstrated the presence or the absence of the cn^+ substance. These tests have shown that

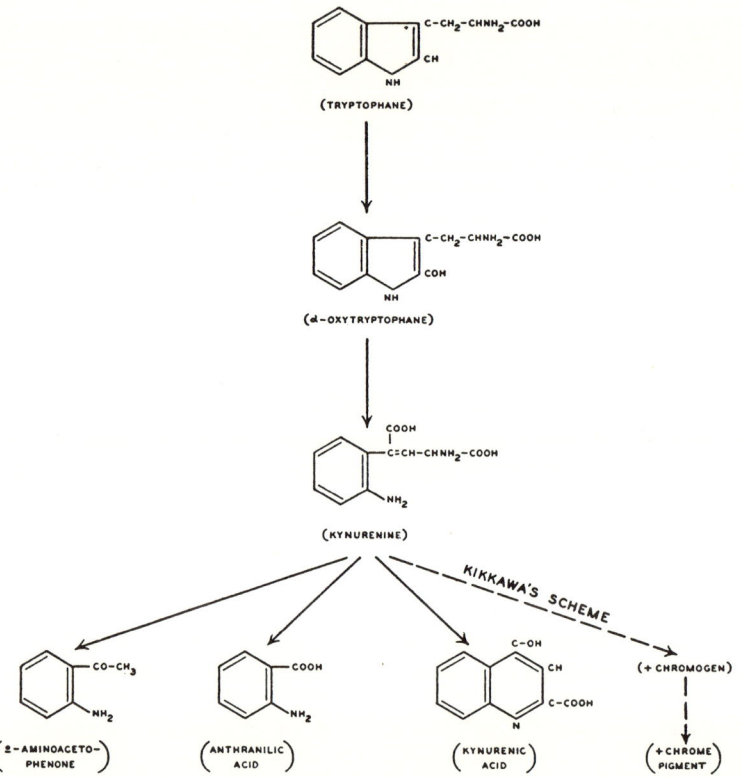

FIG. 2. THE PRODUCTS OF TRYPTOPHANE METABOLISM AND KIKKAWA'S SCHEME OF PIGMENT PRODUCTION IN DROSOPHILA

According to Kotake, in the animal body the transformations of kynurenine most commonly lead, through kynurenine-yellow, to o-amino-acetophenone, anthranilic acid and kynurenic acid (Fig. 2). The solubilities of the three first named substances suggest that they cannot be the cn^+ substance. Injections of anthranilic acid by Kikkawa (*l.c.*) in fact gave negative results. Kynurenic acid, as well as α-decarboxykynurenine have been tested by Butenandt, Weidel and Becker (1940b) and did not show any cn^+ activity.

Kotake has also suggested that the so-called Weiss urochromogen and the Sachs color sub-

there is a clear correlation between the presence of the cn^+ substance and a positive diazo-reaction. Kikkawa further shows that the diazo-reaction, negative in vermilion-brown controls, becomes positive if the larvae are grown on a medium to which kynurenine has been added. In other words there is an identity of distribution of the cn^+ substance and of a factor responsible for a positive diazo-test, and a causal relationship between the latter and kynurenine. From the above identity of distribution Kikkawa concludes that the cn^+ hormone is the substance responsible for the positive diazo-reaction. He calls it the "+

chromogen", thus indicating by the new term that the cn⁺ substance is a direct precursor of the pigment.

PERSPECTIVES

The work of Kikkawa brings the story of the eye-color hormones of *Drosophila* up to date. Starting from the results of transplantation experiments which led to the postulation of the formation in sequence of two diffusible substances representing the first known links in the chain of reactions connecting the genotype with the character, we have now reached the point where a part of this chain can be described in chemical terms (Table 2 recapitulates the obtained results). The task of the future will be to connect these intermediate links with the two opposite ends of the chain.

The relation of kynurenine and of the cn⁺ substance to the pigment is by no means clear, although Kikkawa, as mentioned above, assumes the cn⁺ substance to be a precursor of pigment. Here, however, no major difficulty seems to stand in the way of future progress.

TABLE 2

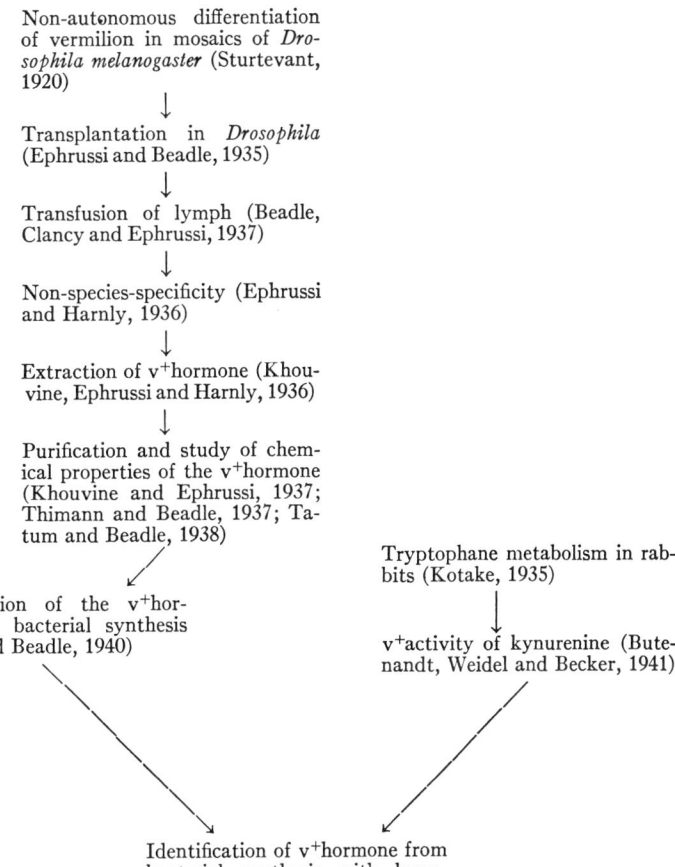

Bridging the gap between the hormone-like substances which, at the present stage, are the first tangible manifestations of certain genes, with these genes, promises to be a much harder task. What is known at present of the action of the v⁺ gene indeed suggests that this gene intervenes in the tryptophane metabolism by producing a specific oxidizing enzyme. Thus the problem of primary gene activity in this case seems to resolve itself into a problem of enzyme chemistry.

In so far as the primary genic action is concerned, however, we are confronted with more than purely practical difficulties. As Sturtevant (1941) puts it in discussing the methods of physiological genetics in a recent article:

"there is no *a priori* reason why the method may not, in certain cases, lead to a knowledge of the primary effect of a given gene; the difficulty is that, at present, there appears to be no way of deciding the point in any given case. That is to say, the chain of developmental reactions may be traced back to the gene, but there is no way of determining when one has reached the gene" (p. 48).

The theoretical difficulty pointed out by Sturtevant is very real indeed. The present writer does not think, however, that the efforts along the same path should be abandoned. Instructed by the lessons of past experience and without underestimating the value of scientific planning, he keeps his optimism unaltered. It is his belief that unpredictable experimental situations often are the seeds of new insight and new ways of attack. Sometimes they do yield a little more than permitted by the theory.

LIST OF LITERATURE

BEADLE, G. W. 1937. The development of eye colors in *Drosophila* as studied by transplantation. *Am. Nat.*, 71: 120–126.

BEADLE, G. W., R. L. ANDERSON, and J. MAXWELL. 1938. A comparison of the diffusible substances concerned with eye color development in *Drosophila*, *Ephestia*, and *Habrobracon*. *Proc. Nat. Acad. Sci.*, 24: 80–85.

BEADLE, G. W., C. W. CLANCY, and B. EPHRUSSI. 1937. Development of eye colours in *Drosophila*: Pupal transplants and the influence of body fluid on vermilion. *Proc. Roy. Soc. Lon.*, Ser. B., 122: 98–105.

BEADLE, G. W., and B. EPHRUSSI. 1936. The differentiation of eye pigments in *Drosophila* as studied by transplantation. *Genetics*, 21: 225–247.

BEADLE, G. W., and B. EPHRUSSI. 1937. Development of eye colors in *Drosophila*: Diffusible substances and their interrelations. *Genetics*, 22: 76–86.

BEADLE, G. W., and L. W. LAW. 1938. Influence on eye-color of feeding diffusible substances to *Drosophila melanogaster*. *Proc. Soc. Exp. Biol. and Med.*, 37: 621–623.

BEADLE, G. W., and E. L. TATUM. 1941. Experimental control of development and differentiation. Genetic control of developmental reactions. *Am. Nat.*, 75: 107–116.

BEADLE, G. W., E. L. TATUM, and C. W. CLANCY. 1938. Food level in relation to rate of development and eye pigmentation in *Drosophila melanogaster*. *Biol. Bull.*, 75: 447–462.

BECKER, E. 1938. Die Gen-Wirkstoff-Systeme der Augenausfärbung bei Insekten. *Naturwiss.*, 26: 433–441.

BRIDGES, C. B. 1925. Elimination of chromosomes due to a mutant (Minute-N) in *Drosophila melanogaster*. *Proc. Nat. Acad. Sci.*, 11: 701–706.

BUTENANDT, A., W. WEIDEL, and E. BECKER. 1940a. Kynurenin als Augenpigmentbildung auslösendes Agens bei Insekten. *Naturwiss.*, 28: 63–64.

BUTENANDT, A., W. WEIDEL, and E. BECKER. 1940b. α-Oxytryptophan als "Prokynurenin" in der zur Augenpigmentbildung führenden Reaktionskette bei Insekten. *Naturwiss.*, 28: 447–448.

CASPARI, E. 1933. Über die Wirkung eines pleiotropen Gens bei der Mehlmotte *Ephestia kuhniella* Zeller. *Arch. Entw. Org.*, 130: 353–381.

EPHRUSSI, B. 1938. Aspects of the physiology of gene action. *Am. Nat.*, 72: 5–23.

EPHRUSSI, B., and G. W. BEADLE. 1935. La transplantation des disques imaginaux chez la Drosophile. *C. R. Acad. Sci. Paris*, 201: 98.

EPHRUSSI, B., and G. W. BEADLE. 1936. A technique of transplantation for *Drosophila*. *Am. Nat.*, 70: 218–225.

EPHRUSSI, B., and G. W. BEADLE. 1937. Development des couleurs des yeux chez la Drosophile: Influence des implants sur la couleur des yeux de l'hote. *Bull. Biol.* 71: 75–90.

EPHRUSSI, B., and M. H. HARNLY. 1936. Sur la présence chez différents insectes des substances intervenant dans la pigmentation des yeux de *Drosophila melanogaster*. *C. R. Acad. Sci.Paris*, 203: 1028–1029.

HARNLY, M. H., and B. EPHRUSSI. 1937. Development of eye colors in *Drosophila*: Time of action of body fluid on cinnabar. *Genetics*, 22: 393–401.

KHOUVINE, Y., and B. EPHRUSSI. 1937. Fractionnement des substances qui interviennent dans la pigmentation des yeux de *Drosophila melanogaster*. *C. R. Soc. Biol. Paris*, 124: 885–887.

KHOUVINE, Y., B. EPHRUSSI, and S. CHEVAIS. 1938. Development of eye colors in *Drosophila*: Effect of yeast, peptones and starvation on their production. *Biol. Bull.*, 75: 425–445.

Khouvine, Y., B. Ephrussi, and M. H. Harnly. 1936. Extraction et solubilité des substances intervenant dans la pigmentation des yeux de *Drosophila melanogaster*. *C. R. Acad. Sci.Paris*, 203: 1542.

Kikkawa, H. 1941. Mechanism of pigment formation in *Bombyx* and *Drosophila*. *Genetics*, 26: 587–607.

Kühn, A., E. Caspari, and E. Plagge. 1935. Über hormonale Genwirkungen bei *Ephestia kühniella* Z. *Nach. Gesell. Wiss. zu Gott., Nach. a.d. Biol.*, 2: 1–30.

Morgan, L. V. 1929. Composites of *Drosophila melanogaster*. *Carn. Inst. Wash., Publ.* 399: 225–296.

Morgan, T. H., and C. B. Bridges. 1919. The origin of gynandromorphs. *Carn. Inst. Wash., Publ.* 278, 1–122.

Morgan, T. H., C. B. Bridges, and A. H. Sturtevant. 1925. The genetics of *Drosophila*. *Bibliogr. Genetica*, 2: 1–262.

Sturtevant, A. H. 1920. The vermilion gene and gynandromorphism. *Proc. Soc. Exp. Biol. and Med.*, 17: 70–71.

Sturtevant, A. H. 1932. The use of mosaics in the study of the developmental effect of genes. *Proc. 6th. Int. Con. Gen.*, 1: 304–307.

Sturtevant, A. H. 1941. Physiological aspects of genetics. *Ann. Rev. Phys.*, 3: 41–56.

Tatum, E. L. 1939a. Development of eye-colors in *Drosophila*: Bacterial synthesis of v^+ hormone. *Proc. Nat. Acad. Sci.*, 25: 486–490.

Tatum, E. L. 1939b. Nutritional requirements of *Drosophila melanogaster*. *Proc. Nat. Acad. Sci.* 25: 490–497.

Tatum, E. L. 1941. Vitamin B requirements of *Drosophila melanogaster*. *Proc. Nat. Acad. Sci.*, 27: 193–197.

Tatum, E. L., and G. W. Beadle. 1938. Development of eye colors in *Drosophila*: Some properties of the hormones concerned. *Jour. Gen. Phys.*, 22: 239–253.

Tatum, E. L., and G. W. Beadle. 1939. Effect of diet on eye-color development in *Drosophila melanogaster*. *Biol. Bull.*, 77: 415–422.

Tatum, E. L., and G. W. Beadle. 1940. Crystalline *Drosophila* eye-color hormone. *Science*, 91: 458.

Tatum, E. L., and A. J. Haagen Smit. 1941. Identification of Drosophila v^+ hormone of bacterial origin. *Jour. Biol. Chem.*, 140: 575–580.

Thimann, K., and G. W. Beadle. 1937. Development of eye colors in *Drosophila*: Extraction of the diffusible substances concerned. *Proc. Nat. Acad. Sci.*, 23: 143–146.

VI
One Gene–One Enzyme

Editor's Comments on Papers 9 Through 15

9 **Beadle and Tatum:** *Genetic Control of Biochemical Reactions in* Neurospora

10 **Sawin and Glick:** *Atropinesterase, A Genetically Determined Enzyme in the Rabbit*

11 **Horowitz:** *On the Evolution of Biochemical Syntheses*

12 **Mitchell and Nyc:** *Hydroxyanthranilic Acid as a Precursor of Nicotinic Acid in* Neurospora

13 **Horowitz and Leupold:** *Some Recent Studies Bearing on the One Gene–One Enzyme Hypothesis*

14 **Wagner and Haddox:** *A Further Analysis of the Pantothenicless Mutants of* Neurospora

15 **Kalckar, Anderson, and Isselbacher:** *Galactosemia, A Congenital Defect in a Nucleotide Transferase: A Preliminary Report*

After concluding that the work with the *Drosophila* eye color mutants had about reached the point at which it appeared new returns would be small, Beadle and Tatum began work with *Neurospora*. The choice was carefully calculated. Neurospora, they knew, was a haploid organism with an easily manipulated sex cycle that grew readily on a simple culture medium containing only inorganic salts, sucrose, and biotin. They started with the assumption that mutants could be induced which would not grow unless specific, known compounds were added to the minimal medium. This discovery changed the whole course of research into the nature of gene action and provided the stimuli for making new discoveries in this area that are being felt even today.

The work of Beadle and Tatum made it clear that a close link existed between what was called genes and specific enzymes. Concrete evidence for this was lacking, however. Sawin and Glick presented such evidence for atropine esterase in the rabbit.

The realization that the biosynthesis of compounds such as amino acids, vitamins, and so forth, proceeded by a series of sequential reactions each of which was under the guidance of a different gene raised some interesting questions of an evolutionary nature. Horowitz considered these questions and offered an explanation that is generally accepted as a reasonable possibility today, and can be no better stated now than it was in 1945.

Soon after the discovery of the method of isolation of biochemical or nutritional mutants of *Neurospora* by Beadle and Tatum, it became apparent that mutants with the same requirement, such as in this instance nicotinamide, were not necessarily identical genetically. It was quickly realized that this probably meant that each step in a series of reactions leading to the end product was under the control of a different gene. This surmise proved to be true.

The 1940s and 1950s were a period in biochemistry when many workers were actively trying to work out biosynthetic pathways for the cell-building blocks such as amino acids, purines, pyrimidines, and coenzymes. It was recognized that the *Neurospora* mutants, and the *Escherichia coli* mutants discovered soon after, might be powerful tools for working out these pathways; this turned out to be correct. Paper 12 is an example of how many metabolic pathways were elucidated by the judicious use of organic chemistry, educated guesswork, and genetics.

The one gene–one enzyme hypothesis dominated the thinking of biochemical geneticists soon after Beadle and Tatum's first discoveries. But could it be proved? Horowitz and Leupold set about doing this, and succeeded to the extent that it became the doubter's role to bring evidence to bear against it. The hypothesis still stands as valid in the form of the one gene–one polypeptide theory.

It was not realized at the time Paper 13 was written that most enzymes are made up of polypeptide subunits some of which are different polypeptides. This matter is considered further in papers on hemoglobins later in this volume.

A notable feature of this paper, aside from the close, logical reasoning, is the use of temperature mutants in some of the experiments described. These have since proved to be indispensible in the study of many phenomena, including development, meiosis and mitosis, and the assembly of phage particles in the host cell. The reader should pay particular attention to this portion of the paper.

Wagner and Haddox consider an important question current at the time it was written. What is a genetic block? Granted that a gene mutates, and therefore a reaction cannot proceeed. Why? Is the enzyme not made at all, or is a protein like it made but altered by the mutation in such a way that it cannot function properly? Paper 13 actually gives a partial answer to this in the results with the temperature mutants. But in this paper it is made quite evident that the pantothenicless mutant of *Neurospora* probably makes an enzyme which is altered in such a way that it does not function under certain environmental conditions other than temperature. That it is present at all times is evident, but it cannot always function.

The early observation of Garrod, described in his book *Inborn Errors of Metabolism* in 1909, pointed the way to making significant findings in human metabolism and inherited alteration of it. The biochemical knowledge and techniques of 1909 were not, however, sufficiently developed to exploit Garrod's findings. Nor were medical people or geneticists sufficiently ready to recognize the importance of the contents of the box that Garrod had opened for them to view.

Kalckar and his associates were among the first to renew the Garrodian quest. In this short note they show that an important inherited human disease, galactosemia, is the result of an inherited inability to produce an active PGal-transferase.

Selected Bibliography

Davis, R. H. 1963. *Neurospora* mutant lacking an arginine-specific carbamyl phosphokinase. *Science* 142: 1652.
Golly, J. A., and G. M. Edelman. 1972. The genetic control of immunoglobulin synthesis. *Ann. Rev. Genetics* 6: 1–46.
Hartwell, L. H. 1970. Biochemical genetics of yeast. *Ann. Rev. Genetics* 4: 373–396.
Levine, R. P., and U. W. Goodenough. 1970. The genetics of photosynthesis and of the chloroplast in *Chlamydomonas Reinhardi. Ann. Rev. Genetics* 4: 397–408.
Mäkelä, P. H., and B. A. D. Stocker. 1900. Genetics of polysacharide biosynthesis. *Ann. Rev. Genetics* 3: 291–322.
Sherman, F., and J. W. Stewart. 1971. Genetics and biosynthesis of cytochrome c. *Ann. Rev. Genetics* 5: 257–296.
Wagner, R. P., and H. K. Mitchell. 1964. *Genetics and Metabolism,* 2nd ed. Wiley, New York.

GENETIC CONTROL OF BIOCHEMICAL REACTIONS IN NEUROSPORA*

By G. W. Beadle and E. L. Tatum

BIOLOGICAL DEPARTMENT, STANFORD UNIVERSITY

Communicated October 8, 1941

From the standpoint of physiological genetics the development and functioning of an organism consist essentially of an integrated system of chemical reactions controlled in some manner by genes. It is entirely tenable to suppose that these genes which are themselves a part of the system, control or regulate specific reactions in the system either by acting directly as enzymes or by determining the specificities of enzymes.[1] Since the components of such a system are likely to be interrelated in complex ways, and since the synthesis of the parts of individual genes are presumably dependent on the functioning of other genes, it would appear that there must exist orders of directness of gene control ranging from simple one-to-one relations to relations of great complexity. In investigating the rôles of genes, the physiological geneticist usually attempts to determine the physiological and biochemical bases of already known hereditary traits. This approach, as made in the study of anthocyanin pigments in plants,[2] the fermentation of sugars by yeasts[3] and a number of other instances,[4] has established that many biochemical reactions are in fact controlled in specific ways by specific genes. Furthermore, investigations of this type tend to support the assumption that gene and enzyme specificities are of the same order.[5] There are, however, a number of limitations inherent in this approach. Perhaps the most serious of these is that the investigator must in general confine himself to a study of non-lethal heritable characters. Such characters are likely to involve more or less non-essential so-called "terminal" reactions.[5] The selection of these for genetic study was perhaps responsible for the now rapidly disappearing belief that genes are concerned only with the control of "superficial" characters. A second difficulty, not unrelated to the first, is that the standard approach to the problem implies the use of characters with visible manifestations. Many such characters involve morphological variations, and these are likely to be based on systems of biochemical reactions so complex as to make analysis exceedingly difficult.

Considerations such as those just outlined have led us to investigate

TABLE 1

Growth of Pyridoxinless Strain of *N. sitophila* on Liquid Medium Containing Inorganic Salts,[9] 1% Sucrose, and 0.004 Microgram Biotin per Cc. Temperature 25°C. Growth Period, 6 Days from Inoculation with Conidia

MICROGRAMS B_6 PER 25 CC. MEDIUM	STRAIN	DRY WEIGHT MYCELIA, MG.
0	Normal	76.7
0	Pyridoxinless	1.0
0.01	"	4.2
0.03	"	5.7
0.1	"	13.7
0.3	"	25.5
1.0	"	81.1
3.0	"	81.1
10.0	"	65.4
30.0	"	82.4

the general problem of the genetic control of developmental and metabolic reactions by reversing the ordinary procedure and, instead of attempting to work out the chemical bases of known genetic characters, to set out to determine if and how genes control known biochemical reactions. The ascomycete *Neurospora* offers many advantages for such an approach and is well suited to genetic studies.[6] Accordingly, our program has been built around this organism. The procedure is based on the assumption that x-ray treatment will induce mutations in genes concerned with the control of known specific chemical reactions. If the organism must be able to carry out a certain chemical reaction to survive on a given medium, a mutant unable to do this will obviously be lethal on this medium. Such a mutant can be maintained and studied, however, if it will grow on a medium to which has been added the essential product of the genetically blocked reaction. The experimental procedure based on this reasoning can best be illustrated by considering a hypothetical example. Normal strains of *Neurospora crassa* are able to use sucrose as a carbon source, and are therefore able to carry out the specific and enzymatically controlled

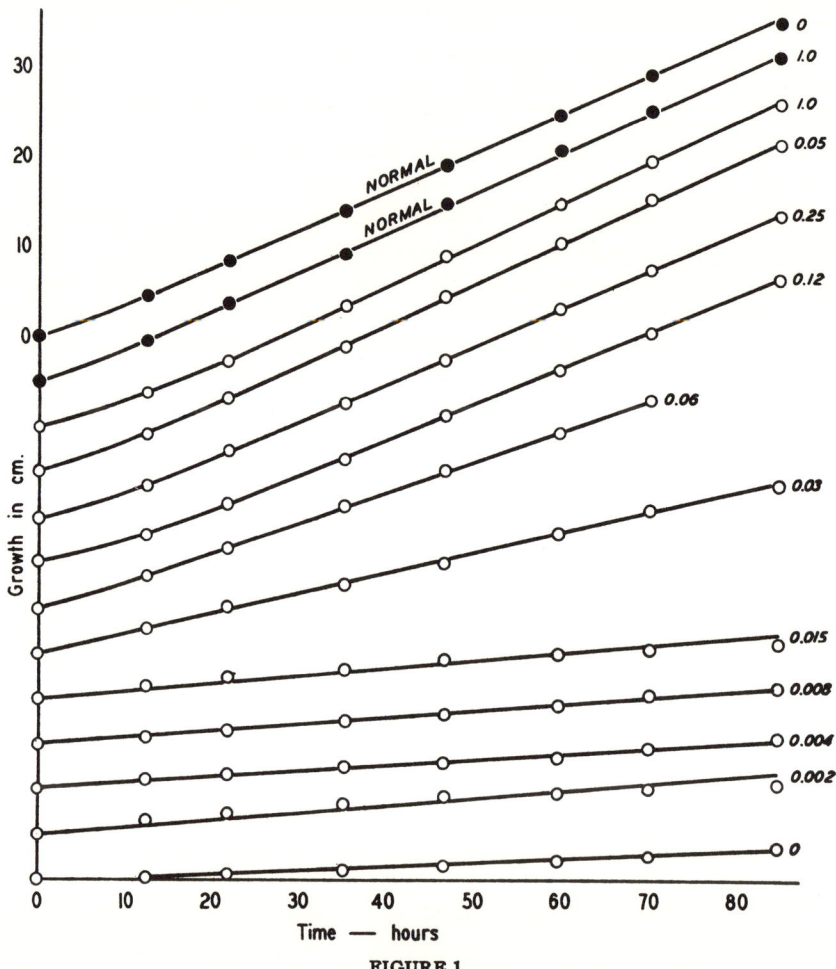

FIGURE 1

Growth of normal (top two curves) and pyridoxinless (remaining curves) strains of *Neurospora sitophila* in horizontal tubes. The scale on the ordinate is shifted a fixed amount for each successive curve in the series. The figures at the right of each curve indicate concentration of pyridoxine (B_6) in micrograms per 25 cc. medium.

reaction involved in the hydrolysis of this sugar. Assuming this reaction to be genetically controlled, it should be possible to induce a gene to mutate to a condition such that the organism could no longer carry out sucrose hydrolysis. A strain carrying this mutant would then be unable to grow on a medium containing sucrose as a sole carbon source but should be able to grow on a medium containing some other normally utilizable carbon source. In other words, it should be possible to establish and maintain such a mutant strain on a medium containing glucose and detect its inability to utilize sucrose by transferring it to a sucrose medium.

Essentially similar procedures can be developed for a great many metabolic processes. For example, ability to synthesize growth factors (vitamins), amino acids and other essential substances should be lost through gene mutation if our assumptions are correct. Theoretically, any such metabolic deficiency can be "by-passed" if the substance lacking can be supplied in the medium and can pass cell walls and protoplasmic membranes.

In terms of specific experimental practice, we have devised a procedure in which x-rayed single-spore cultures are established on a so-called "complete" medium, i.e., one containing as many of the normally synthesized constituents of the organism as is practicable. Subsequently these are tested by transferring them to a "minimal" medium, i.e., one requiring the organism to carry on all the essential syntheses of which it is capable. In practice the complete medium is made up of agar, inorganic salts, malt extract, yeast extract and glucose. The minimal medium contains agar (optional), inorganic salts and biotin, and a disaccharide, fat or more complex carbon source. Biotin, the one growth factor that wild type *Neurospora* strains cannot synthesize,[7] is supplied in the form of a commercial concentrate containing 100 micrograms of biotin per cc.[8] Any loss of ability to synthesize an essential substance present in the complete medium and absent in the minimal medium is indicated by a strain growing on the first and failing to grow on the second medium. Such strains are then tested in a systematic manner to determine what substance or substances they are unable to synthesize. These subsequent tests include attempts to grow mutant strains on the minimal medium with (1) known vitamins added, (2) amino acids added or (3) glucose substituted for the more complex carbon source of the minimal medium.

Single ascospore strains are individually derived from perithecia of *N. crassa* and *N. sitophila* x-rayed prior to meiosis. Among approximately 2000 such strains, three mutants have been found that grow essentially normally on the complete medium and scarcely at all on the minimal medium with sucrose as the carbon source. One of these strains (*N. sitophila*) proved to be unable to synthesize vitamin B_6 (pyridoxine). A second strain (*N. sitophila*) turned out to be unable to synthesize vitamin B_1 (thiamine). Additional tests show that this strain is able to synthesize the pyrimidine half of the B_1 molecule but not the thiazole half. If thiazole alone is added to the minimal medium, the strain grows essentially normally. A third strain (*N. crassa*) has been found to be unable to synthesize para-aminobenzoic acid. This mutant strain appears to be entirely normal when grown on the minimal medium to which p-aminobenzoic acid has been added. Only in the case of the "pyridoxinless" strain has an analysis of the inheritance of the induced metabolic defect been investigated. For this reason detailed accounts of the thiamine-deficient and p-aminobenzoic acid-deficient strains will be deferred.

Qualitative studies indicate clearly that the pyridoxinless mutant, grown on a medium containing one microgram or more of synthetic vitamin B_6 hydrochloride per 25 cc. of medium, closely approaches in rate and characteristics of growth normal strains grown on a similar medium with

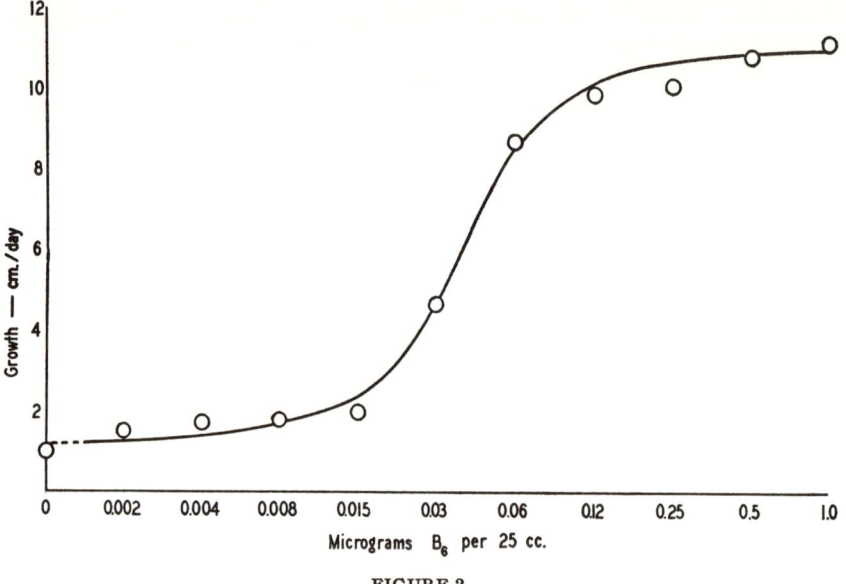

FIGURE 2

The relation between growth rate (cm./day) and vitamin B_6 concentration.

no B_6. Lower concentrations of B_6 give intermediate growth rates. A preliminary investigation of the quantitative dependence of growth of the mutant on vitamin B_6 in the medium gave the results summarized in table 1. Additional experiments have given results essentially similar but in only approximate quantitative agreement with those of table 1. It is clear that additional study of the details of culture conditions is necessary before rate of weight increase of this mutant can be used as an accurate assay for vitamin B_6.

It has been found that the progression of the frontier of mycelia of *Neurospora* along a horizontal glass culture tube half filled with an agar medium provides a convenient method of investigating the quantitative effects of growth factors. Tubes of about 13 mm. inside diameter and about 40 cm. in length are used. Segments of about 5 cm. at the two ends are turned up at an angle of about 45°. Agar medium is poured in so as to fill the tube about half full and is allowed to set with the main segment of the tube in a horizontal position. The turned up ends of the tube are stoppered with cotton plugs. Inoculations are made at one end of the agar surface and the position of the advancing front recorded at convenient intervals. The frontier formed by the advancing mycelia is remarkably well defined, and there is no difficulty in determining its position to within a millimeter or less. Progression along such tubes is strictly linear with time and the rate is independent of tube length (up to 1.5 meters). The rate is not changed by reducing the inside tube diameter to 9 mm., or by

TABLE 2

RESULTS OF CLASSIFYING SINGLE ASCOSPORE CULTURES FROM THE CROSS OF PYRIDOXINLESS AND NORMAL *N. sitophila*

ASCUS NUMBER	1	2	3	4	5	6	7	8
17	—	pdx	pdx	pdx	N	N	N	—
18	—	—	N	N	—	—	pdx	pdx
19	—	pdx	—	—	—	—	—	N
20	—	—	N	—	—	—	—	pdx
22	—	—	N	—	—	—	—	—
23	—	*	*	*	N	N	pdx	pdx
24	N	N	N	N	pdx	pdx	pdx	pdx

N, normal growth on B_6-free medium. *pdx*, slight growth on B_6-free medium. Failure of ascospore germination indicated by dash.

* Spores 2, 3 and 4 isolated but positions confused. Of these, two germinated and both proved to be mutants.

sealing one or both ends. It therefore appears that gas diffusion is in no way limiting in such tubes.

The results of growing the pyridoxinless strain in horizontal tubes in which the agar medium contained varying amounts of B_6 are shown graphically in figures 1 and 2. Rate of progression is clearly a function of vitamin B_6 concentration in the medium.[10] It is likewise evident that there is no significant difference in rate between the mutant supplied with B_6 and the normal strain growing on a medium without this vitamin. These results are consistent with the assumption that the primary physiological difference between pyridoxinless and normal strains is the inability of the former to carry out the synthesis of vitamin B_6. There is certainly more than one step in this synthesis and accordingly the gene differential involved is presumably concerned with only one specific step in the biosynthesis of vitamin B_6.

In order to ascertain the inheritance of the pyridoxinless character, crosses between normal and mutant strains were made. The techniques for hybridization and ascospore isolation have been worked out and described by Dodge, and by Lindegren.[6] The ascospores from 24 asci of the cross were isolated and their positions in the asci recorded. For some unknown reason, most of these failed to germinate. From seven asci, however, one or more spores germinated. These were grown on a medium containing glucose, malt-extract and yeast extract, and in this they all grew normally. The normal and mutant cultures were differentiated by growing them on a B_6 deficient medium. On this medium the mutant cultures grew very little, while the non-mutant ones grew normally. The results are summarized in table 2. It is clear from these rather limited data that this inability to synthesize vitamins B_6 is transmitted as it should be if it were differentiated from normal by a single gene.

The preliminary results summarized above appear to us to indicate that the approach outlined may offer considerable promise as a method of learning more about how genes regulate development and function. For example, it should be possible, by finding a number of mutants unable to carry out a particular step in a given synthesis, to determine whether only one gene is ordinarily concerned with the immediate regulation of a given specific chemical reaction.

It is evident, from the standpoints of biochemistry and physiology, that the method outlined is of value as a technique for discovering additional substances of physiological significance. Since the complete medium used can be made up with yeast extract or with an extract of normal *Neurospora*, it is evident that if, through mutation, there is lost the ability to synthesize an essential substance, a test strain is thereby made available for use in isolating the substance. It may, of course, be a substance not previously known to be essential for the growth of any organism. Thus we may expect to discover new vitamins, and in the same way, it should be possible to discover additional essential amino acids if such exist. We have, in fact, found a mutant strain that is able to grow on a medium containing Difco yeast extract but unable to grow on any of the synthetic media we have so far tested. Evidently some growth factor present in yeast and as yet unknown to us is essential for *Neurospora*.

Summary.—A procedure is outlined by which, using *Neurospora*, one can discover and maintain x-ray induced mutant strains which are characterized by their inability to carry out specific biochemical processes.

Following this method, three mutant strains have been established. In one of these the ability to synthesize vitamin B_6 has been wholly or largely lost. In a second the ability to synthesize the thiazole half of the vitamin B_1 molecule is absent, and in the third para-aminobenzoic acid is not synthesized. It is therefore clear that all of these substances are essential growth factors for *Neurospora*.[11]

Growth of the pyridoxinless mutant (a mutant unable to synthesize vitamin B_6) is a function of the B_6 content of the medium on which it is grown. A method is described for measuring the growth by following linear progression of the mycelia along a horizontal tube half filled with an agar medium.

Inability to synthesize vitamin B_6 is apparently differentiated by a single gene from the ability of the organism to elaborate this essential growth substance.

NOTE: Since the manuscript of this paper was sent to press it has been established that inability to synthesize both thiazole and *p*-aminobenzoic acid are also inherited as though differentiated from normal by single genes.

* Work supported in part by a grant from the Rockefeller Foundation. The authors are indebted to Doctors B. O. Dodge, C. C. Lindegren and W. S. Malloch for stocks and for advice on techniques, and to Miss Caryl Parker for technical assistance.

[1] The possibility that genes may act through the mediation of enzymes has been suggested by several authors. See Troland, L. T., *Amer. Nat.*, **51**, 321–350 (1917); Wright, S., *Genetics*, **12**, 530–569 (1927); and Haldane, J. B. S., in *Perspectives in Biochemistry*, Cambridge Univ. Press, pp. 1–10 (1937), for discussions and references.

[2] Onslow, Scott-Moncrieff and others, see review by Lawrence, W. J. C., and Price, J. R., *Biol. Rev.*, **15**, 35–58 (1940).

[3] Winge, O., and Laustsen, O., *Compt. rend. Lab. Carlsberg, Serie physiol.*, **22**, 337–352 (1939).

[4] See Goldschmidt, R., *Physiological Genetics*, McGraw-Hill, pp. 1–375 (1939), and Beadle, G. W., and Tatum, E. L., *Amer. Nat.*, **75**, 107–116 (1941) for discussion and references.

[5] See Sturtevant, A. H., and Beadle, G. W., *An Introduction to Genetics*, Saunders, pp. 1–391 (1931), and Beadle, G. W., and Tatum, E. L., loc. cit., footnote 4.

[6] Dodge, B. O., *Jour. Agric. Res.*, **35**, 289–305 (1927) and Lindegren, C. C., *Bull. Torrey Bot. Club*, **59**, 85–102 (1932).

[7] In so far as we have carried them, our investigations on the vitamin requirements of *Neurospora* corroborate those of Butler, E. T., Robbins, W. J., and Dodge, B. O., *Science*, **94**, 262–263 (1941).

[8] The biotin concentrate used was obtained from the S. M. A. Corporation, Chagrin Falls, Ohio.

[9] Throughout our work with *Neurospora*, we have used as a salt mixture the one designated number 3 by Fries, N., *Symbolae Bot. Upsalienses*, Vol. 3, No. 2, 1–188 (1938). This has the following composition: NH_4 tartrate, 5 g.; NH_4NO_3, 1 g.; KH_2PO_4, 1 g.; $MgSO_4 \cdot 7H_2O$, 0.5 g.; NaCl, 0.1 g.; $CaCl_2$, 0.1 g.; $FeCl_3$, 10 drops 1% solution; H_2O, 1 l. The tartrate cannot be used as a carbon source by *Neurospora*.

[10] It is planned to investigate further the possibility of using the growth of *Neurospora* strains in the described tubes as a basis of vitamin assay, but it should be emphasized that such additional investigation is essential in order to determine the reproducibility and reliability of the method.

[11] The inference that the three vitamins mentioned are essential for the growth of normal strains is supported by the fact that an extract of the normal strain will serve as a source of vitamin for each of the mutant strains.

ATROPINESTERASE, A GENETICALLY DETERMINED ENZYME IN THE RABBIT

By Paul B. Sawin* and David Glick†

Brown University, Providence, R. I., and the Laboratories of Newark Beth Israel Hospital, Newark, N. J.

Communicated January 12, 1943

Introduction.—Hereditary differences in enzyme activity are well known among both plants and animals, particularly in so far as they have to do with irregularities in pigment formation.[10,13] In animals such differences are not restricted to pigment formation but have to do with many other physiological activities such as the xanthophyllase activity which determines the presence or absence of yellow color of body fat in the rabbit,[1,11] the amylase activity in the digestive juices of silkworm larvae[7] and the uricase activity which determines the amount of nitrogen excreted as uric acid in the dog.[5] In man alkaptonuria, steatorrhea, hematoporphyria, pentosuria and cystinuria are other biochemical differences which are suspected to have a similar background.

Recently interest has been revived in the observation, first reported by Fleischmann[3] and subsequently confirmed by others in various parts of the world, that the blood of certain rabbits can destroy atropine while that of others cannot. Studies of some properties of the enzyme in the serums of those rabbits capable of hydrolyzing atropine have been carried out by Glick,[4] and Glick and Glaubach[5] have investigated the distribution of the atropinesterase among certain tissues in these animals. The possibility that atropinesterase in rabbit serum is an inherited factor was recognized by Levy and Michel[7] but no supporting data have been forthcoming. The present coöperative investigation is an attempt to determine the genetic properties of the enzyme.

Methods.—Blood samples of sufficient volume to provide 0.5 cc. of serum have been obtained from animals at three months of age or older of the proper matings in the laboratory at Brown, centrifuged to remove the cells and placed in the mail for examination within three days at Newark. The enzyme activity was determined by the manometric method employing the Warburg apparatus in the manner previously described.[4,5] The activity was expressed as atropinesterase units per 100 mg. of serum. The unit is defined as the amount of enzyme required to liberate 1 c. mm. CO_2 in 300 minutes at 30° in a total volume of 4 ml. in the bicarbonate-Ringer medium containing a concentration of substrate (0.25% atropine sulphate) sufficient to achieve the maximum rate of hydrolysis.

Data and Discussion.—Preliminary examination of five unrelated families showed that about 55 per cent of the 181 rabbits examined possessed the atropinesterase. This is a considerably higher proportion than previously obtained by other investigators and suggests the existence of racial differences, a fact that becomes more apparent when this population is separated into its constituent parts. Two of the families—III, a New Zealand White race which has been closely bred for more than 10 generations, and the *A* race of Castle, a small-sized multiple-recessive strain—appear to lack the enzyme entirely. The other three families possess it in a high proportion of the individuals. Family V, a pure chinchilla race, also closely bred for 12 generations, and family X, a more heterogeneous race originated by hybridization of the small race with other genetic stocks, produce both enzyme-possessing and non-enzyme-possessing offspring. In the first family, among 50 individuals tested the proportion was equal; in the second, 75 per cent of 87 animals examined were atropinesterase-producing. Only 14 individuals have been examined in family II*c*, of which three did not possess the enzyme. Of these families, III, V and X at least, show significant genetic differences. No entire family was found which does not produce some individuals lacking the enzyme.

The evidence as to the hereditary nature of these differences is as follows. In a total of 69 offspring obtained from parents lacking the enzyme, none have shown any trace of the enzyme. Parents possessing the enzyme, on the other hand, may or may not transmit the character to their offspring. If those which do possess it are paired with mates which do not, one of two alternatives follows. Either the offspring *all* possess the enzyme or they are about equally divided between those which possess it and those which do not. This indicates homozygosity in the first case and heterozygosity in the second case, of the parent possessing the character. Thirty-seven individuals have been obtained from the first type of mating, all of which were positive. In a population of 173 individuals obtained from the second type of mating, 88 possessed the enzyme in their blood and 85 did not, which is a close approximation to the equality expected from a monohybrid backcross. From known hybrids mated with each other, 68 young possessed the enzyme and 30 lacked it. This departure from the expected 3:1 ratio is not significant since it is less than twice the probable error.

From these results it appears that the ability to produce the enzyme is dominant over its absence. Dominance is probably not complete, however, since animals known to be heterozygous show a lower mean value of enzyme production than those which are homozygous. For 25 animals known to be heterozygous the mean value was 107 with a range of 52–174. For 4 animals known to be homozygous the mean value was 271 with a range of 232–348.

It is also interesting to note that the enzyme is not present at birth but first manifests itself at one to two months of age. Forty-five animals have been bled from the heart at birth and at monthly intervals two or three times thereafter. Thirty-nine of these possessed at least one homozygous parent and hence could be expected to have the enzyme in their blood at three months of age. Six were from a heterozygous male and a non-enzyme mother. None of those examined at birth had demonstrable activity in their blood. Twelve out of 16 of the individuals of the first group, examined at the end of the first month, possessed the enzyme in substantial amounts and two others had a trace. All of them were active at subsequent examinations. Of the second group none possessed the enzyme at the end of the first month but one of them did show it at later examinations.

Similar quantitative differences and also differences in the age of production of the enzyme are apparent in the amylase of the digestive juice of the silkworm,[7] and in several enzymes of the pig.[8]

Through information supplied by Dr. C. I. Wright of the National Institute of Health, Bethesda, Md., who had previously discovered the enzy-

TABLE 1

EVIDENCE FOR GENETIC LINKAGE BETWEEN GENES As AND E

Backcross progeny from F^1 double heterozygote × double recessive

	NON-CROSSOVERS		CROSSOVERS		POPULATION
Coupling	$\frac{E\,As}{10}$	$\frac{e\,as}{12}$	$\frac{E\,as}{4}$	$\frac{e\,As}{4}$	30
Repulsion	$\frac{E\,as}{4}$	$\frac{e\,As}{5}$	$\frac{E\,As}{1}$	$\frac{e\,as}{2}$	12
Totals	31		11		42

The deviation from equality, 10, is 4.5 times the P.E., 2.19, and so clearly significant. The indicated crossover percentage is 26.2 ± 5.2.

matic hydrolysis of certain morphine derivatives,[12] there is reason to believe that the enzyme which hydrolyzes atropine may be identical with that which hydrolyzes monoacetylmorphine. Nineteen serums from our laboratory and 8 from his own have been examined by Dr. Wright. Each sample was examined for the hydrolysis of both substrates under the same conditions. All of the 17 individual bloods which hydrolyzed the one compound also hydrolyzed the other, whereas those of 10 individuals hydrolyzed neither. According to Dr. Wright, the order of activity is the same when the serums are arranged as to enzyme concentration for either "monoacetylmorphinease" or "atropinesterase." The initial rate of hydrolysis of the former is somewhat greater than for the latter but the time required for complete hydrolysis is approximately the same, due to the difference in the order of the two reactions.

For the gene responsible for atropinesterase production we shall use the symbol As and for its recessive allele, as. This gene apparently is borne on the same chromosome of the rabbit as is the gene E for the extension of black pigment in the coat, for in matings between a double recessive individual and an F_1 double heterozygote, crossover recombinations are significantly fewer than non-crossovers, the two classes being 11 and 31, respectively (table 1), whereas equality would be expected if no linkage existed. Tentatively the genes As and E are regarded as members of a sixth linkage group of the rabbit. Segregation of both of these pairs of genes in this population is entirely regular when they are considered separately.

We regret that due to the national emergency, which has made it impossible for one of us to continue the work, a more accurate determination of the actual strength of this linkage association cannot be obtained at this time. In comparison, however, similar data have been obtained from crosses involving combinations of As and the agouti gene A and the color gene C. Nine and 8 offspring, respectively, have been obtained from these matings and the offspring are as equally divided between crossover and non-crossover combinations as possible.

Considering the population as a whole there is an obvious tendency for females to manifest greater enzyme activity than males, although no significant difference in the distribution of the character to the two sexes is manifest. Close scrutiny of the individual matings, however, reveals that in the backcross matings ($Asas \times asas$) this tendency is quite pronounced, the average male and female titers being 98 and 129, respectively, and there is a statistically significant tendency for females to possess the character more often than males. A similar situation occurs in the F_2 although less pronounced. The full meaning of this observation is not apparent. It seems probable, however, that possession of atropinesterase is not in itself a sex-linked character since in the same backcross population sufficient numbers have been obtained to indicate that it makes no difference whether the As gene is derived from the mother or the father. It seems more probable that this peculiarity is the result of a difference in the genetic milieu of the two sexes, perhaps acting secondarily through the medium of the sex hormones. It is interesting to compare these results with the similar behavior of cholinesterase in rats and mice in which age and also sex are important factors.[2]

Conclusion.—Rabbits which have in their blood serum an enzyme capable of hydrolyzing atropine (and monoacetylmorphine) inherit that peculiarity in a gene (As) borne in the same chromosome as the gene (E) for the extension of black pigment in the coat. The gene (As) is incompletely dominant, homozygotes producing the enzyme more effectively than heterozygotes. The enzyme is not present at birth but appears first at about one month of age, and tends to occur in greater concentration in females and to be demonstrable in a higher percentage of them than in males.

* Aided by a grant from the Rockefeller Foundation.

† Aided by a grant from the Sidney C. Keller Research Fund. Present address: Chemistry Department, Mount Sinai Hospital, New York, N. Y.

[1] Castle, W. E., *Proc. Nat. Acad. Sci.*, **19**, 947 (1933).
[2] Beveridge, J. M. R., and Lucas, C. C., *Science*, **93**, 356–357 (1941).
[3] Fleischmann, P., *Arch. Exp. Path. Pharm.*, **62**, 518 (1910).
[4] Glick, D., *Jour. Biol. Chem.*, **134**, 617–625 (1940).
[5] Glick, D., and Glaubach, S., *Jour. Gen. Physiol.*, **25**, 197–205 (1941).
[6] Keeler, C. E., and Trimple, H. C., *Jour. Hered.*, **29**, 281–289 (1937).
[7] Levy, J., and Michel, E., *Compt. Rend. Soc. Biol.*, **129**, 820–822 (1938).
[8] Matsumura, S., *Nagairo Sericult. Exp. Sta. Bull.* 28 (1934).
[9] Mendel and Mitchell (cited by H. G. Wells), *Arch. Path.*, **9**, 1055 (1930).
[10] Onslow, H. W., *Biochem. Jour.*, **17**, 334, 567 (1923).
[11] Pease, M. S., *Verh. V. Int. Kong. Ver., Ztschr. Ind. Abst. u. Ver. Suppl.* **2**, 1153 (1927).
[12] Wright, C. I., *Jour. Pharmacol. Exp. Therap.*, **71**, 164–177 (1941).
[13] Wright, S., *Physiol. Rev.*, **21**, 487–527 (1941).

ON THE EVOLUTION OF BIOCHEMICAL SYNTHESES

By N. H. Horowitz

School of Biological Sciences, Stanford University, Calif.

Communicated April 23, 1945

Although it has been recognized for a long time that the biochemistry of the organism is conditioned by its genetic constitution, a more precise definition of this dependence has not been possible until recently. A considerable amount of evidence now exists for the view that there is a one-to-one correspondence between genes and biochemical reactions. This concept, foreshadowed in the work of Garrod[1] on human alcaptonuria, accounts in a satisfactory way for the inheritance of pigment formation in guinea pigs,[2] insects[3] and flowers,[4] and the synthesis of essential growth factors in *Neurospora*.[5] It appears from these studies that each synthesis is controlled by a set of non-allelic genes, each gene governing a different step in the synthesis. As to the nature of this control, it is probable that the primary action of the gene is concerned with enzyme production. That genes can direct the specificities of proteins has been shown in the case of many antigens,[6] while several mutations demonstrably affecting the production of enzymes have been reported.[6] Evidence on the postulated gene-enzyme relationship is in most cases, however, still circumstantial; this is partly because of technical difficulties involved in the study of synthetic, or free-energy consuming reactions *in vitro*, and partly because of the insufficiency of biochemical information on those reactions which happen to be susceptible of genetic analysis.

As a corollary of the above hypothesis, each biosynthesis depends on the direct participation of a number of genes equal to the number of different, enzymatically catalyzed steps in the reaction chain. In attempting to account for the evolutionary development of such a reaction chain one meets in a clear form the problem of explaining macroevolutionary changes in terms of microevolutionary steps. The individual reactions making up the chain are of value to the organism only when considered collectively and in view of the ultimate product. Regarded individually, intermediate substances cannot, in general, be assumed to have physiological significance, and the ability to produce them does not of itself confer a selective advantage. An example from *Neurospora* genetics will serve to illustrate this point. At the present time seven different genes are known to be concerned in the synthesis of arginine by the mold.[7] The inactivation of any one prevents the synthesis from taking place. On the basis of the above hypothesis, at least seven different catalyzed steps must occur in the synthesis. Several of the steps have been identified and controlling genes assigned to each. Two of the intermediates in the chain have been shown to be the amino acids ornithine and citrulline. Unlike arginine, neither of these substances is a general constituent of proteins. Aside from their function as precursors, they are apparently of no further use to the organism.

While the above example probably represents the general case, there are also well-known instances in which precursors serve independent functions. Thus, arginine, glycine and methionine are precursors of creatine in the rat,[8] but the synthesis goes through the non-functional intermediate, glycocyamine. On the other hand, acetylcholine may be synthesized from choline in one step.[9] In cases such as these, the problem is that of accounting for the synthesis of the precursors.

Since natural selection cannot preserve non-functional characters, the most obvious implication of the facts would seem to be that a stepwise evolution of biosyntheses, by the selection of a single gene mutation at a time, is impossible. It will be shown below that this is not a necessary conclusion, but that under special conditions the stepwise evolution of long-chain syntheses may occur. First, however, an alternative to stepwise evolution will be considered; that is, the origin of a new reaction chain through the chance combination of the necessary genes.

Although the probability of the origin of a useful character through the chance association of many genes may be small, it is never zero. Indeed, a consideration of the statistical consequences of the interaction of mutation, Mendelian inheritance, and natural selection has led Wright[10] to the conclusion that such chance associations may be of major importance in evolution. He has analyzed the evolutionary possibilities of various types of breeding structures and has shown that under certain conditions an extensive trial and error mechanism exists, whereby the species can test numerous combinations of non-adaptive genes. The breeding structure which most favors this type of evolution is that of a large population divided into many small, partially isolated groups. Within each group the cumulative effects of the accidents of sampling among the gametes are of major significance in determining gene frequencies, but the penalty of fixation of deleterious genes, ordinarily incurred under inbreeding, is avoided by exchange of migrants with other groups. The pressures of forward and reverse mutations, which between them determine an equilibrium frequency for non-adaptive genes in large, random-breeding populations, become of minor importance. As a consequence, a random drift of gene frequencies occurs. If, by chance, one group finds a particularly favorable combination of genes, a process of intergroup selection comes into play, whereby the favorable combination is spread to the population at large.

This model provides a means for the evolution of a new gene combination in spite of unfavorable mutation rates to active alleles and in the absence of selection of individual genes. It is thus favorable for the evolution of systems of individually non-adaptive, but collectively adaptive, genes. The effectiveness of the process would seem to be strongly dependent on the size of the gene combination required, however, decreasing approximately exponentially with increasing numbers of genes, other factors remaining constant. There would result a tendency toward the evolution of short reaction chains involving the recombination of molecular units already available. There is no doubt that a conservative tendency of this sort actually exists in nature. The wide variety of biologically important compounds built up on the pyrrole nucleus, to mention but one example, is a case in point.

The application of Wright's theory to the particular problem under consideration is limited by the fact that it operates only under biparental reproduction. It is probable that a large number of basic syntheses evolved prior to sexual reproduction. The universal distribution among living forms of certain classes of compounds—viz., the amino acids, nucleotides and probably the B vitamins—identifies them as essential ingredients of living matter. The synthesis of these substances must have evolved very early in geologic time, as a necessary condition for further progress, although loss of certain syntheses may have occurred in the later differentiation of some forms. It is therefore desirable to search for another solution of the problem applicable to compounds of this type, preferably one in which a minimum burden is placed on chance and a maximum one on directed evolutionary forces. It is thought that the following suggestion, while definitely a speculation, offers a possible solution along these lines.

In essence, the proposed hypothesis states that the evolution of the basic syntheses proceeded in a stepwise manner, involving one mutation at a time, but that the order of attainment of individual steps has been in the reverse direction from that in which the synthesis proceeds—i.e., the last step in the chain was the first to be acquired in the course of evolution, the penultimate step next, and so on. This process requires for its operation a special kind of chemical environment; namely, one in which end-products and potential intermediates are available. Postponing for the moment the question of how such an environment originated, consider the operation of the proposed mechanism. The species is at the outset assumed to be heterotrophic for an essential organic molecule, A. It obtains the substance from an environment which contains, in addition to A, the substances B and C, capable of reacting in the presence of a catalyst (enzyme) to give a molecule of A. As a result of biological activity, the amount of available A is depleted to a point where it limits the further growth of the species. At this point, a marked selective advantage will be enjoyed by mutants which are able to carry out the reaction $B + C = A$. As the external supplies of A are further reduced, the mutant strain will gain a still greater selective advantage, until it eventually displaces the parent strain from the population. In the A-free environment a back mutation to the original stock will be lethal, so we have at the same time a theory of lethal genes. The majority of biochemical mutations in *Neurospora* are lethals of this type.

In time, B may become limiting for the species, necessitating its synthesis from other substances, D and E; the population will then shift to one characterized by the genotype $(D + E = B, B + C = A)$. Given a sufficiently complex environment and a proportionately variable germ plasm, long reaction chains can be built up in this way. In the event that B and C become limiting more or less simultaneously, another possibility is opened. Under these circumstances symbiotic associations of the type $(F + G \neq C, D + E = B)(F + G = C, D + E \neq B)$ will have adaptive value.

This model is thus seen to have potentialities for the rapid evolution of long chain syntheses in response to changes in the environment. As has been pointed out by Oparin[11] the hypothesis of a complex chemical environment is a necessary corollary of the concept of the origin of life through chemical means. The essential point of the argument is that it is inconceivable that a self-reproducing unit of the order of complexity of a nucleoprotein could have originated by the chance combination of inorganic molecules. Rather, a period of evolution of organic substances of ever-increasing degree of complexity must have intervened before such an event became a practical, as distinguished from a mathematical, probability. Or, put in another way, any random process which can have produced a nucleoprotein must at the same time have led to the production of a profusion of simpler structures. Oparin has considered in some detail the possible modes of origin of organic compounds from inorganic material and cites a number of known reactions of this type, together with evidences of their large-scale occurrence on the earth in past geologic ages. He concludes that in the absence of living organisms to destroy them highly complex organic systems can have developed. The first self-duplicating nucleoprotein originated as a step in this process of chemical evolution. The origin of living matter by physicochemical means thus presupposes the existence of a highly complex chemical environment.

To summarize, the hypothesis presented here suggests that the first living entity was a completely heterotropic unit, reproducing itself at the expense of prefabricated organic molecules in its environment. A depletion of the environment resulted until a point was reached where the supply of specific substrates limited further multiplication. By a process of mutation a means was eventually discovered for utilizing other available substances. With this event the evolution of biosyntheses began. The conditions necessary for the operation of the mechanism ceased to exist with the ultimate destruction of the organic environment. Further evolution was probably based on the chance combination of genes, resulting to a large extent in the development of short reaction chains utilizing substances whose synthesis had been previously acquired.

[1] Garrod, A. E., *Inborn Errors of Metabolism*, Oxford University Press (1923).

[2] Wright, S., *Biol. Symposia*, **6**, 337–355 (1942).

[3] Ephrussi, B., *Quart. Rev. Biol.*, **17**, 327–338 (1942).

[4] Lawrence, W. J. C., and Price, J. R., *Biol. Rev.*, **15**, 35–58 (1940).

[5] Horowitz, N. H., Bonner, David, Mitchell, H. K., Tatum, E. L., and Beadle, G. W., *Am. Nat.*, in press (1945).

[6] Summarized in Wright, S., *Physiol. Rev.*, **21**, 487–527 (1941).

[7] Srb, A., and Horowitz, N. H., *Jour. Biol. Chem.*, **154**, 129–139 (1944).

[8] Summarized in Schoenheimer, R., *The Dynamic State of Body Constituents*, Harvard University Press (1942).

[9] Lipmann, F., *Advances in Enzymology*, **1**, 99–162 (1941).

[10] Wright, S., *Bull. Am. Math. Soc.*, **48**, 223–246 (1942). Contains summary of earlier papers.

[11] Oparin, A. I., *The Origin of Life*, trans. by S. Morgulis, Macmillan, New York (1938).

HYDROXYANTHRANILIC ACID AS A PRECURSOR OF NICOTINIC ACID IN NEUROSPORA*

By HERSCHEL K. MITCHELL AND JOSEPH F. NYC

THE WILLIAM G. KERCKHOFF LABORATORIES OF THE BIOLOGICAL SCIENCES, CALIFORNIA INSTITUTE OF TECHNOLOGY, PASADENA, CALIFORNIA

Communicated by G. W. Beadle, November 17, 1947

Recent investigations in this laboratory[1] have provided evidence that the biosynthesis of nicotinic acid in *Neurspora* proceeds from tryptophane through the intermediate kynurenine I.

Kynurenine I

Further studies suggested that the pyridine ring of nicotinic acid might arise by ring closure of the α keto acid corresponding to kynurenine to give the naturally occurring compound kynurenic acid II[2] or, if preceded by oxidation, xanthurenic acid III.[3]

Kynurenic acid II   Xanthurenic acid III

In addition to these two compounds a series of nicotinic acid derivatives was synthesized and tested for growth promoting or growth inhibiting properties on *Neurospora* mutant 65001.[1] These compounds included: 3-carboxy-4-hydroxy pyridine, 3-carboxy-4-amino pyridine, 2-hydroxy-3-carboxy pyridine, 3-carboxy-6-hydroxy pyridine, 2,3-dicarboxy pyridine, 3,4-dicarboxy pyridine, 2,6-dimethyl-3,4-dicarboxy pyridine, 2,3,4-tricarboxy-6-methyl pyridine, 3-carboxy-4-chloro pyridine and 2,6-dimethyl-3-carboxy-4-chloro pyridine. In high concentrations, the compound 3-

carboxy-4-amino pyridine promoted a small amount of growth, but the remaining compounds possessed no stimulatory or inhibitory action under the conditions utilized.

From the above facts it was concluded that the pyridine ring of nicotinic acid does not arise from kynurenine through kynurenic acid II or xanthurenic acid III followed by oxidation of the benzene ring. It also appeared evident that the oxidation in position 8 of xanthurenic acid III precedes formation of the pyridine ring, a possible intermediate being 3-hydroxykynurenine IV.

$$\text{benzene ring with } -C(=O)-CH_2-CH(NH_2)-COOH, -NH_2, \text{ and } OH \text{ substituents}$$

Hydroxykynurenine IV

A consideration of this hypothetical compound suggested the possibility of biological oxidation to give 3-hydroxyanthranilic acid (2-amino-3-hydroxybenzoic acid) V instead of xanthurenic acid.

$$\text{benzene ring with } -COOH, -NH_2, \text{ and } OH \text{ substituents}$$

Hydroxyanthranilic acid V

A trimethyl derivative of this compound V is indeed found in nature as the alkaloid damascanine (2-methyl-amino-3-methoxy-methyl benzoate).[4,5] The alkaloid has been isolated from the seeds of two species of *Nigella* (common name of flowers, Love in a Mist).

It is the purpose of the experimental part of the present paper to present evidence that hydroxyanthranilic acid is an intermediate in the biological synthesis of nicotinic acid from tryptophane in *Neurospora*.

Hydroxyanthranilic acid has been synthesized in this laboratory by two independent methods. These methods and the proof of structure of the active compound will be presented elsewhere.

Experimental.—Media and conditions for growth of mutant 65001 have been previously described.[1,6] Growth curves for this mutant in the presence of nicotinamide and hydroxyanthranilic acid (filter sterilized) are presented in figure 1. For these experiments the pH of the medium was adjusted to 4.1 since hydroxyanthranilic acid, like nicotinic acid, is less active at a higher pH where dissociation is greater. In four days at a pH of 5.6 the compound is 50 to 70% as effective as nicotinamide in promoting growth. It is thus more effective than nicotinic acid at pH 5.6.[7]

The growth-promoting activity of hydroxyanthranilic acid on several

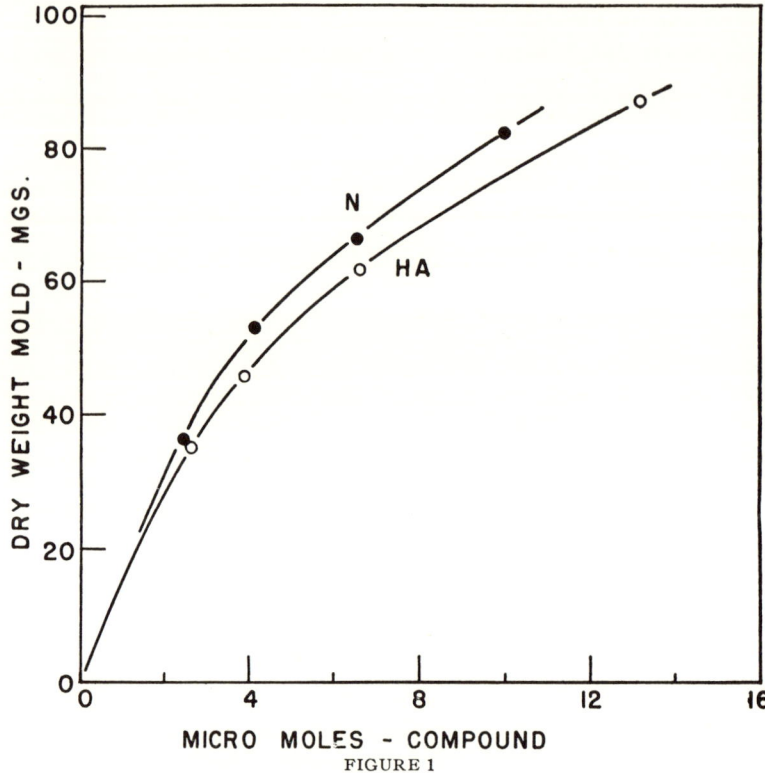

FIGURE 1

Growth curves of mutant 65001 (6½ days) in the presence of nicotinamide (curve N) and hydroxyanthranilic acid (curve HA).

genetically different mutants of *Neurospora* was compared to that of anthranilic acid, indole, tryptophane, kynurenine and nicotinamide, and qualitative data is presented in table 1.

TABLE 1

ACTIVITY OF HYDROXYANTHRANILIC ACID COMPARED TO ANTHRANILIC ACID, INDOLE, TRYPTOPHANE, KYNURENINE AND NICOTINAMIDE ON *Neurospora* MUTANTS

MUTANT STRAIN	ANTHRA- NILIC ACID	INDOLE	TRYPTO- PHANE	KYNURENINE	HYDROXY- ANTHRANILIC ACID	NICOTINA- MIDE
44008	−	+	+	+	+	+
65001	−	+	+	+	+	+
39401	−	+	+	+	+	+
4540	−	−	−	−	−	+
3416	−	−	−	−	−	+

It was previously shown[1] that an excess of a compound with nicotinamide activity is produced by mutant 65001 when it is grown in the presence of an excess of kynurenine. Similar experiments with hydroxyanthranilic acid are summarized in table 2. Nicotinic amide activity was determined by use of strain 3416 which does not utilize hydroxyanthranilic acid. Determinations were made on culture fluids of six-day-old cultures of 65001 and 44008 grown in the presence of various quantities of hydroxyanthranilic acid.

TABLE 2

PRODUCTION OF NICOTINAMIDE ACTIVITY FROM HYDROXYANTHRANILIC ACID BY MUTANTS 65001 AND 44008

HYDROXYANTHRANILIC ACID, μG. PER 20 ML.	DRY WEIGHT		NICOTINAMIDE ACTIVITY, μG. PER 20 ML. OF CULTURE FLUID	
	65001	44008	65001	44008
0	2	0	0	0
20	84	97	0	0
50	97	95	0	0
100	108	94	10	12
200	103	85	12	12

Discussion.—It is evident from the experimental data presented, that for certain *Neurospora* mutants, hydroxyanthranilic acid possesses growth-promoting activity that is quite comparable to the activity of nicotinamide. In addition it has been demonstrated that in the presence of an excess of hydroxyanthranilic acid two of the *Neurospora* mutants produce an excess of a substance with the biological activity of nicotinic acid or nicotinamide. This was determined by use of a mutant that utilizes either of the latter two compounds but does not utilize hydroxyanthranilic acid.

Thus it appears probable that this substance is a natural intermediate in the biological synthesis of nicotinic acid by the mold *Neurospora*. It is of interest to note the complex series of reactions that are required by the mold to convert anthranilic acid to hydroxyanthranilic acid. These reactions are illustrated schematically in figure 2.

No comparison has been made, in this laboratory, between the properties of hydroxyanthranilic acid and those of the nicotinic acid precursor from *Neurospora* described by Bonner and Beadle.[7] From the published data it can be seen that the molecular formula is similar. The isolated precursor, however, is reported to be a pyridine derivative. As such it would be expected to be further along in the series of reactions leading to nicotinic acid synthesis. In this connection it may be suggested that hydroxyanthranilic acid can be converted to nicotinic acid by oxidation and loss of carbon three of the compound, followed by ring closure or rearrangement in the six carbon amino acid residue. If this occurs in animals and in *Neurospora* the 3-carboxy-6-pyridone isolated by Knox and Grossman[8] may well be a by-product of the reaction. Similarly, the occurrence of damascanine in *Nigella* may be accounted for as resulting from a side reaction in the biosynthesis of nicotinic acid in the organism.

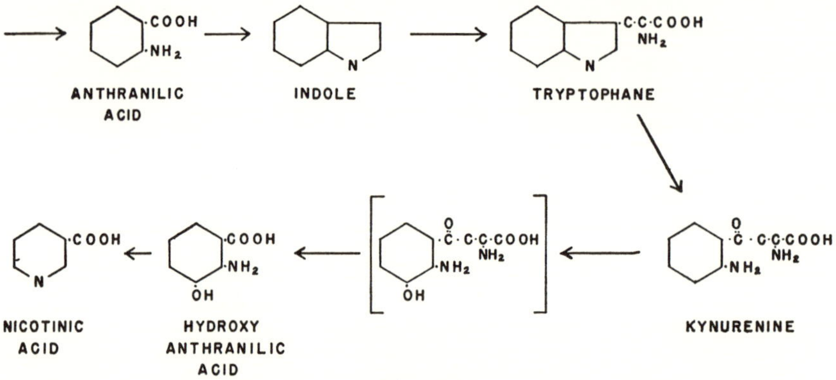

FIGURE 2

A schematic representation of a series of reactions leading to the biosynthesis of nicotinic acid in *Neurospora*.

Summary.—1. Evidence is presented to show that hydroxyanthranilic acid (2-amino-3-hydroxy benzoic acid) is an intermediate in the biosynthesis of nicotinic acid in *Neurospora*.

2. Several nicotinic acid derivatives and other related compounds are shown to lack significant biological activity.

* These investigations were supported by funds from the Rockefeller Foundation and the Williams-Waterman Fund for the Combat of Dietary Diseases.

[1] Beadle, G. W., Mitchell, H. K., and Nyc, J. F., *Proc. Nat. Acad. Sci.*, **33**, 155 (1947).
[2] Ellinger, A., *Ber.*, **37**, 1801 (1904).
[3] Lepkovsky, S., Roboz, E., and Haagen-Smit, A. J., *J. Biol. Chem.*, **149**, 195 (1943).
[4] Pommerehne, F., *Archiv. Pharm.*, **238**, 531 (1900).
[5] Keller, O., *Ibid.*, **246**, 1 (1908).
[6] Beadle, G. W., and Tatum, E. L., *Am. J. Bot.*, **32**, 678 (1945).
[7] Bonner, D. M., and Beadle, G. W., *Archiv. Biochem.*, **11**, 319 (1946).
[8] Knox, W. E., and Grossman, W. I., *J. Biol. Chem.*, **166**, 391 (1946).

13

Copyright © 1951 by The Biological Laboratory, Long Island Biological Association, Inc.

Reprinted from *Cold Spring Harbor Symp. Quant. Biol.*, **16**, 65–74 (1951)

SOME RECENT STUDIES BEARING ON THE ONE GENE-ONE ENZYME HYPOTHESIS[1]

N. H. HOROWITZ AND URS LEUPOLD

Kerckhoff Laboratories of Biology, California Institute of Technology, Pasadena

The assumption that a given gene is involved, in a primary way, in the production of but a single enzyme has been implicit in most speculations on the nature of gene action since Cuénot's time. As a result of the investigations of the last ten years stemming from the discovery of nutritional mutants in *Neurospora* by Beadle and Tatum (1941), one is now in a position to scrutinize this supposition more closely than was previously possible. Specifically, we are in a better position to trace the consequences of the hypothesis and of its various alternatives, and to appraise the evidence which may have a bearing on it. In this paper we propose to examine some of the evidence, deriving from studies on *Neurospora*, and *E. coli*, which relates to this problem.

Before considering the experimental findings, it may be useful to define more explicitly the meaning of the one gene-one enzyme hypothesis. The concept is that of a gene whose sole activity aside from self-duplication is that of functioning in the synthesis of a particular enzyme or enzyme precursor. It is not thereby implied that genes at other loci may not also function directly in the formation of the enzyme. This is a completely independent problem with which we are not concerned here, and regarding which there is little evidence in *Neurospora* one way or the other; all that can be said with assurance is that if two or more genes do, in fact, cooperate in the production of a given enzyme, then their respective contributions must be different. Nor does the one gene-one enzyme hypothesis imply that the final phenotypic expression of a mutation is necessarily restricted to a particular structure or function of the organism. The ultimate effect of a mutation is the result of an enormous magnification of the initial gene change, brought about through a system of reactions which, originating at the gene rapidly branches out in various directions and coalesces with similar networks deriving from other loci to form a reticulum of as yet indeterminate extent and complexity. It is impossible to decide from the end-effects alone whether the gene has one or many primary functions, since on either assumption a complex pattern of effects is expected in most cases. In the biochemical mutants of *Neurospora* and other microorganisms, the end effects would, if they could be analysed, undoubtedly prove to be exceedingly numerous. A mutation which induces a deficiency of an amino acid, for example, must secondarily affect the synthesis of virtually every protein of the cell, and an exhaustive enumeration of the end effects might well include every structure and function of the organism.

It turns out, however, that it is possible in such a case to prevent the secondary damage and the consequent death of the mutant by supplying the lacking amino acid. When given a sufficient quantity of the amino acid the mutant becomes normal in growth rate, morphology, and fertility. It is difficult to escape the conclusion that the sole function of the gene in this case is to play some essential role in the synthesis of the amino acid. When biochemical analysis of the mutant is carried farther, it is discovered that the field of action of the gene is even more circumscribed than might have been supposed: it is restricted to sensibly a single chemical step of the synthesis. Apparently a single reaction is abolished in the mutant, while all others proceed normally. It is inferred that the role of the gene is to function in the synthesis of the enzyme which catalyses this reaction.

It has not yet been possible to analyse all, or even the majority, of the known *Neurospora* mutants in the detail we have just outlined, while in a few cases the analysis has been carried still farther by showing that the mutants are in fact lacking in particular enzymes (Mitchell and Lein, 1948;

---

[1] The studies on *E. coli* reported in this paper were supported by a Grant-in-Aid from the American Cancer Society upon recommendation of the Committee on Growth of the National Research Council; by a grant from the Rockefeller Foundation; and by a contract between the Office of Naval Research, Department of the Navy and the California Institute of Technology (NR 164010).

Fincham, in press). Out of approximately 500 nutritional mutants which are, or have been, in the Pasadena collection, 84 per cent require single, known chemical substances for growth. The remaining 16 per cent have not responded to any of the individual substances tested, but do grow on complex media. It is very likely that many of the strains in this unanalysed group require individual compounds which have not been tested, others may have multiple requirements resulting from multiple mutations, while some may have multiple requirements resulting from the mutation of multifunctional genes.

The one gene-one enzyme hypothesis has been suggested as the simplest interpretation of the large class of mutants whose growth requirement is known to be satisfied by a single growth factor. Are there any grounds for suspecting that these mutants may not, in spite of appearances, represent mutations of unifunctional genes?

One basis for criticism of the one gene-one enzyme interpretation is the difficulty of excluding in every instance the alternative hypothesis that the given gene controls not one, but several sequential steps in the affected pathway. This interesting idea appears rather improbable, however, in view of the cumulative evidence from series of mutants which shows that each gene can be assigned to a particular step in a sequence of reactions; and it is virtually excluded in those cases where it has been possible, by enzyme studies, to define the reaction precisely.

A second ground for suspicion of the one gene-one enzyme interpretation rests on the fact that closer study of the mutants shows that they are not in all cases restored to a fully normal phenotype when supplied with the required growth substance. While all of the lethal consequences of the mutation are avoided, a residue of non-lethal effects may remain. In some cases, these residual effects are readily accounted for—partial sterility, for example, when the mutant carries a chromosomal rearrangement (McClintock, 1945). Others are not so easily understood. One of the commonest residual effects is a sensitivity toward certain natural substances—frequently amino acids. The first reported instance of this phenomenon was that described by Doermann (1944), who found that growth of all of the then known lysine-requiring mutants—a series involving at least three loci—is competitively inhibited by L-arginine in the medium. The growth of wild type *Neurospora* is not affected by arginine. Many similar cases are now known. A significant feature of this phenomenon is that the inhibition may, as in the case cited, extend to a whole class of genetically different, but biochemically related mutants, indicating that the effect is not locus-specific but is inherent in the mechanism of utilization of the exogenously provided growth substance. A third residual effect which has been encountered is sterility in crosses in which both parents carry the same mutant allele. One interpretation is that in these cases the gene performs a specific function, possibly independent of its vegetative biochemical function, in connection with zygote formation or maturation. Some recent preliminary results which have been obtained in our laboratory by Mr. Henry Gershowitz, working with certain methionine-requiring strains, indicate, however, that the sterility can be overcome by supplementing the medium with a large quantity of the amino acid—at least twice as much as is required to produce optimal growth of vegetative cultures. This suggests that the sterility may result from a high metabolic requirement for the growth factor during the sexual process, or to a lowered permeability to it. A fourth, and relatively rare, residual effect is failure of the mutant to attain a normal growth rate. This also can characterize an entire class of mutants, as in a certain group of strains of the cysteine-methionine series now under investigation in our laboratory.

Everything considered, it is perhaps surprising that residual effects are not observed more frequently, since in no event is it possible, even in theory, to avoid all of the consequences of a mutation by supplying the deficient metabolite. The block in the synthetic pathway still remains, and it can have an influence quite apart from the effects of the nutritional deficiency. It has been shown in numerous instances that metabolic intermediates may accumulate behind the block, sometimes in spectacular quantities (for review, see Horowitz, 1950). It would be surprising if the presence of abnormal concentrations of metabolic intermediates in the cells did not at times produce deleterious side-effects. Actually, evidence has been obtained both in *Neurospora* (Bonner, 1946a) and in *E. coli* (Davis, 1950; Umbarger and Mueller, 1951) that accumulated intermediates may exert a lethal action by interfering with reactions in other metabolic pathways. This leads one to suspect that the so-called residual effects are to a large extent the irreparable side-effects of the primary block. (For further discussion, see Emerson, 1950.)

THE SELECTION PROBLEM

At the Cold Spring Harbor Symposium of 1946, Delbrück raised a question as to whether incompatibilities with the one gene-one enzyme hypothesis could be detected even if they occurred (see Discussion following paper by Bonner, 1946b). Delbrück's argument was based on the recognized fact that not all of the mutations which are produced can be detected by the methods usually employed for this purpose. Principally three classes of biochemical mutants are not recoverable: (1) those requiring a substance which is absent from the so-called "complete" medium used for recovering nutritional mutants, (2) those requiring a substance which is unable to diffuse into the cell, and (3) those requiring a substance which, though present and diffusible, is not utilized because of the inclusion in the medium of an inhibitor of the mutant in question. We shall refer to mutants which, for the above, or for any other reasons, are incapable of growing on complete medium as mutants which have lost an *indispensable function*. The point of Delbrück's argument was that if any gene has more than one primary function, it is likely that at least one of these is an indispensable function; in which case mutation of the gene would not be detected.

Now the validity of this argument depends on the relative frequency of indispensable functions. If this frequency is very high, then the probability of recovering a mutation of a gene with several primary functions will be very low. Thus, if 90 per cent of gene functions are indispensable, and if dispensable and indispensable functions are randomly distributed among the genes, then the probability of detecting a mutation in a gene with two primary functions is only one per cent. On the other hand, if the frequency of indispensable functions is low, then the chance of detecting multifunctional genes will be much better. The determination of the proportion of indispensable functions is thus critical for the one gene-one enzyme concept. The question is how this quantity is to be determined. It would seem almost by definition to be unknowable, in which case the one gene-one enzyme idea must be banished to the purgatory of untestable hypotheses, along with the proposition that a blue unicorn lives on the other side of the moon.

The Frequency of Indispensable Functions in Neurospora

What is needed is a method for detecting mutations which result in loss of an indispensable function and for comparing their frequency to that of mutations which cause loss of a dispensable function. It occurred to one of us (Horowitz, 1948, 1950) that the so-called "temperature mutants" of *Neurospora* might form the basis of such a method. Temperature mutants are a class in which the mutant phenotype is fully expressed only in a particular temperature range. Generally, such mutants exhibit a growth factor requirement when cultured at $35°$, but grow in its absence at $25°$; in a few cases this relationship is reversed —i.e., the growth factor is required at the lower, but not at the higher, temperature. In three instances it has been found that particular temperature mutations behave as alleles of mutations of the usual, temperature-independent sort, and it seems not unlikely that this will be found to be generally true.

The usefulness of these mutants for the present problem is based on the expectation that the mutant will be recoverable in the temperature range within which it has no growth factor requirement, regardless of whether a dispensable or an indispensable function has been lost. This expectation is borne out by the fact that a group of temperature mutants which fails to grow on complete medium at the mutant temperature is, in fact, known. Of the 26 temperature mutants known in *Neurospora*, 12 are of this type, while 14 grow on complete medium in the temperature range within which they have a requirement. In other words, roughly one-half of these mutants has lost an indispensable function.

In using the temperature mutants as a sampling device it is assumed that genes controlling indispensable functions are just as likely to yield temperature alleles as those controlling dispensable functions. This assumption is supported by two considerations. In the first place, the two classes of functions are in no sense natural categories, but depend largely on the composition of the particular complete medium which is employed. There is thus no reason to assume that the genes governing these functions differ from one another in any fundamental way. In the second place, among the temperature mutants whose specific requirement is known there is no indication that any one kind of nutritional requirement is favored over others. Mutation to temperature alleles appears to occur at random among genes controlling known biochemical syntheses (Horowitz, 1950).

With the information that the frequency of indispensable gene functions constitutes approximately 50 per cent of the total, it becomes possible to estimate the intensity of the selection which operates against the detection of multifunctional genes. With a random distribution of functions, one-half of genes with a single function will be detectable by the usual methods, one-fourth of bifunctional genes, and, in general, $(½)^n$ of n-functional genes. The original minimal estimate of 84 per cent of unifunctional genes, based on the observation that this fraction of the mutants responds to single growth substances can now be corrected. A sufficiently close approximation is given by neglecting genes with more than two functions, and we obtain 73 per cent as the corrected frequency of unifunctional genes:

$$\text{Observed frequency} = \frac{84}{84 + 16} = 0.84$$

$$\text{Corrected frequency} = \frac{84 \times 2}{84 \times 2 + 16 \times 4} = 0.73$$

The exact value is given by the first term of a Poisson distribution, and is equal to 0.71 (see Appendix).

This value is so high, that in spite of the uncertainties in its determination it may be regarded as strongly supporting the conclusion that at least the majority of genes controlling biosynthetic reactions in *Neurospora* are unifunctional. There are several obvious sources of uncertainty in the calculations. First, they should be based on the number of genetically different mutations, rather than on the total number of occurrences; this cannot be done at the present time. Second, the assumption was made that all of the unanalysed mutants, 16 per cent of the total, represent multifunctional genes; this is almost certainly incorrect and biases the calculations against the one gene-one enzyme theory. Finally, the number of temperature mutants is too small to give an accurate estimate of the frequency of indispensable functions. It is to the last point that we now turn.

The Frequency of Indispensable Functions in *E. coli*

It was clearly desirable to obtain a more reliable estimate of the frequency of indispensable functions, but to even double the existing number of temperature mutants in *Neurospora* would be a formidable operation. We therefore turned to *E. coli* K-12, with the expectation of recovering large numbers of temperature mutants by a modified penicillin technique (Davis, 1948; Lederberg and Zinder, 1948). Providentially, this method proved to be unsuited to our purpose: although temperature-independent mutants were obtained, the yield of temperature mutants was zero. This was a fortunate circumstance, since it forced us to adopt a more direct method, one which introduces fewer uncontrolled selective variables into the experiment than would the penicillin technique. The method is simply that of plating out U.V.-treated cells on minimal medium and incubating them for 48 hours at 40°. The plates are transferred to 25° for an additional 5 days, and the colonies which come up during this second period —so-called secondary colonies—are picked off and tested. This procedure was made feasible by a visual method devised by Dr. Leupold which makes it easier to detect a few secondary colonies on a plate containing hundreds of primary colonies. Altogether 161 temperature mutants were obtained by this method. Of these, only 37, or 23 per cent, were unable to grow on the *Neurospora* complete medium at 40° and therefore represent losses of indispensable functions. The statistics are shown in Table 1.

TABLE 1. STATISTICS OF *E. COLI* STUDY

No. of irradiated cells	1.7×10^9 (approx.)
No. of surviving cells	2.4×10^6 (approx.)
Secondary colonies isolated	2157
No. of temperature mutants	161
Type D40	124
Type I40	37

The remaining 124 mutants, those which grow on complete medium at 40°, were tested by the auxanographic method to determine their growth requirements. Seventy-nine per cent of these mutants were classifiable in this way. A variety of requirements was found (Table 2), indicating again that temperature mutation is random with respect to the classes of syntheses which can be affected. A number of substances are conspicuous by their absence from this list, notably tryptophane and p-aminobenzoic acid. It has not been excluded, however, that requirements for these substances are present among the mutants which were not classifiable in the auxanographic test.

Several other points of interest in connection with the *E. coli* study should be mentioned. These concern the selective forces operating in

TABLE 2. SYNTHESES KNOWN TO BE AFFECTED IN *E. COLI* TEMPERATURE MUTANTS

Amino Acids	Vitamins
Methionine	Biotin
Cystine	Thiamin
Arginine	Pyridoxin
Lysine	Nicotinamide
Histidine	Pantothenic acid
Leucine	
Isoleucine	
Valine	Nucleic Acid Constituents
Threonine	
Aminobutyrate	
Tyrosine	
Glycine	

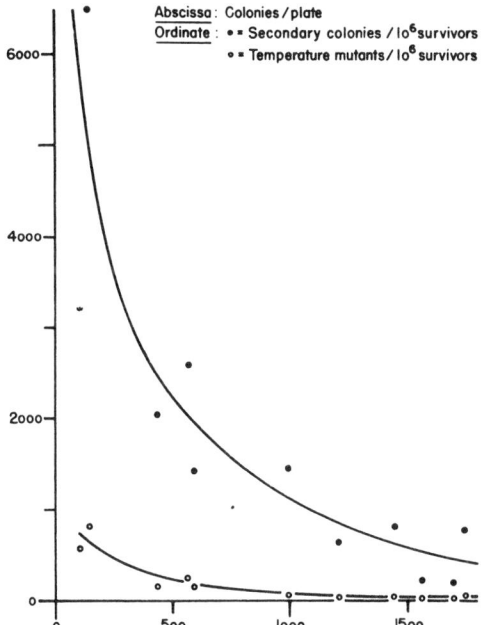

FIG. 1. Frequency of secondary colonies (solid circles) and of temperature mutants (open circles) per 10^6 survivors as functions of population density.

densely populated versus lightly populated Petri plates. It is obvious that in a method like this, in which there may be many hundred of colonies per (15 cm.) plate, we are not actually isolating the mutants in a minimal medium, but in a minimal medium plus or minus whatever the hundreds of wild type colonies add to or subtract from it. This is quite clearly shown in the relative yield of secondary colonies and temperature mutants per number of survivors. In our experiments the total number of colonies (i.e., survivors) per plate varied from 100 to 1,700. In sparsely populated plates the relative yield of both secondary colonies and temperature mutants was much higher than in densely populated plates. In Figure 1 the numbers of secondary colonies and temperature mutants per million survivors are plotted against the total number of colonies per plate. At the lowest densities, with populations of the order of 100 colonies per plate, yields of 3,400 and 6,500 secondary colonies per million survivors have been recorded. At higher densities, the yield decreases systematically and rapidly, reaching values between 200 and 800 secondary colonies per million survivors at population densities of 1,200 to 1,700 colonies per plate. The yield of temperature mutants per million survivors is roughly one-tenth that of all secondary colonies, and it exhibits the same systematic trend. The difference between these two curves expresses the fact that approximately 90 per cent of the secondary colonies are wild types which, for one reason or another, started to grow late; partially blocked, slowly growing biochemical mutants; and completely blocked biochemical mutants which have been fed "syntrophically" by the wild types. Figure 2 shows that the composition of the population of secondary colonies is also influenced by the population density. It is seen that the yield of temperature mutants per hundred secondary colonies decreases with increasing densities, indicating that the yield of temperature mutants decreases even more rapidly than does the yield of secondary colonies in general.

At least three selective forces are at work in these populations: (1) competition for food, (2) probably more important, mutual inhibition by-products of metabolism, and (3) superimposed on these but acting in the opposite direction, cross-feeding, or syntrophism, the mutual exchange of essential growth factors. The first two mechanisms are probably mainly responsible for the rapid decrease in the yield of secondary colonies with increasing plate densities. Their intensity is evidently quite remarkable. They are unspecific forces, however, and cannot be expected to influence systematically the relative frequencies of the two types of temperature mutants which we set out to find. The third influence, cross-

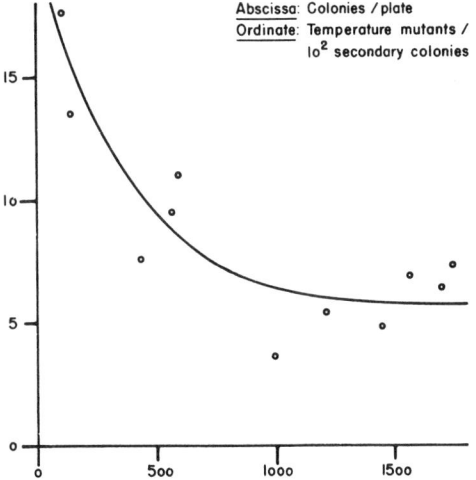

FIG. 2. Frequency of temperature mutants per hundred secondary colonies as a function of population density.

feeding, is much more dangerous in this respect, although its direction and intensity are difficult to predict. The effect of cross-feeding on the selective isolation of the two kinds of temperature mutants depends on both the quantity and quality of the output of growth factors by wild type *E. coli*. For example, let us assume that

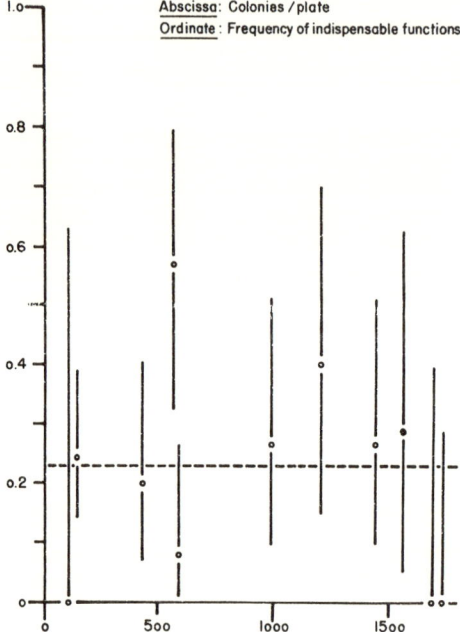

FIG. 3. Frequency of temperature mutants of the indispensable class as a function of population density. Vertical lines show the range within which the experimental points would be expected to fall in 95 per cent of similar experiments. The broken line shows the weighted mean of the distribution.

the growth factor excretion by the wild type is qualitatively similar to the composition of complete medium. Large amounts of such enrichment would tend to support the growth of temperature mutants which had lost a dispensable function, up to the point of visible colony formation, and would thus remove them from the isolation procedure. On the other hand, smaller quantities of the same enrichment might give these same mutants an advantage in the struggle for survival during the first 48 hours at the mutant temperature and during the second period at the lower temperature when they begin to grow against the heavy competition of established wild type colonies. The actual situation would be far more complex, in that the sign and magnitude of the selection would differ from mutant to mutant.

In spite of the indisputable occurrence of cross-feeding on the plates, however, it can be said that no systematic selection favoring either of the two classes of temperature mutants is deducible from our data. In Figure 3 is plotted the fraction of temperature mutants which were found to belong to the indispensable class, together with the ranges within which the values would be expected to fall in 95 per cent of trials, against the population density. With the exception of one experiment, the data are consistent with the assumption that these are random samples drawn from a homogeneous population of temperature mutants in which the frequency of mutants of the indispensable class is between 20 and 25 per cent. In short, with the exception of a single experiment in the middle range of population densities, there is no indication that population density influences the relative frequencies of the two classes of temperature mutants: the various selective forces appear to affect both classes equally.

We may summarize the *E. coli* results, then, by saying that the frequency of indispensable functions, as revealed by the temperature mutant method, is even lower than was indicated by the less extensive series of temperature mutants in *Neurospora*. Actually, the two results—46 per cent of indispensable functions in *Neurospora*, compared to 23 per cent in *E. coli*—are not very different, considering the great differences in the respective organisms. The intensity of the selection which opposes the recovery of multifunctional genes in the usual screening procedures is thus of a rather low order and is incapable of accounting for the fact that the genes detected by these procedures appear to be preponderantly of the unifunctional type.

CONCLUSION

In concluding this paper, we should like to make some brief observations on the significance of the low frequency of indispensable functions and the one gene-one enzyme hypothesis.

Our results indicate that the effect of most lethal mutations in *Neurospora* and *E. coli* is to block the synthesis of metabolites which are replaceable by nutritional means; that is to say, of low molecular weight substances such as might be expected to diffuse into the cell and of which the complete medium is chiefly, if not exclusively, composed. This situation appears to contrast markedly with that encountered in *Drosophila*, where lethal mutations, as well as visibles, result in irreplaceable losses, frequently organ-specific in character. In some measure this may reflect fundamental biological differences between *Drosophila* and *Neurospora* or *E. coli*, but also, and to an indeterminate extent, it reflects the differences in methodological approach to these organisms. It would not be surprising to find that non-diffusible products play a more important role in the development of a highly differentiated organism like *Drosophila* than in relatively undifferentiated ones like *Neurospora* and *E. coli*. But on the other hand, it must be recognized that the methods of *Drosophila* culture are such as virtually to exclude the possibility of detecting nutritional mutants even if they occur.

The results we have obtained from the microorganisms suggest that in the great majority of cases the metabolic function of the gene is to produce an enzyme which catalyzes the formation of a low molecular weight product. If there exists a large class of enzymes whose function is other than this, we must conclude either that they are not individually essential to survival, or else that they are independent of the genes for their production. There is a strong suggestion in this result that the mechanism of biosynthesis of large molecules, especially proteins, is not an enzymatic mechanism in the usual sense. That is to say, the protein molecule may not be built, cathedral-like, by a process of accretion; but rather may be made directly from the amino acids or their simple derivatives by a single catalyst.

It is interesting to note that such a mechanism provides a simple explanation of another essential feature of our findings; namely, the mutual independence of the pathways of synthesis of different enzymes. On the hypothesis of growth by accretion of peptide fragments it seems unlikely that a one gene-one enzyme relationship could be found, even if it existed, since loss or modification of any peptide fragment would be expected to result in loss or modification of a particular group of enzymes; namely, those which contain this fragment in their structures. Unless the fragment could be introduced into the cells, the result would be the frequent occurrence of multiple unrelated biochemical deficiencies among the mutants. This has not been found. We find considerable appeal in the notion that the proposed mechanism of enzyme synthesis may underlie the results we have obtained.

Summary

The one gene-one enzyme hypothesis is supported by the following evidence:

The great majority, at least 84 per cent, of the known nutritional mutants of *Neurospora* require single chemical substances as essential growth factors. Provision of the essential factor results in a normal phenotype in most cases; where a fully normal phenotype is not attained, the evidence indicates that this is to be accounted for on the basis of secondary effects unrelated to the mode of gene action.

Biochemical analysis of the mutants has indicated in many instances that the effect of the mutation is to block sensibly a single step in the pathway of synthesis of the growth factor. In a few cases it has been possible to show that the mutants are lacking in the specific enzyme involved.

The question of whether the known biochemical mutants are a highly selected sample from which multifunctional mutants are excluded by virtue of the screening procedure employed has been attacked by the temperature mutant method. Evidence has been presented which indicates that this method makes possible the recovery of mutants without regard to the nature of the induced biochemical deficiency or the composition of the (complete) medium.

Application of this method has shown that the proportion of biochemical mutants not recoverable by the usual screening tests may be less than 50 per cent in *Neurospora* and less than 25 per cent in *E. coli*. It is calculated that this rate of loss does not produce a sufficiently intense selection of unifunctional mutations to account for the high frequency of such mutations actually found.

Finally, it is suggested that the results can be simply accounted for on the hypothesis that the synthesis of a protein molecule is accomplished by a single catalyst working directly on the constituent amino acids or their simple derivatives.

Appendix

The corrected value of the frequency of unifunctional genes can be computed from a Poisson distribution as follows:

Assume that each gene has one function to begin with and that there is in addition a number of functions, m, randomly distributed among the genes. The fraction i of all gene functions is indispensable. Letting

n = the number of genes

Then $n + m$ = the number of gene functions,

$m/n = c$ = the mean number of additional functions per gene,

And ic = the mean number of indispensable additional functions per gene.

The fraction of genes with no additional functions—i.e., unifunctional—is then given by e^{-c}, the first term of a Poisson distribution. Of these, the fraction $(1-i)$ are recoverable. The fraction of recoverable unifunctional genes is therefore

$$P_a = (1-i)e^{-c}$$

Similarly, the fraction of all genes, unifunctional and multifunctional, which can be recovered is

$$P_{a+b} = (1-i)e^{-ic}$$

The frequency of unifunctional genes among those recovered is therefore

$$P_a/P_{a+b} = e^{-c(1-i)}$$

Equating this expression to the observed value, 0.84, and substituting 0.5 for i, one finds c = 0.34. The corrected frequency of unifunctional genes is then $e^{-0.34} = 0.71$.

REFERENCES

BEADLE, G. W. and TATUM, E. L., 1941, Genetic control of biochemical reactions in *Neurospora*. Proc. Nat. Acad. Sci. Wash. 27: 499–506.
BONNER, D., 1946a, Further studies of mutant strains of *Neurospora* requiring isoleucine and valine. J. Biol. Chem. 166: 545–554.
1946b, Biochemical mutations in *Neurospora*. Cold Spr. Harbor Symposium Quant. Biol. 11: 14–24.
DAVIS, B. D., 1948, Isolation of biochemically deficient mutants of bacteria by penicillin. J. Amer. Chem. Soc. 70: 4267.
1950, Studies on nutritionally deficient bacterial mutants isolated by means of penicillin. Experientia 6: 41–50.
DOERMANN, A. H., 1944, A lysineless mutant of *Neurospora* and its inhibition by arginine. Arch. Biochem. 5: 373–383.
EMERSON, S., 1950, Competitive reactions and antagonisms in the biosynthesis of amino acids by *Neurospora*. Cold Spr. Harbor Symposium Quant. Biol. 14: 40–48.
FINCHAM, J. R. S., J. Gen. Microbiol., in press.
HOROWITZ, N. H., 1948, The one gene-one enzyme hypothesis. Genetics 33: 612–613.
1950, Biochemical genetics of *Neurospora*. Advances in Genetics 3: 33–71.
LEDERBERG, J. and ZINDER, N., 1948, Concentration of biochemical mutants of bacteria with penicillin. J. Amer. Chem. Soc. 70: 4267.
MITCHELL, H. K. and LEIN, J., 1948, A *Neurospora* mutant deficient in the enzymatic synthesis of tryptophan. J. Biol. Chem. 175: 481–482.
UMBARGER, H. E. and MUELLER, J. H., 1951, Isoleucine and valine metabolism of *Escherichia coli*. I. Growth studies on amino acid-deficient mutants. J. Biol. Chem. 189: 277–285.

DISCUSSION

EPHRUSSI: I should like to know what you consider to be convincing evidence that there is genic control of protein synthesis at all?

HOROWITZ: In *Neurospora*, at least two cases are known in which gene mutations lead to deficiency of specific enzymes. These are the tryptophane synthesizing enzyme worked on by Mitchell and Lein and the glutamic dehydrogenase studied by Fincham. If these cases merely involved loss of specific cofactors then it might be expected that the activity of mutant extracts would be restored to some extent by mixing them with extracts from normal cells. This experiment has been performed, and, if I am not mistaken, gave negative results with both enzymes. Similar evidence probably exists in other organisms.

It is more difficult to obtain data bearing on the question of whether genes actually function in the synthesis of the polypeptide chain, or only impress a particular folding on prefabricated protein molecules. The most important evidence is that which Pauling and co-workers have obtained in connection with the electrophoretic mobilities of sickle-cell anemia hemoglobin and normal hemoglobin. They show that these hemoglobins differ by 3 or 4 positive charges in the globin part of the molecule. The most likely interpretation is that the proteins differ in the number or kind of ionizable groups. Thus it appears probable that the sickle cell anemia mutation results in the production of a protein of altered composition.

MAAS: In answer to Dr. Ephrussi's question I would like to mention some recent experiments which indicate that a biochemical mutation can affect synthesis of an enzyme protein. The biochemical reaction under investigation was the coupling of β-alanine and pantoic acid through a peptidic linkage to form pantothenate in *E. coli*. Two types of mutants blocked at this reaction were obtained: an absolute one which requires pantothenate at all temperatures, and another which requires it only above 32°. Acetone dried powder of the wild type readily yielded stable and highly active cell-free preparations of the coupling enzyme. In contrast, no enzyme activity was found in extracts of the absolute mutant; the system was sensitive enough to detect 1/2000 of the usual wild type activity. From the temperature sensitive mutant active but extremely heat labile extracts were obtained. Extraction had to be carried out in the cold and activity measured at 15°C, since at 25°C the enzyme is rapidly and irreversibly inactivated. These results, in accordance with the one gene: one enzyme hypothesis, show the enzymatic activity of the cell-free extracts of the several strains to parallel their growth behavior.

In this connection, I would like to point out that a previous apparent exception to this parallelism has now been clarified. Wagner had found pantothenate synthesizing enzyme activity in extracts of two *Neurospora* mutants whose growth requirements under ordinary conditions implied absence of this enzyme. However, he has shown recently (Genetics 35, 697, 1950) that under appropriate conditions the mutants can grow without pantothenate and can synthesize pantothenate as rapidly as the wild type. As in Dr. Bonner's tryptophane/niacin mutants the biochemical block is incomplete; it is therefore not surprising to find enzyme activity in the extracts.

DAVIS: Supplementing the demonstrations by Pauling and by Maas that mutation can result in the production of a qualitatively altered protein, I should like to present evidence that an enzyme of altered specificity is present in a drug-resistant mutant.

The argument is based on the following facts. In *E. coli* p-nitrobenzoic acid (PNBA) is a simultaneous competitor of two metabolites, p-hydroxybenzoic acid (POB) and p-aminobenzoic acid (PABA). In the presence of an excess of PABA, PNBA inhibits growth only by competition with POB; mutants resistant to this PNBA/POB inhibition are readily obtained. Similarly, in the presence of POB, mutants resistant to PNBA/PABA inhibition can be obtained. These two types of mutants show no cross resistance with each other. This observation excludes decreased permeability or increased destruction of the drug as the mechanism of resistance. Furthermore, resistance to a sulfonamide (e.g., sulfathiazole) competing with PABA is quite different from resistance to PNBA/PABA. Similarly, resistance to a sulfonamide analogue of POB does not entail resistance to PNBA/POB. These observations exclude mechanisms that could not distinguish different analogues of a metabolite, such as increased concentration of the metabolite antagonizing the drug, increased efficiency of utilization of the product of this metabolite, or an alternative metabolic pathway bypassing the metabolite. Furthermore, if we assume that both sulfonamide and PNBA inhibit by direct competition with the corresponding metabolite at the same enzyme surface, the specificity of the resistance to each type of inhibitor could not be explained by an increased concentration of the enzyme. If this assumption is correct, and unless some other explanation is being overlooked, these results appear to imply a specific change in the enzyme: i.e., an alteration in configuration that results in decreased affinity for the inhibitor in comparison with the metabolite. The type of alterations required in the present case can readily be visualized since the sulfonamide analogues differ from the metabolite at the 1-position of the ring, and PNBA at the 4-position (see diagram).

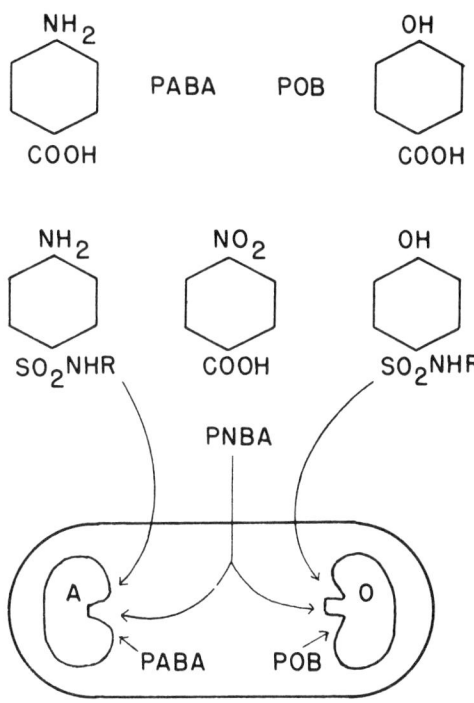

FOX: I should like to relate some observations which have a bearing on Dr. Muller's question regarding antigens, or Dr. Ephrussi's question regarding the role of genes in protein synthesis, and at the same time on the mechanism of protein synthesis proposed by Dr. Horowitz. In *Neurospora crassa*, the two mating types of strain 15,300 are distinct antigenically, each possessing an antigenic component not possessed by the other. These antigens are proteins, and chemical fractionation indicates that their specificity is not haptenic (?) in nature but resides in the protein itself. Genetic analysis of this difference, so far as it has gone, indicates that at least two, but probably more, loci are concerned with the production of these specific antigens. The data do not exclude the possibility that the mating-type locus is one of those concerned. We have here a case where genes are demonstrated to be involved in protein synthesis, and which, since more than one locus is concerned, would seem to argue against the single-step hypothesis of protein synthesis.

VILLEE: Many of the enzymes whose substrate specificities have been studied biochemically have been shown to be not completely specific for a single substrate but rather specific for a certain type of reaction; they will catalyze reactions involving any one of a class of compounds with a common reactive group. Have any biochemical mutants been obtained in which the enzyme involved has been shown to be the catalyst for a type of reaction rather than for one involving a single specific substrate as most of these stepwise gene-controlled syntheses appear to be?

(Question answered by Horowitz that he didn't know of any. Lederberg then cited a lactose mutant discovered in *E. coli* in which the enzyme involved was a general B-D-galactosidase.)

I think Dr. Horowitz's conclusion that his results suggest that protein synthesis is accomplished by a single step with a single catalyst working directly on amino acids is perhaps an unwarranted extrapolation from the data. His data show that there are still a considerable fraction of mutants for indispensable functions, some of which may be for protein synthesis. I would like to cite some evidence bearing on this obtained by Drs. Anfinsen and Steinberg at Harvard. They incubated minced chick oviduct in a saline medium containing $C^{14}O_2$. After a three or five hour incubation period, the synthesis was stopped and the ovalbumin was extracted, purified, and finally crystallized with added carrier. Then, using the *B. subtilis* enzyme discovered by Linderstrøm-Lang which splits off a hexapeptide from ovalbumin, leaving a distinctly different, crystallizable protein called plakalbumin, the radio-active ovalbumin was split into peptide and plakalbumin parts. When aspartic acid was isolated from the two parts, the specific activity of the peptide aspartic was twice that of the plakalbumin aspartic. This suggests that the ovalbumin molecule is not put together in one fell swoop by the accumulation of amino acids against some sort of template, but rather that the molecule is assembled bit by bit. More recent experiments show that a similar relation is true for the alanine isolated from the two parts of the ovalbumin molecule.

HOROWITZ: The *E. coli* experiment indicates that less than 25 per cent of temperature dependent lethals involve loss of indispensable functions. Preliminary experiments of Dr. Leupold have already shown that some of these mutants actually require low-molecular weight substances present in the complete medium; they appear in the indispensable class because their growth is inhibited by other substances in this medium. Still other mutants may require small molecules which are absent from the medium. The point is that the fraction of mutants whose requirement is for large molecules is even smaller than the fraction which have lost indispensable functions. It should be mentioned that the enzymatic synthesis of peptides with special metabolic functions—for example, glutathione—is not excluded by our hypothesis. I agree that our suggestion is an extrapolation, but not an unwarranted one, since it accounts for our results in a straightforward way. For the present it is to be regarded as a working hypothesis and nothing more.

In connection with the interesting experiment by Anfinsen and Steinberg, cited by Villee, I wonder if exchange at certain preferred or exposed sites of the ovalbumin molecule was excluded.

(Dr. Villee stated that he thought exchange was excluded.)

WALLACE: Your tests were made on the survivors that were but 1/1000 of the original population. Aren't you disturbed by the fact that you know nothing about the other 999 in every 1000?

HOROWITZ: Every reasonable precaution was taken to insure that the temperature mutants isolated represented a random sample of those present immediately following the irradiation. After our experiments had been begun, however, Weigle and Delbrück discovered a mechanism which might lead to selective killing by ultraviolet in cultures of *E. coli* K-12. They found that ultraviolet induces lysis of K-12 by activating the phage lambda which, as shown by Mrs. Lederberg, is carried in a latent form by K-12. It seemed possible, therefore, that the survivors in our experiments had been selected on the basis of non-lysogenicity. With this in mind, our mutants have been tested by Dr. Margaret Lieb, and only five of them were found to be non-lysogenic. It appears unlikely that lysogenicity was an important factor in determining the types of temperature mutants recovered.

Another point raised by the finding of Weigle and Delbrück concerns the actual genetic effectiveness of the U.V. dose used in our experiments. It seems probable from the results of these workers that induced lysis was a major cause of death of the irradiated cells, in which case the mutagenic effectiveness of the irradiation would have been considerably less than might be inferred from the high mortality. It is quite possible that a large fraction of our mutants were, in fact, of spontaneous origin.

A FURTHER ANALYSIS OF THE PANTOTHENICLESS MUTANTS OF NEUROSPORA

R. P. WAGNER AND C. H. HADDOX

The University of Texas[1]

During the last few years emphasis in the work on Neurospora biochemical mutants has been shifting from the use of these mutants as tools in determining pathways of biosynthesis to a study of the mutants themselves. Where previously it had been primarily a question of where the genetic blocks were located, it now seems more germane to inquire about the nature of these blocks. Since it has been assumed for many years, with good reason, that genes mediate chemical reactions through enzymes which they in turn control, the most feasible approach to the problem of genetic blocks is through a study of the enzyme systems in the mutant and wild type strains.

Initial attempts to distinguish enzyme differences between wild type and mutant, such as in the case of the adenineless mutant, 44206 by McElroy and Mitchell (1946), yielded inconclusive results. However, Mitchell and Lein (1948) investigated a tryptophanless mutant blocked at the reaction which couples serine and indole to produce tryptophane, and were able to show that cell-free preparations of wild type were active in this synthesis whereas those from the mutant were not. This result clearly indicated that the genetic block in question was due to the lack of an enzyme. It therefore provides support for the simple hypothesis that genetic blocks are due to the loss of enzymes *per se*. That is, loss of enzyme activity due to gene mutation is accompanied by the absence of the enzyme.

On the other hand, the work of Wagner and Guirard (1948) and Wagner (1949) on the pantothenicless mutants, 5531 and 34556 showed that the explanation of genetic blocks might not be quite so simple in all cases. It was found that in these mutants (which appear to be allelic or identical) the apparent genetic block is at the point of synthesis of pantothenic acid from β-alanine and pantoyl-lactone. The wild type is able to synthesize pantothenic acid from these precursors at a high rate *in vivo* in resting cell preparations but the mutant is not. When acetone powders are prepared from the mutant and wild type mycelia, these are found in both cases to be active in this synthesis. The mutant cannot carry out the synthesis *in vivo* under ordinary conditions of culture, therefore, but is able to do so under certain *in vitro* conditions. Obviously, the interpretation that genetic blocks are produced by simple loss of enzymes is not applicable in the

[1] The Genetics Laboratory of the Department of Zoology.

case of the pantothenicless mutants, unless it is postulated that the acetone treatment of the mutant produced the enzyme artificially, as suggested by Horowitz (1949).

It is the purpose of this present communication to report the results of additional work on the pantothenicless mutants, and to discuss their implications in the analysis of how genetic blocks are produced. It was assumed that if the mutant mycelium possessed the enzyme for pantothenic acid synthesis, it should be possible to demonstrate its activity under certain *in vivo* conditions. This assumption proved to be correct as is shown below.

EXPERIMENTAL

Synthesis of pantothenic acid in vivo:

The initial approach to the problem of inducing synthesis of pantothenic acid *in vivo* was to modify the culture conditions. It had been previously shown that a number of Neurospora mutants such as the riboflavinless and adenineless mutants could be caused to grow in the absence of riboflavin and adenine by changing the temperature (Mitchell and Houlahan, 1946a, b). The vitamin B_6-requiring mutant, 299, grows in the absence of B_6 at basic pH's (above 6.0) but not at lower pH values (Stokes, Foster and Woodward, 1943). Under acid conditions the B_6-requiring mutant, 44602, will grow on minimal medium provided the ammonium content of the medium is increased (Strauss, 1951). Attempts were made to stimulate growth of the pantothenicless mutants, 5531 and 34556, in the absence of pantothenic acid by varying the pH, temperature and nitrogen content of the medium. No stimulation was observed even when suboptimal amounts of pantothenic acid were added. Therefore, it was assumed that no pantothenic acid was synthesized. None of the conditions was so drastic as to prevent the growth of the mutants in the presence of pantothenic acid.

Following these negative results an attempt was made to determine if pantothenic acid is synthesized by the mutant after growth has been initiated by the addition of pantothenic acid. There are a number of cases reported in which biochemical mutants have responded by synthesizing the required nutrilite after growth had been started in the presence of the nutrilite. The riboflavinless mutant of Neurospora has been shown to be one of these (Mitchell and Houlahan, 1946a).

The mutants, 34556 and 5531, and wild type, 5256, were tested in flasks containing 50 ml of standard Neurospora minimal medium, some containing in addition the pantothenic acid precursors, 0.004M β-alanine and pantoyllactone. The mutant culture media were made complete by the addition of 18γ of pantothenic acid per flask. The flasks were incubated at 25°C and 37°C and pantothenic acid assays were made each day on a set of flasks using *Lactobacillus arabinosus*. The results of these assays, which were essentially the same for both mutants at both temperatures, are given in fig. 1 for 34556 and for wild type over an eight-day period at 25°C. The dotted line in this graph represents the average dry weight increase over

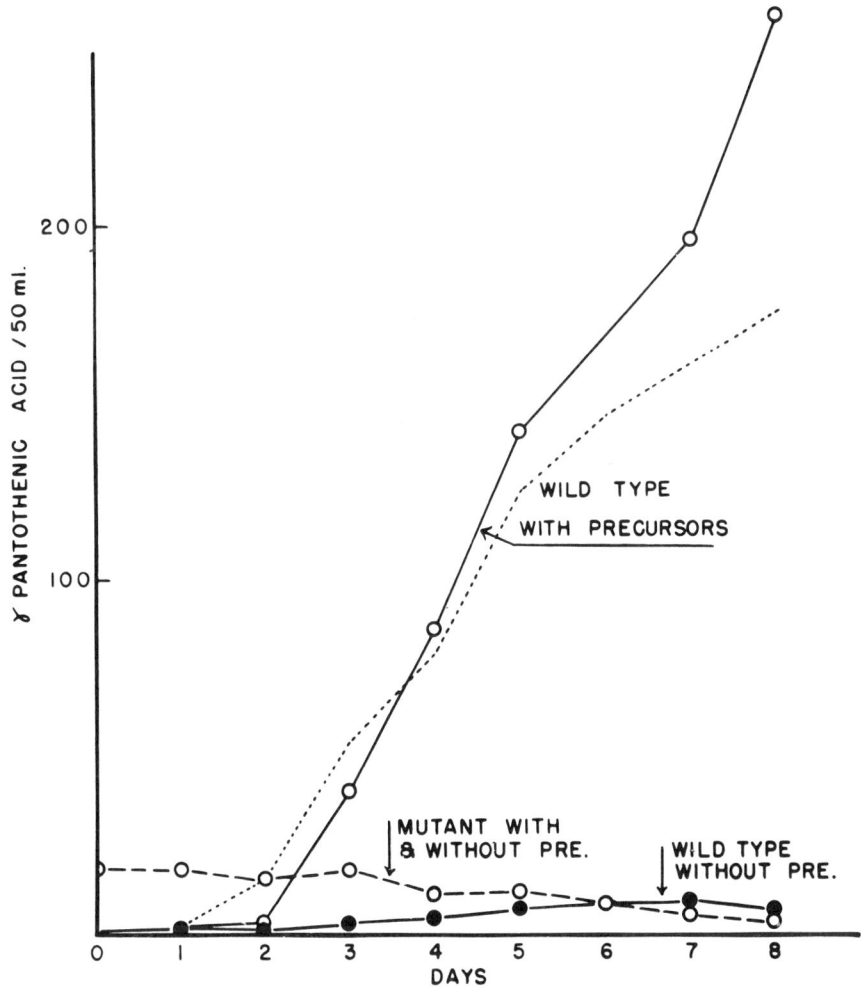

FIGURE 1. Production of pantothenic acid by wild type and mutants in standing culture.

the period observed. Actually assays were made each day for 13 days, but no change was noted in the amount of pantothenic acid present in the mutant culture media, whereas the wild type growing in the presence of the precursors continued to synthesize pantothenate, producing up to as much as 6 mg in 50 ml of medium. Wild type in the absence of the precursors produced no more pantothenate after 13 days than indicated in fig. 1 for an 8-day period.

To test the possibility that substances excreted into the medium by the pantothenicless strains inhibited the synthesis of pantothenic acid, mycelial pads were removed from some of the flasks on certain days, washed and incubated in the presence of 0.004 M concentrations of both precursors and phosphate buffer at ph 6.5 for 24 hours. The results are given in table 1. Included in this table are the values for pantothenic acid synthesis by the wild type mycelium. It is evident that the intact mutant mycelium has no enzyme activity. The intact wild type mycelium decreases in activity for pantothenic acid synthesis on about the third day. In addition, comparable amounts of pantothenic acid are synthesized by wild type whether or not the mycelium is grown in the presence of the precursors.

TABLE 1

PANTOTHENIC ACID SYNTHESIZED BY UNTREATED RESTING MYCELIUM OF VARIOUS AGES AFTER GROWTH IN STANDING CULTURE.
γ PANTOTHENIC ACID/MG DRY WEIGHT MYCELIUM PER 24 HOURS.

	Age of mycelium in days				
	3	6	7	8	13
Wild type grown in absence of precursors	8.7	4.1	2.4	3.0	1.8
Wild type grown in presence of precursors	8.9	5.1	3.0	4.9	1.5
34556 grown in absence of precursors	0.02	0.02	0.01	0.01	0.01
34556 grown in presence of precursors	0.02	0.01	0.02	0.03	0.01

Attention was next directed to the possibility of synthesis under conditions of vigorous aeration. It had been found that a culture of mutant mycelium grown in a five liter volume of medium under constant aeration for five days had accumulated 51 mg of pantothenic acid. This meant that 41 mg of the vitamin had been synthesized in five days, since 10 mg were used as a supplement to start growth. Subsequent experiments under controlled temperature conditions verified this synthesis.

The mutants and wild type were cultured in 500 ml of medium in one liter round bottom flasks immersed in a water bath set at 30°C. A constant stream of sterile air was bubbled through the medium in each flask throughout the course of the experiments by means of a glass tube which opened near the bottom of the flask. The culture medium for the mutant was supplemented with 1γ of pantothenic acid per ml. Pantothenic acid assays were made on a 10 ml aliquot of each culture every day for eight days. Fig. 2 presents the data obtained for 5531, giving the amount of pantothenic acid present per ml of medium. Flasks were also incubated at 25°C with essentially the same results. The volume of medium in the flasks did not vary more than 10 to 20 ml from the 500 ml original volume during the course of the experiment. It was kept within this range by saturating the entering air with water and replacing the assay aliquots with an equal amount of sterile distilled water after each withdrawal. On the eighth day the mycelium in each flask was removed, squeezed dry, weighed, and a small

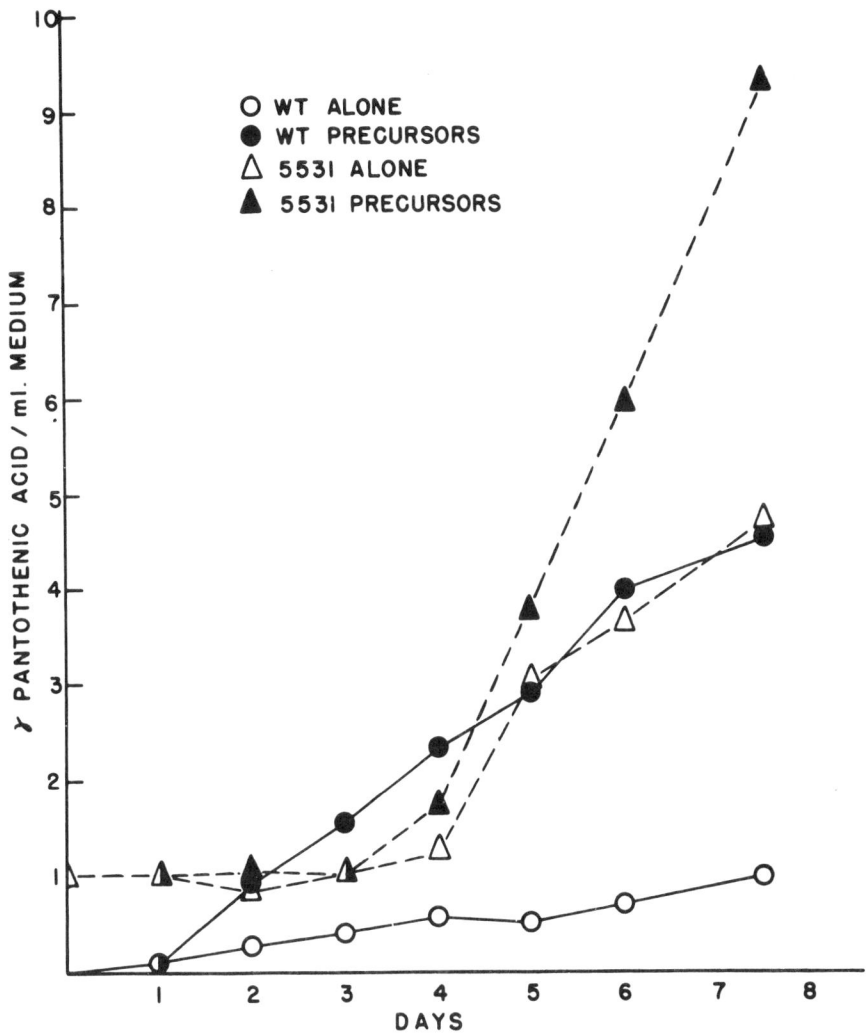

FIGURE 2. Synthesis of pantothenic acid by wild type and 5531 in an aerated medium.

sample dried and weighed. The remainder was blended with 0.1M potassium phosphate buffer at pH 6.5 and allowed to autolyze for 72 hours under toluene. The residue was discarded and the autolysate assayed for pantothenic acid. From this the pantothenic acid content of the mycelium was determined. The total pantothenic acid determined in the culture filtrate was added to this, after subtracting the pantothenic acid added originally to the mutant cultures. The total dry weight of the mycelium was determined and the amount of pantothenic acid synthesized per mg dry weight of mycelium was calculated. These results are given in table 2 for two independent runs using the above procedure. These values may be some-

TABLE 2

SYNTHESIS OF PANTOTHENIC ACID UNDER CONDITIONS OF AERATION. PANTOTHENIC ACID GIVEN AS γ PER DRY WEIGHT OF MYCELIUM AFTER 8 DAYS GROWTH.

5531 (mutant)			5256 (wild type)		
A Minimal only	B with precursors	B/A	A Minimal only	B with precursors	B/A
0.985	3.12	3.14	0.649	2.95	4.55
0.717	2.63	2.92	0.498	2.30	4.60

what low since pantothenic acid in bound form, particularly in the form of coenzyme A, is not assayable with *Lactobacillus arabinosus,* unless freed by the proper hydrolytic procedure. It is not known whether such hydrolysis occurred during autolysis of the mycelium.

Small amounts of resting mycelium grown under aeration were tested for activity in the presence of precursors and buffer by the procedure described previously (Wagner and Guirard, 1948). The results are given in table 3 for wild type, 5531 and 34556 cultured in the presence and absence of precursors. The activity of the mutant mycelium is evidently similar to that of the wild type. The activity of the wild type mycelium in this experiment is consistently higher than for the 8 day old wild type mycelium grown in standing culture. (See table 1.)

With regard to the observed activity for synthesis of pantothenic acid by the mutant cultures, the question arises as to whether there had been a back mutation to the normal at the pantothenicless locus, or a suppressor mutation. This possibility was tested by transferring small bits of mycelium from the mutant cultures into fresh minimal medium. No growth was observed in the inoculated flasks. On the other hand, growth was obtained if pantothenic acid was added as a supplement to the medium. It seems firmly established therefore that back mutation is not the explanation for the activity of the mutants. In this connection it is interesting to note that the pantothenicless gene seems to be a particularly stable one. Giles and Lederberg (1948) have performed extensive experiments to determine the back mutation rate of a number of biochemical mutants, and were unable to detect any for pantothenicless, although they found that other loci such as inositolless back mutated readily. The present authors have done similar experiments with pantothenicless and obtained similar negative results.

In the experiments previously reported (Wagner, 1949) in which pantothenic acid synthesis by the mutants was described, the evidence for the identity of the pantothenic acid synthesized by the mutant with synthetic pantothenate had been based entirely on the growth response of the mutants themselves and *Lactobacillus arabinosus*. While this is very good presumptive evidence that the two are identical it is not proof. Additional information as to their relationship was obtained by means of paper chromatography using a butyl alcohol-acetic acid-water mixture. The mutant-synthesized pantothenic acid has an R_f of 0.76 compared to 0.75 for synthetic pantothenate (converted to the free acid with oxalate before chromatography). The close similarity between the two R_f values makes it almost certain that the mutant produces pantothenic acid and not some other compound with similar activity.

COMBINATION MUTANTS INVOLVING PANTOTHENICLESS

Attention was next directed to the possibility of modifying the *internal* conditions of the mutant in order to suppress the phenotypic expression of the pantothenicless gene. The assumption was made that perhaps the enzyme necessary for uniting pantoyl lactone and β-alanine is inhibited by a compound or compounds produced by some other metabolic process.

In order to test the possibility that some naturally occurring compound synthesized by Neurospora might be responsible, a number of amino acids, vitamins, purines and pyrimidines were tested for inhibitory activity on the synthesis of pantothenic acid *in vitro*. None proved to be consistently

TABLE 3

SYNTHESIS OF PANTOTHENIC ACID BY 8 DAY OLD, RESTING, AERATED MYCELIUM

Strain	γ pantothenate per mg. dry weight mycelium per 24 hrs.
Wild type	6.8
5531	7.5
34556	7.7

inhibitory with the exception of cystine and betaine. Betaine was investigated further to determine its effect on the synthesis of pantothenic acid *in vivo* by wild type. In addition, tryptophane which definitely inhibits the growth of both wild type and pantothenicless was tested in the same way. Betaine proved to have no effect on wild type in concentrations up to 5 mg per 25 ml, but tryptophane showed a decided inhibition. In the absence of tryptophane, actively growing wild type mycelium produced 0.88 ± 0.06 γ of pantothenic acid per mg dry weight of mycelium from precursors in eight days, whereas in its presence (5 mg per 25 ml) only 0.54 ± 0.06 γ was synthesized per mg dry weight.

The pantothenicless mutant 5531 was crossed to UT 1, a mutant which when originally discovered in this laboratory grew slightly on minimal but required tryptophane for growth equivalent to wild type. Ascospores from this cross were picked at random and tested for growth on minimal, minimal supplemented with pantothenic acid or tryptophane and minimal with both. Ascospore isolations from single asci were not feasible since there was too large a number of inviable spores. Wild type, pantothenicless and presumed double mutant segregants were identified. However, no spores of the parent UT 1 strain were recovered. It was assumed that these were possibly represented among the inviable spores. Cultures from viable spores which produced mycelium on minimal medium, but showed a stimulatory effect of pantothenic acid or tryptophane were assumed to be double mutants of UT 1 and 5531 and were tested by backcrossing to wild type. In a number of these both wild type and 5531 segregated in the same ascus sac but again UT 1 was not recovered. The results clearly showed, however, that 5531 was included in the double mutant strains. One of the double mutants, here designated as B3, was tested most extensively with respect to its growth requirements and ability to synthesize pantothenic acid. It is not only capable of synthesizing pantothenic acid, but its capacity for synthesis in the presence of the precursors and under conditions of aeration is from five to six times greater than the wild type and 5531.

UT 1, unfortunately, was practically sterile in crosses to wild type; only about 10 per cent of the spores resulting were viable, and all these proved to be wild type. It may be that UT 1 is not a simple point mutation, but a chromosome aberration. In addition to the aberrant behavior in crosses, UT 1 reverted to the wild type condition in so far as tryptophane no longer had any stimulatory influence on its growth. Because of this, and its sterility in crosses, experiments on this particular combination were discontinued.

The pantothenicless mutant was also crossed to UT39, an ultra-violet induced mutant which has been characterized in detail by Haddox (1951). UT39 grows only slightly (trace to 3 mg in 25 ml) on minimal medium during the first 72 hours, but after 72 hours its rate approaches that of wild type. In the presence of both phenylalanine and tyrosine together in equimolar concentrations of about 2×10^{-4} M, growth during the first 72 hours is almost identical to wild type. Tyrosine or phenylalanine stimulate growth alone, but not to as great an extent as in combination. Repeated backcrosses of this mutant to wild type have always resulted in a type of segregation characteristic of a single gene difference.

The double mutant obtained from the cross 5531 × UT39 was tested by backcrossing to wild type, and its constitution proven by finding that the mutant strains 5531, UT39 and the wild type and combination segregated in the expected ratios. The three types of ascus arrangements which occurred out of 16 dissected are as follows:

Ascus type	Spores							
	1	2	3	4	5	6	7	8
1	C	C	C	C	wt	wt	wt	wt
2	5531	5531	5531	5531	39	39	39	39
3	5531	5531	39	39	5531	5531	39	39

This double mutant has the following growth characteristics: on minimal medium it grows like the parent strain UT39; it is not stimulated by pantothenic acid, tryosine, or phenylalanine alone, but equimolar amounts of the latter two combined show definite stimulation (see tables 4, 5). Panto-

TABLE 4

GROWTH OF THE DOUBLE MUTANT (UT39, 5531) AND UT39 ON MINIMAL AND VARIOUS SUPPLEMENTS. MG DRY WEIGHT MYCELIUM AFTER 72 HOURS

Supplement	UT39	5531, UT39
none	5.0	1.5
tyrosine		
2×10^{-6} M	17	1.9
2×10^{-5} M	31	1.5
2×10^{-4} M	34	1.6
2×10^{-3} M	16	1.5
phenylalanine		
2×10^{-6} M	22	2.2
2×10^{-5} M	39	1.5
2×10^{-4} M	49	1.5
2×10^{-3} M	39	2.2
tyrosine and phenylalanine*		
2×10^{-6} M	19	1.5
2×10^{-5} M	30	6.5
2×10^{-4} M	40	2.7
2×10^{-3} M	21	3.6
pantothenic acid	4.1	1.6

*Range between 4×10^{-5} to 4×10^{-4} gives 20 mg or more growth of double mutant in 72 hours.

thenic acid and tyrosine combined have no effect on either UT39 or the double mutant, other than the expected tyrosine stimulation of UT39. The double mutant, then, shows a complete suppression of the pantothenicless phenotype, and in part the tyrosineless and phenylalanineless, since it does not respond to either of the stimulatory amino acids alone.

DISCUSSION

The original interpretation of the pantothenicless mutants of *Neurospora* as being due to the absence of the enzyme catalyzing the synthesis of pantothenic acid from β-alanine and pantoyl lactone (Beadle, 1945) is no longer tenable. It is now established that such an enzyme is present *in vivo* as

TABLE 5

GROWTH OF THE DOUBLE MUTANT (5531, UT39) AND UT39 ON MINIMAL AND VARIOUS SUPPLEMENTS. MG DRY WEIGHT MYCELIUM AFTER 84 AND 120 HOURS

Supplement	UT39		(5531, UT39)	
	84 hrs.	120 hrs.	84 hrs.	120 hrs.
none	10	40	3.2	40
20 γ/25 ml. pantothenic acid	8.2	40	6.2	39
2×10^{-4} M tyrosine	45	69	8.0	53

well as *in vitro* preparations, but is inactive under certain *in vivo* conditions. Alternative explanations for the genetic block in these mutants are therefore in order.

Since there is no other rational basis for hypotheses dealing with the primary or nearest observable action of genes evident at the present time, one must proceed on the assumption that genes mediate chemical reactions through enzymes. Starting then with the observation that the enzyme *per se* is present in this case one may interpret the results reported in this and previous papers by the following alternatives:

I. The mutation involves a gene directly concerned with the enzyme coupling β-alanine and pantoyl-lactone.

II. The mutation involves a gene controlling a reaction having nothing to do directly with the synthesis of pantothenic acid. This reaction results in the production of a compound which inhibits the enzyme catalyzing the synthesis of pantothenic acid.

Applying these hypotheses to the observed facts results in the following conclusions:

Considering hypothesis I first, we may conclude that any change brought about in the enzyme is not a simple quantitative change, since the synthesis of pantothenic acid by wild type and the mutant under proper conditions proceed at approximately equal rates. Nor, is there a simple qualitative change involving only an enzyme changed in its affinity for its substrate. The extremely low activity of the mutant *in vitro* preparations at 25°C compared to the wild type, and to mutant activity at 30°C (Wagner, 1949) indicates a qualitative change, but there is no effect of increasing the temperature on the growth of the mutant in the absence of pantothenic acid. Furthermore, the mutant enzyme is active *in vivo* under aeration at 25°C. The change must involve the relationship between the enzyme and substances in the cell other than the pantothenic acid precursors. The enzyme in the mutant must be so altered that even though it is capable of carrying out the synthesis of pantothenic acid, it is prevented from doing so by the inhibitory activity of other substances. The affinity of the altered enzyme for the pantothenic acid precursors may be affected under certain conditions, but the more important result is that it is made subject to complete inhibition by other metabolites in the cell. These may be natural compounds such as proteins, amino acids, etc., which are involved in other phases of metabolism.

Using the first hypothesis, then, the activity of the *in vitro* preparations could be explained on the basis that an inhibitor or inhibitors are destroyed or removed in the process of preparation with acetone. The former would be more likely since concentrates prepared from the acetone washings do not seem to have any inhibitory activity on the synthesis *in vitro*. The demonstrated activity *in vivo* is more difficult to interpret, but one may assume that after an initial period of growth in aerated medium made possible by added synthetic pantothenate, the excess of oxygen or perhaps changes in CO_2 content bring about changes in certain phases of metabolism which result in part in a decrease of the inhibitor. Dagley, Dawes and Morrison (1950) have reported that *E. coli* and *Aerobacter aerogenes*, when grown in unaerated medium, excrete amino acids (tentatively identified as histidine, aspartic acid, glutamic acid and alanine) into the medium which are easily detectable by paper chromatography without recourse to concentration of the filtrate. Cells cultured in aerated medium, however, only produce these amino acids in traces. The filtrate must be concentrated ninety-fold in order to get visible spots on paper. Accompanying this difference in amino acid excretion it has been found that the concentration of pyruvic acid increases during the active phase of growth of an aerated culture, but falls rapidly to zero after the stationary phase begins. In unaerated cultures the concentration of pyruvic acid attained is higher and the fall in concentration is not pronounced after the stationary phase sets in. Comparable but not necessarily identical changes in metabolism might be expected in the case of Neurospora as a result of aeration. Preliminary experiments conducted to test this have indicated that such is the case, but more work along this line is in order before any definite conclusion can be drawn.

The double mutant (5531, UT39), which grows in the absence of pantothenic acid, would by orthodox genetic terminology be described as a case of suppression of the 5531 gene by UT39. This is a good description of a condition, but not an analysis. A complete analysis of the suppression would require a more or less complete understanding of the inherited metabolic upsets in each one of these mutants, and we do not have this understanding at the present time. However, if one accepts the hypothesis that 5531 is incapable of producing pantothenic acid due to an inhibition, and that this inhibition is relieved in the presence of the UT39 gene by the reduction in concentration of the active inhibitor, a working hypothesis may be set up which may be of value for future investigations.

The second hypothesis given above is related to the first in so far as an inhibition is postulated. The essential difference is that the pantothenate enzyme is not altered, but that a natural inhibitor produced in excessive amounts due to mutation at another locus completely inhibits the enzyme at the higher concentrations. Both hypotheses, therefore, center around the relationship between an enzyme and its inhibitor. Altering the enzyme qualitatively or the inhibitor quantitatively should produce the observed effects.

It is not possible with the present data to differentiate between the two alternatives, or even to state that they are the only two possible explanations. Other hypotheses may be formulated which will explain the above described facts, but in the opinion of the authors the two given are the simplest, and most susceptible to further experimental investigation.

It must be noted in conclusion that the results obtained here with double mutants are not the only ones reported which are in conflict with the enzyme absence hypothesis as a general explanation of the inability of an organism to produce a particular chemical compound. Houlahan and Mitchell (1948),

Emerson (1948) and Mitchell and Mitchell (1950) have all described Neurospora double mutants of various types in which suppressors are involved. It is our opinion that proper combinations of mutant genes made by selected crosses between previously characterized mutants will result in a clarification of the concept that many biochemical mutants are due to internal upsets in the *balance* of metabolic systems. The object should now be to investigate the validity of this concept rigorously by combining unbalanced systems which complement one another and thus reestablish new balances. This is not heterocaryosis, but the synthesis of new balanced systems at a somewhat different level from the old. The evolutionary and biochemical implications of this concept are obvious.

SUMMARY

The pantothenicless mutants, 5531 and 34556, are active in the synthesis of pantothenic acid *in vivo* when the cultures are aerated. When 5531 is combined with the mutants UT 1 and UT39, the double mutants do not require pantothenic acid for growth. The bearing of these results on the analysis of genetic blocks is discussed.

LITERATURE CITED

Beadle, G. W., 1945, Biochemical genetics. Chem. Rev. 37: 15–96.

Dagley, S., E. A. Dawes and G. A. Morrison, 1950, Production of amino-acids in synthetic media by *Escherichia coli* and *Aerobacter aerogenes*. Nature 165: 437–438.

Emerson, S., 1948, A physiological basis for some suppressor mutations and possibly for one gene heterosis. Proc. nat. Acad. Sci. 34: 72–74.

Giles, Norman H., Jr., and Ester Z. Lederberg, 1948, Induced reversions of biochemical mutants in *Neurospora crassa*. Amer. J. Bot. 35: 150–157.

Haddox, C. H., 1951, Phenylalanine and tyrosine mutants of *Neurospora crassa*. Thesis. University of Texas.

Horowitz, N. H., 1949, Biochemical genetics of Neurospora. Advances in Genetics 3: 33–71.

Houlahan, M. B., and H. K. Mitchell, 1948, Evidence for an interrelation in the metabolism of lysine, arginine, and pyrimidines in Neurospora. Proc. nat. Acad. Sci. 34: 465–470.

McElroy, W. D., and H. K. Mitchell, 1946, Enzyme studies on a temperature sensitive mutant of Neurospora. Federation Proceedings 3: 376–379.

Mitchell, H. K., and M. B. Houlahan, 1946a, Neurospora IV. A temperature-sensitive riboflavinless mutant. Amer. J. Bot. 33: 31–35.

    1946b, Adenine-requiring mutants of *Neurospora crassa*. Federation Proceedings 3: 370–375.

Mitchell, H. K., and J. Lein, 1948, A Neurospora mutant deficient in the enzymatic synthesis of tryptophane. J. Biol. Chem. 175: 481–482.

Mitchell, M. B., and H. K. Mitchell, 1950, The selective advantage of an adenineless double mutant over one of the single mutants involved. Proc. nat. Acad. Sci. 36: 115–119.

Stockes, J. L., J. W. Foster, and C. R. Woodward, 1943, Synthesis of pyrodoxin by a "pyridoxinless" X-ray mutant of *Neurospora sitophila*. Arch. Biochem. 2: 235–245.

Strauss, B. S., 1951, Studies on the vitamin B_6, pH sensitive mutants of *Neurospora crassa*. Arch. Biochem. 30: 292–305.

Wagner, R. P., 1949, The *in vitro* synthesis of pantothenic acid by pantothenicless and wild type Neurospora. Proc. nat. Acad. Sci. 35: 185–189.

Wagner, R. P., and B. M. Guirard, 1948, A gene-controlled reaction in Neurospora involving the synthesis of pantothenic acid. Proc. nat. Acad. Sci. 8: 398–402.

GALACTOSEMIA, A CONGENITAL DEFECT IN A NUCLEOTIDE TRANSFERASE: A PRELIMINARY REPORT

By Herman M. Kalckar, Elizabeth P. Anderson,*
and Kurt J. Isselbacher

NATIONAL INSTITUTE OF ARTHRITIS AND METABOLIC DISEASES
NATIONAL INSTITUTES OF HEALTH, UNITED STATES PUBLIC
HEALTH SERVICE, BETHESDA, MARYLAND

Communicated by Gerty T. Cori, December 22 1955.

Galactosemia is a disease of childhood which manifests itself biochemically as a disorder in the metabolism of galactose (cf. Hartmann[1]) as well as of α-galactose-1-phosphate (Gal-1-P).[2] The latter ester has recently been shown to accumulate in the blood of galactosemic infants after the administration of galactose.[2]

According to Leloir,[3] the conversion of Gal-1-P to α-glucose-1-phosphate (G-1-P) involves the following steps, in which a nucleotide, uridinediphospho-glucose (UDPG) is required:

$$\text{Gal-1-P} + \text{UDPG} \rightleftharpoons \text{G-1-P} + \text{UDPGal}, \tag{1}$$
$$\text{UDPGal} \rightleftharpoons \text{UDPG}. \tag{2}$$

Step (2) is catalyzed by an enzyme discovered by Leloir[4] in 1951 and called "galacto-waldenase." Step (1) was found to be catalyzed by an enzyme which has been found in galactose-adapted yeast[5] and mammalian liver.[6] We call this enzyme "PGal-transferase." It has been postulated at various times that galactosemia is a defect due to a block in the synthesis of galacto-waldenase. The fact that Gal-1-P accumulates in the erythrocytes of galactosemic subjects when galactose is administered, together with the fact that the clinical symptoms completely disappear if the patients are put on a galactose-free diet, would be compatible with a block in step (1), i.e., PGal-transferase, rather than with a block in step (2) catalyzed by galacto-waldenase.

Methods and Materials.—Hemolyzates were used as the enzyme source. The incubation time was 30 minutes. One sample was incubated with Gal-1-P alone and the second with UDPG alone. The missing substrate was added after deproteinization. The third sample was incubated with both substrates. For the determination of Gal-1-P, UDPG, and UDPGal specific enzymatic methods were used. The indicator in these methods is one of the pyridine nucleotides (TPN or DPN), both of which upon reduction develop an increase in absorption at 340 mμ. The Gal-1-P was determined by means of its liberation of G-1-P from UDPG, using purified PGal transferase and a TPN indicator system.[7]

UDPG was determined by means of a specific, purified UDPG dehydrogenase,[8] using DPN as indicator. UDPGal was determined by the same principle, except that galacto-waldenase (fractionated from liver[6]) was also added.[9] Alpha Gal-1-P was kindly made available to us through Drs. Hewitt G. Fletcher, Jr., and Elizabeth Maxwell. UDPG was a commercial product from the Sigma Chemical Company, St. Louis, Missouri. UDPGal was prepared by enzymatic techniques.[6]

Results.—We have confirmed and extended the observation of Schwartz *et al.*[2] that galactose added to erythrocytes from galactosemic children gives rise to accumulation of Gal-1-P. The erythrocytes were incubated in vitro with galactose for three hours at 37° and the Gal-1-P determined enzymatically.[7] Blood from normal children showed no accumulation of Gal-1-P under these conditions (less than 0.01 μM per milliliter of red blood cells per hour), whereas, in three cases of galactosemia, about 0.1 μM of Gal-1-P accumulated per milliliter of red blood cells per hour.

The presence of PGal transferase was studied by means of incubating Gal-1-P and UDPG together with hemolyzates and subsequently measuring by enzymatic techniques the conversion of UDPG to UDPGal.[8, 9]

In order to insure that the hemolyzates were not fortuitously varying in activity with respect to this class of enzymes, another nucleotide transferase which we call "PP transferase,"[10] was also measured by the same type of method and at the same time as the PGal transferase. The PP transferase catalyzes the following reaction, which involves inorganic pyrophosphate (PP) and uridine triphosphate (UTP): G-1-P + UTP ⇌ PP + UDPG. The PP transferase does not play a direct role in the metabolism of Gal-1-P.[7] It brings about a conversion of G-1-P to UDPG, which by subsequent enzymatic dehydrogenation is converted to UDP-glucuronic acid,[11] or, alternatively, without oxidation can be converted into UDPGal by the above-mentioned enzyme, galacto-waldenase.[4]

In Table 1, average values of PGal and PP transferases for three groups of

TABLE 1

AVERAGE RATES FOR PGAL AND PP TRANSFERASES IN HUMAN HEMOLYZATES FROM NONGALACTOSEMIC AND GALACTOSEMIC SUBJECTS

No. of Cases	Types of Cases	Diet	μM UDPG Exchanged per Milliliter Red Blood Cells per Hour	
			PGal Transf.	PP Transf.
11	Normal	Ordinary	0.75	1.10
3	Milk allergies	Galactose-free	0.88	1.35
8	Galactosemic	Galactose-free	<0.02	1.85

subjects are presented. The first group comprises normal subjects of various ages (male and female); the second group deals with individuals on galactose-free diets on account of milk allergy (three cases); and the third group deals with galactosemic subjects (eight cases). It can be seen that blood from the galactosemic subjects is devoid of PGal transferase (less than 3 per cent, which means essentially undetectable). The PP transferase, however, is constantly present not only in the group of normal and milk-allergy subjects but also in the galactosemic subjects.

We therefore propose that galactosemia is an inborn defect, presumably of genetic origin,[12] in the production of PGal transferase.

Summary.—Galactosemia seems to furnish an example of a congenital human metabolic disease in which a specific enzyme is missing. The enzyme which catalyzes the exchange of α-galactose-1-phosphate with uridinediphospho-glucose, forming α-glucose-1-phosphate and uridinediphospho-galactose, is absent in blood from galactosemic subjects. It is known that this enzymatic exchange is an important step reaction by which administered galactose is used in general carbohydrate metabolism. Several of the metabolic manifestations of the disease might readily be explained on the basis of this enzymatic defect.

Our thanks are due to Drs. G. T. Cori, V. O'Donnell, J. Kety, H. H. Mason, R. Harris, and A. Hartmann for their kind help in obtaining the galactosemic cases, and to Miss Bodil Waage-Jensen, who rendered valuable assistance as a trainee under the Scandinavian-American Foundation through a grant-in-aid generously made available to one of us (H. M. K.) by the Eli Lilly Laboratories.

[*] Fellow in Cancer Research of the American Cancer Society.

[1] A. Hartmann, *J. Pediat.*, **47**, 537, 1955.

[2] V. Schwartz, L. Golberg, G. M. Komrower, and A. Holzel, *Biochem. J.*, **59**, xxii, 1955.

[3] L. F. Leloir, in W. D. McElroy and B. Glass (eds.), *Phosphorus Metabolism* (Baltimore: Johns Hopkins Press, 1951), **1**, 67.

[4] L. F. Leloir, *Arch. Biochem. and Biophys.*, **33**, 186, 1951.

[5] H. M. Kalckar, B. Braganca, and A. Munch-Petersen, *Nature*, **172**, 1039, 1953.

[6] E. Maxwell, H. M. Kalckar, and R. Burton, *Biochim. et biophys. acta*, **18**, 389, 1955.

[7] A. Munch-Petersen, H. M. Kalckar, and E. E. B. Smith, *Kgl. Danske Videnskab. Selskab Biol. Medd.*, Vol. **22**, No. 7, 1955.

[8] J. L. Strominger, E. Maxwell, and H. M. Kalckar, in S. P. Colowick and N. O. Kaplan (eds.), *Methods in Enzymology*, Vol. **3** (in press).

[9] H. M. Kalckar and E. P. Anderson, *Biochim. et biophys. acta* (in press).

[10] A. Munch-Petersen, H. M. Kalckar, E. Cutolo, and E. E. B. Smith, *Nature*, **172**, 1036, 1953.

[11] J. L. Strominger, H. M. Kalckar, J. Axelrod, and E. Maxwell, *J. Am. Chem. Soc.*, **76**, 6411, 1954.

[12] A. Holzel and G. M. Komrower, *Arch. Disease Childhood*, **30**, 155, 1955.

VII
Mutant Genes and Mutant Proteins

Editor's Comments on Papers 16 Through 18

16 Suskind, Yanofsky, and Bonner: *Allelic Strains of* Neurospora *Lacking Tryptophan Synthetase: A Preliminary Immunochemical Characterization*

17 Bonner, Suyama, and DeMoss: *Genetic Fine Structure and Enzyme Formation*

18 Ingram: *Gene Mutations in Human Haemoglobin: The Chemical Difference Between Normal and Sickle Cell Haemoglobin*

During the pursuit of the gene–enzyme problem, it became evident that a method was needed to recognize altered proteins which did not have enzyme activity. The important contribution of Paper 16 was to supply one such method, which proved to be a powerful tool. By making antibodies to tryptophan synthetase, these workers were able to establish that many *Neurospora* mutants which did not make active tryptophan synthetase did indeed make a protein immunologically similar to it. This essentially proved that mutations may result in the formation of altered proteins, which may have no enzymatic activity but sufficient similarity to the unaltered wild-type protein to have the same immunological specificity.

The tryptophan synthetase mutants of *Neurospora*, which Bonner and his group started to work with in the 1950s, proved to be a storehouse of information. Paper 17, which is in part a review and synopsis of work done over a period of time in Bonner's laboratory, makes several important points. First, building upon the Watson–Crick model of DNA, which by 1960 was generally accepted, if the base sequence of DNA dictates the amino acid sequence in a protein, then changes in base sequence by mutation should result in altered amino acid sequence. Second, the large number of tryptophan synthetase mutants, for which detailed characteristics are given, can be grouped into eight different categories depending on the effect of the mutation on the enzyme (see Table 1 in paper). Third, there is a reasonable correlation between the location of the mutational sites and phenotype of the enzyme. That is, similar mutant effects map together.

The ideas in this paper presaged the events that followed in the next 10 years, primarily in the laboratory of Yanofsky. His findings on the tryptophan synthetase of *Escherichia coli* are described in Papers 27, 28, 29, and 33.

In 1949, Pauling (see the paper by Pauling et al. on sickle-cell anemia reproduced in Schull's volume, *Medical Genetics,* in this series) and his coworkers reported that hemoglobin from persons with sickle-cell anemia had an electrophoretic mobility different from normal hemoglobin. Coupled with Neel's demonstration made in the same year (1949) that sickle-cell anemia is inherited as a single gene trait, Pauling's finding revealed, in his words, "a clear case of a change produced in a protein molecule by an allelic change in a single gene involved in synthesis."

So far so good, but it remained to show precisely what this change is in chemical terms. This next step was taken by Ingram, who shows in Paper 18 that, out of nearly 300 amino acids in hemoglobin, sickle-cell hemoglobin has only one difference from normal hemoglobin: one peptide obtained by hydrolysis has valine in place of the glutamic acid in the corresponding hemoglobin of the normal. This was a finding of profound significance. It established that what appeared to be a single mutation in a gene could result in a single amino acid substitution. Hence for the first time the gene–protein relationship was tied down at the molecular level.

Selected Bibliography

Dintzis, H. M. 1961. Assembly of the peptide chains of hemoglobin. *Proc. Natl. Acad. Sci. U.S.* 47: 247–261.
Naughton, M. A., and H. M. Dintzis. 1962. Sequential synthesis of the peptide chains of hemoglobin. *Proc. Natl. Acad. Sci. U.S.* 48: 1822–1830.
Neel, J. V. 1949. The inheritance of sickle cell anemia. *Science 110:* 64.
Stamatoyannopoulos, G. 1972. The molecular basis of hemoglobin disease. *Ann. Rev. Genetics 6:* 47–70.

ALLELIC STRAINS OF NEUROSPORA LACKING TRYPTOPHAN SYNTHETASE: A PRELIMINARY IMMUNOCHEMICAL CHARACTERIZATION*

By Sigmund R. Suskind,† Charles Yanofsky,‡ and David M. Bonner

DEPARTMENT OF MICROBIOLOGY, YALE UNIVERSITY, NEW HAVEN, CONNECTICUT

Communicated by J. S. Fruton, June 6, 1955

The action of genetic material and its relation to enzyme biosynthesis has only in recent years received attention from an experimental standpoint. A variety of enzyme changes has been found to be associated with genetic alterations, including the formation of altered enzymes, the modifications in various quantitative aspects of enzyme formation, and the complete loss of ability to form specific enzymes (see review[1]). The first two categories have received some attention in previous investigations. The last category, which might be the most informative, has received scant attention in the past, largely because the lack of enzyme activity severely limits a direct experimental approach. However, the absence of detectable enzyme activity need not abolish all hope of examining such a condition. Even if the ability to form a specific enzyme were lost, there might remain the synthesis of proteins antigenically similar to that enzyme, and the presence of these could be investigated by immunochemical methods. This approach has recently been employed in studies of the β-galactosidase system of *Escherichia coli*[2] and of the tyrosinase of *Glomerella*.[3]

The present paper is a preliminary report on the application of immunochemical techniques to problems of gene action and of the control of tryptophan synthetase formation in *Neurospora crassa*. This enzyme was selected for study because a series of well-characterized alleles affecting its formation was available,[4,5] as well as detailed information about the enzyme itself.[6,7,8]

Materials and Methods.—The various tryptophan independent (T^+) and tryptophan requiring (T^-) strains of *N. crassa* used in this work are listed in Table 1.

TABLE 1

STRAINS OF *N. crassa* EMPLOYED IN THE IMMUNOLOGICAL EXPERIMENTS

Strain	Tryptophan Requirement
Wild type (5256 A)	Tryptophan independent (T^+)
td_1, td_2, td_3, td_6, td_7	Tryptophan requiring (T^-)
$td_1 su_2$	Tryptophan requiring (T^-)
$td_2 su_2$, $td_2 su_3$, $td_6 su_6$	Tryptophan independent (T^+)

The tryptophan requiring strains employed, td_1, td_2, td_3, td_6, and td_7,[4,5] represent independent mutations affecting the ability to convert indole to tryptophan. These mutants specifically lack tryptophan synthetase.[4,5,7] Genetic tests have indicated that these strains represent mutations of the same locus or of closely linked interdependent genetic material.[4,5,9] The tryptophan independent T^+ strains employed were of two types: parental strains and suppressed mutant strains (td su). As previously reported,[4,5] suppressor mutations restore to specific td mutants the ability to grow in the absence of tryptophan and to form detectable amounts of tryptophan synthetase.

T+ cultures were grown with aeration in 20 liters of glucose (1 per cent) minimal medium[10] for 60–70 hours at 30° C.; T- cultures were grown in the presence of a supplement of 3 gm. of DL-tryptophan. The preparation, purification, and method of assay of extracts have previously been described.[8] The tryptophan synthetase unit (TSU) employed is the amount of enzyme which will catalyze the conversion of 0.1 μ mole of indole to tryptophan in a 60-minute incubation at 37° in the presence of saturating concentrations of indole, serine, and pyridoxal phosphate. Protein was determined by the methods of Lowry[11] and of Robinson et al.[12]

For immunization purposes an eightfold purified preparation of tryptophan synthetase was used as antigen. Four rabbits received approximately the same total dose of enzyme, 150 tryptophan synthetase units in about 45 mg. of N. crassa protein. All the rabbits were bled before receiving enzyme and again 6 weeks after the first injection. Inactivation of tryptophan synthetase by normal serum does not occur during incubation periods up to 1 hour. Normal serum inactivation of tryptophan synthetase does become a problem, however, if long incubation periods are used. This problem can be overcome by the use of sera which have been heat-treated and fractionated by the ammonium sulfate method of Tiselius.[13] Thus, for incubation periods up to 1 hour, whole untreated sera could be used. For long incubation periods, fractionated, heat-treated (56° C. for 30 minutes) sera were employed.

Precipitin reactions were ordinarily carried out at 5° C. At various times aliquots were removed, centrifuged in the cold, and carefully decanted. Enzyme determinations were routinely made on the supernatant solutions and occasionally on the precipitates. Enzyme assays were also run on uncentrifuged samples.

Absorption experiments were performed in a similar manner; the absorbing antigen was allowed to react with the serum at 5° C. for several hours before the addition of active enzyme. The removal of antibody against the active enzyme by this procedure was considered evidence for the presence of cross-reacting antigens in the absorbing extract. Inhibition values given are based on differences in activity between control serum and antiserum.

Results.—The antitryptophan synthetase activity of whole and fractionated immune serum was determined by measuring the reduction in activity of tryptophan synthetase preparations incubated with control serum and antiserum. From the inhibition values shown in Table 2, it can be seen that immunization with tryptophan synthetase preparations effectively elicits the formation of antibodies against this enzyme.

TABLE 2

INHIBITION OF TRYPTOPHAN SYNTHETASE ACTIVITY BY ANTISERUM, AS A FUNCTION OF TIME

Time at 5° C. (Hours)	Per Cent Inhibition by Antiserum	Time at 5° C. (Hours)	Per Cent Inhibition by Antiserum
0.25	65	48	72
1	66	72	69
24	74	168	53

Two tubes were tested at each time interval, one containing nonimmune (control) serum, and the other, antitryptophan synthetase antiserum. To each tube were added 1.0 ml. of a 1:1 dilution of the appropriate fractionated serum in phosphate buffer (0.1 M, pH 7.8), 0.2 ml. enzyme (14 TSU), 0.04 ml. pyridoxal phosphate (40 μgm.), and 0.1 ml. glutathione (5 × 10^{-3} M). At the indicated times the tubes were centrifuged, and the supernatants tested for enzyme activity. Inhibition values are based on difference in activity between control serum and antiserum.

It was of interest to determine two characteristics of this enzyme–anti-enzyme reaction: first, the time course of the reaction and, second, whether the antiserum neutralized the enzyme's activity or merely precipitated the enzyme. From Table 2 it can be seen that the reaction, measured in terms of tryptophan synthetase inhibition, is complete in less than 1 hour. It may be seen from Table 3 that the antiserum neutralizes enzyme activity, since there is no difference in percentage inhibition of enzyme activity between mixtures tested before centrifugation and those tested after centrifugation. Furthermore, no enzyme activity could be detected in the precipitate.

TABLE 3

NEUTRALIZATION OF TRYPTOPHAN SYNTHETASE ACTIVITY BY ANTISERUM

	Per Cent Enzyme Activity Remaining in		
Antiserum	Uncentrifuged Sample	Supernatant Solution after Centrifugation	Precipitate
1	70	74	...
2	52	46	...
3	0	0	0

Determinations were carried out in nonimmune serum and antitryptophan synthetase antiserum, and activity values are based on the difference in activity between these two sera. Antiserum from three rabbits was tested. To each tube were added 1.0 ml. of a 1:1 dilution of the appropriate fractionated serum in phosphate buffer (0.1 M, pH 7.8), 0.8 ml. enzyme (14 TSU), 0.04 ml. pyridoxal phosphate (40 μgm.), and 0.1 ml. glutathione (5 × 10^{-3} M). Tubes were stored at 5° C. for 40 hours before assay.

With a constant amount of antiserum and increasing amounts of enzyme, it is possible to standardize a given immune serum in terms of its antitryptophan synthetase activity. Absorption experiments were then carried out with extracts of various mutants to test for the presence in them of proteins antigenically related to tryptophan synthetase.

Extracts of strains td_1 and td_2 were employed in the initial absorption experiments. Aliquots of these extracts were incubated with antiserum of known antitryptophan synthetase titer for 1 hour at 5° C., the mixture centrifuged, and the supernatant solutions tested for their antitryptophan synthetase activity. As can be seen from the data in Table 4, strain td_2 contains proteins sufficiently similar to

TABLE 4

ABSORPTION OF ANTITRYPTOPHAN SYNTHETASE ANTISERUM WITH CRUDE AND FRACTIONATED PREPARATIONS OF STRAINS td_1 AND td_2

Absorbing Antigen	Test Antigen	Per Cent Inhibition of Tryptophan Synthetase Activity by Absorbed Antiserum
None	Tryptophan synthetase	100
td_1—crude extract	Tryptophan synthetase	100
td_1—dialyzed extract	Tryptophan synthetase	100
td_1—fractionated extract	Tryptophan synthetase	100
td_2—crude extract	Tryptophan synthetase	10
td_2—dialyzed extract	Tryptophan synthetase	12
td_2—fractionated extract	Tryptophan synthetase	3

To each tube were added 0.5 ml. of the appropriate fractionated serum, 0.5 ml. absorbing antigen, 0.04 ml. pyridoxal phosphate (40 μgm.), and 0.05 ml. glutathione (2.5 × 10^{-3} M). Tubes were adjusted to equal volume with phosphate buffer (0.1 M, pH 7.8). Absorption was continued for 2 hours at 5° C. before the test antigen (0.05 ml. enzyme [7 TSU]) was added. After an additional 30 minutes at 5° C., the tubes were assayed for enzyme activity. Determinations were carried out in nonimmune serum and antitryptophan synthetase antiserum, and activity values are based on difference.

tryptophan synthetase to react with and remove antibody to the enzyme, thereby reducing the anti-enzyme activity of the serum. Furthermore, the same results are obtained regardless of whether the antiserum-extract mixtures are centrifuged after absorption or not. This observation indicates that the antigenically similar material inactivates or blocks the activity-neutralizing site of the anti-enzyme.

Of particular interest is the finding that extracts of strain td_1 do not contain cross-reacting protein.

Extracts of td_2 were fractionated in a manner similar to that used for purifying tryptophan synthetase preparations, and the final fraction tested for cross-reacting material. It can be seen in Table 4 that the cross-reacting material in extracts of strain td_2 is similar to tryptophan synthetase in its behavior during fractionation. Similar treatment of td_1 extracts failed to yield a preparation containing cross-reacting material.

In view of the differences observed between extracts of strains td_1 and td_2, extracts of other td mutants were examined for antigens which would absorb antitryptophan synthetase antibodies. As may be seen from Table 5, strains td_3, td_6, and td_7 all form proteins which cross-react with antitryptophan synthetase.

TABLE 5

ADSORPTION OF ANTITRYPTOPHAN SYNTHETASE ANTISERUM WITH EXTRACTS OF VARIOUS td MUTANTS AND OF SEVERAL td MUTANTS CARRYING SPECIFIC SUPPRESSOR GENES

Absorbing Antigen	Test Antigen	Per Cent Inhibition of Tryptophan Synthetase Activity by Absorbed Antiserum
td_1su_2—crude extract (0 TSU)	Tryptophan synthetase	100
td_2su_2—crude extract (0.3 TSU)	Tryptophan synthetase	19
td_2su_2—dialyzed extract (0 TSU)	Tryptophan synthetase	0
td_2su_2—fractionated extract (0.4 TSU)	Tryptophan synthetase	4
td_6su_6—crude extract (1 TSU)	Tryptophan synthetase	38
td_6—crude extract (0 TSU)	Tryptophan synthetase	36
td_3su_3—crude extract (0.65 TSU)	Tryptophan synthetase	16
td_3—crude extract (0 TSU)	Tryptophan synthetase	5
td_7—crude extract (0 TSU)	Tryptophan synthetase	13

To each tube were added 0.5 ml. of the appropriate fractionated serum, 0.5 ml. absorbing antigen, 0.04 ml. pyridoxal phosphate (40 μgm.), and 0.1 ml. glutathione ($5 \times 10^{-3} M$). Absorption was continued for 2 hours at 5° C. before the test antigen was added (0.2 ml. enzyme [9 TSU]). After an additional 30 minutes at 5° C., the tubes were assayed for enzyme activity. Determinations were carried out in nonimmune serum and antitryptophan synthetase antiserum, and activity values are based on difference.

Several experiments were carried out with extracts of strains carrying suppressor genes. As may be seen in Table 5, extracts of td_1su_2 (strain td_1 is unaffected by su_2) do not contain cross-reacting proteins. Thus td_1su_2 and td_1 are similar in this respect. It can also be seen in this table that extracts of td_2su_2 *do* remove antibodies to tryptophan synthetase. This is not unexpected, since strain td_2su_2 forms small amounts of tryptophan synthetase.[4] However, what is surprising is that td_2su_2 extracts remove antibody out of proportion to the amount of active tryptophan synthetase they contain. This would suggest that this strain forms the antigen or antigens characteristic of td_2 in addition to the normal enzyme. The results obtained with extracts of strains td_6su_6 and td_3su_3 appear to be similar to those found with td_2su_2. Thus it would appear that suppressed strains possessing suboptimal tryptophan synthetase levels continue to form another protein or proteins, antigenically related to tryptophan synthetase but having no tryptophan synthetase activity.

Discussion.—The various tryptophan requiring strains used in these investigations are noteworthy in that they possess a number of properties in common. All these strains have an absolute requirement for tryptophan which cannot be satisfied by either indole or anthranilic acid; they have similar requirements for tryptophan; they are characterized by their complete lack of tryptophan synthetase activity; and they all appear to be allelic. In view of these similarities, it may be said that these strains represent mutations of the same or of closely linked interdependent genetic material at the same biochemical locus.[4,5]

These same mutants, however, do show some very striking differences, particularly in their response to specific suppressor genes;[4,5] a given suppressor restores tryptophan synthetase formation only in combination with a particular td allele.

The results of the present investigation further demonstrate that differences exist between these allelic strains. Some of the mutants form material closely related antigenically to tryptophan synthetase, but this property is not characteristic of all the alleles. Whether the antigens formed by the various strains differ is not as yet known. It may well be that we are dealing with differences other than the presence or absence of a single cross-reacting antigen. Hence these experiments further strengthen the conclusion that a locus controlling the formation of a specific enzyme can be exceedingly complex and subject to numerous changes. The elucidation of differences between alleles would thus appear to be dependent upon the number of criteria that are applied.

The fact that td_1su_2 does not form cross-reacting material demonstrates that it is the mutant locus and not the suppressor which controls the production of the antigen or antigens. Of particular interest in this regard is the observation that td_2su_2, td_3su_3, and td_6su_6 appear to form both cross-reacting material and tryptophan synthetase. This would suggest that suppressor genes do not completely overcome defects in tryptophan synthetase formation, a conclusion consistent with the fact that suppressed strains form considerably less tryptophan synthetase than the wild-type strain.

One additional point of interest is that all attempts to obtain a suppressor for strain td_1, the strain which does not form cross-reacting material, have been unsuccessful.[4,5] Td_2 and all the alleles studied except td_7 form cross-reacting material and are suppressed by specific suppressor genes. The suppressibility of strain td_7 has not as yet been demonstrated. These data seem to suggest that the ability to form cross-reacting material and suppressibility may be related. The validity of this correlation must await further study, particularly in cases such as that of td_7.

What is the relationship of the cross-reacting antigen or antigens to tryptophan synthetase? Certainly it is not known at this time, although a great many possibilities exist. The cross-reacting material could be enzyme precursor, altered or incomplete forms of the enzyme, inactivated tryptophan synthetase, or a protein or proteins formed along the same biosynthetic pathway as tryptophan synthetase. Further study is necessary to answer the question.

In any event, the presence or absence of proteins antigenically related to tryptophan synthetase in different members of a group of very similar mutants does suggest that these differences may be a reflection of defects in specific and separate phases of the synthesis of tryptophan synthetase.

Summary.—Preliminary data have been presented on the immunochemical characterization of the tryptophan synthetase–antitryptophan synthetase reaction. Antibody against partially purified tryptophan synthetase preparations from *N. crassa* was found to neutralize enzyme activity. Extracts of several allelic tryptophan requiring mutants, which lack the enzyme tryptophan synthetase, were examined for the presence of proteins which cross-react with antitryptophan synthetase antibody. All but one of the mutants studied were found to possess such material. The cross-reacting material in the mutants behaved as did tryptophan synthetase during the course of purification. Mutants which carry specific suppressor genes, with the attendant formation of low levels of tryptophan synthetase activity, were found to form both active enzyme and cross-reacting material.

The authors wish to thank Dr. A. M. Pappenheimer, Jr., for his helpful advice on certain phases of this work.

* The work reported in this paper was supported in part by the Atomic Energy Commission (Contract AT [30-1] 1017) and in part by the American Cancer Society on recommendation by the Committee on Growth.

† Present address: Department of Microbiology, School of Medicine, New York University.

‡ Present address: Department of Microbiology, School of Medicine, Western Reserve University, Cleveland, Ohio.

[1] D. M. Bonner, *Cold Spring Harbor Symposium Quant. Biol.*, **16**, 143, 1951.

[2] M. Cohen and A. M. Torriani, *J. Immunol.*, **69**, 471, 1952; *Biochim. et Biophys. Acta*, **10**, 280, 1953.

[3] C. L. Markert and R. D. Owen, *Genetics*, **39**, 818, 1954.

[4] C. Yanofsky, these PROCEEDINGS, **38**, 215, 1952.

[5] C. Yanofsky and D. M. Bonner, *Genetics* (in press).

[6] W. W. Umbreit, W. A. Wood, and C. I. Gunsalus, *J. Biol. Chem.*, **165**, 731, 1946.

[7] H. K. Mitchell and J. Lein, *J. Biol. Chem.*, **175**, 481, 1948.

[8] C. Yanofsky, *Methods in Enzymology*, ed. S. P. Colowick and N. O. Kaplan (New York: Academic Press, 1955).

[9] D. Newmeyer, *Genetics*, **39**, 604, 1954.

[10] G. W. Beadle and E. L. Tatum, *Am. J. Bot.*, **32**, 678, 1945.

[11] C. H. Lowry, N. J. Rosebrough, A. L. Farr, and R. J. Randall, *J. Biol. Chem.*, **193**, 265, 1951.

[12] H. W. Robinson and C. G. Hogden, *J. Biol. Chem.*, **135**, 727, 1940.

[13] A. Tiselius, *Biochem. J.*, **31**, 313, 1464, 1937.

Copyright © 1960 by the Federation of American Societies for Experimental Biology

Reprinted from Fed. Proc., 19(4), 926–930 (1960)

Genetic fine structure and enzyme formation[1]

D. M. BONNER[2], Y. SUYAMA[2] AND J. A. DEMOSS[2]
Department of Microbiology, Yale University, New Haven, Connecticut

DURING THE PAST TEN YEARS clear proof that enzyme formation is genetically controlled has been obtained. This proof has come largely from the study of microorganisms and stems from the efforts of many investigators (1, 2). Extensive evidence has also been obtained indicating that the structural characteristics of the formed enzyme are genically determined (3, 4). Recent work additionally points to the conclusion that quantitative aspects of enzyme formation is an action of genetic material and is distinct from the action controlling the qualitative characteristics of the formed enzyme (5). It then appears that both the qualitative and the quantitative characteristics of enzyme formation are gene determined. The knowledge that enzyme formation is under the control of the genetic elements of the cell and that there exists an apparent one-to-one relation of gene to enzyme, viewed with the Watson-Crick model of DNA and the observed amino acid difference of genetically distinct hemoglobins (6), has led to the widespread belief that the base sequence of the DNA of the gene serves as a code, specifying the amino acid sequence of the formed enzyme. Alteration of the base sequence, in turn, is thought to alter the amino acid sequence and so results in the formation of an altered enzyme. This theory is of great interest at the present time and experimental evidence is in general agreement with such a postulation. It is known that mutation leads to formation of altered enzymes, though it has not been shown that the alteration results from an amino acid sequence alteration. While there is much that is in agreement there are odd bits and pieces of information which suggest that the theory may well be modified with time. It is obvious that a knowledge of the base sequence of a gene and the amino acid sequence of its enzyme would be of great interest. Progress in this field is being made. An understanding of gene and enzyme relationships, however, requires knowledge of sequences, and of genetic fine structure. Genetic fine structural analysis can serve as an excellent tool for directly relating gene structure to the enzyme, when a number of criteria characterizing diverse functions of the latter are available. It is the study of one gene and its product enzyme which will be reviewed here.

The genetic control of tryptophan synthetase formation of *Neurospora crassa* will be reviewed. These studies have been in progress for a number of years and our present knowledge of this system stems from the investigation of a number of different investigators. We might first consider the enzyme. The enzyme, tryptophan synthetase, is required for the catalysis of the terminal step in the formation of tryptophan, i.e. the substitution reaction between indole-glycerol-phosphate (InGP) and serine. This reaction requires as a cofactor pyridoxal phosphate (7, 8). Tryptophan synthetase, however, can carry out two additional reactions (see fig. 1). It will catalyze the conversion of (InGP) to indole (reaction 3) and the conversion of indole to tryptophan in the presence of serine and pyridoxal phosphate (reaction 2) (7, 8). All three activities are associated with a single enzyme as fractionation of the enzyme does not alter the relative activity of the three reactions in question (4, 8, 9; and S. E. Ensign, unpublished observations). Thus, from purification studies there has been no indication of the relative enrichment or loss of one activity, as contrasted to the second or third. Mutational evidence gives additional strong support to the view that the three activities involve a single enzyme species.

Mutations affecting tryptophan synthetase are readily obtained. When tryptophan requiring mutants are isolated, one class is found to lack enzymatic activity for reaction 1. A large number of independently isolated mutants all of which are unable to form tryptophan and all of which lack detectable tryptophan synthetase activity, judged in terms of reaction 1, have been isolated. With such a group of mutants we can determine whether all of the mutations which have resulted in loss of this enzymatic activity are mutations in the same genetic region or are mutations at other regions as well. This can be determined by use of the inheritance test. Such experiments permit the clear conclusion that all of the mutations which have been isolated involve alterations of a restricted region of the genetic material of Neurospora (10). Thus, all of the mutations which have been isolated appear to be clustered in a single region of chromosome number 2 as indicated in figure 2. How complex is this region? Is it a region containing a single mutational site or is it a region containing many mutational sites, or, phrased in experimental terms, are all of the mutant strains which have been isolated identical or are functional differences observed between them? As mentioned earlier all of the mutant strains appear

---

[1] These investigations supported in part by Atomic Energy Commission Contract (AT(30-1)-1017) and the American Cancer Society.

[2] Present address: Dept. of Biology, University of California, La Jolla, Calif.

similar in requiring tryptophan and in lacking enzymatic activity for reaction 1. If one scans these mutants using additional criteria, many differences are found. For instance, it is possible to determine whether a mutation, which has resulted in loss of enzymatic activity, has at the same time resulted in loss of ability to form a protein which is antigenically related to the parental enzyme. By preparing antibodies against the parental enzyme, mutant extracts can be scanned for the presence of material which will react with these antibodies. Such investigations reveal that mutant strains can be subdivided into two major groups on the basis of whether or not they lack a protein which will react with tryptophan synthetase antibodies. Strains lacking a cross reacting material are called CRM−, those having such a material are called CRM+. Detailed investigation of a CRM+ strain has indicated that the material found in at least this one mutant is serologically identical with the enzyme of the parental strain (12). Still other methods of comparison can be employed. Let us consider those strains which are CRM+. As was mentioned earlier all of the mutations which have been investigated to date result in loss of enzymatic activity judged in terms of reaction 1. However, what about reactions 2 and 3? Can mutations occur which result in the loss of reaction 1, but permit retention of reaction 2 or 3? The CRM's of various strains have been tested for their ability to catalyze reactions 2 or 3 and on this basis the CRM+ strains can be subdivided (see table 1). The majority of the strains which form a cross reacting material are found to lack enzymatic activity for reactions 1, 2, or 3. This is not true of all strains, however, since certain strains are found which while having lost the ability to catalyze reactions 1 and 2 retain enzymatic activity judged in terms of reaction 3 (8, 13). The CRM of these strains is active in catalyzing the conversion of InGP to indole. As might be expected, such mutant strains are characterized by their accumulation of indole during growth, in contrast to strains which accumulate the compound indole-glycerol. Thus, we have a second group of CRM+ mutants, a group which might be called indole accumulators. The indole accumulator strains in turn can be subdivided. Purified tryptophan synthetase does not show cofactor requirements for reaction 3, i.e. the InGP → indole reaction. Surprisingly this is not true of all the indole accumulators. If extracts of these mutant strains are purified, one finds that the CRM's of certain mutants are similar to the parental enzyme, i.e. are capable of catalyzing the conversion of InGP to indole in the absence of additional cofactors. However, a second group is found which is capable of catalyzing this reaction only if pyridoxal phosphate is present. Still a third group is found which is capable of carrying out this reaction only if pyridoxal phosphate and serine are both present in the reaction mixture (8). The role of pyridoxal phosphate and serine in this reaction is not known at present. It is possible to compare the affinity constant for serine in the mutant reaction with the affinity constant of serine for the indole serine reaction catalyzed by the parental enzyme. The affinity constants are found to be identical (8). It would appear that mutation has not affected serine affinity, and one can only suggest that serine may be required to permit InGP access to its active site. A novel mutant was discovered by Catcheside (15) which led to the realization that a third class of CRM+ mutants can be found. This group has lost enzymatic activity judged in terms of reactions 1 and 3, but has retained the ability to catalyze reaction 2 (14, 15). Such strains are able to grow on indole in place of tryptophan and are referred to as indole utilizing mutants.

One therefore finds that mutant strains, all of which have lost the ability to catalyze the conversion of indole-

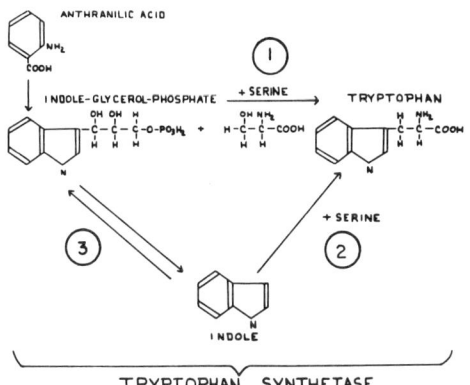

FIG. 1. Reactions catalyzed by tryptophan synthetase.

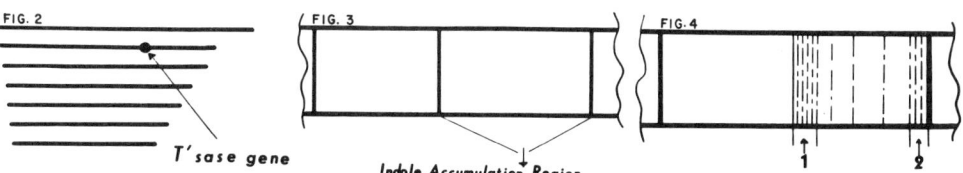

FIG. 2. The chromosomes of *Neurospora*.
FIG. 3. Tryptophan synthetase gene showing regions in which mutations result in indole accumulation.
FIG. 4. Genetic fine structure of the indole accumulation region; *1* requires B_6PO_4; *2* requires B_6PO_4 + serine.

TABLE I.

Functional Differences Found in Various Tryptophan Synthetaseless Mutants

I. CRM−
II. CRM+
  a) Indoleless—reaction 1⁻, 2⁻, 3⁻
  b) Indole accumulator—reaction 1⁻, 2⁻, 3⁺

	B_6PO_4	Serine
1	−	−
2	+	−
3	+	+

  c) Indole utilization—reaction 1⁻, 2⁺, 3⁻

glycerol phosphate to tryptophan, show many differences. Mutation may result in loss of ability to form a protein which is antigenically related to the parental strain. Mutation, however, may also result in the retention of ability to form a protein which is antigenically related to the parental enzyme, but a protein which is unable to catalyze any of the 3 tryptophan synthetase reactions. Mutation can also result in the loss of ability to catalyze reactions 1 and 2 but permit survival of reaction 3, or mutation may result in the loss of the ability to catalyze reactions 1 and 3, but permit the survival of reaction 2. Still other differences are found if mutants are compared in terms of temperature sensitivity or suppressibility (10, 16, 17). In any event, such investigations clearly point to the fact that the genetic region controlling the formation of tryptophan synthetase must be mutationally complex. It must contain many mutational sites, alteration of any one of which will result in a different functional alteration of the formed enzyme. We can enlarge the region of the chromosome controlling the formation of tryptophan synthetase and in the absence of other knowledge simply subdivide it indicating it as a region consisting of a large number of mutational sites. Obviously one would like to know something about the distribution of these sites within the genetic region as a whole. For instance, are the mutational sites arranged in linear array within the region, and do we find a clustering of mutational sites in terms of function, or do we find mutational sites which result in similar functional loss scattered at random throughout the genetic area as a whole? This problem can also be tackled by means of the inheritance test. Two tryptophan synthetase mutants of independent origin (alleles), if crossed with each other may form occasional tryptophan independent, or prototrophic, progeny. One can, therefore, prepare a recombination map in which these strains are mapped in terms of the relative frequency of prototroph formation in intra-allelic crosses. Inasmuch as the frequency of tryptophan independent progeny is low, a genetic analysis of this sort is based solely on the relative frequency of all recombinational events in diverse intra-allelic crosses.

We can now determine whether mutations which yield indole accumulators, i.e. mutations which have permitted the survival of reaction 3, appear to be distributed at random throughout the entire functional genetic region or whether they appear to be clustered. As contemporary theory suggests that mutation results in alteration of the primary structure and since folding characteristics are thought to be determined by the primary structure of the enzyme itself, one might a priori predict that mutations which result in similar functional alterations in a given end product should be distributed throughout the genetic area as a whole. This is not found to be the case (Y. Suyama and D. M. Bonner, manuscript in preparation). Genetic fine structure indicates that mutations which permit survival of reaction 3 are grouped in roughly one half of the genetic area as a whole, and thus do not appear to be at random (fig. 3). A more interesting relationship is found when we study recombination frequency within this group. Mutations which have resulted in a requirement for both pyridoxal phosphate, and serine appear to be clustered at one end of the tryptophan synthetase locus (Y. Suyama and D. M. Bonner, manuscript in preparation). Mutations which have resulted in a requirement for only pyridoxal phosphate also appear to be clustered at the other side of the indole accumulation region (fig. 4). Thus we find a region in which mutations result in retention of ability to catalyze reaction 3 but within this area mutations in one region impose a cofactor requirement for pyridoxal phosphate and serine, while in the second region they impose a requirement for pyridoxal phosphate alone. Scattered between these two major groups are those mutations which show no additional cofactor requirement.

A similar conclusion holds for mutations which have resulted in loss of reactions 1 and 3, but survival of reaction 2, i.e. indole utilizing mutations. The maximum frequency of recombination which has been observed in any intra-allelic cross involves an indole utilizing strain and an indole accumulator strain which requires both pyridoxal phosphate and serine. Mutations which have interfered with the ability of the enzyme to convert InGP to tryptophan but which have not interfered with the ability of the enzyme to catalyze the indole-serine reaction appear to be located at one end of the tryptophan synthetase locus (fig. 5). This observation was initially made with a single mutant strain, as this mutation has appeared only once. However, in reversion experiments as will be mentioned later, additional indole utilizing mutant strains have been obtained. These strains are clustered with the original mutant strain and like it show maximal rates of recombination in crosses with indole accumulator strains which require both pyridoxal phosphate and serine. Thus, as indicated in figure 5, we can delineate an additional region in our genetic map, a region which we can simply indicate as an indole utilization region.

What about mutations which are CRM+ but which have lost the ability to carry out reactions 1, 2, and 3? Such mutations are the majority class and are ones which might perhaps most logically be expected to be scattered at random. Surprisingly enough preliminary

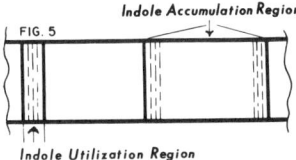 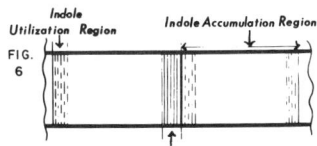 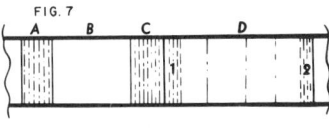

FIG. 5. Indole utilization region.　　FIG. 6. The profound region.　　FIG. 7. Map of t'sase gene, April, 1960. *A*. Indole utilization-triose/InGP site. *B*. Quiet. *C*. Profound-indole site. *D*. Indole accumulation: *1*. B6PO4 site. *2*. Serine site.

genetic analysis indicates that mutations of this group are clustered and are clustered in about the middle of the locus (fig. 6), a region called the profound region, as alteration has a profound effect. The last class that we have then to locate is the class which has not only lost all three enzymatic activities, but which at the same time has lost the ability to form a detectable CRM. Mutant strains of this type show aberrant recombination values characteristic of deletions. This aberrant recombination has been found characteristic of all of the CRM− strains. Not all of these strains result from deletion however, for were this the case, reversion to tryptophan independence should not occur, yet many of the CRM− strains which have been investigated are capable of reverse mutations to tryptophan independence. Recombination characteristics might then suggest that the reversible strains result from multisite mutation, i.e. in contrast to CRM+ strains, these strains might involve alteration at more than one mutational site. This suggestion is supported from reverse mutation studies of CRM− strains which appear to overlap the indole utilization region. Reverse mutation of such CRM− strains yield not only tryptophan independent strains, but also an equal number of strains which are still tryptophan requiring but which can now utilize indole in place of tryptophan. The frequency with which these two types of reverse mutations occur, however, make a multisite explanation unlikely. In fact at the present it is not possible to answer satisfactorily in all cases the nature of the genetic change which has resulted in the formation of the CRM− mutants. However, it is of interest to note that the recombinational characteristics of CRM− strains are distinct from those of the CRM+ and so suggest that alteration of a single mutational site may not yield a CRM− strain.

These data as a whole suggest numerous interesting possibilities. It is clear, that the genetic region controlling the formation of tryptophan synthetase can be divided into two major areas: a genetic area in which mutations appear to affect catalytic activity in reactions involving serine and pyridoxal phosphate, and a second region in which mutations appear to affect the enzyme affinity for InGP. Such a conclusion is not unreasonable in view of the finding by Crawford and Yanofsky (18) that the tryptophan synthetase of *Escherichia coli* will dissociate into two protein moieties, an A and a B protein. Neither the A nor the B protein has substantial activity for any of the three reactions by itself, but the A, B complex does show all three enzymatic activities. However, one of the two proteins again appears to be involved primarily in InGP affinity while the second protein appears to be primarily involved in those reactions involving serine and pyridoxal phosphate. At present we have no indication that the enzyme of *Neurospora* will dissociate into two independent protein fractions. In fact, all of the data at present is consistent with the view that this enzyme is a single moiety (9; S. E. Ensign, unpublished observations; and C. J. Wust, manuscript in preparation). This conclusion is of interest in view of the phenomenon of complementation in *Neurospora* (19–21). The tryptophan synthetase region of *Neurospora* consists of 5–6 complementation groups (15, 21). These complementation groups can be mapped, but inconsistencies between the complementation and recombination map are found. Complementation may not therefore measure discrete functional areas of the genetic region, but may rather measure interaction at the enzyme forming site itself.

The genetic map, however, is of interest for other reasons. This map suggests that mutational sites do not occur at random within the locus. Thus, mutations which can affect the ability of serine to react with indole or InGP are clustered. In view of the presumed dependence of the secondary and tertiary structures of an enzyme on its primary structure, such observations emphasize the need for further knowledge concerning protein structure. Of additional interest is the observation that while mutational sites are linearly arranged in the region they are not uniformly placed, gaps appear to exist (fig. 7). For instance there appears to be a genetically quiet region between the region in which mutations can result in loss of ability to carry out reactions 1 and 2, but permit survival of reaction 3, and the region in which mutation results in loss of the ability to carry out all three reactions. The relative size of this area, judged in terms of recombination, can only be guessed at, for the size of the region is a function of the number of genetic analyses which have been carried out. There is a clear possibility that what is indicated now as a mutationally quiet region may simply reflect the fact that insufficient mutants have been analyzed to date. However, the fact that there is this apparent gap does pose the possibility of there being regions within the gene in which mutations can occur and yet have no substantial effect on the activity of the formed enzyme. This perhaps would be expected in view of the observed clustering of mutational sites. The very fact that mutations affecting specific catalytic activities are not found randomly distributed suggests that regions may well be found in which mutation and amino acid alteration could occur without affecting the catalytic characteristics of the formed enzyme. The region has been indicated as a mutationally quiet region rather than silent, for it is known that mutations which result in CRM— strains can occur within this region and thus mutational events can occur in this region and result in loss of enzymatic activity. Mutationally quiet regions pose a problem in terms of amino acid differences found between presumed mutant and parental enzymes. If mutationally quiet or silent regions do exist, the possibility of having substantial amino acid differences between diverse enzymes of identical enzymatic activity clearly exists. This could pose a thorny problem when comparing the amino acid composition of two catalytically similar enzymes of diverse genetic origins.

SUMMARY

Combined genetic and enzymatic study of tryptophan synthetase formation in *Neurospora crassa* indicate the following: The enzyme is capable of catalyzing the three reactions, *1*) a substitution reaction involving InGP and serine in the presence of pyridoxal phosphate, *2*) a reaction involving indole and serine in the presence of pyridoxal phosphate, and *3*) the conversion of InGP to indole. Mutation in the genetic region controlling the formation of this enzyme results in loss of catalytic activity for reaction 1. Such mutations, however, may result in the formation of an altered enzyme which may be active in reaction 2 or 3. Study of these mutants suggests that the genetic area concerned with tryptophan synthetase formation may be subdivided into two major regions. One of these two regions is concerned with the catalytic activities involving serine and pyridoxal phosphate and the second region is concerned with the catalytic activities involving InGP and indole. The former region may be subdivided into two major areas judged in terms of reaction 3, one in which mutations result in a co-factor requirement for pyridoxal phosphate and serine, and a second area in which mutations result in a cofactor requirement for pyridoxal phosphate. The second major region may be subdivided into two areas, one area in which mutations result in loss of enzymatic activity in all three reactions, and a second area in which mutations result in retention of catalytic activity in reaction 2.

These studies suggest that mutational sites are not randomly distributed in the genetic area which controls tryptophan synthetase formation. Rather mutations which are functionally similar appear to be clustered. Additionally, mutationally quiet regions in the locus appear to exist. These observations permit the construction of a genetic map of the tryptophan synthetase locus. Such a map poses problems of interest in the field of molecular coding and points to avenues of fruitful exploration for the future.

REFERENCES

1. TATUM, E. L. *Science* 129: 1711, 1959.
2. BEADLE, G. W. *Science* 129: 1715, 1959.
3. BONNER, D. M. *Genetics and Cancer.* Univ. of Texas Press, 1959, p. 207.
4. YANOFSKY, C. *Bact. Rev.* 24: 221, 1960.
5. COHEN, G. AND F. JACOB. *Compt. rend.* 248: 3490, 1959.
6. INGRAM, V. M. *Nature* 180: 326–328, 1957.
7. YANOFSKY, C. AND M. RACHMELER. *Biochem. et biophys. acta* 28: 640, 1958.
8. DEMOSS, J. A. AND D. M. BONNER. *Proc. Natl. Acad. Sci.* 45: 1405, 1959.
9. MOHLER, W. C. AND S. R. SUSKIND. *Biochem. et biophys. acta* 43: 288, 1960.
10. YANOFSKY, C. AND D. M. BONNER. *Genetics* 40. 761, 1955.
11. SUSKIND, S. R., C. YANOFSKY AND D. M. BONNER. *Proc. Natl. Acad. Sci.* 41: 577, 1955.
12. SUSKIND, S. R. *J. Bacteriol.* 74: 308, 1957.
13. SUSKIND, S. R. AND E. JORDAN. *Science* 129: 1614, 1959.
14. RACHMELER, M. AND C. YANOFSKY. *Bacteriol. Proc.* 30, 1959.
15. CATCHESIDE, D. G. *Microbial Genetics.* Cambridge Univ. Press, 1960, p. 181.
16. YANOFSKY, C. AND D. M. BONNER. *Rec. Genet. Soc. Amer.* 24: 602, 1955.
17. SUSKIND, S. R. AND L. I. KUREK. *Proc. Natl. Acad. Sci.* 45 193, 1959.
18. CRAWFORD, I. P. AND C. YANOFSKY. *Proc. Natl. Acad. Sci.* 44: 1161, 1958.
19. WOODWARD, D. O., C. W. H. PARTRIDGE AND N. H. GILES. *Proc. Natl. Acad. Sci.* 44: 1237, 1958.
20. FINCHAM, J. R. S. *J. Gen. Microbiol.* 21: 600, 1959.
21. LACY, A. M. Thesis, Yale University, 1959.

GENE MUTATIONS IN HUMAN HÆMOGLOBIN: THE CHEMICAL DIFFERENCE BETWEEN NORMAL AND SICKLE CELL HÆMOGLOBIN

By Dr. V. M. INGRAM

Medical Research Council Unit for the Study of the Molecular Structure of Biological Systems, Cavendish Laboratory, University of Cambridge

I REPORTED recently[1] that the globins of normal and sickle cell anæmia human hæmoglobins differed only in a small portion of their polypeptide chains. I have now found that out of nearly 300 amino-acids in the two proteins, only one is different; one of the glutamic acid residues of normal hæmoglobin is replaced by a valine residue in sickle cell anæmia hæmoglobin. The latter is an abnormal protein which is inherited in a strictly Mendelian manner; it is now possible to show, for the first time, the effect of a single gene mutation as a change in one amino-acid of the hæmoglobin polypeptide chain for the manufacture of which that gene is responsible.

In previous experiments[1], tryptic digests of the two proteins had been prepared; the resulting mixtures of small peptides were separated on a sheet of paper, using electrophoresis in one direction and partition chromatography in the other. These paper chromatograms derived from the two proteins, which I had called 'finger-prints', showed all peptides to have identical electrophoretic and chromatographic properties, except for one spot, peptide No. 4. This occupied different positions in the 'finger-prints' of normal (Hb A) and sickle cell anæmia (Hb S) hæmoglobins, indicating that the difference between the two proteins was located there. I have now determined the chemical constitution of these No. 4 peptides derived from both hæmoglobin A and S.

The hæmoglobin A No. 4 peptide was prepared by elution from the neutral fraction of many 'finger-prints', followed by cooled paper electrophoresis in pyridine/acetic acid/water (pH 3·6 [2]) on Whatman No. 3 MM paper at 30 V./cm. for 75 min. The peptide was obtained as a well-separated band and eluted. The hæmoglobin S No. 4 peptide could be produced in a fairly pure state by eluting the slowest positively charged band from an extended one-dimensional paper electrophoresis of the peptide mixture in the pH 6·4 buffer[1]. In both cases qualitative amino-acid analysis by paper chromatography showed the presence of histidine, valine, leucine, threonine, proline, glutamic acid and lysine. There was more glutamic acid in the hæmoglobin A peptide, but more valine in the hæmoglobin S peptide. In view of the known specificity of trypsin[3], it was to be expected that these peptides, obtained by tryptic hydrolysis, had lysine at the C-terminal end. This agreed with all the results from the partial acid hydrolysis studies, as reported below.

Partial hydrolysis in 12 N hydrochloric acid at 37° C. for two or three days, followed by 'finger-printing'[1], gave the peptides indicated in Fig. 1, and

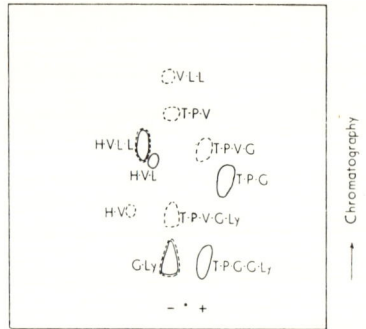

Fig. 1. Acid degradation and structure of the No. 4 peptides from hæmoglobins A and S. Hæmoglobin A (full lines): His—Val—Leu—Leu—Thr—Pro—$\overset{++}{Glu}$—Glu—Lys. Hæmoglobin S (broken lines): $\overset{+}{His}$—Val—Leu—Leu—Thr—Pro—$\overset{+-}{Val}$—$\overset{+-}{Glu}$—Lys

also free amino-acids, which are omitted from the figure. The N-terminal amino-acids of most of these peptides were determined by the fluoro-2,4-dinitrobenzene method[4]. Together with the amino-acid compositions, these fragments indicated the sequences of the No. 4 peptides of hæmoglobins A and S shown in Fig. 1. The only ambiguity was the amino-acid following threonine. Here the relevant products from the hydrochloric acid splitting—threonyl-prolyl-glutamyl-glutamyl-lysine and threonyl-prolyl-valyl-glutamyl-lysine—were subjected, on paper strips, to a stepwise Edman degradation for two cycles[4,5]. The results indicated the sequence threonyl-prolyl- in both cases. The charge distribution of the two No. 4 peptides shown in Fig. 1 was deduced from the electrophoretic behaviour of the two peptides, especially in relation to the behaviour of the smaller split peptides. The only difference found between the two No. 4 peptides is that the first glutamic acid residue of the hæmoglobin A peptide is replaced by valine in the hæmoglobin S peptide.

It is known from X-ray crystallographic[6] and from chemical[7] studies that the human hæmoglobin molecule of molecular weight 66,700 is composed of two identical half-molecules, each approximately 33,000. It is believed that this substitution, which occurs in each of the two identical half-molecules, constitutes the only chemical difference between normal and sickle cell anæmia hæmoglobins. Certainly the hæm groups of the two proteins are the same[8]. The fact that in each half-molecule a glutamic acid is replaced by the neutral amino-acid valine agrees with previous findings that the whole hæmoglobin S molecule has two to three carboxyl groups fewer than the normal protein[9,10]. All the other peptides of the tryptic digest occupy identical, and characteristic, positions in the two 'finger-prints'. Qualitative amino-acid analyses of these peptides have now been carried out, but have failed to reveal any differences between them. It would seem probable, therefore, that they have identical structures, leaving the two No. 4 peptides as the only ones that differ.

About 30 per cent of the hæmoglobin molecule is not susceptible to attack by trypsin and does not appear on the 'finger-print'. To eliminate the possibility that an additional difference resides in these large hæmoglobin A and S fragments, they were digested with chymotrypsin, which attacks them readily. Again two peptide mixtures were obtained, which were examined both by 'finger-printing' and by careful paper chromatography of the neutral peptides. No differences between them could be detected.

We owe to Pauling and his collaborators[9] the realization that sickle cell anæmia is an example of an inherited 'molecular disease' and that it is due to an alteration in the structure of a large protein molecule, an alteration leading to a protein which is by all criteria still a hæmoglobin. It is now clear that, per half-molecule of hæmoglobin, this change consists in a replacement of only one of nearly 300 amino-acids, namely, glutamic acid, by another, valine—a very small change indeed.

Differences between closely related proteins, involving only a very small number of amino-acids, are known; the clearest examples are the differences between horse, whale, sheep, pig and cattle insulins, which show changes in only one sequence of three amino-acids[11]. However, since these are inter-species differences, the genetic mechanism underlying them is by no means clear and cytoplasmic inheritance has not yet been ruled out. The abnormal human hæmoglobins, on the other hand, are a group of very closely related proteins within the same species. It is certain that the inheritance of these proteins is Mendelian in character and occurs through the chromosomal genes. Neel[12] has shown that a single mutational step of such a gene, the one responsible for making hæmoglobin, produces the new abnormal sickle cell anæmia hæmoglobin. Previous investigations on the normal and the sickle cell anæmia protein could not decide whether the difference between them is due to a difference in folding of identical polypeptide chains or to a difference in the amino-acid sequences of the two chains. While there may also be changes in folding, it has now been definitely established that the amino-acid sequences of the two proteins differ, and differ at only one point. Thus it can be seen that an alteration in a Mendelian gene causes an alteration in the amino-acid sequence of the corresponding polypeptide chain. In the case of sickle cell anæmia hæmoglobin, this is the smallest alteration possible —only one amino-acid is affected—reflecting, presumably, a change in a very small portion of the hæmoglobin gene. It is not known, but it may well be that this involves a replacement of no more than a single base-pair in the chain of the deoxyribonucleic acid of the gene.

It is well known that mutations lead to very small chemical differences between, for example, flower pigments[13]. It seems likely that these mutations produce first a change in a protein, in this case probably an enzyme, which in turn causes the production of a changed flower pigment. These enzymes, which have not yet been investigated for differences, stand in closer relationship to the gene than do the flower pigments themselves. The protein hæmoglobin is just as close. It has therefore been called the first gene product, and is probably the first protein to be made by the gene.

The divisibility of genes in a virus was shown previously in bacteriophage by Benzer[14] and Streisinger[15], who studied the effects of many different mutations of a gene on the behaviour of the virus. Such sub-units in genes have also been shown in *Aspergillus*[16] and *Neurospora*[17]. The results presented in this communication are certainly what one would expect on the basis of the widely accepted hypothesis of gene action; the sequence of base-pairs along the chain of nucleic acid provides the information which determines the sequence of amino-acids in the polypeptide chain for which the particular gene, or length of nucleic acid, is responsible. A substitution in the nucleic acids leads to a substitution in the polypeptide.

The abnormally low solubility of reduced hæmoglobin S, which causes the sickling of the erythrocytes in the anæmia, is presumably a function of the charge distribution on the surface of the molecule. The replacement of two charged glutamic acid residues for two uncharged valines is presumably enough to alter this distribution towards one favouring abnormally easy crystallization.

It is hoped that similar studies of other abnormal human hæmoglobins[18] will provide further insight into the effects of gene mutations.

Full details of this work will be published shortly elsewhere. I am indebted to Drs. S. Brenner and G. Seaman, Cambridge, for supplying blood from patients with homozygous sickle cell anæmia. It is a pleasure to acknowledge the constant interest and encouragement shown by Drs. M. F. Perutz and F. H. C. Crick. Some of the enzymatic digestion and 'finger-prints' were done by Mr. J. A. Hunt; Miss Rita Prior rendered invaluable assistance.

[1] Ingram, V. M., *Nature*, **178**, 792 (1956).
[2] Michl, H., *Monatsh. Chem.*, **82**, 489 (1951).
[3] Sanger, F., and Tuppy, H., *Biochem. J.*, **49**, 463, 481 (1951). Sanger, F., and Thompson, E. O. P., *Biochem. J.*, **53**, 353, 356 (1953).
[4] Fraenkel-Conrat, H., Harris, J. I., and Levy, A. L., "Methods of Biochemical Analysis" **2**, 359 (1955).
[5] Edman, P., *Acta Chem. Scand.*, **4**, 277, 283 (1950).
[6] Perutz, M. F., Liquori, A. M., and Eirich, F., *Nature*, **167**, 929 (1951).
[7] Schroeder, W. A., Rhinesmith, H. S., and Pauling, L. (in the press).
[8] Havinga E., and Itano, H. A., *Proc. U.S. Nat. Acad. Sci.*, **39**, 65 (1953).
[9] Pauling, L., Itano, H. A., Singer, S. J., and Wells, I. C., *Science*, **110**, 543 (1949).
[10] Scheinberg, I. H., Harris, R. S., and Spitzer, J. L., *Proc. U.S. Nat. Acad. Sci.*, **40**, 777 (1954).
[11] Harris, J. I., Sanger, F., and Naughton, M. A., *Arch. Biochem. Biophys.*, **65**, 427 (1956).
[12] Neel, J. V., *Science*, **110**, 64 (1949).
[13] Haldane, J. B. S., "Biochemistry of Genetics" (Allen and Unwin, London, 1954).
[14] Benzer, S., in "The Chemical Basis of Heredity", edit. by McElroy, W. D., and Glass, B., 70 (Johns Hopkins Press, Baltimore, 1957).
[15] Streisinger, G., and Franklin, N. C., *Cold Spring Harbor Symp. Quant. Biol.*, **21**, 103 (1956).
[16] Pritchard, R. H., *Heredity*, **9**, 343 (1955).
[17] Giles, N. H., Partridge, C. W. H., and Nelson, N. J., *Proc. U.S. Nat. Acad. Sci.*, **43**, 305 (1957).
[18] Itano, H. A., *Ann. Rev. Biochem.*, **25**, 331 (1956).

VIII
Complementation and the Nature of Proteins

Editor's Comments on Papers 19 Through 22

19 Giles, Partridge, and Nelson: *The Genetic Control of Adenylosuccinase in* Neurospora crassa

20 Catcheside and Overton: *Complementation Between Alleles in Heterocaryons*

21 Woodward: *Enzyme Complementation* in Vitro *Between Adenylosuccinaseless Mutants of* Neurospora crassa

22 Fincham and Coddington: *The Mechanism of Complementation Between am Mutants of* Neurospora crassa

Prior to the publication of Paper 19 it had been generally assumed that allelic mutants could not complement to give a wild phenotype in heterozygotes. Giles and his coworkers here show that this is not so. By forming heterocaryons between *ad-4* mutants of *Neurospora,* they were able to show that those mutants at different sites, as established by obtaining wild types from crosses between them, may indeed complement to give heterocaryotic strains which grow in the absence of adenine required by the *ad-4* mutants. This was a finding of considerable significance as will become evident from reading the following three papers.

The discovery of interallelic complementation at the *ad-4* locus in *Neurospora* was followed by numerous other reports showing that this phenomenon was rather general. This paper by Catcheside and Overton supplies further examples, and, more important, a discussion which makes the highly significant observation that the presence of complementation indicates the presence and interaction of two or more polypeptide chains in the enzyme controlled by that gene locus.

In Paper 20 the authors make the important observation that one might expect complementation between two allelic mutant genes, if they have mutations at different sites, and if the protein formed is a multimer made up of a number of polypeptide units. Assuming that each allele coded for a polypeptide altered in a different way, one might expect to obtain in certain cases "hybrid" multimers that have some of the catalytic function of the wild-type enzyme.

Work with hemoglobins had also formed a basis for this kind of reasoning, as mentioned in the introduction of this paper. It had been shown that the various hemoglobin polypeptides could be disassociated and then reassociated *in vitro* to give various different combinations of the mutant and normal polypeptides.

Using this work as a basis, Woodward succeeded in showing that one could take extracts from *A-4* mutants which complement *in vivo* in heterocaryons, mix the extracts which alone have no enzymatic activity, and obtain an active adenylo-succinase.

The *am* mutants of *Neurospora* are deficient in NADP-linked glutamic dehydrogenase, which results in their having a requirement for α-amino nitrogen. After finding that there was a considerable degree of complementation within the *am* locus of the type found with the *ad-4* and other loci, and building upon the findings of Woodward discussed in Paper 21, Fincham and Coddington confirmed this work with a different enzyme. Additionally, they performed experiments with different ratios of the mutant protein extracts, and obtained evidence that various kinds of aggregates can be formed between the two mutant forms.

The work on complementation of the kind discussed in this and the previous papers had a powerful influence on shaping ideas about the nature of functional proteins. It can be considered another important example of the contribution of biochemical genetics to a better understanding of general biochemistry.

Selected Bibliography

Creighton, T. E., and C. Yanofsky. 1966. Association of the α and β_2 subunits of the tryptophan synthetase of *E. coli. J. Biol. Chem. 241:* 980–990.

Doy, C. H. 1966. Anthranilate synthetase and the allosteric protein model. *Biochim. Biophys. Acta 118:* 173–188.

Fincham, J. R. S. 1966. *Genetic Complementation.* W. A. Benjamin, Menlo Park, Calif.

*THE GENETIC CONTROL OF ADENYLOSUCCINASE IN NEUROSPORA CRASSA**

By Norman H. Giles, C. W. H. Partridge, and Norma J. Nelson

DEPARTMENT OF BOTANY, JOSIAH WILLARD GIBBS RESEARCH LABORATORIES,
YALE UNIVERSITY, NEW HAVEN, CONNECTICUT

Communicated by Karl Sax, February 14, 1957

In order to elucidate the mechanisms by which genes control enzyme formation, an experimental system favorable for both biochemical and genetical investigations is of paramount importance. Although an increasing number of instances are now known in which mutational changes have given rise to specific enzymatic deficiencies (Horowitz[1]), few of these systems have yet been characterized genetically in detail. The present paper will report initial results with a system in *Neurospora crassa* which appears to be particularly well suited for combined biochemical and genetical studies. A series of over twenty adenine-requiring mutants of independent origin has been obtained in one wild-type strain as a result of mutation at a single locus which is quite favorable for genetical studies, having suitable closely linked markers on either side. These mutants are all characterized by being deficient for a single enzyme involved in the terminal step in adenine biosynthesis—adenylosuccinase[2]—for which a convenient assay procedure is available. Furthermore, certain of the mutants are capable of reversion, and the characteristics of the resulting adenine-independent types can be investigated both biochemically and genetically.

Additionally, since all mutants were produced in a single **wild-type strain**, it has been possible to test for heterocaryon formation and growth in the absence of adenine with the various strains. The unexpected result has been that certain combinations of mutants are able to form adenine-independent heterocaryons (bicaryons)† in which appreciable adenylosuccinase activity can be demonstrated, even though both components lack detectable adenylosuccinase activity. Thus the initial studies with this system are already proving to be of exceptional interest in relation to the problem of gene action in enzyme formation. Brief reports of certain of these results have been presented previously (Giles, Partridge, and Nelson[3]; Giles and Partridge[4]).

Production and Preliminary Classification of Mutants.—The adenine-requiring mutants used in these studies were obtained in wild-type strain 74A (of St. Lawrence) by application of the filtration-concentration technique of Woodward, deZeeuw, and Srb.[5] Untreated (control) macroconidia and macroconidia exposed to either X-rays or ultraviolet were utilized to obtain mutants. Following filtration, platings were made on minimal agar supplemented with adenine (and other supplements in certain experiments in which additional types of mutants were being sought), and single colonies were isolated for testing. The resulting adenine-requiring mutants (606 in number) were next screened for their response to hypoxanthine. The present studies are concerned with the resulting 47 adenine-specific mutants which do not grow on hypoxanthine.

Further characterization of these mutants was sought by means of heterocaryon tests for complementation resulting in growth in the absence of adenine. Utilizing a single, arbitrarily chosen strain as a standard, all other adenine-specific mutants were tested for heterocaryon complementation with this strain by mixing conidia of the two mutants on a minimal agar plate. On the basis of these tests, two distinct groups of mutants were detected: group E (27 mutants), which gave abundant growth with the tester within 3 days or less after inoculation in the absence of adenine, and group F (20 mutants) which gave no growth with the tester (Table 1).

Additional evidence for two distinct groups of mutants has been obtained from genetic studies. In crosses to wild type, individual mutants from the two groups

TABLE 1

ORIGIN OF ADENINE-SPECIFIC MUTANTS

| Group | Mutants Derived from Conidia Exposed to: | | | Totals |
	X-rays	Ultraviolet	No Treatment	
E	25	0	2 (E11 and E12)	27
F	17	1 (F19)	2 (F6 and F10)	20*

* Does not include mutant 44206, of ultraviolet origin (Barratt *et al.*, *Advances in Genetics*, **6**, 1, 1954).

exhibit generally regular 1:1 segregation in asci. Crosses between mutants in the two groups give approximately 25 per cent wild-type segregants, demonstrating that the two types are not linked. Further studies with the F mutants have shown that these constitute a group of allelic mutants in linkage group III located between *prol*-1 and *leu*-1 (Barratt *et al.*[6]). Heterocaryon and crossing tests have also shown that the F mutants are allelic with the previously known *ad-4* mutant (44206) reported as adenine-specific (and temperature-sensitive) by Mitchell and Houlahan.[7] Group E mutants constitute a new group of allelic, adenine-specific mutants not as yet located in any particular linkage group.

Biochemical Characterization of Mutants.—Biochemical studies of the two groups of mutants were initiated in an effort to determine what step in adenine synthesis is blocked in each type. Culture filtrates of E mutants were found to be active in supporting growth of all other groups of adenine mutants except F, a result which appeared to place E prior to F in the biosynthetic sequence. Ultraviolet absorption spectra of the culture filtrates of both types were obtained, and characteristic accumulation products were then separated by ion-exchange chromatography. The results of these studies, especially with F filtrates, have been described in a preliminary note (Partridge and Giles[8]) and are similar to those obtained independently for the F mutant 44206 by Whitfeld.[9] The evidence indicates that F mutants are blocked in the terminal step in adenine synthesis, involving the splitting of adenosine monophosphate succinate (AMP-S) to adenosine monophosphate (AMP).

Comparative studies of E and F mutants, including examination of the absorption spectra of filtrates from the single and double mutants, as well as identification of the accumulation product (hypoxanthine) in an E mutant responsible for the biological activity of its filtrate, led to the tentative conclusion that E mutants were blocked in the step prior to F in biosynthesis, namely, the conversion of inosine monophosphate (IMP) to adenosine monophosphate succinate (AMP-S). More recent observations (Partridge and Giles, unpublished) indicate that a revision of this view may be required. It now appears that F mutants are blocked in not one, but two, reactions, the second involving the splitting of 5-amino-4-imidazole-(N-succinylo-carboxamide) ribotide (SAICAR) to 5-amino-4-imidazole carboxamide ribotide (AICAR) (Buchanan[10]). The accumulation of hypoxanthine by E-F double mutants appears to place the E block prior to the two F blocks and hence to require a modification of the previous view[3,4] regarding the step blocked in E mutants. Additionally, preliminary investigations of a third group of mutants (J) indicate that these may be considered basically adenine-specific, although they grow quite well on hypoxanthine when histidine is also present. The absorption spectrum of J-mutant filtrates is much like that of F mutants and suggests a close relationship between the substrates of the reactions under their respective control. Further studies of these relationships are under way. A study of the properties of double mutants involving J, E, and F types should help to decide the question of the relative positions of the compounds whose metabolism they affect, in the path of adenylic acid synthesis or as by-products thereof. All present evidence, however, supports the previous conclusion that F mutants are indeed blocked in the splitting of AMP-S to AMP.

The present experiments have been concerned principally with an analysis of the genetic control of this reaction, based on studies of the AMP-S splitting enzyme, adenylosuccinase, in wild type, F mutants, and revertants induced in F mutants.

Adenylosuccinase in Wild Type, Mutants, and Revertants.—On the basis of the evidence just discussed, indicating that F mutants accumulate AMP-S, it appeared that they might be defective with respect to the AMP-S splitting enzyme, adenylosuccinase (Carter and Cohen[11]). Hence tests for this enzyme were performed on mycelial extracts of wild type, F mutants, and certain other adenine mutants.

The following assay procedure was utilized: Mutant strains were grown with adenine sufficient for submaximal growth; prototrophic cultures were grown routinely on Fries minimal medium; mycelial pads were harvested by filtration, washed with water at 5° C., dried in the frozen state, weighed, and aliquot portions taken. The aliquots were crushed to powder, extracted by shaking in 0.05 M phosphate buffer, pH 7.0, for one hour at 5°, and centrifuged with refrigeration to remove visible sediment. Supernates were separated from lipid layers and sediment and were assayed either immediately or after one or two days' storage in a deep-freeze. Activity was not removed by filtration through a membrane of pore diameter >1 μ.

For assay, a fraction of 1 ml. of extract was brought to 3.3 ml. with the extraction buffer, containing AMP-S at a final concentration of $30\mu g/ml$, in a 10-mm. silica cell. The absorption of the substrate-product mixture was followed continuously or at fixed intervals by comparison with an identical mixture, lacking substrate, in a matched reference cell. Activity was recorded as rate of change of absorbance at 280 mμ during the period of constant reaction rate, on a mycelial dry-weight basis.

A study of variation of activity in wild-type and E-mutant extracts with culture times from 32 to 60 hours at 35° indicated a relatively constant activity for both strains, with a maximum between 36 and 44 hours (approximately 30 per cent greater than earlier or later samples). Times of harvesting in subsequent experiments were chosen to correspond to a stage of development in the range of optimal activity. Similar activities were found in 25° and 35° cultures at equivalent stages of development, except in temperature-sensitive strains. High levels of activity were found in wild type (equal in both mating types) and in mutant strains of groups E, J, and B. However, the B strain, when grown on hypoxanthine in place of adenine, developed an extractable purple pigment in the mycelium and was found to be inactive. When the pigmented extract was mixed with the colorless adenine-grown extract of the same strain, it removed the activity of the latter to a large extent. No activity (not more than 0.1 per cent that of wild type) could be detected in most of the nineteen F strains tested, even under conditions of double the usual extract concentration and a fivefold increase in substrate concentration with prolonged incubation. The exceptions were a strain (F16) which showed a trace of growth on minimal medium and a barely detectable activity and strain 44206, which is prototrophic and enzymically active at temperatures up to about 32°. No significant inhibition of wild-type activity was noted upon addition of crude extracts of any of the F strains. Therefore, the enzyme is presumed to be absent or grossly altered in F. Tests with extracts of both wild type and 44206 showed that activity was proportional to extract concentration over the experimental range, giving no evidence of intrastrain inhibition in crude extracts. It is not impossible that fractionation or special treatments of the mutant extracts, or growth of mutants under different conditions, will reveal latent activity (Wagner[12]).

On the basis of the results just discussed, it can be concluded that forward mutation at the *ad-4* locus in the wild type to give adenine-requiring mutants results in the loss of activity of the enzyme adenylosuccinase. It thus became of interest to determine whether reversions in such mutants to adenine independence would result in the restoration of adenylosuccinase activity. For these studies a series of 15 revertants was produced by ultraviolet irradiation of F12 macroconidia. The revertants were backcrossed to an F12 strain of opposite sex, and homocaryotic adenine-independent isolates were obtained for genetical and biochemical analysis. On the basis of backcrosses to wild type, all 15 revertants test as reverse mutations rather than as suppressor mutations.

Tests were next performed for adenylosuccinase activity, and these demonstrated that activity is present in all 15 revertants. Preliminary quantitative comparisons indicate, however, that the level of enzyme activity is markedly less (averaging about 50 per cent) in the revertant extracts than in those of wild type (Fig. 1). Further tests will be necessary to establish the reproducibility of these

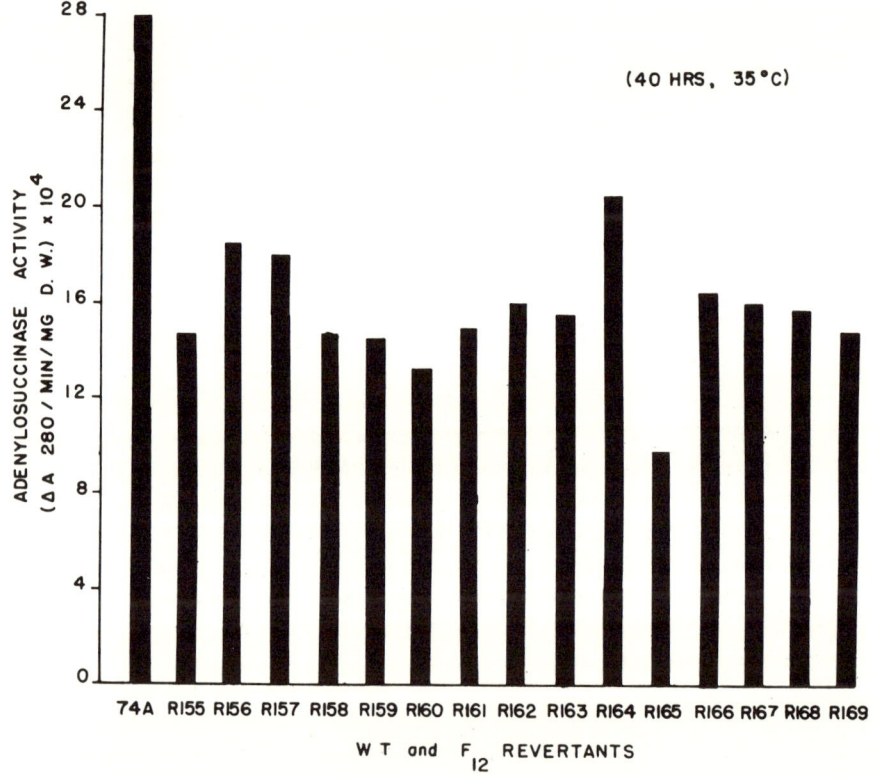

FIG. 1.—Comparative adenylosuccinase activity in wild type (74A) and 15 ultraviolet-induced revertants in the adenylosuccinase-deficient mutant F12.

quantitative comparisons, and to determine whether the observed quantitative differences among revertants arise exclusively from differences in the kinds of reverse mutational changes occurring at the F locus. A brief report of these studies has already been made (Giles and Partridge[4]), and additional detailed observations with these and other revertants will be published subsequently. Certain very recent observations indicate that the lower enzyme activities in at least some of the revertants may result from a more rapid loss of activity after extraction as compared with wild type.

Preliminary studies have also been made of adenylosuccinase in the temperature-sensitive mutant 44206 which grows in the absence of adenine at 25° but has an absolute requirement at 35°. Mycelial extracts of this strain grown at 25° have about 25 per cent of wild-type activity, whereas those grown at 35° have substantially no activity. When the strain is grown at 32°, the upper limit of growth in the absence of adenine, extract activity is about 2 per cent that of wild type. An examination has been made of the relative thermostabilities of adenylosuccinase activity in extracts obtained from the temperature-sensitive mutant and from wild type. On the basis of comparative residual activities following incubation at 35° and at 48°, adenylosuccinase activity is much less stable in extracts from the temperature-sensitive mutant than in those from wild type. Neither the activities nor the stability in frozen storage of extracts of wild type or of mutant 44206 were altered significantly by insertion of biochemical mutant markers with their concomitant nutritional supplements in growth at 25°, nor was there any effect upon activity of any prototrophic strain upon supplementation of growth medium with adenine.

Further detailed comparisons of the enzymes from the two allelic strains are in progress and will be reported later. In recent experiments temperature-sensitive revertants have been obtained in one F strain, and these are also being studied.

Heterocaryon Complementation between Adenylosuccinase-deficient Mutants.—The previously reported results indicate that all F mutants are biochemically similar in being blocked in the same step in adenine biosynthesis, and all either lack or have modified adenylosuccinase activity. Hence, on the basis of previous studies in *Neurospora* which have indicated that prototrophic bicaryons are regularly formed (in heterocaryon-compatible strains) between biochemically unlike (nonallelic) mutants, but not between biochemically similar (allelic) mutants (Beadle and Coonradt[13]), it was anticipated that F mutants would form adenine-independent heterocaryons with other biochemically dissimilar (nonallelic) adenine-requiring mutants, but not with one another. However, tests of this hypothesis have indicated that it is incorrect. Although F mutants do regularly form prototrophic heterocaryons with biochemically and genetically distinct groups of mutants (as, for example, with E), certain combinations of F mutants are also capable of forming bicaryons which grow in the absence of adenine. Regular complementation with the prompt initiation of growth is obtained in combinations involving mutant F12 with three other mutants, F1, F4, and F5, although the latter three mutants do not complement one another. The growth rates in the absence of adenine of such heterocaryons compared to those of wild type and of a heterocaryon between F12 and an E mutant (E4), as measured at two temperatures, are shown in Table 2. It will be noted that at both 25° and 35° all bicaryons grow at rates comparable to wild type and are not markedly stimulated by adenine, with one exception, F12 + F1, which grows much more slowly at 35° and responds to exogenous adenine. Instances of complementation involving F mutants other than F12 have also been obtained, but these are characterized by delayed initiation of growth and much slower rates on minimal. For the present, attention has been centered on the three heterocaryons cited. In certain of the numerous combinations of F mutants which do not complement, forced heterocaryons have been made, and in all instances these have been found to require exogenous adenine for growth. Attempts to modify these negative results by employing heterocaryons containing various nuclear ratios have been unsuccessful.

The occurrence of adenine-independent heterocaryons between adenine-requiring, adenylosuccinase-deficient mutants immediately raised the question whether enzyme activity is present in such heterocaryons. Tests of mycelial extracts have indeed shown that adenylosuccinase activity is present. As shown in Figure 2, heterocaryons between F12 and F4 and F12 and F5 have about one-quarter the activity of wild type and about one-half that of a heterocaryon between F12 and E4.

TABLE 2

GROWTH RATES OF HETEROCARYONS BETWEEN F MUTANTS COMPARED WITH THOSE OF WILD TYPE AND AN E + F HETEROCARYON*

Mutant Combination	Growth Rate (Mm/Hr)			
	25°		35°	
	Minimal	Adenine	Minimal	Adenine
Wild type (74A)	3.0	3.2	4.0	4.2
F12 + E4	3.2	3.5	4.5	4.9
F12 + F1	3.0	3.5	2.0	4.4
F12 + F4	3.1	3.6	3.9	4.5
F12 + F5	2.8	3.5	4.3	4.6

* Tubes in duplicate; adenine supplement, 50 µg/ml.

A later experiment has shown that the heterocaryon between F1 and F12, which is temperature-sensitive, provides an extract only about 6 per cent as active as that of wild type, when grown at 25° either with or without exogenous adenine. Its activity relative to that of the non-temperature-sensitive intra-F heterocaryons is then 25 per cent, a ratio similar to that of strain 44206 compared to wild type. The fact that a heterocaryon of F1 and F12 grows at wild-type rate on minimal medium, whereas it produces only 6 per cent of wild-type activity in its extract, might suggest that at least one step (the terminal) in the production of an essential cellular constituent (AMP) may be greatly oversupplied in terms of its specific enzyme in a normal organism.

These results indicate that the heterocaryons, even those within the F group, can synthesize AMP by the same route as wild type. The spectra of the heterocaryon filtrates are indistinguishable from those of wild type and thus provide no indication of an alternative pathway in these special cases.

Recent evidence that F mutants also lack activity for another reaction in adenine biosynthesis, the splitting of SAICAR to AICAR (Buchanan[10]; Partridge and Giles, unpublished), appeared to reopen the possibility that complementation might arise on the basis of these two reactions—one mutant (e.g., F12) lacking activity for one, the other three mutants lacking activity for the other. However, tests of mycelial extracts of the four mutants indicate that all lack enzyme activity for this second substrate, an activity which can be demonstrated in wild type, in F revertants, and also in heterocaryons of the complementary F mutants. No significant differences in relative activities for the two reactions have been found in any instance.

Genetic tests have been made to determine whether the adenine-independent cultures obtained from the various F-mutant combinations are indeed heterocaryons with mutant nuclei only and do not contain wild-type nuclei arising as a result of reversion or some other genetic change. In all instances tested, such

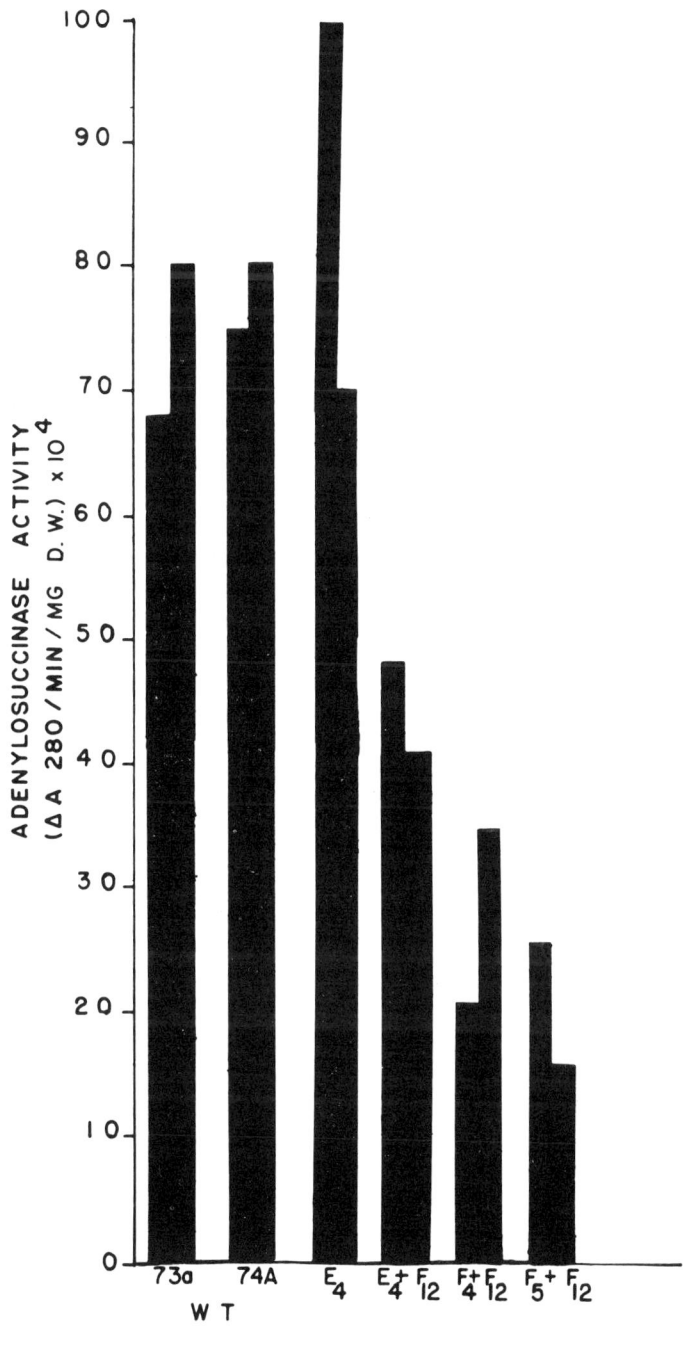

FIG. 2.—Comparative adenylosuccinase activities in mycelial extracts of wild types (74A and 73a), an E mutant (E4), and adenine-independent heterocaryons between F mutants (F4 + F12; F5 + F12). Results of duplicate tests are indicated.

presumptive heterocaryons can be resolved by hyphal-tip isolations and conidial platings to recover the two original mutant types, which can be identified by their pattern of heterocaryon formation with F testers. Despite extensive tests, no evidence has been obtained for the presence of other than mutant nuclei in these heterocaryons. In addition, preliminary tests using genetically marked strains have given no indication of mitotic recombination between the two types of mutant nuclei in such heterocaryons.

In view of the complementation in vivo between certain F mutants, resulting in the restoration of adenylosuccinase activity in heterocaryons, preliminary attempts have been made to detect such activity in an in vitro system in which mycelial extracts from two complementary mutant types were mixed. To date, such tests have given only negative results.

Genetic Relationships of Adenylosuccinase-deficient Mutants.—As already indicated, all F mutants are located in linkage group III between the same two markers. Intensive genetic analyses of these apparently allelic mutants have only recently been initiated, now that satisfactorily marked stocks are available. Analyses of crosses between various mutants are proving to be difficult because of sterility and poor ascospore viability, especially in serial isolations. However, it is already clear that crosses of certain mutants yield a low frequency of adenine-independent progeny, whereas selfings of the parents give only adenineless progeny.

To date, most adenine-independent segregants have proved to be pseudo-wild[14] rather than true wild types. Pseudo-wild types have resulted only from crosses of mutants which complement in heterocaryons. Only a few instances of true wild types have been detected as yet. In crosses where markers have been present, the origin of these adenine prototrophs has been associated with marker recombination. However, present crossing results are not extensive enough to determine what type of recombination (whether "orthodox" or involving "conversion") occurs between various mutants. It is expected that current studies will clarify further the genetic relationships of the various F mutants.

A particular study has been made of the inheritance of the pattern of heterocaryon complementation in F mutants exhibiting this behavior. This analysis involved crosses of such F mutants to wild type and tests of the heterocaryon responses of these F mutants in the first and subsequent generations. In all tests to date, F segregants which are heterocaryon-positive (capable of forming heterocaryons on minimal with compatible, biochemically unrelated strains) have shown the same pattern of heterocaryon response as the parental strain. Further tests have involved crosses of heterocaryon-positive strains of two F mutants (capable of complementation), one strain carrying a closely linked marker gene. Tests of segregants from such crosses indicate that a given pattern of heterocaryon complementation is linked with the marker, as is expected if this response is a regular characteristic of a particular adenine mutant.

In summary, in all tests to date, it has not been possible to separate the pattern of heterocaryon complementation from the particular adenine mutant with which it was originally associated. Hence it appears highly probable that this pattern is a result of the particular type of mutational event producing a given adenine mutant.

Discussion.—The preceding evidence indicates that one group of allelic, adenine-specific mutants in *N. crassa* (group F—or *ad-4*—containing 21 mutants of independent origin) can be characterized biochemically as being blocked in the terminal step in adenine biosynthesis, involving the splitting of AMP-S to AMP, as a result of the absence or impairment in these mutants of the AMP-S splitting enzyme, adenylosuccinase. This enzyme is present in wild type and in other biochemically distinct mutants. One of the *ad-4* mutants is temperature-sensitive and, when grown at 25° in the absence of adenine, contains appreciable adenylosuccinase activity. There is preliminary evidence that the enzyme from this mutant is much more thermolabile than that from wild type.

Furthermore, it has been demonstrated that certain of the F mutants lacking detectable adenylosuccinase activity can revert either spontaneously or following ultraviolet irradiation to give adenine-independent isolates testing as reverse mutations. Fifteen such revertants have been shown to possess adenylosuccinase activity, although their average activity is only about 50 per cent that of wild type, and none is equal to wild type in activity, at least after frozen storage of their extracts.

The foregoing results establish that in this instance forward mutation in the wild type to adenine requirement results in the loss of activity of a specific enzyme involved in adenine biosynthesis, and that reverse mutation results in the restoration of this activity. Additionally, it appears, on the basis of qualitative (temperature-sensitivity) and quantitative enzyme activity tests, that changes at the *ad-4* locus arising from mutation in either the forward or the reverse direction may produce diverse mutant and revertant types. Hence this particular situation provides an exceptionally favorable system for further combined biochemical and genetical studies of mutants arising by both forward and reverse mutational events at a single gene locus.

Of particular interest in the present investigation has been the unexpected finding that certain combinations of *ad-4* mutants are able to form heterocaryons (bicaryons) which can grow in the absence of adenine, and that adenylosuccinase activity, although absent from the mutants, is present in the heterocaryons. This type of complementation takes place between certain mutant combinations only. It is also clear that the absence of complementation does not result from a failure of heterocaryon formation, since heterocaryons have been forced with several such combinations, and these require adenine for growth. Several possible interpretations of these observations have been considered.

The occurrence between *ad-4* mutants of complementation essentially similar to that observed between biochemically unrelated types, in which different reactions are known to be blocked, raises the question as to whether the presumed single-step reaction from AMP-S to AMP may actually consist of two steps. However, all presently available evidence tends to refute this possibility. Biochemical studies with yeast and pigeon liver favor the view that a single-step reaction is involved, catalyzed by a single enzyme (Carter and Cohen[11]). In the present material no qualitative differences were detected among the various *ad-4* mutants, such as the accumulation of intermediates other than AMP-S. Furthermore, in vitro mixing of extracts of the F mutants capable of complementation in heterocaryons failed to produce any activity for the over-all reaction. In fact, no spectral changes were observed in AMP-S in the presence of such extracts, either separate or mixed, whereas it might have been anticipated that, since the enzyme is soluble in active strains, an indication of one or both of the hypothetical component reactions would have been observed in the presence of separate extracts, and certainly an indication of the over-all reaction should have appeared in a mixture of such hypothetical soluble component enzymes.

The possibility that complementation might arise on the basis of differential activities for the splitting of SAICAR and AMP-S also appears to have been eliminated by the failure to demonstrate any activity for either substrate in any of the four F mutants concerned. At present it appears likely that a single enzyme, deficient in all the *ad-4* mutants, is able to catalyze both splitting reactions, which are chemically quite similar.

Additional general evidence against the view that two reactions may be involved comes from the fact that the *ad-4* mutants constitute not two classes, but three, with respect to their complementation. It is thus necessary to assume that the majority of the mutants lack both postulated activities. However, it is clear that the majority of these mutants must have arisen by single mutational events, which would require a remarkably close juxtaposition of the loci presumably controlling the distinct reactions. In the absence of any evidence to the contrary, it seems reasonable to assume that the splitting of AMP-S involves a single reaction catalyzed by a single enzyme. Furthermore, all available data support the view that mutation at only one locus results in the absence or modification of adenylosuccinase.

Another possibility considered was that heterocaryons able to grow in the absence of adenine could do so because they contained wild-type nuclei. However, the following evidence is against this view. (1) Such heterocaryons can be resolved by hyphal-tip isolations and conidial platings, with the recovery of adenine-requiring colonies containing either one or the other of the two original types of mutant nuclei as determined by their heterocaryon responses with the parental tester stocks. Adenine-independent colonies are also recovered from macroconidial platings, but all such colonies tested have proved to be heterocaryons, undoubtedly arising from conidia containing both types of mutant nuclei. Extensive tests have failed to produce any evidence that true wild-type nuclei are present in these heterocaryons. (2) Preliminary tests for mitotic recombination involving linkage group III markers have given negative results. (3) Quantitative comparisons of adenylosuccinase activity in E-F heterocaryons with that in intra-F heterocaryons would appear to require that about one-quarter of the nuclei in the latter heterocaryons be wild type in genotype, which is obviously not the case. Thus these results appear to eliminate the possibility that mitotic recombination or some type of nuclear interaction analogous to transformation or transduction is occurring to produce permanently wild-type nuclei in these adenine-independent heterocaryons.

The final hypothesis to be considered, and the one that seems most probable on the basis of present evidence, is that heterocaryon complementation arises from an interaction involving two different types of mutant nuclei present in a common cytoplasm, resulting in the synthesis of adenylosuccinase, and that this complementation occurs without any permanent genetic change having taken place in the nuclei involved. As yet there is no evidence as to what the nature of such an interaction may be. It is possible that the presence of one type of nucleus may so modify cytoplasmic conditions that the other nucleus becomes capable of effecting adenylosuccinase synthesis. A more likely possibility would appear to be that some type of nuclear product interaction is involved. The present genetic evidence indicates that a given pattern of complementation is a genetic characteristic of a particular mutant and inseparable from the mutant locus. Thus the presence of a suppressor in one adenine strain, specific for another strain, and hence permitting growth in a heterocaryon, is ruled out (unless the suppressor locus is exceedingly closely linked with the ad-4 locus). A reasonable possibility would seem to be that, in cases where complementation occurs, the two mutants involved have arisen as a result of mutational changes at different sites in the ad-4 locus. Where complementation fails, the two mutants may have a certain portion of the locus damaged in common. Thus the mechanism could be visualized as involving cytoplasmic interaction between two types of imperfect templates resulting in the formation of perfect templates for enzyme formation, or a co-operative functioning of two imperfect templates to produce a normal enzyme. It will be of interest to determine whether this hypothesis can be correlated with the behavior of these mutants on crossing, especially with respect to the origin of wild types. It is evident, of course, that other interaction mechanisms can also be hypothesized.

The occurrence of prototrophic heterocaryons between ad-4 mutants provides further evidence that the heterocaryon test is not a consistent one for detecting allelism (Mitchell[15]). Furthermore, although it has not been possible to obtain stable diploid heterozygotes in *Neurospora*, the evidence that adenine-independent pseudo-wild types arise from crosses of mutants which complement, strongly supports the view that the characteristics of such heterozygotes would be similar to those of heterocaryons with respect to enzyme formation. Thus the present results appear to make more difficult the general application of the *cis-trans* position effect test to delimit a locus as a functional unit (Benzer[16]), at least if functional unity is defined as control over the synthesis of a single enzyme.

Summary.—Combined genetic and biochemical studies have been performed on a series of adenine-specific mutants in *N. crassa*. Twenty-one mutants of independent origin have arisen as a result of mutation at a single locus (ad-4) in linkage group III between two closely linked marker genes. Initial crossing analyses with certain of these mutants indicate that adenine-independent isolates arise with a low frequency in some crosses, and that the majority of these types are pseudo-wilds, although true wild types occur in certain combinations. The recombination mechanism associated with the origin of wild types has not yet been established.

Biochemically, all the mutants have been found to be blocked in the terminal step in adenine biosynthesis, involving the splitting of adenosine monophosphate succinate (AMP-S) to adenosine monophosphate (AMP), and to lack or have impaired activity for the AMP-S splitting enzyme, adenylosuccinase. One of the mutants is temperature-sensitive and produces an adenylosuccinase which in crude extracts is much more thermolabile than that from wild type. Certain of the ad-4 mutants are capable of reverse mutation to adenine-independent phenotypes, and these have been shown to possess restored adenylosuccinase activity, although at levels (or of stabilities) below that of wild type.

These results establish that in this instance forward mutation at a single locus in the wild type to adenine requirement results in the loss of activity of a specific enzyme involved in adenine biosynthesis and that reverse mutation to adenine independence results in the restoration of this activity. Additionally, changes at this locus, arising from either forward or reverse mutational events, may produce diverse mutant and revertant types as judged by qualitative and quantitative tests of enzyme activity.

Certain combinations of *ad-4* mutants have been found to form heterocaryons (bicaryons) able to grow in the absence of adenine. Although the individual mutants lack detectable adenylosuccinase activity, this enzyme is synthesized by the bicaryons. Possible mechanisms for this unexpected type of complementation between alleles in enzyme formation are discussed, along with its implications for biochemical genetic theory in general.

The authors wish to express their appreciation to Miss Mary Case and Mrs. Norman Giles for technical assistance with these experiments; to Dr. C. E. Carter for the sample of AMP-S, and to Dr. J. M. Buchanan for the sample of SAICAR.

* This work was supported in part by a research contract with the Atomic Energy Commission, AT(30-1)-872, and by Institutional Grant 47F from the American Cancer Society.

† This term, rather than dicaryon which generally has a more restricted meaning, is used to designate a heterocaryon having only two kinds of mutant nuclei. Heterocaryons having more than two kinds of mutant nuclei, e.g., tricaryons, have not yet been investigated in these studies.

[1] N. H. Horowitz, *Federation Proc.*, **15**, 818, 1956.

[2] C. E. Carter, *Ann. Rev. Biochem.*, **25**, 123, 1956.

[3] N. H. Giles, C. W. H. Partridge, and N. J. Nelson, *Proc. Japan. Genetics Symposia*, 1956 (in press).

[4] N. H. Giles and C. W. H. Partridge, *Genetics*, **41**, 645, 1956 (abstr.).

[5] V. W. Woodward, J. R. deZeeuw, and A. M. Srb, these PROCEEDINGS, **40**, 192, 1954.

[6] R. W. Barratt, D. Newmeyer, D. D. Perkins, and L. Garnjobst, *Advances in Genetics*, **6**, 1, 1954.

[7] H. K. Mitchell and M. B. Houlahan, *Federation Proc.*, **5**, 370, 1946.

[8] C. W. H. Partridge and N. H. Giles, *Arch. Biochem. and Biophys.* (in press).

[9] P. Whitfeld, *Arch. Biochem. and Biophys.*, **65**, 585, 1956.

[10] J. M. Buchanan, personal communication.

[11] C. E. Carter and L. H. Cohen, *J. Biol. Chem.*, **222**, 17, 1956.

[12] R. P. Wagner and C. H. Haddox, *Am. Naturalist*, **85**, 319, 1951.

[13] G. W. Beadle and V. L. Coonradt, *Genetics*, **29**, 291, 1944.

[14] M. B. Mitchell, T. H. Pittenger, and H. K. Mitchell, these PROCEEDINGS, **38**, 589, 1952.

[15] M. B. Mitchell, *Compt. rend. trav. Lab. Carlsberg, Sér. physiol.*, **26**, 285, 1956.

[16] S. Benzer, these PROCEEDINGS, **41**, 344, 1955.

Complementation between Alleles in Heterocaryons

D. G. CATCHESIDE AND ANNE OVERTON

Department of Microbiology, University of Birmingham, Birmingham, England

Two cases have been described recently in *Neurospora crassa* in which mutants, each lacking a particular function, show partial manifestation of the function when combined in a heterocaryon. These cases concern, respectively, mutants affecting glutamic dehydrogenase (Fincham and Pateman, 1957) and adenylosuccinase (Giles, Partridge and Nelson, 1957). For each, there is a number of mutants, every one of which completely lacks the specific function. Each set of mutants is divisible into three groups, A, B and C, on the basis of heterocaryon complementation. Primarily this is detected by the ability of heterocaryons to grow on minimal medium, in the absence of a supplement of the growth factor specifically required by the mutants. Group A shows no complementation with either group B or group C, but B and C complement one another. Of course, there is no complementation between members of the same group.

The degree of complementation varies considerably, but the heterocaryons usually grow much less well on minimal medium than wild type does on minimal or either of the mutants do on minimal plus an optimal supply of growth factor. Secondarily, the heterocaryons show some of the enzyme activity which the mutants have lost. Again the degree of this, as a proportion of wild type activity, varies. In the case of glutamic dehydrogenase, it does not exceed 20–30 per cent of wild type. In the adenylosuccinase case, one mutant in group B and three in group C have been described. Of the three heterocaryons, two have about 25 per cent of the specific enzymic activity of the wild type, while the third has only about 6 per cent.

Several more cases of such interallelic complementation have been discovered. One has been reported at the *pan-2* locus by Case and Giles (1958). While the analysis of none of them has so far extended to a study of their enzyme systems, they are important in two respects. First, they show the rather frequent occurrence of the phenomenon, since it occurs at about a third of the loci at which any considerable number of mutations has been tested. Secondly, at two loci a more complex system has been found. These loci are *arg-1* (arginine-1) and a *lys* whose identity with one of the previously known lysine loci is unsettled, except that it is not allelic with *lys-1* (33933), *lys-3* (4545) or *lys-4* (15069).

All of the mutants used in this study have been derived from Emerson *a* wild type, following irradiation of conidia with ultraviolet light. Enrichment of the proportion of mutants among the conidia was secured, before plating on supplemented media, by means of the filtration technique (Catcheside, 1954; Woodward, De Zeeuw and Srb, 1954). Some improvements were introduced, including incubation in a water-bath and continuous agitation of the conidial suspension by means of a stream of sterile air bubbled through it. In some experiments, citrulline was added to the filtration medium in order to eliminate the rather abundant arginine mutants which have blocks at earlier steps than *arg-1*.

Altogether 40 *arg-1* mutants have been obtained and tested in all combinations for their ability to form heterocaryons which will grow on minimal medium. This has usually been done by superimposing small, but dense, inocula of conidia of the pairs of mutants to be tested on the surface of minimal medium in petri dishes. The medium was usually VSM, *i.e.* Vogel's medium made up with 0.5% sorbose and 0.1% sucrose. On this medium, if incubated at 25°C., the formation of a heterocaryon able to grow on minimal medium is easily detected in two to three days. In some cases, forced heterocaryons have been made between mutants which were negative by the previous method.

The 40 *arg-1* mutants are divisible into six groups, distinguished by the ability or not of their heterocaryons to grow on minimal medium (Fig. 1a). The distribution of the mutants among the six groups is shown in Table 1, the laboratory numbers of all being listed for ease of later reference. It will be noticed that by far the largest class is that containing the mutants which are negative with all other groups, as well as with other mutants in the same group. It should be mentioned that some of the combinations do not grow as vigorously as the others. This variation is encountered in homologous heterocaryons. Thus, of the heterocaryons made up of the group C mutant (K287) and the four group E mutants, two (those involving K248 and K406 respectively) are quite vigorous, one (that with K272) grows rather feebly, and the fourth (that with K236) grows only moderately on agar slopes.

On slopes of minimal Vogel's medium (VM), most of the heterocaryons appear to be as vigorous as wild type, but when grown in liquid VM in flasks none of them attain a dry weight as great as that of wild type. In liquid medium,

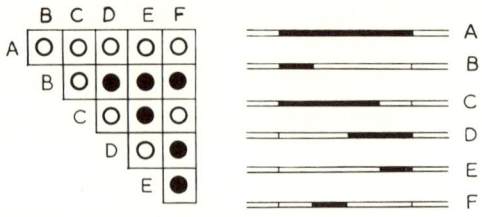

FIGURE 1. Patterns of presence (closed circles) or absence (open circles) of complementation in heterocaryons of arginine-1 mutants and their representation on a one-dimensional scheme of ranges of defective action.

all show at least some response to a supplement of arginine.

The two-dimensional pattern of reactions can be represented by a one-dimensional diagram, in which each mutant exerts a range of influence. In Fig. 1b these ranges are drawn in such a way that overlapping ranges occur where the heterocaryons do not grow on minimal medium, and non-overlapping ranges where complementation occurs in the heterocaryons. A completely consistent representation of the data is obtained in this way. The problem now arises as to whether this linear diagram has anything more than a formal meaning. That the diagram is linear is highly suggestive that the data on which it is based disclose some kind of linear differentiation within the gene, or its product.

The diagram could be interpreted as indicating that the arginine-1 locus is complex and made up of four physiologically distinct blocks, say four cistrons in Benzer's (1956) sense. If so further groups could be expected, viz. one affecting only block 3, one affecting blocks 1 and 2, one affecting blocks 2 and 3, and one affecting blocks 2, 3 and 4. If these are found in further samples of mutants they will constitute support for this interpretation, but could hardly be regarded as decisive. On this interpretation, the members of group A would have to be regarded as deletions encompassing all four cistrons. The fact that

TABLE 1. DISTRIBUTION OF arginine-1 MUTANTS AMONG THE GROUPS DISTINGUISHABLE BY HETEROCARYON COMPLEMENTATION

Group A: K159, K209, K262, K275, K279, K286, K305, K312, K334, K340, K348, K357, K366, K367, K400, K403, K404, K407, K408, K409, K412, K415, K416, K417, K418, K419, K420, K423, 46004, (36703?)
B: K166
C: K287
D: K311, K359, K401
E: K236, K248, K272, K406
F: K337, K351, K414

the largest injury is the one that occurs most frequently could, perhaps, be reconciled with the fact that all of the mutations were induced by ultraviolet light. It would imply that the ultraviolet light had usually induced large deletions. Nevertheless, there is definite evidence that at least one group A mutant cannot be a large deletion. Both K334 (group A) and K311 (group D) have several times back mutated to wild type. Tests have shown that these are back mutations at the locus itself, rather than the occurrence of suppressor mutations. Thus, the interpretation as a complex locus made up of four distinct physiological units encounters difficulties which render it untenable.

A second possibility is that some of the mutants carry suppressors which are specific in their action on other mutants, the one-dimensional interpretation being merely a chance result. Thus K166 (group B) might carry a suppressor of mutants in groups D, E and F; K287 (group C) a suppressor of mutants in group E; and so on. If this were so, it should be possible to separate the presumed suppressor from the mutant with which it was at first associated. These tests may be carried out in either of two ways, namely (1) crossing an arg-1 mutant to some other stock, recovering arginine segregates and testing their ability to complement in heterocaryons and (2) crossing two complementary arginine-1 mutants together and examining their progeny for evidence of whether any will grow on minimal medium. Various difficulties have prevented these tests being as extensive and complete as is desirable. The arg-1 locus is rather close to the mating type locus, so rather many progeny must be analyzed to secure suitable stocks. Secondly, the stocks which have been outcrossed must then be freed of genetic factors leading to inability to form heterocaryons at all. Thirdly, stocks of opposite mating type do not form heterocaryons very readily; the rather poor growth of some of the heterocaryons, mentioned previously, is accentuated when the two components are of opposite mating type. Fourthly, arginine mutants require relatively large supplements of arginine. In the standard Westergaard and Mitchell crossing medium, the addition of arginine often so disturbs the carbon:nitrogen ratio as to lead to sterility. Frequently, abundant perithecia are formed, but the asci within abort completely or else only a few of the ascospores mature. Some improvement has been achieved by adjustments in the medium, such as a compensating reduction in the amount of KNO_3 in the basal medium or by raising the sugar concentration. These devices have given sporadic success, but in the case of crosses between arg-1 mutants satisfactory fertility has been achieved so far only in the case of pairs of mutants which will complement one another in heterocaryons.

The crosses which have been analyzed show a

total absence of suppressors except such as may be fairly, or very, closely linked to *arg-1*. In other words, there is no evidence that any gene, other than *arg-1* itself, is concerned in the complementation in the heterocaryons. Specifically, it has been shown that K166 (group B) carries no suppressors of mutants in groups E and F; K359 (group D) carries no suppressor for group F; K248 and K272 (group E) carry no suppressors for groups B, C and F; and K351 (group F) carries no suppressors for groups D and E. This leaves open the possibility that group B could carry a suppressor for group D, group C a suppressor for group E, group D a suppressor for group B, and group F a suppressor for group B. Thus while the occurrence of some suppressors, specific to particular alleles, cannot be excluded, it is evident that the whole of the data cannot be explained by means of a system of specific suppressors.

Another kind of information about the mutants might be expected from the interallelic crosses, namely recombination data permitting the mapping in order along the chromosome of the points or blocks of defect which characterize the mutants. No progress towards this has been made, though one cross has given recombinants which grow on minimal medium. The numbers of recombinants from these crosses are: (1) 0 out of 1700 progeny of the cross K359 × K351 (group D × group F); (2) 2 out of 7749 progeny of K248 × K351 (group E × group F); (3) 0 out of 436 from K248 × K166 (E × B); (4) 0 out of 761 from K272 × K351 (E × F). However, it is possible that these apparent recombinants are pseudowild types (PWT).

Disappointment at the lack of recombination data is somewhat mitigated by the reflection that in other cases of interallelic crosses in Neurospora, and indeed yeast, the recombinants appear to arise by a process different from the normal reciprocal crossing over. Moreover, they lead to data which do not permit an unambiguous linear order of positions of defect. Nevertheless, knowledge of the detailed map is so important for an understanding of these mutants that a strenuous effort must be made to secure proper data, or to know why it is unattainable.

We are led therefore to consider other possible modes of interaction between differently altered genes. It must be remembered that the interaction occurs in a heterocaryon and therefore that the site of the interaction is probably in the cytoplasm. In several cases, the component nuclei have been recovered from heterocaryons and found to be unaltered by the association. Hence, the interaction must be between gene products, either at the level of the formation of the units from which potential enzymes are built or at the level of the final shaping of the enzymes. In either case it must involve the cooperation or aggregation of two or more homologous, though somewhat different, gene products or derivatives.

The first of these possibilities envisages some mechanism of the formation of units, presumably polypeptide chains, from a template, perhaps RNA. It is presumed that the template is compounded of two or more gene derivatives. In the heterocaryon it may be formed from two complementarily different derivatives. Hence in a heterocaryon compounded of, say, B and D type mutants, there could be BB, DD and BD templates. The last of these would be capable of producing some normal products (polypeptide chains), like those formed by wild type templates. How this may be done is a matter of speculation. It might be by a form of copy-choice or it could be that the formation of one product always requires the double unit in the template. However, if either of these possibilities were true, it is difficult to see why any two different alleles, which are not actual deletions, should not complement one another. With this kind of mechanism it is difficult to see how reasonable restrictions could be placed upon cooperation between alleles, consistent with the observed pattern of interaction. One possible way would be a restriction upon the heterologous units which could aggregate to form templates, but the mechanism by which this could be secured is obscure.

The second of the possibilities presumes that the templates are unitary in quality and that, in a B plus D heterocaryon, two types of template, B and D, exist, each determining its characteristic product, a B- or a D-type polypeptide chain. It is further presumed that one stage in the shaping of the enzyme is the aggregation of at least two of these products. In some cases, by chance, the aggregation will involve heterologous products. Other stages in the final shaping of the enzyme are likely to involve specific folding. Presumably, any one mutant has a defect which results in a fault in folding and so in the formation of a protein sufficiently different from the enzyme as to lack its specific activity. However, the difference is often small enough for the mutant protein to be serologically homologous with the enzyme.

There are various ways in which interaction at

TABLE 2. OCCURRENCE OF INTERALLELIC COMPLEMENTATION IN HETEROCARYONS BETWEEN *Neurospora crassa* MUTANTS

Locus	Number of mutants	Number of classes	Distribution among classes					
			A	B	C	D	E	F
arg-1	40	6	28	1	1	3	4	3
arg-10	13	3	11	1	1	—	—	—
orn-2	40	3	29	2	9	—	—	—
his-1	13	3	10	2	1	—	—	—
his-2	14	3	12	1	1	—	—	—

this level might lead to complementation, *i.e.* to restoration of activity. One is a cooperation in the direction of the process of folding in a mixed aggregate. This could be competitive between the normal and the abnormal at each point, or perhaps random. The variation in degree of complementation could be explained by differences in competitive power. On this interpretation, total lack of complementation would correspond to complete dominance of the abnormal, over the normal, control at a given point. A second way could lie in the patterning of the folds, but it is difficult to understand on this interpretation why so few pairs of mutants should complement one another.

The occurrence of complementation is fairly widespread but not universal. The absence of complementation could be taken to indicate that the enzyme is shaped from a single polypeptide chain, whereas the presence of complementation indicates the aggregation of two or more chains. The present information is given in Table 2. In addition one large class of lysine mutants, 46 in number, shows a complex pattern of complementation, similar to that of *arg-1*. The number of groups is at least six and may be seven, the number of apparent "cistrons" being three or four. The tests of these mutants are still incomplete. Cases of absence of complementation are: *arg-5*, 19 mutants tested; *his-3*, 15 mutants; *his-1*, 13 mutants; *his-2*, 18 mutants; *lys-1*, 8 mutants; *lys-3*, 5 mutants; *td*, 64 mutants incompletely tested; *cys-1*, 22 mutants; *cys-2*, 21 mutants. Information about lysine and tryptophane mutants is due to Professor M. Ahmad, and about cysteine mutants to Miss N. E. Parker.

References

BENZER, S., 1957, The elementary units of heredity. In: The Chemical Basis of Heredity. (W. D. McElroy and B. Glass, eds.,) pp. 70–93. Baltimore, Johns Hopkins Press.

CASE, MARY E. and GILES, N. H., 1958, Evidence from tetrad analysis for both normal and aberrant recombination between allelic mutants in *Neurospora crassa*. Proc. Natl. Acad. Sci., Wash. *44*: 378–390.

CATCHESIDE, D. G., 1954, Isolation of nutritional mutants of *Neurospora crassa* by filtration enrichment. J. Gen. Microbiol. *11*: 34–36.

FINCHAM, J. R. S. and PATEMAN, J. A., 1957, Formation of an enzyme through complementary action of mutant alleles in separate nuclei in a heterocaryon. Nature *179*: 741–2.

GILES, N. H., PARTRIDGE, C. W. H. and NELSON, N. J., 1957, The genetic control of adenylosuccinase in *Neurospora crassa*. Proc. Natl. Acad. Sci., Wash. *43*: 305–317.

WOODWARD, V. W., DE ZEEUW, J. R., and SRB, A. M., 1954, The separation and isolation of particular mutants of Neurospora by differential germination of conidia, followed by filtration and selective plating. Proc. Natl. Acad. Sci., Wash. *40*: 192–200.

DISCUSSION

GILES: Extensive further complementation studies have been performed by Mr. Dow Woodward with the *ad-4* mutants lacking adenylosuccinase (referred to by Catcheside and Overton). These results indicate a much more complex relationship than was originally reported. In general, the pattern of complementation is remarkably similar to that found by Catcheside and Overton at the *arg-1* locus. In all, 122 mutants at the *ad-4* locus have been tested for heterocaryon complementation. In addition to primary mutants derived from wild type 74A, these include secondary *ad-4* mutants derived from reverse mutants of certain primary mutants. Fifty of the 122 mutants gave at least one positive response. Based on the complementation pattern, it is possible to arrange these *ad-4* alleles in a linear sequence. This can be done because of apparent overlap between non-functional regions characteristic of the individual alleles. The relationships of the 15 different complementing types obtained indicate a total of seven distinct functional units (cistrons) in this "complementation map" of the *ad-4* locus.

In all cases tested in which complementation occurs, partial restoration of adenylosuccinase activity has resulted. Enzyme activities of crude and partially purified extracts (tested by following the splitting reaction of adenosine monophosphate succinate to adenylic acid in a recording spectrophotometer) range from less than 1% to 25% of that found in the parental wild type strain. In general, there is a striking direct correlation between the relative distance apart of any two mutants on the complementation map and the level of enzyme activity in a heterocaryon between them. Thus heterocaryons between mutants in adjacent cistrons have low activities, and activities increase with increasing distances to the maximum value of 25% for mutants from 4 to 6 cistrons apart. A similar correlation can be made by comparing heterocaryon growth rates and/or the length of time required for growth to commence when conidial inocula from complementing types are mixed on minimal agar.

Meiotic recombination occurs between some of the alleles, giving rise to wild type progeny. Based on the complementation map, these appear to be the more widely separated alleles. Furthermore, the order of the alleles deduced from such crosses utilizing linked markers on either side of the locus does not conflict with the order in the complementation map. To date, infertility in many of the crosses has made a complete recombinational analysis impossible.

Additional instances of complementation, some exhibiting moderate degrees of complexity, have been detected in two other groups of adenine mutants, group E and group J (*ad-5*) by Mr. Tatsuo Ishikawa, and in two groups of histidine mutants (*hist-1* and *hist-3*) by Mr. Brooke Webber.

ENZYME COMPLEMENTATION IN VITRO BETWEEN ADENYLOSUCCINASELESS MUTANTS OF NEUROSPORA CRASSA*

By Dow O. Woodward

DEPARTMENT OF BOTANY, JOSIAH WILLARD GIBBS RESEARCH LABORATORIES, YALE UNIVERSITY

Communicated by Edmund W. Sinnott, April 24, 1959

Previous studies[1] have shown that certain mutant strains of *Neurospora crassa* lacking detectable adenylosuccinase activity, and hence unable to grow in the absence of exogenous adenine, complement one another in certain combinations when grown as heterocaryons. Such complementation results in adenine-independent growth, and certain of the resulting heterocaryons possess a maximum of 25 per cent of wild-type adenylosuccinase activity. The pattern of complementation has been depicted as a complementation map of the *ad-4* locus and suggests a structural correlation between gene and gene product.[2-4] These results led to the hypothesis that complementation involves a cytoplasmic exchange mechanism operating at either the RNA or polypeptide level. The work of Singer and Itano,[5] demonstrating the formation of hybrid protein *in vitro*, lends further support to the suggestion that exchange at the polypeptide level could provide a mechanism for complementation. In addition, such a mechanism could also explain the restoration of enzyme activity in heterocaryons to a maximum level not exceeding 25 per cent of wild type. Other recent studies[6-9] suggest related types of exchange mechanisms that could also yield hybrid protein or protein complexes.

The above hypothesis has received additional support with the demonstration of *in vitro* complementation. By mixing mycelial homogenates from certain pairs of complementing adenylosuccinaseless *Neurospora* mutants grown separately, adenylosuccinase activity as high as 20 per cent of that obtained from *in vivo* complementation has been recovered. The same procedure performed with complementing mutants, singly, or with mixtures of non-complementing mutants, has failed to restore detectable enzyme activity. Earlier attempts[1] to detect enzyme activity in mixtures of extracts from complementing *ad-4* mutants gave negative results.

Materials and Methods.—Data on the origin and other characteristics of the *ad-4* mutants employed in these studies have been presented.[4]

The following general procedure has been found to be the most effective in demonstrating *in vitro* complementation in mixing experiments. Mutant cultures are grown separately for 3–4 days at 25°C on Fries minimal medium containing 100 μg/ml of adenine sulfate. Mycelial pads are removed, washed in chilled distilled water, pressed dry on absorbent paper and quick frozen. Frozen samples are mixed, and the cells are disrupted in a Hughes Press at 5000 psi. The frozen powder from the mixed samples is taken from the press and diluted to a liquid paste condition with $M/20$ tris HCl buffer containing 10^{-3} M reduced glutathione (pH 7.0) and placed on a shaker for 10–15 min or until the frozen extract thaws. The pH is adjusted to approximately 8.0 and the mixture diluted further (usually a two-fold dilution); the mixture is agitated for another hour at 4°C. Cell debris is removed by centrifugation at about $50,000 \times g$ for 30 min. The supernatant solution is further purified by column chromatography on DEAE cellulose.[10] The column purification is not essential to the recovery of enzyme activity, but some of the ultraviolet-absorbing material that would otherwise interfere with the assay is removed. The assay used is a measure of decrease in absorption at 280 mμ as the reaction proceeds from adenosine monophosphate succinate (AMP-S) to adenosine monophosphate (AMP).[1]

The assumption that the decrease in absorption observed at 280 mμ is a measure of the rate of conversion of AMP-S to AMP is based on spectral scans during the reaction. These scans show a change in maximum absorption from 267 mμ (AMP-S peak) to near 260 mμ (AMP peak). This is the same spectral change observed when the wild-type enzyme is used, but a change that does not occur when single mutant extracts are used. A maximum change in absorption occurs at 280 mμ when either wild-type enzyme or enzyme recovered as a result of *in vitro* or *in vivo* complementation is reacted with the AMP-S substrate. The rate of change in absorption is measured on a recording spectrophotometer and is linear for more than 30 min; thus quantitative determinations can be quite precise.

Results and Discussion.—The results of various mixing experiments performed to date are presented in Table 1. At this stage of the investigation, it appears premature to draw many conclusions about the *in vitro* restoration of enzyme activity except that it is accomplished by the procedure described. Other attempts using variations of this procedure and other extraction procedures have failed to restore enzyme activity. Negative results were obtained whenever mycelia were lyophilized before mixing in the extraction buffer; similarly, when the mycelia were ruptured by grinding in sand with mortar and pestle, no enzyme activity was recovered. From the mixing experiments that have been performed, it appears that the process through which restoration of enzyme activity is accomplished is short-lived under the particular experimental conditions used. Mixing must take place within 5–10 min after the extracts have thawed or no enzyme activity is restored. It does not seem likely that the procedure used exerts its influence through the activation of an inactive enzyme since mutants treated alone exhibit no enzyme activity.

The presence of 10^{-3} M reduced glutathione during extraction appears to improve the recovery of enzyme activity, and there is some indication that ATP enhances the recovery, but highly refined quantitative measurements are not possible at this time.

Additional data have also been obtained from further studies involving *in vivo* restoration of enzyme activity. These data support earlier findings[4] of a correlation between complementation map distance and the levels of adenylosuccinase

TABLE 1
ADENYLOSUCCINASE ACTIVITY RESTORED THROUGH IN VITRO COMPLEMENTATION

Mutants	Additives During Extraction	Specific Activity (ΔA/mg. P/min/ml)	% of in Vivo Activity
F4 and F39 (heterocaryon)	None (no significant variation with additives present)	1.90	100.0
F4 and F39 (mixture)	None	0.19	10.0
		0.20	10.6
	ATP (Ave. of 3 experiments)	0.30	15.8
	ATP + GSH + NaF	0.31	16.4
		0.28	14.7
F4 and F48 (heterocaryon)	None (no significant variation with additives present)	1.60	100.0
F4 and F48 (mixture)	None	0.15	9.4
	GSH	0.33	20.6
F39 and F31 (heterocaryon)	None	1.17	100.0
F39 and F31 (mixture)	None	0.20	17.1
F4 and F31 (heterocaryon)	None	0.80	100.0
F4 and F31 (mixture)	None	0.04	5.0
F4	All variations	0.00	0.0
F39	All variations	0.00	0.0
F48	All variations	0.00	0.0
F31	All variations	0.00	0.0
F4 and F2* (mixture)	GSH	0.00	0.0
F39 and F2* (mixture)	GSH	0.00	0.0

* Non-complementing combination *in vivo*.

activity found in interallelic heterocaryons. The results plotted in Figure 1 indicate a steady increase of enzyme activity with increasing map distance up to a separation of about 4 cistrons, at which point no further significant increase in activity is observed. According to the hypothesis,[4] if the sites of damage in the two different defective gene products produced by the two complementing mutant nuclei are sufficiently widely separated such that random exchange can occur, then such a heterocaryon should yield 25 per cent of wild-type enzyme activity. This theoretical maximum of 25 per cent was in fact the maximum adenylosuccinase yield observed. However, an attempt to establish the precise nature of this correlation between map distance and enzyme activity does not yet appear justified in view of the limited data and the variability that exists.

Obvious irregularities in the plot of map distance vs. enzyme activity are limited to one or two cases. The one adjacent cistron combination that yields 10 per cent of wild-type enzyme activity may be such an exception. However, this discrepancy may be more apparent than real. For example, it is not yet certain that all of the cistrons have been located at the *ad-4* locus. If a cistron not yet detected exists between the two presumably adjacent cistrons, the 10 per cent value observed would fit into the plot of Figure 1 very well. This particular heterocaryon combination presents an additional complication since one of the mutant components (F23) is leaky to the extent of almost one per cent of wild-type adenylosuccinase activity even though it is unable to grow on minimal medium. The leakiness may contribute to the unexpectedly high 10 per cent recovery. The only other adjacent cistron combination studied (F12 × F14) possessed less than one per cent of wild-type enzyme activity. Because of poor growth in other adjacent cistron combinations, assays for enzyme activity were not possible. Other variables involved in making specific activity determinations are either inherent in the assay procedure or are brought about by unknowns present in the crude extract.

Heterocaryons have also been made between F12 reversions that possess only

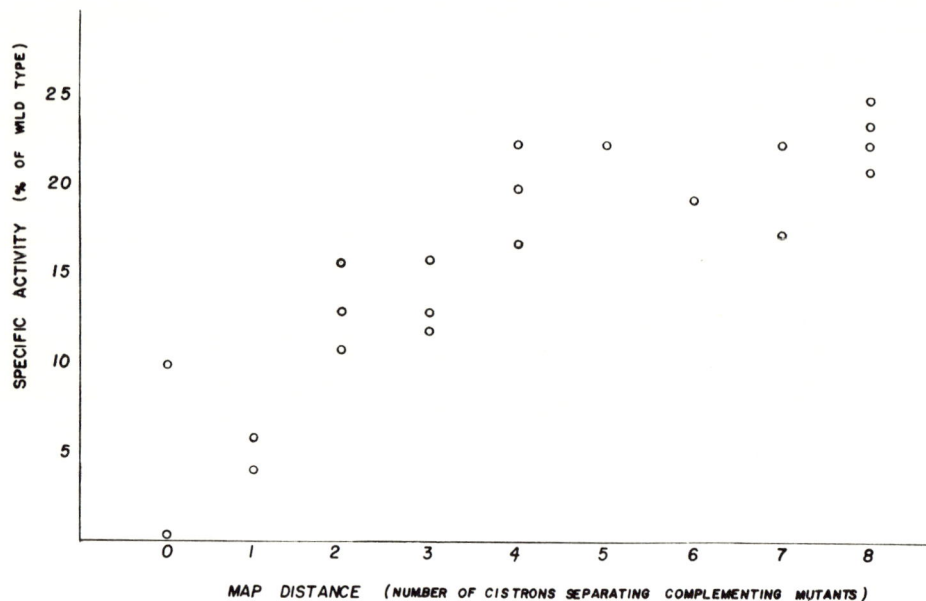

FIG. 1.—Relationship between complementation map distance and adenylosuccinase activity in heterocaryons. Each point plotted represents a heterocaryon composed of different pairwise combinations of adenylosuccinaseless mutants. Many of these figures are averages from several assays of separately grown cultures of the heterocaryons. The specific activity is a measure of the ΔA/28 mg dry wt/20 min/0.77 ml. expressed as a per cent of wild-type activity.

3 per cent of wild-type enzyme activity and mutants that complement F12 and are widely separated from F12 on the complementation map. Since there is now evidence[11] indicating qualitative differences between the enzymes from these reversions and the wild-type enzyme, the assumption can be made that the low level of enzyme activity in the reversions is not due to any quantitative change in the amount of protein synthesized but only to such qualitative changes.[12] On this basis, the hypothesis of random exchange for the wide separation involved predicts a maximum of 25.75 per cent of wild-type enzyme activity. The actual figures obtained from these heterocaryons were 24.3, 24.0, 22.0, and 22.4 per cent of wild-type adenylosuccinase activity.

Until more is understood about *in vitro* complementation, definitive statements concerning the mechanism involved cannot be made. Nevertheless, the present system appears to provide an excellent opportunity for studying this problem.

Summary.—*In vitro* complementation between homogenates from *Neurospora* adenylosuccinaseless mutants resulting in restored adenylosuccinase activity is described. As much as 20 per cent of the activity recovered from *in vivo* complementation in heterocaryons has been restored by the *in vitro* mixing procedure. If the homogenates from two complementing mutants are not mixed immediately after the mycelia have been disrupted, enzyme activity is not recovered. No enzyme activity has been detected when homogenates from non-complementing mutants are mixed. *In vivo* complementation results in the restoration of enzyme activity to a maximum level not exceeding 25 per cent of wild-type. There is a direct correlation between enzyme activity recovered from interallelic heterocaryons and distance between the mutant components on the complementation map. These results support the hypothesis previously proposed that, in general, interallelic complementation in enzyme formation involves a cytoplasmic exchange mechanism resulting in the formation of hybrid protein.

The author is greatly indebted to Dr. Norman H. Giles for his guidance throughout this investigation. The author is also grateful to Drs. C. W. H. Partridge, S. J. Singer and V. W. Woodward for helpful suggestions concerning this manuscript.

* This research was performed during the tenure of a pre-doctoral fellowship, MF-7405, of the National Institute of Mental Health, Public Health Service, and was supported in part by a research contract with the Atomic Energy Commission, AT (30-1)-872, administered by Norman H. Giles. This paper is a part of a dissertation presented for the degree of Doctor of Philosophy in Yale University.

[1] Giles, N. H., C. W. H. Partridge, and N. J. Nelson, these PROCEEDINGS, **43**, 305 (1957).
[2] Woodward, D. O., *Proc. X Intern. Cong. Genetics*, **2**, 321 (1958) (abstr.).
[3] Giles, N. H., *Proc. X Intern. Cong. Genetics*, **1**, (in press) 1958.
[4] Woodward, D. O., C. W. H. Partridge, and N. H. Giles, these PROCEEDINGS, **44**, 1237 (1958).
[5] Singer, S. J., and H. A. Itano, these PROCEEDINGS, **45**, 174 (1959).
[6] Crawford, I. P., and C. Yanofsky, these PROCEEDINGS, **44**, 1161 (1958).
[7] Richards, F. M., these PROCEEDINGS, **44**, 162 (1958).
[8] Bearn, A. G., and E. C. Franklin, *Science*, **128**, 596 (1958).
[9] Harris, H., E. B. Robson, and M. Siniscalco, *Nature*, **182**, 1324 (1958).
[10] Sober, H. A., F. J. Gutter, M. M. Wyckoff, and E. A. Peterson, *J. Am. Chem. Soc.*, **78**, 756 (1956).
[11] Woodward, D. O., unpublished Ph.D. thesis, Yale University, 1959.
[12] Yura, T., these PROCEEDINGS, **45**, 197 (1959).

The Mechanism of Complementation between *am* Mutants of *Neurospora Crassa*

J. R. S. FINCHAM AND A. CODDINGTON

The John Innes Institute, Hertford, England

INTRODUCTION

Inter-allelic or *intra-cistronic* (Schlesinger and Levinthal, 1963) complementation is now recognized as a very general phenomenon. Wherever there is an extensive series of mutants structurally defective with respect to some particular enzyme, it is likely that some pairs of them will be found to be capable of complementing one another (in heterocaryon, heterozygote or heterogenote, depending on the organism) to form more enzyme activity than either could produce by itself. Not to find examples of such complementation seems to be more the exception than the rule.

When this sort of complementation first came into prominence in 1957, there was a tendency to attempt to explain it in terms of some kind of interaction at the level of template RNA. However, the early demonstration by Woodward (1959), that in the Neurospora *ad-4* series of mutants complementation with the formation of active adenylosuccinase could occur in mixtures of cell-free extracts, made a protein–protein interaction theory seem much more likely. Because of the finding in most series of mutants showing complementation of a large class of point mutants which do not complement in any combination, and also because of the existence within the same series of several classes of mutant which will complement each other in *all* combinations, it seems unlikely that complementation is due generally to complementary mutants being defective in different *kinds* of polypeptide chain. Such an explanation would imply that there are as many different kinds of polypeptide in the protein as there are mutually complementing classes of mutants, and that many mutations affect all the chains simultaneously. It seems much simpler to postulate interaction between different mutant forms of the same kind of polypeptide. On this view the normal enzyme would be an aggregate of identical subunits and the complementation activity would be due to the formation of mixed aggregates containing mutant chains with defects at different points. This kind of explanation was offered both by Fincham (1959b), and by Brenner (1959), but it is only fair to point out that essentially the same idea had already been put forward by Catcheside (Catcheside and Overton, 1958). There are a number of models, most of them involving effects on conformation, which can be suggested to account for an active, or partially active, mixed aggregate, when the two homogeneous mutant aggregates are both inactive. Many of us will have seen the unpublished manuscript by Crick and Orgel (1960) in which a number of such models are described.

This paper describes the results of experiments with the *am* series of mutants in *Neurospora crassa*, which are structurally defective in the NADP-linked glutamate dehydrogenase (GDH). (NADP = nicotinamide-adenine dinucleotide phosphate = triphosphopyridine nucleotide.) The results seem to lend strong support to the identical subunit hypothesis, and show a number of complications which seem to be interpretable in terms of this hypothesis.

PROPERTIES OF THE ENZYME

If the identical subunit hypothesis is correct, *Neurospora* GDH must consist of identical subunits. That this is so has seemed probable since 1961, when Barratt (1961) presented a preliminary report on the chemistry of the enzyme. He obtained an S value for the purified enzyme protein of about 11.5, indicating a molecular weight probably in the range 200,000 to 250,000. On the other hand, peptide mapping of tryptic digests of the protein showed only 26 strong spots, 9 or 10 of which contained arginine. Amino acid analysis indicated 9 arginine and 15 lysine residues per 28,000 molecular weight. The obvious conclusion is that the enzyme molecule contains 8 identical subunits, though perhaps 6 to 10 would be a more suitably cautious estimate.

Our own ultracentrifuge and fingerprinting studies have done little more than confirm Barratt's conclusions. Although our peptide maps have commonly shown considerably more spots than the theoretical number, the number of strong spots, and the numbers staining specifically for tyrosine, histidine, arginine, and sulphur are in agreement with a subunit of around 30,000 molecular weight. We had hoped by this time to have been able to present a definitive estimate of the number of

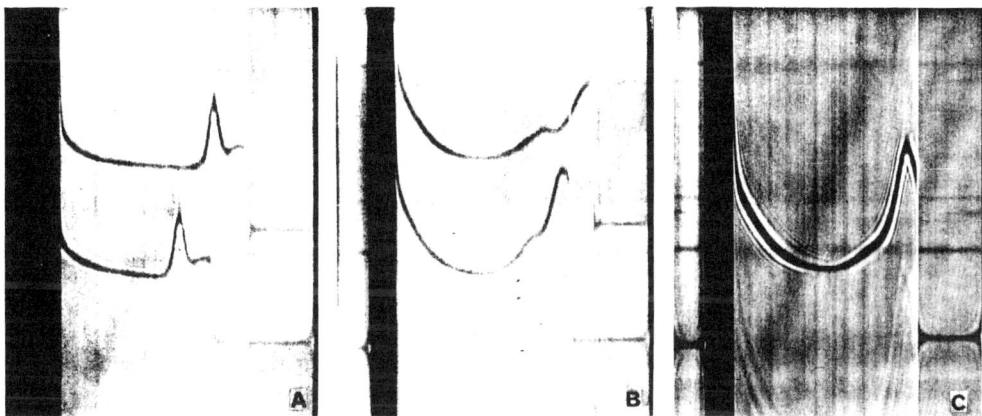

FIGURE 1. Pictures taken in the Spinco Model E analytical ultracentrifuge of GDH dialyzed against urea at various concentrations in 0.1 M phosphate buffer at pH 7.06. A: No urea. Upper picture wild type GDH, lower picture am^1 GDH, after 9 min at 59,780 rpm. B: Wild type GDH. Upper picture 4 M-urea; lower picture 5 M-urea, after 25 min at 59,780 rpm. C: Wild type GDH. 6 M-urea, after 42 min at 59,780 rpm. Protein concentrations about 4–5 mg/ml.

subunits based on quantitative end-group determinations, but we still have no certain answer on this point, largely because of difficulties in preparing large amounts of sufficiently pure material.

However, we do have a certain amount of physical evidence for the dissociability of the enzyme into relatively small subunits. Studies with the analytical ultracentrifuge have shown that 5 M or 6 M urea has a spectacular effect on the sedimentation properties of the protein. In 6 M urea practically all the material is converted to a form sedimenting with an S_{20w} of 2.5 (Fig. 1), while at lower urea concentrations a component with an intermediate sedimentation rate is seen. Even taking account of the fact that some decrease in S value may be due to unfolding of the molecule, a considerable decrease in molecular weight must, presumably, also be taking place to account for an effect of this magnitude. It has not been found possible to recover activity from these apparently dissociated preparations by dialyzing out the urea. The partial inactivation brought about by *brief* treatments with 4.5 M urea is reversible, but it is not certain that such milder urea treatments do, in fact, cause any dissociation.

THE *AM* MUTANTS

The *am* mutants of *Neurospora* need to be supplied with a source of α-amino nitrogen for normal growth and produce no normal NADP-linked GDH. *Neurospora* also produces an NAD-linked GDH which is not affected by *am* mutations and is not relevant to the present work (Sanwal and Lata, 1961). The presence of this second enzyme is probably the reason why *am* mutants are able to grow after a lag on unsupplemented medium, and this somewhat "leaky" property makes the mutants rather difficult to isolate in large numbers. Nevertheless, 19 independently isolated *am* mutants have now been studied in some detail. Of these am^{12} seems to be a recurrence of am^3, and am^{13} a recurrence of am^7; all the others (with the possible exceptions of am^{15}–am^{18}, which are not yet fully tested) are genetically distinct as judged by ability to form wild type recombinants in crosses (Pateman, personal communication) or by complementation properties. The origins of all the mutants up to am^{11} have been given previously (Fincham, 1959a, b). Two more, am^{12} and am^{13} were isolated following ultraviolet irradiation of conidia, and a further six, am^{14} to am^{19} were obtained during the past year by D. R. Stadler following treatment of conidia with nitrous acid.

The recently isolated nitrous acid mutants have added considerably to the complexity of the complementation relationships at the *am* locus. The present state of affairs is illustrated in Fig. 2. In addition, there is now a large family of abnormal revertants from am^2 and am^3 which produce their own varieties of partially active GDH and have their own complementation relationships, necessitating a more complex complementation map (Pateman and Fincham, 1963). These secondary mutants will not be considered here. We might expect, on the basis of our protein–protein interaction hypothesis, that mutants capable of complementation should each produce some variety of the GDH subunit. Roberts, working in Pateman's laboratory in Cambridge, has recently made a

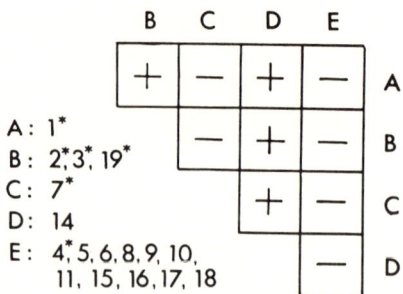

FIGURE 2. Chart showing the complementation relationships of *am* mutants; + indicates complementation, − no complementation. The mutants are divided into five classes: A–E, and the mutants falling into each class are shown. Asterisks indicate the formation of cross-reacting material (CRM) by the mutant concerned. Mutants am^{12} and am^{13} are not included; am^{12} is a recurrence of am^3 and am^{13} of am^7. Data on am^{14} and am^{19} complementation by courtesy of Dr. D. R. Stadler. Data on CRM by courtesy of Mr. D. Roberts.

careful investigation of this point (Roberts, D. B., in press). Using antiserum prepared against wild-type GDH, he has looked for cross-reacting material (CRM) in all the *am* mutants using both immuno-electrophoresis and the Ouchterlony double-diffusion technique. He finds clear evidence of CRM in am^1, am^2, am^3, am^4, am^7, am^{12}, am^{13}, and am^{19}, but in none of the others. Thus, there is a clear correlation between ability to complement and formation of CRM. The exceptions are am^4, which produces CRM but does not complement significantly; and am^{14}, which is our most versatile complementer, but is CRM-negative. Examples of complementing mutants which fail to form CRM are known in other systems (e.g. Garen and Siddiqi, 1962).

From several of the CRM-forming mutants the protein related to GDH has actually been isolated (Fincham, 1962; Fincham and Coddington, 1963). In am^2 and am^3, which are rather similar in their complementation properties (Fig. 2), the related protein has latent GDH activity. The protein from am^3, which we will call GDH-3, gives almost no activity when tested in either of two assay systems customarily used for wild type enzyme. These are (system A) α-oxoglutarate, ammonium chloride and $NADPH_2$ in tris-HCl buffer at pH 8.5, and (system B) 33 mM sodium L-glutamate and NADP in the same buffer. However, when added to a modified glutamate-NADP system with the glutamate concentration increased to 0.14 M (system C), GDH-3 catalyzes a rapid reduction of the coenzyme. The rate of the reaction is enhanced for some preparations by incubation of the enzyme for 2 min with the glutamate-Tris buffer mixture before addition of the NADP. The reaction is rather short-lived, generally slowing down and almost stopping well before equilibrium is attained. GDH-3 will also catalyze the reductive amination of α-oxoglutarate at something approaching a wild type rate if 0.14 M glutamate *and* NADP are added to the usual system A components. Evidently GDH-3 is activated by a high concentration of glutamate in the presence of NADP. Some preliminary evidence indicates that the activating effect of glutamate is not a specific one, since succinate, malate, and even acetate at similar concentrations will also work. The properties of the protein from am^2 (GDH-2) are rather similar to those of GDH-3, but more extreme. GDH-2 shows activity in the presence of 0.14 M glutamate and NADP, but the activity is much less than in GDH-3 and even shorter-lived. Somewhat better activity can be obtained from GDH-2 at a higher pH. In experiments reported in this paper we have used an assay system D, similar to system C, but buffered at pH 8.7 instead of pH 8.5. This pH value is above the optimum for wild type enzyme. The protein produced by am^1 (GDH-1) has no enzyme activity under any conditions yet tried. However, in its fractionation characteristics, sedimentation coefficient, and electrophoretic mobility at alkaline pH in both starch and acrylamide gels, it cannot be distinguished from wild type GDH.

Peptide maps have been prepared from both trypsin and trypsin-plus-chymotrypsin digests of wild type GDH and the three mutant varieties just mentioned. No clear and reproducible differences between the four proteins have been seen. This, though disappointing, was not totally unexpected, since all four are electrophoretically similar (Fincham, 1962; Fincham and Coddington, 1963).

Two attempts to isolate a mutant form of GDH from am^{14} have failed. This failure, which was at first quite surprising, is in line with Roberts' (1963) more recent demonstration that am^{14} contains no CRM.

THE COMPLEMENTATION REACTION IN VITRO

PROTEIN PREPARATIONS

Most of our more recent work has been concerned with reactions between mutant varieties of GDH in vitro. The mutant forms of the enzyme have been purified by the same method as is used for the wild type (Fincham and Coddington, 1963), involving ammonium sulphate fractionation, adsorption on to, and elution from, calcium phosphate gel, and two successive column fractionations using diethylaminoethylcellulose (DEAE). (There is an error in the original description of this method [Fincham, 1962]. The quantities of ammonium sulphate given for 500 ml should be for 1 liter.) In the case of GDH-1 the course of the fractionation cannot be followed by assays of enzyme activity, but it is possible to find out where the GDH-1 is at critical stages by electrophoresis on acrylamide gel (Ornstein and Davis). Even though it has no enzyme activity, the GDH-1 protein band is easily recognizable in somewhat purified preparations by its characteristic position. We have also used acrylamide gel electrophoresis routinely to check the purity of our final preparations. In most cases they appeared to be contaminated by two proteins showing as faint bands when relatively high concentrations of protein were run. GDH-3 preparations were more often free of these contaminants because of the slightly different behavior of this GDH variety on DEAE. A few of the GDH-1 preparations used were more heavily contaminated. The gels showed GDH-1 protein to be the major component, but with several minor bands. Because of these differences in degree of purity, the absolute amounts of the various GDH proteins used in *in vitro* mixing experiments are not precisely known, and the values given can only be regarded as approximations erring on the side of overestimation. Except where otherwise stated, however, the various protein preparations used were very similar in degree of purity by the electrophoresis criterion. The best preparations were about 100-fold purified on a protein basis, as compared with crude extracts.

CONDITIONS FOR IN VITRO COMPLEMENTATION

In two published cases, adenylosuccinase of *N. crassa* (Woodward, 1959) and imidazole glycerol phosphate dehydrase of *Salmonella* (Loper, 1961), extracts of enzyme-deficient mutants have been shown to form enzyme activity by simple mixing in buffered solutions at pH values at, or slightly above, neutrality. Our preparations of mutant proteins are usually obtained in 0.05 M sodium phosphate — 0.001 M ethylenediaminetetracetic acid at pH 7.4 (standard buffer). Simple mixing of GDH-1 and GDH-3 in this buffer did not produce any new enzymic activity. However, adjustment of the pH either up with sodium hydroxide, or down with acetic acid led to the appearance of activity in assay system A (reductive amination of α-oxoglutarate). Increase in ionic strength by increasing the phosphate concentration without change in pH had no effect. The best pH range for the complementation reaction in vitro was found to be 5.0–6.0. For the activity to appear, however, it was necessary to readjust the mixture to pH 7.4; a sample added directly to the assay system from a pH 5.8 mixture gave little or no activity. At pH 5.8 the reaction is evidently rapid at 20°; only small differences in yield of activity were obtained by varying the incubation period at pH 5.8 from 10 min to 1 hr (Fig. 5). Studies on the time course of the reaction have not been seriously attempted. Usually a 30 min period at pH 5.8 was used. The appearance of activity following readjustment to pH 7.4 was also rapid and not appreciably time-dependent in the range 15–50 min. At least 15 min were allowed to elapse after pH readjustment before mixtures were assayed. Studies on other conditions of pH which are somewhat effective in bringing about complementation have only been made in a perfunctory way. From pH 10 very little activity was recovered; but pH 9.4 gave (after 30 min followed by readjustment to pH 7.4) about 40% of the system A activity observed after treatment at pH 5.8 for a similar period of time. Between pH 7.0 and pH 6.0 the yield of activity from a 30 min treatment decreased with increasing pH, but even pH 6.8 gave a little activity. The optimum conditions for complementation between GDH-1 and GDH-3 had no activating effect on either protein by itself. The same procedure yielded activity from GDH-1 + GDH-2 mixtures, but GDH-2 and GDH-3 did not complement in vitro.

Dependence on the Proportions of the Interacting Proteins

The GDH-1 + GDH-2 system. An experiment on the effect of varying the concentration of GDH-1 with constant concentration of GDH-2, and vice versa, gave results which suggest that not only the *amount* of complementation product but also the *kind* of complementation enzyme formed depends on the proportions of the two proteins. As Fig. 3 shows, GDH-2 has negligible activity by itself in the reductive amination system (system A), and fairly low activity in the high glutamate — NADP system at pH 8.7 (system D). On reacting GDH-1 with the GDH-2, activity appears in system A and greatly increases in system D. With relatively small amounts of GDH-1 the system A activity is very low in comparison with the *increase* in the system D activity. As the quantity of GDH-1 is increased a range of concentration is reached in which the system A activity is increasing in proportion to more than the first power of GDH-1 concentration, while the system D activity is not increasing proportionately. With about a 1:1 ratio of the two proteins the increment of system D activity is as much as five times greater than the system A activity, while with a large excess of GDH-1 the same ratio is down to about 0.75. This result

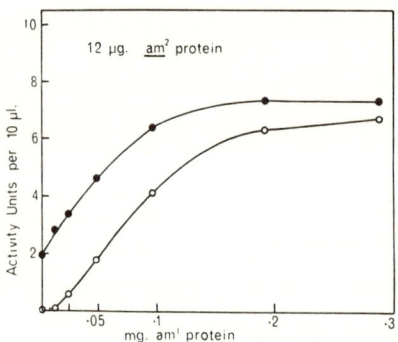

FIGURE 3. Complementation between a constant amount of GDH-2 and varying amounts of GDH-1. The quantities indicated were mixed in 0.5 ml 0.05 sodium phosphate-0.001-M EDTA buffer at pH 7.4. The pH was adjusted to 5.8 with N acetic acid and after 30 min at 20° readjusted to 7.4 with N-sodium hydroxide. Enzyme activities were assayed starting 15 min later. Open circles: assay system A (at 35°: 0.15 ml 0.2 M disodium α-oxoglutarate; 0.10 ml M-ammonium chloride; 0.10 ml 0.1% $NADPH_2$; 2.55 ml 0.05 M-Tris-HCl buffer, pH 8.5; 0.1 ml complementation mixture at zero time). Solid circles: assay system D (at 35°: 2.7 ml 0.15 M-monosodium glutamate in 0.05-M-Tris-HCl buffer pH 8.7; 0.1 ml, complementation mixture at zero time; 0.2 ml 0.2% NADP at 2 min). The activity unit is a rate of change of absorbancy at 280 mμ of 0.001 per min. The GDH-1 preparation used in this experiment was probably no more than about 50% pure, judging by minor bands visible after electrophoresis.

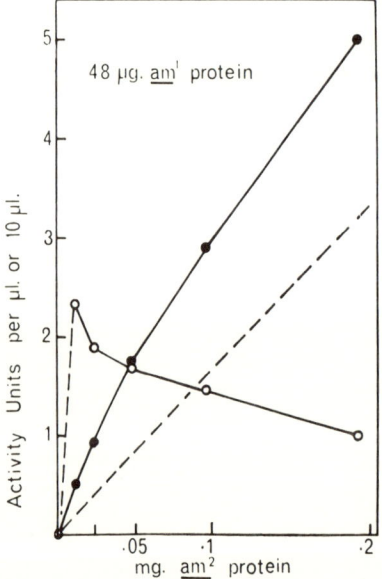

FIGURE 4. Complementation between a constant amount of GDH-1 and varying amounts of GDH-2. All conditions as for Fig. 3. The lower dashed line shows the amount of activity in system D which would have been expected from the same quantities of GDH-2 in the absence of GDH-1 (c.f. Fig. 3).

strongly suggests that at least two kinds of hybrid aggregate are being formed, with the one with the most system A activity containing a higher proportion of GDH-1 subunits. The reciprocal experiment, illustrated in Fig. 4, in which increasing amounts of GDH-2 were reacted with a constant amount of GDH-1, is also consistent with this idea. Here the system A activity actually decreased as greater amounts of GDH-2 were added, though the absolute amount of system D activity, and the excess system D activity over what could have been due to the GDH-2 alone, both increased. So far as system A activity is concerned, the optimum ratio of GDH-1 to GDH-2 is evidently quite an extreme one with a large excess of GDH-1. Considering specific activity in system D, however, a one-to-one

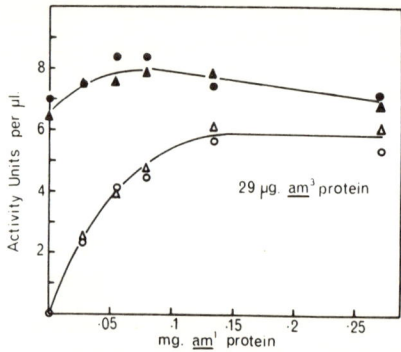

FIGURE 5. Complementation between a constant amount of GDH-3 and varying amounts of GDH-1. Complementation reaction conditions as for Fig. 3 except that the mixtures were kept at pH 5.8 for either 17 min (circles) or 50 min (triangles), and the amounts of protein indicated were in a volume of 3.2 ml. Open symbols: assay system A; solid symbols: assay system C, which is similar to assay system D except the pH is 8.5 instead of 8.7 (see legend to Fig. 2); 10 μl. of each complementation reaction mixture were assayed in each system. Redrawn from Fincham and Coddington (1963) with a few of the points altered slightly as a result of recalculation.

ratio is not far from the optimum. Unfortunately, GDH-2 and GDH-1 show very little difference in their chromatographic properties on DEAE, and it has not been possible to fractionate GDH-1 + GDH-2 reaction mixtures to any useful extent.

The GDH-1 + GDH-3 system. The results obtained with this pair of mutant proteins are quite analogous to those just described for GDH-1 + GDH-2. The main difference is that GDH-3 is much more readily activated than GDH-2 and gives specific activity approaching that of wild type enzyme in a system containing a high concentration of glutamate with NADP. The assay system C, with 0.14 M glutamate and pH 8.5, has been used in experiments involving GDH-3. The system C

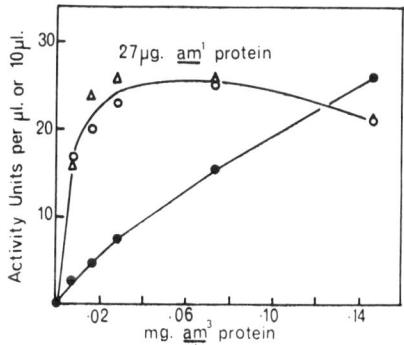

FIGURE 6. Complementation between a constant amount of GDH-1 and varying amounts of GDH-3. All conditions as for Fig. 5.

activity of GDH-3 can be somewhat enhanced by reaction with GDH-1 but the effect is much less marked than in the analogous situation with GDH-2 (cf. Figs. 3 and 5). Like GDH-2, GDH-3 has a negligible activity in assay system A unless it has previously been reacted with GDH-1. The system A activity for a constant amount of GDH-3 increases linearly, or at low GDH-1 concentrations rather more than linearly, with GDH-1 concentration (Figs. 5 and 7). At higher levels of GDH-1 the A activity tends to approach a maximum, although the ratio of A activity to C activity probably continues to increase up to the highest concentrations of GDH-1 tested since a large excess of GDH-1 seemed to cause some decline in C activity (Fig. 5).

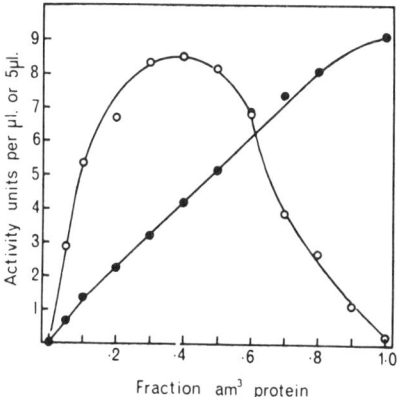

FIGURE 7. Complementation between GDH-1 and GDH-3, varying the ratio of the two proteins at a constant total protein concentration of 0.12 mg/ml. Complementation conditions as in legend to Fig. 2. Solid circles assay system C (see legend to Fig. 5); open circles assay system A (see legend to Fig. 2). The activities in system A are expressed in units per 5 μl. of complementation mixture, and in system C in units per μl.

Reacting increasing amounts of GDH-3 with a constant amount of GDH-1 gives a system A activity optimum with about a two-fold excess of GDH-3, after which there is a definite decline in activity (Fig. 6).

Figure 7 shows the variation in activity in both assay systems as the ratio of GDH-1 to GDH-3 is varied, the total protein concentration being kept constant. The curve for the system A activity shows very clearly the asymmetrical nature of the relationship of the two proteins.

FRACTIONATION OF THE GDH-1 + GDH-3 COMPLEMENTATION MIXTURE

Quite unexpectedly it was found that GDH-3 can be separated with very little overlap either from wild type GDH or from GDH-1 by chromatography on a DEAE column. This opened up the possibility of separating the complementation product or products from the proteins of the original types. Figure 8 shows the effect of the "hybridization" procedure on the chromatographic behavior of a mixture of about 4 parts of GDH-1 to 1 part of GDH-3. This is a ratio of the two proteins close to the optimum for obtaining system A activity. When the mixture without the pH 5.8 treatment is fractionated on a DEAE column with a gradient of sodium phosphate, an almost complete separation is achieved between an early protein peak, corresponding to the amount of GDH-3 added to the mixture, and having all the system C activity, and a late peak of electrophoretically identical, but enzymically inactive, protein which must be the GDH-1 component. The same mixture after the pH 5.8 treatment shows a significantly different elution profile with respect to protein and a markedly different distribution of activities. The early peak of GDH-3 is now reduced to a shoulder and the existence of two protein components is no longer clear. In the region which, in the analysis of the unhybridized mixture, had been between the two peaks, there is now a well-defined peak of system A activity. The protein eluted at higher phosphate concentrations shows decreasing enzymic activity; that is, it takes on more and more of the character of GDH-1. As we have shown elsewhere (Fincham and Coddington, 1963) a similar analysis of the GDH protein formed in $am^1 + am^3$ heterocaryons shows a very similar picture.

In another experiment it was shown that the material with system A activity could be rechromatographed without any change in either its enzymic or chromatographic properties. Two successive additional column fractionations yielded enzyme which appeared chromatographically homogeneous and with the system A and C activities running roughly parallel to each other and

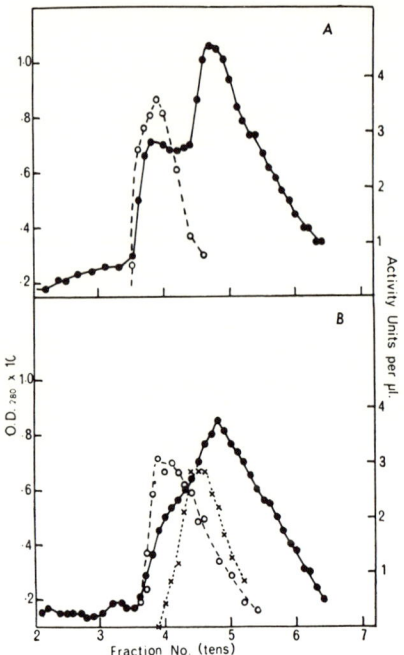

FIGURE 8. Fractionation on identical DEAE-cellulose columns of mixtures of 5 mg of GDH-1 and 1.25 mg of GDH-3, mixed (A) in 14 ml 0.05 M phosphate-0.001 M EDTA, pH 7.4 without any adjustment of pH; and (B) in the same volume of the same buffer, then adjusted to pH 5.8 with N acetic acid and readjusted after 35 min to pH 7.4 with N sodium hydroxide. The columns were eluted with sodium phosphate buffer at pH 7.4, increasing linearly in concentration from 0.02 M (fraction 4) to 0.07 M (fraction 64). EDTA at 0.001 M was present throughout. Fraction volumes were 4 ml. Solid circles: absorbancy (OD) at 280 mμ. Open circles: enzyme activity in assay system C (see legend to Fig. 5). Crosses: enzyme activity in assay system A (see legend to Fig. 2).

found that the active enzyme was eluted at nearly the same point in the phosphate gradient, irrespective of a wide variation in the ratio of GDH-1 and GDH-3 in the mixture. The chromatographic system, however, is probably hardly sensitive enough to distinguish hybrids not grossly different in composition.

MECHANISM OF THE REACTION AND NATURE OF THE PRODUCT

ARE THE IN VIVO AND IN VITRO COMPLEMENTATION PRODUCTS THE SAME?

It may well be that the detailed mechanism of the complementation reaction in an intact heterocaryon is not the same as that of the process in vitro. In vitro, if our hypothesis is correct, we first have homogeneous mutant aggregates and then form mixed aggregates from them. In vivo the possibility exists that the mixed aggregates are formed primarily from newly-formed polypeptide chains. Nevertheless, there is good reason to think that the nature of the active molecules which we can form in the in vitro reaction is the same as that

to protein concentration over the whole peak. In this relatively pure complementation product the activity for system A was about double that for system C, which is not very different from the ratio of activities found with the wild type enzyme. However, the complementation enzyme is rather less heat-stable than is the wild type (half-lives at 60° in standard buffer of 12.5 and 20 min were determined for the two enzyme varieties in one experiment) and has a Michaelis constant for glutamate which is higher than the normal value by a factor of about three (Fig. 9, cf. Fincham, 1959b, where the same difference was observed with cruder preparations).

It should be noticed that chromatographic fractionation gives little indication of more than one kind of complementation product. In an experiment designed to check this point it was

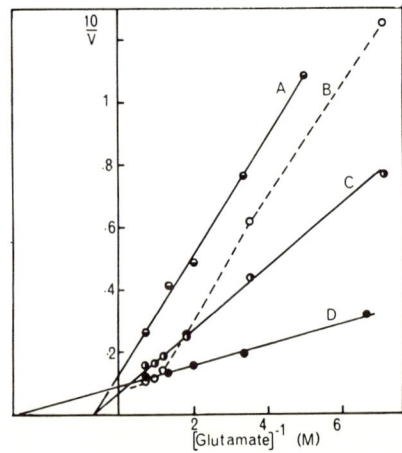

FIGURE 9. Lineweaver-Burk plots for the calculation of Michaelis constants for glutamate of: (A) an in vitro complementation fraction with high system A activity from a column fractionation similar to that shown in Fig. 8b; (B) fraction with properties resembling those of GDH-3 from a column fractionation of protein from an $am^1 + am^3$ heterocaryon; (C) fraction with high system A (complementation) activity from the same column fractionation of heterocaryon protein as B; (D) wild type GDH. The amounts of protein used per assay were 0.8 μg, 1.2 μg, 0.6 μg and 0.7 μg in A, B, C, and D respectively. All four fractions were substantially homogeneous electrophoretically. The concentrations of glutamate indicated were present in reaction mixtures containing TPN and Tris-HCl buffer as in assay system C (pH 8.5). Note the disproportionately low activity in B at low glutamate concentration.

of the complementation enzyme formed in the corresponding heterocaryon. As we have shown in a previous paper (Fincham and Coddington, 1963) chromatography on a DEAE column of GDH protein isolated from an $am^1 + am^3$ heterocaryon gives an active fraction which is indistinguishable in activity, and in position relative to the GDH-3 and GDH-1 fractions, from the corresponding product in the in vitro reaction mixture. Furthermore, complementation enzyme preparations from the two sources are indistinguishable in K_m for glutamate, both deviating from the wild type value to the same extent (Fig. 9). In the discussion which follows we feel justified in assuming that the in vivo and in vitro products are molecules of the same type.

THE SIZE OF THE COMPLEMENTATION PRODUCT

Two kinds of evidence virtually rule out the possibility that complementation between am mutants is due to the addition of one kind of mutant protein to the other to form a double-size molecule. Firstly, fractions of GDH-1 + GDH-3 reaction mixtures with high system A activity are indistinguishable from GDH-1, GDH-3, or wild type GDH by acrylamide gel electrophoresis. A similar conclusion can be drawn with regard to the complementation enzymes from all the other complementary pairs of am mutants in heterocaryons. Even though the complementation product has not been purified from most of these heterocaryons, it is possible to locate the enzymically active bands by immersing the acrylamide gels in a glutamate-NADP-phenazine methosulphate-neotetrazolium chloride mixture, buffered with Tris (Markert and Møller, 1959, and Markert, personal communication). In every case the active enzyme formed by complementation was similar to wild type in electrophoretic mobility in acrylamide gel. Since this gel should separate molecules on the basis of size in the molecular weight range we are concerned with here (Ornstein and Davis), it seems that the complementation enzymes must resemble normal GDH in the number of subunits which they contain. The same conclusion was drawn (Fincham and Coddington, 1963) from an ultracentrifugal analysis of a mixture of GDH-1 and GDH-3 "hybridized" in vitro. The sedimentation characteristics of the mixture, containing high system A activity, were no different from those of the original proteins which were mixed.

POSSIBLE MECHANISMS OF THE IN VITRO REACTION

A recent paper by Schlesinger and Levinthal (1963) has shown that complementation between different mutant varieties of *E. coli* alkaline phosphatase, which is a dimer, can be achieved by procedures which cause more or less total dissociation of the enzyme into subunits, followed by reassociation. At first such a mechanism also seemed probable in our system, especially in view of the need for two steps in the procedure, namely acidification followed by readjustment of pH. Analysis in the analytical ultracentrifuge of purified GDH-1 and GDH-3, separately and in mixture, at pH 7.4 and pH 5.8 showed, however, that any dissociation into subunits at the lower pH was at the most only very slight. These data have recently been published elsewhere (Fincham and Coddington, 1963). It is, of course, not at all excluded that at pH 5.8 there is an equilibrium between subunits and aggregate molecules heavily favoring the latter, but there is no evidence for this at present. We have no data on which to base any further discussion on this point. It may be questioned, however, whether the formation of hybrid aggregates need necessarily depend on a dissociation-reassociation mechanism. It seems quite possible that the opposite mechanism might be responsible; that is, the transitory formation of higher aggregates with the possibility of recombination on subsequent dissociation, a process faintly analogous to zygote formation followed by meiosis in genetics. If this were true, free subunits need never exist in the complementation reaction.

EXPLANATIONS IN TERMS OF MULTIPLE HYBRIDS

In spite of the doubt about the mechanism of their formation, it is difficult to explain the data except on the basis that hybrid aggregates are, in fact, formed. Given this assumption it is easy to think of likely explanations for many of the results, especially if several kinds of hybrid are possible. One of the most striking observations is that, in both the in vitro systems we have studied, the optimum composition of the mixture is heavily weighted on the side of the GDH-1 component. One might attempt to explain this on the basis that the GDH protein could split into halves for purposes of hybrid formation, so that the reaction mixture contained GDH-1, GDH-3, and one kind of 1-3 hybrid. The higher requirement for GDH-1 would then have to be explained as due to a greater stability of GDH-1 as compared to GDH-3 and a corresponding reduced availability of its subunits for hybrid formation. Alternatively, one could go to the other extreme and suppose that the 8 subunits which are suggested by the chemical studies could become re-assorted freely with the possible formation of seven kinds of hybrid. A model for this kind of situation in vivo is provided by Wilson, Cahn, and Kaplan (1963) who have shown that lactic dehydrogenase in bird muscle is a tetramer composed of two broadly similar but electrophoretically distinct kinds of subunit, and that all three possible kinds of hybrid (3:1, 2:2, and 1:3) can be distinguished. From this view, the higher requirement for GDH-1 could be because the hybrids which contribute most of the complementation activity contain a predominance of GDH-1 subunits. The attraction of this kind of picture is that a number of other features of the data can be accommodated within it. To avoid having to write too many digits, let us consider that the GDH is effectively a tetramer for purposes of hybrid formation. Then we can denote GDH-1 as 1.1.1.1, GDH-2 as 2.2.2.2, and GDH-3 as 3.3.3.3. If the only hybrids with appreciable activity in system A are 1.1.1.2 and 1.1.1.3, then the high requirement for GDH-1 in the complementation reactions is accounted for without the need to postulate any differential stability of the different aggregates. Also explained is the depressing effect of an excess of either GDH-2 or GDH-3 on active product formation. On our model this could be due to many of the GDH-1 subunits getting tied up in 1.2.2.2 and 1.1.2.2 (or 1.3.3.3 and 1.1.3.3) hybrids which would have low reductive amination activity. The rather strong indication of two kinds of complementation activity in the GDH-1 + GDH-2 system (Fig. 3) is also simply explained on the same basis. We need only suppose that 1.1.2.2, and perhaps also 1.2.2.2, have enhanced system D activity per GDH-2 subunit as compared with 2.2.2.2 without having appreciable system A activity. This type of model is obviously still more flexible if we take into account the possibility of hybrids with 7:1 and 5:3 subunit compositions.

STABILITIES OF COMPLEMENTATION ENZYMES

It would be expected that hybrid aggregates of the type envisaged here would differ from wild type enzyme in different degrees. One of the earliest reasons for favoring the subunit hypothesis of interallelic complementation was that enzyme formed by complementation was, in several instances, clearly different from the corresponding wild type enzyme (Fincham, 1959a; Partridge, 1961). In fact all the complementary pairs of am mutants produce varieties of active GDH with more or less abnormal properties. The criterion which we have employed more than any other is stability to heat. The $am^1 + am^3$ complementation enzyme approaches most nearly to wild type stability, but is still consistently less stable than normal GDH, while the $am^1 + am^2$ product is less stable still (Fincham, 1959b). The least stable complementation enzymes we have tested are those produced by $am^2 + am^{14}$ and $am^3 + am^{14}$. Dr. Stadler, who kindly permitted us to quote his results, finds that these enzymes are completely inactivated by 5 min heating at 60° at pH 8.0 in 0.05 M phosphate buffer, a treatment which has no appreciable effect on wild type activity. These differences in stability seem quite consistent with the following picture of the nature of complementation. We know that GDH-2 and GDH-3 have all the necessary component groups of the enzyme active center. However, unless the protein, in either case, is activated by incubation with the high glutamate-NADP mixture, the active center cannot function. Activation by substrate mixtures of various kinds or, in one case, simply by mild heating, is common among the imperfectly functional GDH varieties encountered in partial revertants from am^2 and am^3 (Fincham, 1957; Fincham and Bond, 1960; Pateman and Fincham, 1963). It seems very likely that all these activation effects are due to changes in the conformation of the protein, and that the nature of the mutational alterations in GDH-2 and GDH-3 are such as to lead to a "faulty" conformation so that the groups of the active center no longer have the correct spatial relationships. The defect in the GDH-1 protein could be something more radical causing unconditional loss of activity, but with the conformation of the protein still normal. Thus we can readily imagine that the function of GDH-1 subunits in complementation is to correct and stabilize the conformation of the GDH-2 or GDH-3 subunits. The higher the proportion of GDH-1 subunits in the mixed aggregates, the better may the conformation be stabilized in the active state. A minority of GDH-1 subunits could be insufficient for activation, although they might well render the aggregate more easily and completely activable by glutamate-NADP mixtures (cf. Fig. 3). If the basis for complementation activity in mixed aggregates is a unilateral (or perhaps in some cases mutual) modification of the conformation of one subunit by another, one might expect complementing aggregates to be under a certain amount of internal strain which could hardly fail to be reflected in reduced heat stability. Complementation involving am^{14} may represent the extreme case where one of the participating subunits is conformationally so abnormal that it is unable to form stable aggregates at all by itself. This would seem to be the most probable reason for the failure of am^{14} to form CRM. Since GDH-2 and GDH-3 subunits are themselves, according to our hypothesis, inclined to fall into incorrect conformations, extreme heat lability of the products of am^{14} complementation with either am^2 or am^3 is not surprising. The $am^1 + am^{14}$ enzyme is considerably more stable.

ATTEMPTS TO DEHYBRIDIZE THE COMPLEMENTATION PRODUCT

If the complementation reaction consists of a re-assortment of subunits one might expect to be able to make the reaction go in either direction. Starting with complementation product from the GDH-1 + GDH-3 reaction, and freeing it so far as possible from unchanged GDH-1 and GDH-3, it seemed likely that the same conditions as were effective in hybridization would result in dehybridization, with the regeneration of some GDH-1 and GDH-3 from the hybrid. This would result in a drop in the activity in assay system A with relatively little effect on the activity in assay system C. The experiment was tried, both with complementation enzyme isolated from an $am^1 + am^3$ heterocaryon and with the corresponding product of the in vitro reaction. In neither case did the pH 5.8 treatment result in any significant change in enzyme properties. In an experiment with the heterocaryon enzyme the fractions with complementation activity were treated at pH 5.8 and re-chromatographed on a DEAE column; they showed no indication whatever of the presence of regenerated GDH-3. Thus we seem to be forced to the conclusion that the complementation product is quite stable under conditions in which GDH-1 and GDH-3 react with each other very rapidly. It may be that ability to exchange subunits under our complementation conditions is not a general property of all the GDH varieties but is characteristic of certain mutant forms of the protein. The negative result given by the dehybridization experiment has deprived us of one possible means of clinching the argument that complementation enzyme molecules are indeed aggregates of different mutant subunits. Other experimental approaches to the same end will have to be tried.

The failure to reverse the complementation reaction raises some puzzling questions relating to its kinetics. If the reaction is irreversible why does it seem not to go to completion? In the reaction mixture whose analysis is shown in Fig. 8, considerable protein with the properties of GDH-3 remains, in spite of the initial 4-fold excess of GDH-1. A large excess of GDH-1 should cause the complete conversion of GDH-3 to complementation product with system A activity considerably higher than the system C activity. Yet Fig. 3 shows that this situation is not readily achieved even with something approaching a ten-fold excess of GDH-1. In all these experiments the formation of active enzyme was, if not complete, at least very slow at the time it was stopped. It is as if the reaction, after a very rapid start, proceeds to completion only very slowly. It is clear that more knowledge of the basic chemistry of the system will be necessary before we can begin to understand what is happening.

SUMMARY AND CONCLUSIONS

Complementation between *am* mutants of *Neurospora crassa* in heterocaryons results in the formation of active glutamate dehydrogenase (GDH). Mutants capable of complementation produce, with one exception, proteins closely resembling normal GDH in physical and chemical properties, but either lacking activity, or possessing it only after special activating treatments.

Attention has been focussed on the pairs of mutants $am^1 + am^2$ and $am^1 + am^3$. The GDH-like proteins have been isolated from these mutants in nearly pure form. The combinations of proteins 1 + 2 and 1 + 3, but not 2 + 3, will react, under appropriate conditions of pH, to form active GDH. The GDH variety from the 1 + 2 combination differs slightly in properties from the 1 + 3 variety and both differ slightly from wild type GDH (Fincham, 1959b,). The 1 + 3 GDH formed in vitro is probably identical to the enzyme in the corresponding heterocaryon.

Although the chemical evidence is not yet complete, *Neurospora* GDH is almost certainly composed of 6 to 10 identical subunits (Barratt, 1961). Experiments presented here on the effects of variation in the ratio of the two reacting proteins are, on the whole, readily interpreted on the basis of the hypothesis that various kinds of mixed aggregate can be formed in the complementation reaction mixtures. In the case of the 1 + 3 reaction, the active interaction product can be to a large extent separated from the original mutant proteins by column chromatography.

REFERENCES

BARRATT, R. W. 1961. Studies on gene-protein relations with glutamic dehydrogenase in *Neurospora crassa*. Genetics, *46:* 849–850.

BRENNER, S. 1959. The mechanism of gene action, p. 304–316. *In:* Biochemistry of Human Genetics, Ciba Foundation Symposium. Churchill, London.

CATCHESIDE, D. G. and A. OVERTON, 1958. Complementation between alleles in heterocaryons. Cold Spring Harbor Symp. Quant. Biol., *23:* 137–140.

FINCHAM, J. R. S. 1957. A modified glutamic dehydrogenase as a result of gene mutation in *Neurospora crassa*. Biochem. J., *65:* 721–738.

———. 1959a. The role of chromosomal loci in enzyme formation. Proc. X Int. Congr. Genetics, I. p. 355–363. Univ. of Toronto Press.

———. 1959b. On the nature of the glutamic dehydrogenase formed by inter-allele complementation at the *am* locus of *Neurospora crassa*. J. Gen. Microbiol., *21:* 600–611.

———. 1962: Genetically determined multiple forms of glutamic dehydrogenase in *Neurospora crassa*. J. Mol. Biol., *4:* 257–274.

FINCHAM, J. R. S. and P. A. BOND, 1960. A further genetic variety of glutamic acid dehydrogenase in *Neurospora crassa*. Biochem. J., *77:* 96–105.

FINCHAM, J. R. S. and A. CODDINGTON, 1963. Complementation at the *am* locus of *Neurospora crassa*: a reaction between different mutant forms of glutamate dehydrogenase. J. Mol. Biol., *6:* 361–373.

GAREN, A. and O. SIDDIQI, 1962. Suppression of mutations in the alkaline phosphatase structural cistron of *Escherichia coli*. Proc. Natl. Acad. Sci., *48:* 1121–1127.

LOPER, J. C. 1961. Enzyme complementation in mixed extracts of mutants from the *Salmonella* histidine B locus. Proc. Natl. Acad. Sci., *47:* 1440–1450.

MARKERT, C. L. and F. MØLLER, 1959. Multiple forms of enzymes: tissue, ontogenetic and species specific patterns. Proc. Natl. Acad. Sci., *45:* 753.

ORNSTEIN, L. and B. J. DAVIS. Disc electrophoresis. Reprinted by Distillation Products Industries, Kodak Ltd., Kirkby, Liverpool.

PARTRIDGE, C. W. H. 1961. Altered properties of the enzyme, adenylosuccinase, produced by interallelic complementation at the *ad-4* locus in *Neurospora crassa*. Biochem. Biophys. Res. Commun., *3:* 613.

PATEMAN, J. A. and J. R. S. FINCHAM, 1963. Partial revertants at the *am* locus in *Neurospora crassa*. (Abstract). Heredity, in press.

ROBERTS, D. B. 1963. Immunological studies of amination deficient mutants of *Neurospora crassa*. J. Gen. Microbiol., in press.

SANWAL, B. D. and M. LATA, 1961. The occurrence of two different glutamic dehydrogenases in *Neurospora*. Canad. J. Microbiol., *7:* 319–328.

SCHLESINGER, M. J. and C. LEVINTHAL, 1963. Hybrid protein formation of *E. coli* alkaline phosphatase leading to *in vitro* complementation. J. Mol. Biol., *7:* 1.

WILSON, A. C., R. D. CAHN, and N. O. KAPLAN, 1963. Functions of two forms of lactic dehydrogenase in the breast muscle of birds. Nature, *197:* 331–334.

WOODWARD, D. O. 1959. Enzyme complementation *in vitro* between adenylosuccinaseless mutants of *Neurospora crassa*. Proc. Natl. Acad. Sci., *45:* 846–850.

IX
One Gene–One Polypeptide

Editor's Comments on Paper 23

23 Itano and Robinson: *Genetic Control of the α- and β-Chains of Hemoglobin*

At the time this paper was written it was recognized that the normal adult hemoglobin A of humans consisted of two different polypeptide chains, α and β. Sickle-cell hemoglobin S was known to be due to a mutation affecting the β chain so that individuals with HbS had normal α chains, but abnormal β chains. A new abnormal hemoglobin (Ho-2) was discovered, which appeared to be caused by the mutation of a different gene than the gene for the β chain. The question then was did the Ho-2 mutation affect the α chain? Itano and Robinson show here that it did. Therefore, two different loci determine the structure of HbA; one determines the α chain, and the other the β chain. This finding resulted in the demise of the one gene–one enzyme (protein) hypothesis and the substitution of a modified one gene–one polypeptide hypothesis, which is now generally accepted. Soon after this work, other proteins, such as vertebrate lactic dehydrogenase, were found to be heteromultimers like hemoglobin consisting of two different peptides each under the control of a different gene locus.

Selected Bibliography

Bradley, T. B., S. H. Boyer, and F. H. Allen, Jr. 1961. Hopkins-2 hemoglobin: a revised pedigree with blood and serum data. *Bull. Johns Hopkins Hosp. 108:* 75–79.

Itano, H. A., and E. Robinson. 1959. Formation of normal and doubly abnormal haemoglobins by recombination of haemoglobin I with S and C. *Nature 167:* 1799–1800.

Rhinesmith, H. S., W. A. Schroeder, and L. Pauling. 1957. A quantitative study of the hydrolyses of human dinitrophenyl (DNP) globin; the number and kind of polypeptide chains in normal adult hemoglobin. *J. Amer. Chem. Soc. 79:* 4682–4686.

Vinograd, J. R., W. D. Hutchinson, and W. A. Schroeder. 1959. C^{14}-hybrids of human hemoglobins: II. The identification of the aberrant chain in human hemoglobin S. *J. Amer. Chem. Soc. 81:* 3168–3169.

*GENETIC CONTROL OF THE α- AND β-CHAINS OF HEMOGLOBIN**

BY HARVEY A. ITANO AND ELIZABETH A. ROBINSON

THE NATIONAL INSTITUTE OF ARTHRITIS AND METABOLIC DISEASES†

Communicated by Bentley Glass, September 21, 1960

Identical pairs of two chemically different polypeptide chains, the α-chains and the β-chains,[1,2] make up the protein portion of human adult hemoglobin. Hemoglobin A, in which both types of chains are normal, is designated $\alpha_2^A \beta_2^A$. Either type may carry the inherited defect of an abnormal hemoglobin; the specific amino acid substitution of hemoglobin S or C occurs in the β-chain, and the substitution of hemoglobin I occurs in the α-chain.[3,4] These abnormal adult hemoglobins are designated $\alpha_2^A\beta_2^S$, $\alpha_2^A\beta_2^C$, and $\alpha_2^I\beta_2^A$, respectively. The biosynthetic and structural bases for stable pairing of identical chains is as yet unexplained. Free exchange of chain-pairs occurs when a mixture of hemoglobins is dissociated and recombined to produce, for example, $\alpha_2^A\beta_2^A$ and $\alpha_2^I\beta_2^S$ from $\alpha_2^A\beta_2^S$ and $\alpha_2^I\beta_2^A$;[5] on the other hand hybrid molecules of the type $\alpha^A\alpha^I\beta_2^A$ or $\alpha_2^A\beta^A\beta^S$ have not been detected in hemolysates or in recombined mixtures.

Allelic genes control the synthesis of hemoglobins A, S, and C,[6] which differ from each other only in their β-chains,[3, 4] homozygosity at this locus resulting in the presence of one adult hemoglobin, and heterozygosity, in two. The absence of hemoglobin A in sickle cell-hemoglobin C disease, in fact, furnished the initial clue to allelic control.[7] Whether or not other loci participate in the genetic control of hemoglobin synthesis has been the subject of much discussion. Recent speculation has centered about the more specific question of whether or not different loci control the α-chains on the one hand and the β-chains on the other; however, critical data bearing on this point have heretofore been unavailable.

The present study stems from the discovery by Smith and Torbert[8] that hemoglobin Hopkins-2 (Ho-2) is determined at a locus different from the one which determines hemoglobin S. The simultaneous presence of sickle cell trait and hemoglobin Ho-2 trait in the same individual (an abnormal gene and its normal allele at each locus) is associated with the presence of three hemoglobin components, the two abnormal hemoglobins and a component with the electrophoretic mobility of hemoglobin A. The findings of Smith and Torbert raise questions pertinent to the mode of genetic control of synthesis of the chains of hemoglobin. Are hemoglobins Ho-2 and S defective in different chains? If so, why does double heterozygosity result in only three electrophoretic components? Formation of four molecular species, normal and doubly abnormal as well as two singly abnormal, would be expected to result from independent synthesis of a normal and an abnormal α-chain and a normal and an abnormal β-chain in the same red cell. The present paper reports studies of these problems by the technique of asymmetric recombination of of hemoglobin.[9] Further familial studies on the kindred reported by Smith and Torbert[8] have been conducted by Bradley, Boyer, and Allen.[10]

Materials and Methods.—Hemoglobin solutions: Carbonmoxyhemoglobin (HbCO) was prepared from the blood of members of the kindred with hemoglobins S and Ho-2. Washed red cells were lysed by dialysis at 5° in a CO atmosphere against several changes of water. The hemolysate was electrodialyzed, clarified by centrifugation, and saturated with CO. Previously prepared stock solutions of HbCO S and C and of a mixture of HbCO A and I were also used in the present studies.

Chromatographic separations: Fractions for recombination experiments were prepared by chromatography on 5 × 35 cm and 5 × 45 cm columns of the cation exchange resin, Amberlite CG-50, Type 2,[11] at 5°. The columns were prepared and the HbCO samples applied as described by Allen, Schroeder, and Balog.[12] 1–2 g of HbCO in 20–30 ml of water was chromatographed in each experiment. The stock solution of developer was a sodium phosphate buffer at 0.5 M Na$^+$ concentration, which contained 27.6 g NaH$_2$PO$_4$·H$_2$O, 21.3 g Na$_2$HPO$_4$, and 0.65 g KCN per liter. The developer was diluted to 0.1 M Na$^+$, at which concentration its pH is 6.7, for initial equilibration of the columns. The samples were eluted at this concentration until the first steady minimum of HbCO in the effluent was reached. The second main hemoglobin zone was eluted with stock developer diluted to 0.15 M Na$^+$, and the last zone was eluted with developer diluted to 0.20 M Na$^+$. The flow rate was 20–40 ml/hr, and the volume of each fraction was 20 ml. Concentrations of the fractions were determined from absorbancy readings at 538 mμ. Fractions corresponding to the major zones of elution were pooled and dialyzed against water or dilute neutral phosphate buffer and then concentrated by pervaporation. The concentrates were treated with $Na_2S_2O_4$ and CO to convert the ferrihemoglobin produced during pervaporation to HbCO, dialyzed against water, and clarified by centrifugation.

Recombination experiments: Samples were analyzed before and after recombination by moving boundary electrophoresis at 0.5° in potassium phosphate buffer of pH 6.8, μ 0.02 for 5 hr at a field strength of 9.3 v/cm. Recombination of HbCO was carried out by dissociation of mixtures at 1 g/100 ml concentration at pH 4.7 or 11.0 for either 4 or 24 hr followed by dialysis against three changes of the phosphate buffer used for electrophoretic analysis.[9, 13]

Results and Discussion.—Samples of HbCO from hemoglobin Ho-2 trait (HbCO A and Ho-2), double heterozygosity for the genes that control hemoglobins S and Ho-2 (HbCO S, Ho-2, and A-like component), and the mixture of HbCO A and I used in an earlier study[5] were analyzed by column chromatography. HbCO Ho-2 and I were eluted with developer at 0.1 M Na$^+$, HbCO A and A-like component of double heterozygotes at 0.15 M Na$^+$, and HbCO S at 0.20 M Na$^+$. Although completely homogeneous preparations were not obtained,[14] the proportion of

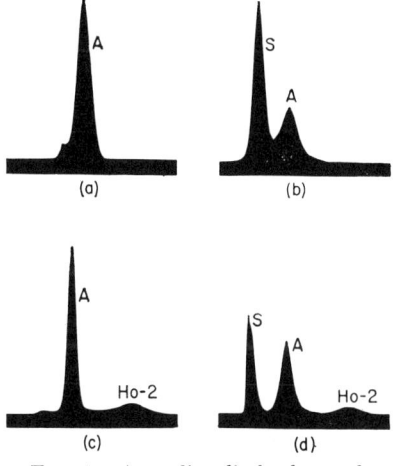

FIG. 1.—Ascending limb electrophoretic patterns of 1% HbCO solutions of the four genotypes of the Fuller kindred in pH 6.8, μ 0.02 phosphate buffer. The direction of migration is from right to left. (*a*) Normal, (*b*) Strait, (*c*) Ho-2 trait (*d*) Ho-2-S double heterozygote.

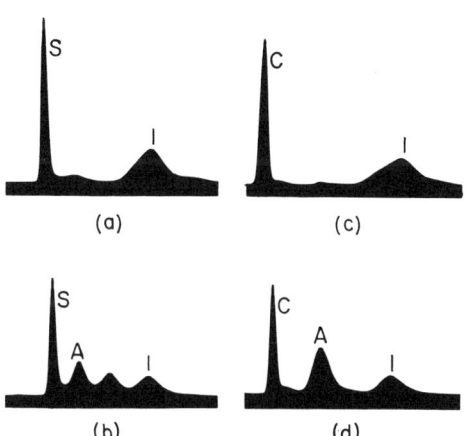

FIG. 2.—Ascending limb electrophoretic patterns of 1% HbCO in pH 6.3, μ 0.02 phosphate buffer. The direction of migration is from right to left. (*a*) I + S control (*b*) I + S recombined (*c*) I + C control (*d*) I + C recombined.

desired component was in each case sufficiently high to permit satisfactory recombination experiments.

Samples from each of the four genotypes in this kindred were analyzed by moving boundary electrophoresis (Fig. 1). The discovery of Smith and Torbert, that the hemoglobin of the S-Ho-2 double heterozygote contains an intermediate component with the mobility of hemoglobin A, was confirmed. In the buffer of pH 8.6 used by Smith and Torbert, in which the hemoglobins are negatively charged, Ho-2 migrates more rapidly than A, and A migrates more rapidly than S. At pH 6.8 each of the hemoglobins carries a positive charge. HbCO C has the greatest electrophoretic mobility, and HbCO S, A, Ho-2, and I follow in order.

When HbCO S and I were recombined, two components, one with the mobility of A and the other with mobility between those of A and I, were produced; when HbCO C and I were recombined, one component with the mobility of A was produced (Fig. 2). The amount of the single product of the C-I recombination was, however, as great as the sum of the amounts of the two products of the S-I recombination (Table 1). Recombination of HbCO S and Ho-2 resulted in net formation of one intermediate electrophoretic component, and recombination of HbCO C and Ho-2 resulted in net formation of two intermediate components which were poorly resolved (Fig. 3). One of the two products of the C-Ho-2 recombination has the mobility of HbCO A, and the other has a slightly higher mobility. The total increase of the single product of the S-Ho-2 recombination was approximately equal to that of the two products of the C-Ho-2 recombination taken together (Table 2). Net formation of a component with the mobility of HbCO S always resulted from recombination of the intermediate component of the S-Ho-2 double heterozygote (Fig. 4 and Table 3). A component with the approximate mobility

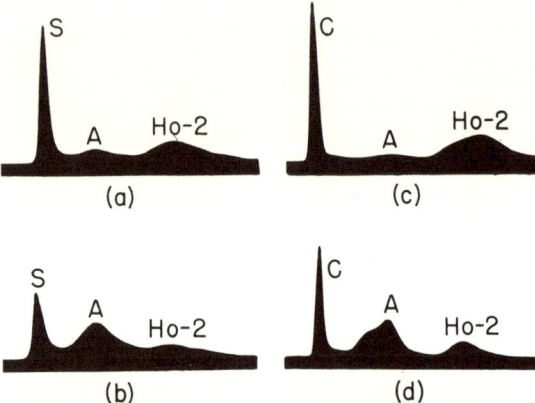

Fig. 3.—Ascending limb electrophoretic patterns of 1% HbCO in pH 6.8, μ 0.02 phosphate buffer. The direction of migration is from right to left. (a) Ho-2 + S control (b) Ho-2 + S recombined (c) Ho-2 + C control (d) Ho-2 + C recombined. Note component migrating ahead of A.

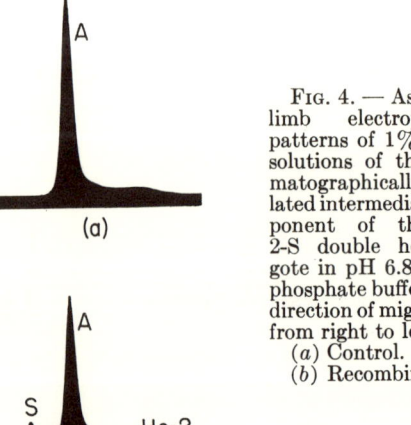

Fig. 4. — Ascending limb electrophoretic patterns of 1% HbCO solutions of the chromatographically isolated intermediate component of the Ho-2-S double heterozygote in pH 6.8, μ 0.02 phosphate buffer. The direction of migration is from right to left.
(a) Control.
(b) Recombined.

of HbCO Ho-2 remained as a contaminant of the intermediate component after chromatographic separation. This component either increased or decreased in different experiments; however, net change in amount was never large. No significant change was observed on recombination of the A-Ho-2 mixture of

TABLE 1

RECOMBINATION OF HbCO I WITH HbCO S AND C

Mixture		Electrophoretic composition: Per cent of total				
		I	$\alpha_2^I \beta_2^S$	A	S	C
$I^1 + S$	Control	46	..	6^2	48	..
	Recombined	23	19	22^2	36	..
$I^1 + C$	Control	47	..	3	..	50
	Recombined	22	..	42^3	..	36

[1] Hb-I was prepared from an A-I sample by column chromatography resulting in a mixture of 94% I plus 6% A. 1 part of this mixture was used to 1 part S or C.
[2] This includes the trailing shoulder of S.
[3] This presumably includes $\alpha_2^I \beta_2^C$.

TABLE 2

RECOMBINATION OF HbCO Ho-2 WITH HbCO A, S, AND C

Mixture	Dissociation[1] time	Electrophoretic composition: Per cent of total			
		Ho-2	A	S	C
Ho-2 trait	Control	21	79	..	..
	24 hr	19	81	..	..
Ho-2 + A + S[2]	Control	42	17	41	..
	4 hr	33	32	35	..
	23 hr	26	47	27	..
Ho-2 + A + C[3]	Control	46	9	..	45
	24 hr	20	47^4	..	33

[1] All dissociations were done at pH 4.7.
[2] The sample was a 50–50 mixture of S and Ho-2 + A. The latter was prepared from Ho-2 trait by column chromatography, resulting in a mixture of 75% Ho-2 plus 25% A.
[3] The sample was a 55–45 mixture of C and Ho-2 + A. The latter was prepared as above (2), giving a mixture of 90% Ho-2 plus 10% A.
[4] This includes approximately 16% of a faster component, presumably $\alpha^{Ho-2}\beta_2^C$.

TABLE 3

RECOMBINATION OF INTERMEDIATE COMPONENT OF Ho-2-S DOUBLE HETEROZYGOTE

Dissociation pH	Dissociation time	Electrophoretic composition: Per cent of total		
		Ho-2	A	S
Control	..	15	84	1
4.7	4 hr.	17	73	10
Control	..	13	87	..
4.7	4 hr.	12	77	11
4.7	24 hr.	9	79	12
Control	..	14	86	..
11.0	4 hr.	18	73	9
11.0	24 hr.	14	76	10

hemoglobin Ho-2 trait.

The results on the recombination of nearly homogeneous HbCO I with S and C confirm our earlier conclusions based on experiments conducted in the presence of HbCO A.[5] The observed net changes as shown in Figure 2 and Table 1 are consistent with asymmetric recombination of a hemoglobin abnormal in the α-chain with a hemoglobin abnormal in the β-chain to form hemoglobin A (normal adult hemoglobin) and a hemoglobin abnormal in both chains. The net reactions may be written

$$\alpha_2^A \beta_2^S + \alpha_2^I \beta_2^A = \alpha_2^A \beta_2^A + \alpha_2^I \beta_2^S$$

and

$$\alpha_2^A \beta_2^C + \alpha_2^I \beta_2^A = \alpha_2^A \beta_2^A + \alpha_2^I \beta_2^C.$$

Our inference from electrophoretic patterns that the abnormalities in mobilities of HbCO C and I are equal and opposite and that $\alpha_2^A \beta_2^A$ and $\alpha_2^I \beta_2^C$ therefore have the same mobility is consistent with recent results of Murayama,[15] which have shown that the inherited defect of hemoglobin I is the presence of an aspartic acid residue in a position of the α-chain normally occupied by a lysine residue. With respect to alteration of net charge this substitution is equal and opposite to the substitution in hemoglobin C of a lysine residue for a glutamic acid residue in the β-chain. The negative defect in net charge of hemoglobin I is greater than the positive defect of hemoglobin S; consequently $\alpha_2^I \beta_2^S$ migrates between HbCO A and I.

The above experiments with hemoglobins known to be abnormal in the α-chain and the β-chain facilitate interpretation of results of experiments with hemoglobin Ho-2. Indeed the suggestion that the intermediate component of S-Ho-2 double heterozygotes might be a mixture of two electrophoretically equivalent species was based on the C-I recombination pattern.[16] Asymmetric recombination of a HbCO mixture does not result in change in electrophoretic pattern when the components of the mixture have a common chain, as do HbCO A, S, and C.[17] The results of Figure 3 and Table 2 thus indicate that hemoglobin Ho-2 differs from S and C in both chains and must be abnormal in the α-chain, the normal chain of S and C. The β-chain of hemoglobin Ho-2 must then be normal, i.e., the same as that of hemoglobin A, since recombination of Ho-2 with A did not produce any new species. Further support for the latter conclusion was furnished by the appearance of a component with mobility of HbCO A after recombination of HbCO Ho-2 with S and C. Since S and C furnished the normal α-chain, Ho-2 must have furnished the normal β-chain. We therefore represent hemoglobin Ho-2 as $\alpha_2^{Ho-2} \beta_2^A$ and write its recombinations with S and C, respectively, as

$$\alpha_2^A \beta_2^S + \alpha_2^{Ho-2} \beta_2^A = \alpha_2^A \beta_2^A + \alpha_2^{Ho-2} \beta_2^S$$

and

$$\alpha_2^A \beta_2^C + \alpha_2^{Ho-2} \beta_2^A = \alpha_2^A \beta_2^A + \alpha_2^{Ho-2} \beta_2^C.$$

Net formation of only one electrophoretic component when HbCO S and Ho-2 were recombined is consistent with the approximately equal and opposite abnormalities in the mobilities of these hemoglobins (Figs. 1 and 3) and lends support to the hypothesis that the intermediate component of S-Ho-2 double heterozygotes is a mixture of $\alpha_2^A \beta_2^A$ and $\alpha_2^{Ho-2} \beta_2^S$. The positive abnormality in mobility of HbCO C is greater than the negative abnormality of Ho-2; therefore $\alpha_2^{Ho-2} \beta_2^C$ migrates between A and C.[18]

Presence of a component with the mobility of HbCO Ho-2 in the chromatographically prepared material complicated the results of recombination experiments with the intermediate component of the S-Ho-2 double heterozygote. The same component appeared when a solution of HbCO A was merely diluted and reconcentrated. This component, which migrates more slowly than A at acid pH, corresponds to rapidly migrating material at pH 8.6, which at least in part is derived from hemoglobin A.[19] While the results of Table 3 are conflicting with respect to formation of HbCO Ho-2, they show net formation of material with the mobility of HbCO S in every experiment. Such a component did not appear after recombination of unaltered HbCO A, a mixture of HbCO A and Ho-2, or diluted and reconcentrated HbCO A. The most likely source of the chains that cause its high mobility is the postulated doubly abnormal molecule, $\alpha_2^{Ho-2}\beta_2^S$, which has the mobility of HbCO A and can recombine with the latter to produce HbCO S and Ho-2. The postulated net reaction is the reverse of the recombination of HbCO S and Ho-2. The small increases and decreases in proportion of HbCO Ho-2 or a Ho-2-like component observed under different conditions of recombination of the intermediate component probably represent the net effect of two competing reactions. Material was lost on account of irreversible denaturation of the Ho-2-like component derived from HbCO A, and HbCO Ho-2 was produced from the recombination of HbCO A and $\alpha_2^{Ho-2}\beta_2^S$.

The erythrocytes of sickle cell trait contain more hemoglobin A than S;[7] accordingly the net synthesis of β^A-chains must be greater than that of β^S-chains in heterozygous cells. Likewise the fraction of hemoglobin Ho-2 in Ho-2 trait of 15–20 per cent[8] must signify that the net synthesis of α^A-chains is greater than that of α^{Ho-2}-chains. It follows that the S-Ho-2 double heterozygote would have more normal hemoglobin than doubly abnormal, an inference supported by the results of recombination of the intermediate component, in which the maximum proportion of hemoglobin S produced was 12 per cent.

The familial data of Smith and Torbert and of Bradley, Boyer, and Allen show that nonallelic loci are responsible for the defects of hemoglobins S and Ho-2 and that probably these loci are not closely linked. Our recombination data show that different chains carry the abnormalities of these hemoglobins, the β-chain of S and the α-chain of Ho-2, and that double heterozygosity probably results in production of four hemoglobins that are the four possible combinations of α_2^A, α_2^{Ho-2}, β_2^A, and β_2^S. From these data we conclude that a pair of alleles at one locus controls the synthesis of α^A-chains and α^{Ho-2}-chains and that a pair of alleles at a different locus controls the synthesis of β^A-chains and β^S-chains (Table 4). Furthermore each of

TABLE 4
Products of Genotypes with Respect to α-Chain and β-Chain Loci

Genotype	Components		
	-2	0	$+2$
$\dfrac{\alpha^A\beta^A}{\alpha^A\beta^A}$	...	$\alpha_2^A\beta_2^A$	...
$\dfrac{\alpha^A\beta^A}{\alpha^A\beta^S}$	...	$\alpha_2^A\beta_2^A$	$\alpha_2^A\beta_2^S$
$\dfrac{\alpha^A\beta^A}{\alpha^{Ho-2}\beta^A}$	$\alpha_2^{Ho-2}\beta_2^A$	$\alpha_2^A\beta_2^A$ $\alpha_2^A\beta_2^A$	...
$\dfrac{\alpha^A\ \beta^A}{\alpha^{Ho-2}\beta^S}$	$\alpha_2^{Ho-2}\beta_2^A$	$\alpha_2^A\beta_2^A$ $\alpha_2^{Ho-2}\beta_2^S$	$\alpha_2^A\beta_2^S$

the four chains apparently is synthesized independently and in identical pairs. According to available familial studies the genes for hemoglobins C and E, which, like S, are abnormal in their β-chains,[4] behave as alleles of the gene for hemoglobin S.[6, 20] All these findings are consistent with the hypothesis that one locus determines the structure of the α-chain and a different locus, of the β-chain, and that mutation at either locus results in an abnormal hemoglobin.[21] An individual heterozygous for two abnormal genes at either of these loci would not have any hemoglobin A; conversely the absence of hemoglobin A in the presence of two abnormal hemoglobins would, according to this hypothesis, indicate that these abnormal hemoglobins carry their respective defects in the same chain and are controlled by allelic genes.[22] A doubly heterozygous individual would have four species of adult hemoglobin, which would separate electrophoretically into three or four electrophoretic components.[5]

Hemoglobins D_α and I are defective in the α-chain, and hemoglobins D_β and J are defective in the β-chain.[4, 16] Genetic control of these hemoglobins has not been established by familial studies; however, if the theory of control of α- and β-chains by nonallelic loci is correct, D_α, I, and Ho-2 are controlled by alleles at one locus, and D_β, J, and S, by alleles at another. Test of this theory requires familial as well as chemical data, for reliance on the latter alone may lead to an ambiguous result. For example, the presence of three or four electrophoretic components in the same individual probably would indicate double heterozygosity but would not distinguish between linkage or independence. Adjacent or closely linked loci would yield the same set of products as independent loci if the sites of synthesis of the chains are independent.

Atwater, Schwartz, and Tocantins[23] have recently reported familial and electrophoretic data consistent with nonallelic genetic control of hemoglobins C and G. The abnormal chain of hemoglobin G was not ascertained; however, a defect in the α-chain, as postulated by the authors, is in accord with their findings. Presumed double heterozygotes have four hemoglobins, A, C, G, and a hemoglobin with a mobility that corresponds to the composition, $\alpha_2^G\beta_2^C$. Nonallelic control of hemoglobins S and G was postulated earlier in a different kindred because an individual with hemoglobins S and G had a child with only hemoglobin A.[24] Neither this parent nor another member of the kindred with hemoglobins S and G had any hemoglobin A, and the defect of hemoglobin G was found in its β-chain.[25] These inconsistencies with the hypothesis of separate genetic control and independent synthesis of α-chains and β-chains apparently have been resolved, since re-examination has revealed that the parent in question actually does not have hemoglobin G.[26] It appears therefore that the two examples of hemoglobin G cited above are different molecular species with the same electrophoretic mobility. Shooter and co-workers[27] have reported a minor component, hemoglobin G_2, in a presumed homozygote for hemoglobin G. Hemoglobins G and G_2 of the sample examined by Shooter probably have the same abnormal α-chain and are analogues of hemoglobins A and A_2, which have the same normal α-chain.[28] The Ho-2 analogue of A_2 is not detectable in our electrophoretic patterns. If such a component were present, it probably would comprise less than 1 per cent of the hemoglobin of Ho-2 trait and would migrate between hemoglobins A and Ho-2. According to Jones and co-workers and Hunt hemoglobins A and F (fetal hemoglobin) have the same α-chain.[29, 30] The other chain of fetal hemoglobin is the γ-chain.[31] It is likely that a newborn infant with hemoglobin Ho-2 trait would have four hemoglobins, $\alpha_2^A\beta_2^A$, $\alpha_2^A\gamma_2$, $\alpha_2^{Ho-2}\beta_2^A$, and $\alpha_2^{Ho-2}\gamma_2$.

Summary.—We have shown that hemoglobin Ho-2 is abnormal in its α-chain and normal in its β-chain, in contrast to hemoglobin S, which is normal in its α-chain and abnormal in its β-chain. Familial data have shown that the loci which determine these two abnormal hemoglobins are genetically distinct and probably not closely linked. These results are consistent with the hypothesis that different loci determine the structure of α-chains on the one hand and β-chains on the other. Presence of a normal and an abnormal allele at each locus probably results in four hemoglobins in the same individual, normal, doubly abnormal, and two singly abnormal. The fact that these four hemoglobins represent the four possible combinations of an abnormal and a normal pair of each of the two types of chains suggests that the chains are synthesized in identical pairs at independent sites.

We are indebted to Drs. Ernest W. Smith and Thomas B. Bradley for the samples of blood, and to Dr. T. Viswanatha for his collaboration on the chromatographic separations.

* Presented in part at the meeting of the American Society of Biological Chemists in Chicago, Illinois, April 11–15, 1960. H. A. Itano and E. Robinson, *Federation Proc.*, **19**, 193 (1960).

† National Institutes of Health, Public Health Service, U.S. Department of Health, Education, and Welfare, Bethesda, Maryland.

[1] Rhinesmith, H. S., W. A. Schroeder, and L. Pauling, *J. Am. Chem. Soc.*, **79**, 4682 (1957).
[2] Rhinesmith, H. S., W. A. Schroeder, and N. Martin, *J. Am. Chem. Soc.*, **80**, 3358 (1958).
[3] Vinograd, J. R., W. D. Hutchinson, and W. A. Schroeder, *J. Am. Chem. Soc.*, **81**, 3168 (1959).
[4] Ingram, V. M., *Nature*, **183**, 1795 (1959).
[5] Itano, H. A., and E. Robinson, *Nature*, **183**, 1799 (1959).
[6] Ranney, H. M., *J. Clin. Invest.*, **33**, 1634 (1954).
[7] Itano, H. A., *Am. J. Human Genet.*, **5**, 34 (1953).
[8] Smith, E. W., and J. V. Torbert, *Bull. Johns Hopkins Hosp.*, **101**, 38 (1958).
[9] Singer, S. J., and H. A. Itano, these PROCEEDINGS, **45**, 174 (1959).
[10] Bradley, T. B., S. H. Boyer, and F. H. Allen, submitted for publication.
[11] Chromatographic grade, 200–400 mesh, of Amberlite IRC–50. Rohm and Haas Company, Philadelphia.
[12] Allen, D. W., W. A. Schroeder, and J. Balog, *J. Am. Chem. Soc.*, **80**, 1628 (1958).
[13] Robinson, E., and H. A. Itano, *Nature*, **185**, 547 (1960).
[14] The minor components A_1[12] and A_2[19] are eluted at 0.10 M Na$^+$ and 0.20 M Na$^+$, respectively. Their presence does not interfere with recombination of the major components but makes it impossible to determine accurately the percentage composition of each mixture chromatographically.
[15] Murayama, M., *Federation Proc.*, **19**, 79 (1960).
[16] Itano, H. A., and E. Robinson, *Nature*, **184**, 1468 (1959).
[17] Itano, H. A., and S. J. Singer, these PROCEEDINGS, **44**, 522 (1958).
[18] According to their amino acid abnormalities the abnormalities in net charge of hemoglobins S, C, and I are $+2$, $+4$, and -4, respectively, per molecule. The abnormality in net charge of hemoglobin Ho-2 is estimated to be -2 per molecule. The correspondence between estimated net charge and mobility is not, however, exact, presumably because of interactions with other charged groups in the neighborhood of the amino acid substitution.
[19] Kunkel, H. G., and A. G. Bearn, *Federation Proc.*, **16**, 760 (1957).
[20] Aksoy, M., *Blood*, **15**, 610 (1960).
[21] Itano, H. A., *Advances in Protein Chem.*, **12**, 215 (1957).
[22] These conclusions would not hold in the event of an interallelic recombination within a locus. Presumably the number of amino acids in each polypeptide chain is the maximum number of detectable mutational sites within each locus.
[23] Atwater, J., I. R. Schwartz, and L. M. Tocantins, *Blood*, **15**, 901 (1960).
[24] Schwartz, H. C., T. H. Spaet, W. W. Zuelzer, J. V. Neel, A. R. Robinson, and S. F. Kaufman, *Blood*, **12**, 238 (1957).
[25] Hill, R. L., and H. C. Schwartz, *Nature*, **184**, 681 (1959).
[26] Hill, R. L., R. T. Swenson, and H. C. Schwartz, *J. Biol. Chem.* (in press).
[27] Shooter, E. M., E. R. Skinner, J. P. Garlick, and N. A. Barnicot, *Brit. J. Haemat.*, **6**, 140 (1960).

[28] Stretton, A. O. W., and V. M. Ingram, *Federation Proc.*, **19,** 343 (1960).
[29] Jones, R. T., W. A. Schroeder, and J. R. Vinograd, *J. Am. Chem. Soc.*, **81,** 4729 (1959).
[30] Hunt, J. A., *Nature*, **183,** 1373 (1959).
[31] Schroeder, W. A., and G. Matsuda, *J. Am. Chem. Soc.*, **80,** 1521 (1958).

X
Origin of New Genes and Proteins

Editor's Comments on Papers 24 Through 26

24 **Ingram:** *Gene Evolution and the Haemoglobins*

25 **Smithies, Connell, and Dixon:** *Chromosomal Rearrangements and the Evolution of Haptoglobin Genes*

26 **Baglioni:** *The Fusion of Two Peptide Chains in Hemoglobin Lepore and Its Interpretation as a Genetic Deletion*

Soon after the important observation of Itano and Robinson described in Paper 23 was made, it was established that at least four different polypeptides participate in the formation of functional human hemoglobins. These were designated as the $\alpha, \beta, \gamma,$ and δ hemoglobin chains; each was believed to be controlled by a different gene.

In Paper 24 Ingram points out that the four different chains are actually quite closely related in their amino acid sequences, which raises the question of whether they are not, in fact, related by having evolved from a common ancestral polypeptide. He suggests that this process may have proceeded by the duplication of the original α-polypeptide gene and the subsequent divergence of the duplicates by mutation into the present β and δ determining genes. These speculations were the forerunners of an important new area in biochemical genetics—macromolecular evolution.

The duplication of genes by such processes as unequal crossing over or by translocation of small chromosome fragments receives active support to this day from numerous observations such as that in Paper 24 on the different polypeptides in hemoglobin. Smithies and colleagues extend the principle as a result of an analyses of the different kinds of haptoglobins segregating in the human population. The data presented strongly indicate that proteins may evolve not only by point mutation, i. e., base changes resulting in missense mutations, but by rearrangements of the genetic material, principally by unequal crossing over *within* genes, resulting in all or part of the original gene being duplicated, or reciprocally, deleted. The code may be changed, and hence amino acid sequences altered, by internal rearrangements not involving point mutation.

In Paper 26, published almost simultaneously with the previous one, Baglioni points out that human hemoglobin Lepore may have arisen in the same way that certain of the α chains of human haptoglobin arose—by unequal crossing over. In this case it is postulated that two adjacent genes, the β- and α-polypeptide genes for human hemoglobin have had an adjacent region of each deleted. This results in the two genes being joined together in such a way that they produce a single polypeptide instead of two separate ones, β and α.

Selected Bibliography

Buettner-Janusch, J. 1970. Evolution of serum protein polymorphisms. *Ann. Rev. Genetics* 4: 47–68.

Ohno, S. 1970. *Evolution by Gene Duplication*. Springer-Verlag, New York.

Schroeder, W. A., J. H. J. Huisman, J. R. Shelton, J. B. Shelton, and E. F. Shelton. 1968. Evidence for multiple structural genes for the γ-chain of human fetal hemoglobin. *Proc. Natl. Acad. Sci. U.S.* 60: 537–544.

GENE EVOLUTION AND THE HÆMOGLOBINS

By Prof. VERNON M. INGRAM

Division of Biochemistry, Department of Biology, Massachusetts Institute of Technology, Cambridge, Mass.

THE four types of polypeptide chain which go to make up the molecules of the three normal human hæmoglobins are believed to be controlled by four independent genes[1]. The following article is an attempt to discuss the chemical and genetic relationship between these chains from an evolutionary point of view; to a lesser extent, because much less is known, the hæmoglobins of other vertebrates will also be mentioned. It should be emphasized that the main purpose of this discussion, and of the evolutionary scheme to be proposed, is to provide a basis for the discussion of the evolution of genes in general and hæmoglobin genes in particular. Questions are raised which will, in part, be answered soon by the chemical work on vertebrate hæmoglobins which is now proceeding in various laboratories (see, for example, ref. 2).

The study of the chemistry of the human hæmoglobins has already provided evidence of the kind of phenotypic effects to be expected from gene mutations as, for example, in the known single amino-acid substitutions in the peptide chains of abnormal hæmoglobins[1]. It is on such changes in protein structure that the forces of natural selection are assumed to act. From the evolutionary and the practical point of view it is convenient to study a commonly occurring protein, such as hæmoglobin, which is found in all vertebrates. Hæmoglobin is not only accessible for chemical study but is also sure to have played an important part in vertebrate evolution, because of its vital physiological function as the carrier of oxygen.

Most of the discussion which follows will centre around the chemical findings in the human hæmoglobins, since they are by far the best studied of vertebrate hæmoglobins. However, the basic similarities in chemical structure between man's hæmoglobins and those of other mammals and lower vertebrates are very striking[3]. All known vertebrate hæmoglobins—except those of the lamprey and the hagfish—are built on the same molecular pattern: they consist of four polypeptide chains of roughly 17,000 molecular weight, with one hæm group each. The hæmoglobin of the lamprey is peculiar in consisting of a single polypeptide chain of molecular weight 17,000. The hagfish hæmoglobin appears to be similar, or possibly a dimer of 34,000 molecular weight[3].

The peptide chains in the other vertebrates are of two different types; for example, human adult hæmoglobin consists of two so-called α-chains and two β-chains. There is usually interaction between the four hæm groups in their reaction with molecular oxygen; in other words, a plot of the degree of oxygenation of the hæmoglobin molecules against the partial pressure of oxygen is sigmoid in shape[1]. This fact enhances enormously the physiological efficiency of the hæmoglobin molecule as our oxygen carrier[5], since it favours complete saturation with oxygen in the lungs and complete discharge in the tissues. This great similarity which characterizes most vertebrate hæmoglobins suggests that we are studying that aspect of the evolution of a particular protein molecule which is concerned with detailed development of an already well-defined molecule. It is true that the hæmoglobin quadruples its size during this evolution (if we include the lamprey hæmoglobin), but the change is due to the aggregation of four fairly similar protein sub-units (peptide chains), rather than to an actual lengthening of a molecule. In the evolutionary scheme to be discussed it is suggested that the increase in complexity, and in diversity, of the hæmoglobins is an illustration of a more general process of gene evolution which results in an increase of the number of genes.

Human Hæmoglobins

The following striking situation is found in the human hæmoglobins:

(1) The three normal human hæmoglobins all possess a common half-molecule which may be written $-\alpha_2 A$. This formulation indicates that this half-molecule unit is composed of two normal α-peptide chains as first described for normal adult hæmoglobin A. However, the other half of each of the three hæmoglobins consists of different types of peptide chain, β, γ or δ, one type for each hæmoglobin. All four chains are of roughly equal size with a chain molecular weight of about 17,000. These three hæmoglobins may be formulated as follows:

Adult	= hæmoglobin A	= $\alpha_2{}^A\beta_2{}^A$
Fœtal	= hæmoglobin F	= $\alpha_2{}^A\gamma_2{}^F$
Kunkel's minor component[6]	= hæmoglobin A_2	= $\alpha_2{}^A\delta_2{}^{A_2}$

This list may have to be extended as the study of other minor hæmoglobins continues[7].

I postulate[1] that four genes are involved in the manufacture of these peptide chains, so that the *genotype* of a normal individual may be written as:

$$\alpha A/\alpha A,\ \beta A/\beta A,\ \gamma^F/\gamma^F,\ \delta A_2/\delta A_2.$$

In this view the products of the α-genes are common to all the human hæmoglobins.

(2) In total amino-acid composition[8] the four types of chains compare as follows:

αA and βA	— differ in perhaps 21 out of nearly 140 amino-acids
γ^F and βA	— differ in perhaps 21 out of nearly 140 amino-acids
δA_2 and βA	— differ in less than 10 amino-acids.

In addition, γ^F is the only one of these chains to contain the amino-acid *iso*leucine—four residues per chain.

(3) The αA-chain begins with the amino-acid sequence[7] Val-Leu- . . . ; the βA-chain and the δA_2-chain begin with Val-His-Leu . . . . The γ^F-chain begins with the sequence Gly-His-Phe-, which although different from the β-chain shows a clear affinity in the type of amino-acid with the beginning of the βA-chain. In both cases, the first amino-acid is neutral and aliphatic, the second is the basic amino-acid histidine, and the third has a non-polar side-chain.

The peptide chain beginning with the Val-Leu-sequence (α-chain) has a counterpart[3] not only in all known mammalian hæmoglobins but also in the hæmoglobins of all the lower vertebrates so far examined, except for the hæmoglobin of the lamprey. This is not to say that the Val-Leu- chains of other animals resemble the human chain precisely, but rather that there might be a strong 'family resemblance'. It is of interest to attempt to explain the repeated occurrence and strange similarity of the Val-Leu- (the α) chain in the vertebrates.

The following postulates are used in developing a scheme for the evolution of the hæmoglobin genes.

Postulate 1. Mutations of a gene result in either single or multiple amino-acid substitutions in the peptide chain which that gene controls, or inversions of part of the amino-acid sequence, or a combination of these possibilities. The new hæmoglobin peptide chains produced by such mutations are then either favoured or discarded in the course of natural selection.

Postulate 2. At several points in the course of evolution a gene for a particular hæmoglobin peptide chain has undergone *duplication*, followed by, or simultaneous with, translocation. The two initially equivalent genes have then evolved independently, governed by the selective pressure of their environment on their protein products. Such mechanisms have been previously postulated; for example, ref. 9. The role of gene duplication[9,10] in evolution has been discussed by Stephens[11]. His conclusion was that the case for duplication as an important factor in evolution was so far neither proved nor disproved.

It is of course equally possible to postulate the occurrence of chromosome duplication, with subsequent independent evolution of the initially identical chromosomes. Such a scheme would fit into the proposed hypothesis equally well.

It seems likely that α-, β- and γ-genes are located on different chromosomes, since they segregate independently, whereas β- and δ-genes appear to be linked[12].

Postulate 3. Once the α-chain, or rather the α_2-dimer, is required to fit precisely with at least two other dimers—β_2 and γ_2—the α-chain is no longer as free to be varied by mutation. It has become 'conservative' and its rate of evolution will be less than that of the other chains, as is perhaps seen in the persistence of the Val-Leu- beginning of the α-chain of the vertebrates.

In addition, one can postulate that mutations of the α-gene produce mutant α-chains which are more severely selected against in the later stages of evolution than are mutants of other chains. This is likely to be so, because in the later stages of evolution the α-chains participate in the formation of fœtal hæmoglobin, a vulnerable point in the life-cycle of the animal. It is a fact that mutants of the human α-chain never reach a high frequency of distribution, in contrast to some β-chain mutants, although we know of as many different kinds of α-chain mutants as of β-chain mutants.

Scheme for the Evolution of the Hæmoglobin Genes

We might suppose that originally the hæmoglobin molecule was rather like the present-day myoglobin molecule, that is, that it had a single peptide chain with a single hæm group, and that therefore it could not show hæm–hæm interaction. The size of this molecule might vary, but eventually it might be stabilized at a value of around 17,000 molecular weight.

At this stage of evolution, presumably earlier than the teleost fishes, the hæm protein inside the muscle cells is assumed to be the same as that in the circulation. The muscle hæm protein became myoglobin in the course of evolution; it retained a molecular weight of 17,000 and a complexity of only one hæm group and one peptide chain per molecule. It was, of course, still subject to mutational changes, as can be seen from the fact that its present-day amino-acid composition and sequence in a given animal often differs considerably from that of any of the analogous hæmoglobin chains[13]. For example, human myoglobin contains *iso*leucine, but no cysteine, whereas the reverse is true of human adult hæmoglobin.

On the other hand, one can foresee limits to the kind of mutations which would be tolerated. The X-ray studies of Kendrew[14] and of Perutz[15] show that the three-dimensional arrangement of chains in myoglobin and the hæmoglobin sub-units (also of

17,000 molecular weight) are remarkably similar, though not identical. This statement applies also to the two kinds of sub-units found in hæmoglobin itself. Presumably, mutational alterations in the course of evolution which would drastically affect the three-dimensional structure were not tolerated. Such considerations imply that the configuration of the peptide chains in myoglobin and hæmoglobin became stabilized early in evolutionary history, at least in its most important features.

During the evolution of the primitive hæmoglobin chain—provisionally called the 'α'-chain—there occurred a gene duplication followed or accompanied by translocation. From now on the two duplicate 'α'-chain genes could evolve independently—one to become the modern myoglobin gene, the other to become the α-chain gene of present-day hæmoglobin. Eventually, according to the scheme, the α-chain gene would evolve in such a way that its product, the α-chain, had the property of dimerization in solution to form α_2 molecules. Such a property would be favoured strongly, if it entailed, in addition, the possibility of hæm–hæm interaction between the two hæm groups of the new dimer molecule and therefore the possibility of more efficient oxygenation and deoxygenation. Once produced, such a mutation is unlikely to be lost in the further evolution of hæmoglobin.

At this stage, the sequence of the 'α'-chains is still variable within the dictates of structural requirements, since there is nothing yet to put additional restrictions on it.

We might next postulate that the genes of the α_2-chains duplicated again. After this gene duplication two types of dimer—α_2 and γ_2—would evolve side by side. Sooner or later, these chains would have evolved sufficiently to be able to form tetramers with even greater selective advantage because of the increased hæm–hæm interaction likely to be found in such tetramers. The characteristics of the genes responsible for the ability of the chains to form tetramers would certainly be fixed. This stage of hæmoglobin evolution seems to have been reached already in some teleost fishes, because they already possess a four-chain hæmoglobin[3].

The third gene duplication and translocation is pictured as occurring with the γ-chain gene, giving rise to a new γ-gene destined to evolve into the β-chain gene. At this gene duplication the property of forming tetramers is already firmly established. The new gene can develop along its own line to provide a hæmoglobin tetramer—$\alpha_2\beta_2$—particularly adapted for the adult body. On the other hand, the old γ-chain continues to develop and to provide half the molecule of the fœtal hæmoglobin ($\alpha_2\gamma_2$). It is the γ-chain gene, rather than the α-chain gene, which is said to duplicate here, because the γ-chain dimers, γ_2, have already the necessary complementariness for forming tetramers with α_2. This complementariness will be automatically a property of the product of the new gene. In addition, we shall see later that β- and γ-chains are more closely related to one another than either is to the α-chain. Therefore we might consider them to have diverged at a later stage.

At this point of evolution, three independent genes—α, β, γ—are assumed to be present, each one capable of forming chains which dimerize and which aggregate to the tetramers $\alpha_2\beta_2$ or $\alpha_2\gamma_2$. Hæm–hæm interaction is strongly present in the tetramers. Such a situation has an important effect on the further

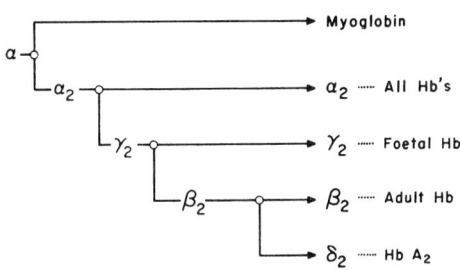

Fig. 1. Evolution of the hæmoglobin chains. The α-chain is the ancestral peptide chain. —O— indicates a point of gene duplication followed by translocation of the new gene

evolution of the α-chain. This chain, or rather its dimer, α_2, is required to fit with two different partners, β_2 and γ_2. As a result, less variation is allowed to the α-chain; it has become 'conservative'. Perhaps such conservation is in part responsible for the apparently universal presence of α-chains beginning with Val-Leu- in the hæmoglobins of the higher vertebrates. There is an alternative explanation for the apparently greater stability of the α-genes; this gene controls also the fœtal hæmoglobin and therefore may not undergo extensive mutational alterations, since the fœtus is a much more delicate organism. The very fact that any alteration in the α-chain gene, and therefore in the α-chain, seems to affect all types of hæmoglobin may be sufficient to explain the apparent 'conservatism'.

On the other hand, there is no *a priori* reason why different parts of a molecule as complex as a protein should develop at the same rate. The difference in apparent stability between the α-chain and the others could be no more than what would be expected as normal variation. It will be interesting to see just how similar the Val-Leu- chains of different vertebrates are.

At the fourth and last gene-duplication in the scheme we suppose that it is the β-gene which becomes duplicate, leading to the δ-chain genes controlling the δ-chains of hæmoglobin A_2. The origin of this δ-chain is placed near the end of the evolutionary scheme, because of its great chemical similarity to the β-chain[16]. Furthermore, the presence of a hæmoglobin A_2-like component seems to be confined to the higher primates[17]. In this view, hæmoglobin A_2 is a new hæmoglobin rather than an archaic one, as has often been supposed. It has been reported to have a higher affinity for oxygen[18] and perhaps it is a more efficient hæmoglobin, destined to replace eventually $\alpha_2\beta_2$ (ref. 19). Its proportion is certainly doubled in some thalassæmias[17], perhaps as a compensating mechanism. There is genetic evidence that the genes for δ- and for β-chains are linked[12], indicating perhaps that the process of translocation of the new δ-gene has not yet occurred.

Where does the lamprey fit into this scheme? One form, *Petromyzon planeri*, has two fœtal and two (different) adult hæmoglobins[20]. The molecular weight of the adult mixture is given as 17,000 (ref. 3). Perhaps these lamprey hæmoglobins are the result of an independent evolution scheme similar to the one discussed in the present article, but which has never included the mutations which led to the formation of dimers and tetramers. We might regard the lampreys as having branched off before or just after

Table 1. COMPARISON OF THE N-TERMINAL SEQUENCES OF THE α- AND β-PEPTIDE CHAINS OF HUMAN HÆMOGLOBIN A.[21]

α-chain Val-Leu-Ser-Pro-Ala-Asp-**Lys**-Thr-Asp-Val-**Lys**-Ala-Ala-Try-Gly-**Lys**-Val-Gly-Ala-His-Ala-Gly-Glu-Tyr-Gly-(Ala,Glu)-Ala-Leu-Glu-**Arg**-
  1   2   3   4   5   6   7   8   9 10 11 12 13 14 15 16 17 18 19 20 21 22 23 24 25   26   27 28 29 30 31
       Met-Phe-Leu-Ser-Phe-Thr-Pro-Thr-**Lys**-
        32 33 34 35 36 37 38 39 40

β-chain Val-His-Leu-Thr-Pro-Glu-Glu-**Lys**-Ser-Ala-Val-Thr-Ala-Leu-Try-Gly-**Lys**-Val-Asp-Val-Asp-Glu-Val-Gly-Gly-Glu-Ala-Leu-Gly-**Arg**-Leu-
  1   2   3   4   5   6   7   8   9 10 11 12 13 14 15 16 17 18 19 20 21 22 23 24 25 26 27 28 29 30 31

the first gene duplication in the scheme of Fig. 1. Unfortunately, nothing seems to be known about the presence or absence of a separate myoglobin in the lampreys.

Discussion

The proposed scheme expresses a desire to explain the striking chemical similarity between α, β, γ and δ human hæmoglobin chains in terms of their evolution from a common precursor.

Braunitzer *et al.*[21] have published some preliminary findings of the amino-acid sequences in the first 30–40 positions of the human α- and β-chains (Table 1). Between the two sequences there is a strong 'family resemblance' of the degree expected for two gene products which have evolved independently, but from a common ancestor. Also Gratzer and Allison[3] have very recently discussed the possibility that the hæmoglobin chains might have evolved from a common ancestor.

It has been suggested to me that the four genes controlling the four chains might have evolved from four unrelated genes which originally controlled the synthesis of quite unrelated proteins; thus there would have been no increase in the total number of genes. By a process of parallel evolution—successive and independent mutations—each of these four genes eventually changed so that it made one of the hæmoglobin peptide chains. The last stages of evolution, when pairs of genes were involved in producing parts of a complex molecule, must then be called on to produce similarity in chemical structure and in configuration. It is easy to conceive of a selective mechanism which would favour changes in various members of a molecular aggregate, such that the members might fit together better. However, it is hard to think of such a mechanism also selecting for chemical and structural similarity of the monomers. In addition, the ability to dimerize would have to arise *de novo* four times, an unlikely situation.

Although we cannot at the moment completely reject the idea of parallel evolution of the four genes from unrelated genes, it is not a very palatable one. The similarity between the human α-, β-, γ- and δ-chains is so great: the C-terminus of the chains ends with the sequence -tyrosyl-histidine or -tyrosyl-arginine[16,22]; their common ability to incorporate an active hæm group; their similar folding and chain length[14,15]; absence of *isoleucine* in α, β and δ[8,16]; and so forth. Altogether, an evolution from a common ancestor is more attractive.

Chemical Evolutionary Relationship Between the Chains

How many mutational steps occurred between the original 'α'-gene and the present-day human α-gene? Or between the original 'α'-gene and the human β-gene? It is more meaningful, perhaps, to ask how many mutational events might separate the present-day human α- and β-genes. Out of 140 amino-acids, we know from the composition that approximately 20 are different. This gives a minimum of 20 mutations of the kind effecting single amino-acid substitutions.

It is believed that the evolution of the vertebrates began some 5×10^8 years ago. Let us assume a mean generation life of 5 years among the vertebrates, which implies some 10^8 generations for the evolution of vertebrate hæmoglobins. A generally acceptable rate[24] per generation is 10^{-5}–10^{-6}, which leads to a figure of 100–1,000 mutations in the evolutionary history of a vertebrate hæmoglobin. This figure, which is itself a minimum one, is very considerably higher than the (minimum) number of mutations which appear to distinguish α-chains from β-chains and α-chains from γ-chains.

The only conclusion which can be drawn from such a calculation is that a more than sufficient number of mutational events have passed in the history of the vertebrates to account for present-day differences between the various parts of the hæmoglobin molecule in the human species.

Comparison of the Fingerprints of the Human α-, β-, γ- and δ-Chains

Rather than deduce the chemical relationship between the chains from their amino-acid composition, it is better to compare their 'fingerprints'[25]. The two-dimensional maps, or fingerprints, of the tryptic peptides derived from each of these four peptide chains give at least a crude idea of the similarity and dissimilarity in their amino-acid sequences and hence of the closeness of the chemical relationship between them.

Human α-Chains and β- or γ-Chains

With the exception of peptide number 21 (ref. 25), all the peptides (some 14 for each chain) occupy different positions on the fingerprints[23] of α-chains compared with β- or γ-chains, or they can be distinguished by such methods as extended paper ionophoresis. They are therefore different in their amino-acid sequences. Peptide 22 is free lysine. It does not follow that the differing peptides of the α- and of the β- or γ-chains are totally altered. They could contain short amino-acid sequences in common[21]. In any event, the indications are that the amino-acid differences between the α- and the β- or γ- chains are distributed throughout their length.

The general comment might be made that there is a familial likeness between the distribution of peptides on the fingerprints of α-, β- and γ-chains. Surely this will reflect a structural relationship (Table 1). We would suppose that portions of their amino-acid sequences are common to all three, as is the tertiary structure of the two types of chain of horse hæmoglobin, which correspond to the human α- and β-chains.

Human β-Chains and γ-Chains

Apart from peptides number 21 (a pentapeptide) and 22 (free lysine), the following peptides[25] occupy very similar positions[23] and would be closely related. if not identical: peptides 12, 14, 15β, 19, 20β, 26. The number is an impressive proportion of the whole chain. If it is correct that these peptides are indeed identical in the β- and γ-chains, it would mean that perhaps 9 peptides out of some 14 are shared by the chains. Some of these peptides are quite long. Their position along the chains is not known, except that peptide 26 is the third one from the N-terminus[21]. These conclusions underline the chemical similarity between β- and γ-chains, which can also be deduced from their amino-acid compositions.

Human β-Chains and δ-Chains

The recent work of Stretton[16] has shown that the β-chains and the δ-chains are very closely similar in their fingerprints. In fact, out of a total of some 14 tryptic peptides, only 4 peptides are different; they are numbers 5, 12, 25 and 26. The similarity between other pairs of peptides from the β- and δ-chains has been ascertained much more carefully by amino-acid analysis than has so far been done with the γ-chains. Moreover, the differences between peptides 12, 25 and 26 from β- and from δ-chains have been shown to involve only 4 single amino-acid differences, for example, serine-threonine (in two places), threonine-asparagine, glutamic acid-alanine. These four single amino-acid differences, together with an unconfirmed fifth one, are so far the only ones definitely known between this pair of chains, each of which is some 140 residues long—a very close chemical similarity indeed.

Apart from the tryptic peptides of the α-, β-, γ- and δ-chains, there is in each case the trypsin resistant 'core' to be considered which amounts to rather more than a quarter of the chain. Here the differences between the chains are far from clear and cannot yet be usefully discussed, except to say that the 'cores' all appear to be similar.

Fingerprints of hæmoglobins from other mammals (ref. 2, and Muller, C. J., unpublished) are recognizable as 'hæmoglobin'. It will be interesting to compare the degree of similarity between the mammalian hæmoglobins on one hand with, say, chicken or teleost hæmoglobin.

Recently, Zuckerkandl[2] has shown that fingerprints of hæmoglobin from the gorilla and chimpanzee are indistinguishable from those of human hæmoglobin. A greater number of differing peptide spots are observed in the less closely related orangutan and rhesus monkey. These extraordinarily interesting findings give added proof of the validity of using a chemical study of vertebrate hæmoglobins to discover evolutionary relationships.

Conclusion

It appears that a sufficient number of generations has passed since the beginning of vertebrate evolution to allow for the known or suspected number of mutations which separate the α-, β- and γ-chains of human hæmoglobin.

The similarity in the fingerprints between β- and γ-chains of human hæmoglobin supports the suggestion that these chains diverged from one another at a later stage in evolution than either from the α-chain gene.

The close similarity between β- and δ-chains gives weight to the idea that the δ-chains of hæmoglobin A_2 are a recent evolutionary development from the β-chains.

The suggestion is made that a single primitive myoglobin-like hæm protein is the evolutionary forerunner of all four types of peptide chain in the present-day human hæmoglobins, and of the corresponding peptide chains in other vertebrate hæmoglobins. Such a scheme involves an increase in the number of hæmoglobin genes from one to five by repeated gene duplication and translocations; the scheme may thus illustrate a general phenomenon in gene evolution.

I acknowledge the many stimulating discussions with Drs. C. Levinthal, S. E. Luria, C. Baglioni, A. O. W. Stretton, J. V. Neel and P. S. Gerald. I am also grateful to Drs. W. B. Gratzer and A. C. Allison for allowing me to read their review paper[3] before publication.

Two books have been particularly stimulating: J. B. S. Haldane, *The Biochemistry of Genetics*; and C. B. Anfinsen, *The Molecular Basis of Evolution*.

No serious attempt has been made to survey critically the literature on gene evolution or duplication. This is partly in order to present a clearer and more provocative hypothesis and partly because I do not feel qualified to evaluate the numerous contributions to that field. Recent experimental work on the animal hæmoglobins has been admirably summarized in the Gratzer and Allison article[3].

This work was supported by grants from the National Science Foundation and the National Institutes of Health, U.S. Public Health Service.

Note added in proof. Further work by G. Braunitzer and his colleagues (*Z. physiol. Chem.*, in the press) shows that the similarity of the α- and β-chains continues throughout these chains.

[1] Ingram, V. M., in *Genetics*, Macy Found. Symp., **141**, 147 (1959); *Hemoglobin and Its Abnormalities* (C. C. Thomas, in the press).
[2] Zuckerkandl, E., Jones, R. T., and Pauling, L., *Proc. U.S. Nat. Acad. Sci.*, **46**, 1349 (1960).
[3] Gratzer, W. B., and Allison, A. C., *Biol. Rev.*, **35**, 459 (1960).
[4] Edsall, J. T., in *Hemoglobin* (National Research Council, Wash., 1958).
[5] Riggs, A., *Nature*, **183**, 1037 (1959).
[6] Kunkel, H. G., and Wallenius, G., *Science*, **122**, 288 (1955).
[7] Schroeder, W. A., *Prog. Chem. Org. Nat. Prod.*, **17**, 322 (1959).
[8] Stein, W. H., Kunkel, H. G., Cole, R. D., Spackman, D. H., and Moore, S., *Biochim. Biophys. Acta*, **24**, 640 (1957). Hill, R. J., and Craig, L. C., *J. Amer. Chem. Soc.*, **81**, 2272 (1959).
[9] Lewis, E. B., *Cold Spring Harbor Symp. Quant. Biol.*, **16**, 159 (1951).
[10] Bridges, C. B., *Science*, **83**, 210 (1936).
[11] Stephens, S. G., *Adv. Genetics*, **4**, 247 (1951).
[12] Cepellini, R., in *Biochemistry of Human Genetics*, 135 (CIBA Symp., 1959).
[13] Rossi-Fanelli, A., Cavallini, D., and de Marco, C., *Biochim. Biophys. Acta*, **17**, 377 (1955).
[14] Kendrew, J. C., Dickerson, R. E., Strandberg, B. E., Hart, R. G., Davies, D. R., Phillips, D. C., and Shore, V. C., *Nature*, **185**, 422 (1960).
[15] Perutz, M. F., Rossmann, M. G., Cullis, A. F., Muirhead, H., Will, G., and North, A. C. T., *Nature*, **185**, 416 (1960).
[16] Ingram, V. M., and Stretton, A. O. W. (submitted to *Nature* for publication).
[17] Kunkel, H. G., Cepellini, R., Muller-Eberhard, U., and Wolf, J., *J. Clin. Invest.*, **36**, 1615 (1957).
[18] Meyering, C. A., Israels, A. L. M., Sebens, T., and Huisman, T. H. J., *Clin. Chim. Acta*, **5**, 208 (1960).
[19] Stretton, A. O. W., Ph.D. Thesis, Univ. Cambridge (1960).
[20] Adinolfi, N., Chieffi, G., and Siniscalco, M., *Nature*, **184**, 1325 (1959).
[21] Braunitzer, G., Liebold, B., Muller, R., and Rudloff, V., *Z. physiol. Chem.*, **320**, 170 (1960).
[22] Guidotti, G., *Biochim. Biophys. Acta*, **42**, 177 (1960).
[23] Hunt, J. A., *Nature*, **183**, 1373 (1959), and Ph.D. Thesis, Univ. Cambridge (1959); Hunt, J. A., and Lehmann, H., *Nature*, **184**, 872 (1959).
[24] Haldane, J. B. S., *Biochemistry of Genetics*, 15 (Allen and Unwin, London, 1954).
[25] Ingram, V. M., *Biochim. Biophys. Acta*, **28**, 539 (1958).

Copyright © 1962 by Macmillan Journals Ltd.

Reprinted from *Nature*, **196**(4851), 232–236 (1962)

CHROMOSOMAL REARRANGEMENTS AND THE EVOLUTION OF HAPTOGLOBIN GENES

By O. SMITHIES, G. E. CONNELL and G. H. DIXON

Department of Medical Genetics, University of Wisconsin, and Department of Biochemistry, University of Toronto

THE three common types of haptoglobin (Hp 1–1, 2–1 and 2–2)[1-3] have recently been divided into a total of six types (Hp 1F–1F, 1F–1S, 1S–1S, 2–1F, 2–1S and 2–2)[4] on the basis of electrophoretic differences in their component polypeptide chains dissociated by treatment with a thiol in the presence of urea. Family studies[5] indicate that these types are the expression of the combinations of three alleles, Hp^{1F}, Hp^{1S} and Hp^2, each of which gives rise to the production of a corresponding polypeptide chain (hp 1Fα, hp 1Sα and hp 2α). All haptoglobins contain an additional polypeptide chain, the β-chain, which appears to be unaffected by the genotype at the Hp locus.

Structure of haptoglobin α-polypeptide chains. The results of our investigations[6] of the differences between the three common haptoglobin α-polypeptide chains may be summarized in part as follows: Amino-acid analyses of hp 1Fα and hp 1Sα suggest the replacement of a single lysine residue in hp 1 Fα by an acidic amino-acid (or its amide) in hp 1Sα. In agreement with these analyses we find that the peptides obtained from hp 1Fα and hp 1Sα after chymotryptic digestion are indistinguishable by finger-printing[7] except for the replacement of one peptide (F) in hp 1Fα by another peptide (S) in hp 1Sα. The effect of pH on the mobilities of the α-chains and the native haptoglobins is also consistent with the substitution of a basic amino-acid in hp 1Fα by an amidated acidic amino-acid in hp 1Sα. On the other hand, although the amino-acid composition of hp 2α is very similar to the compositions of hp 1Fα and hp 1Sα, the peptides obtained from hp 2α differ from those obtained from hp 1Fα and hp 1Sα (Fig. 1) more extensively than can be accounted for by amino-acid substitutions. Thus, all peptides present in digests of hp 1Fα and hp 1Sα are found in digests of hp 2α together with an extra peptide (J); two peptides (N and C), which are common to all three α-chains, are relatively deficient in hp 2α. The amino-terminal amino-acid of peptide N and of the whole α-chains is valine, and the carboxy-terminal amino-acid of peptide C and of the whole of α-chains is glutamine, which suggests that these peptides originate from the ends of the α-chains. Amino-acid

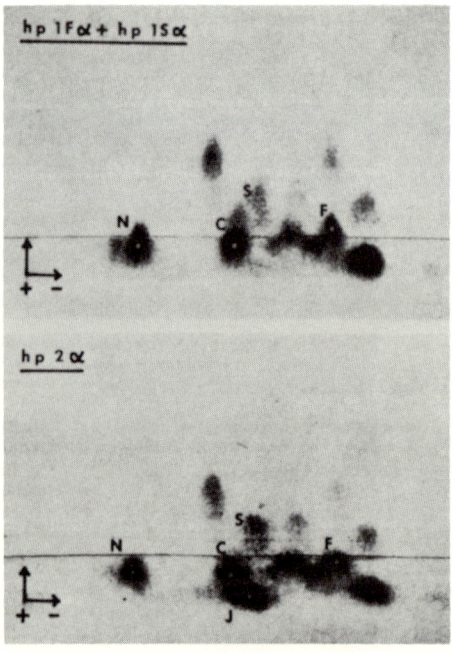

Fig. 1. Peptides from α-chymotrypsin digests of hp 1Fα + hp 1Sα, and hp 2α. The peptides labelled F and S, characteristic of hp 1Fα and hp 1Sα, both occur in hp 2α. Peptides N and C are relatively deficient in hp 2α; J is additional

hp 1Fα   Val──Tyr────(Lys)────Ileu──Glu
          └──┬──┘ ↑ └──┬──┘↑ └──┬──┘
             N   F   C

hp 1Sα   Val──Tyr────(Asp or Glu)────Ileu──Glu
          └──┬──┘ ↑ └──────┬──────┘↑ └──┬──┘
             N   S   C

hp 2α Val──Tyr──(Lys)──Ileu──Tyr──(Asp or Glu)──Ileu──Glu
       └─┬─┘↑ └─┬─┘↑ └─┬─┘↑ └─┬──────┬──────┘↑ └─┬─┘
         N   F   J   S   C

Fig. 2. Outlines of the proposed structures for the three common haptoglobin α-polypeptide chains showing their characteristic features. The order of peptides F and S in hp 2α is arbitrary and could be reversed. The arrows indicate points of attack of chymotrypsin. Amino-acids at the ends of the chymotryptic peptides have been determined by the dinitrofluorobenzene and carboxypeptidase methods; amino-acids in parentheses are present in the corresponding peptides but their positions are not known

analyses of the peptides N, C and J and examination of the peptides obtained from them by further enzymatic digestion with subtilisin or trypsin, suggest that J is slightly smaller than the peptide which would be obtained by joining C and N with a peptide bond. Peptide J has the same amino-terminal amino-acid (isoleucine) as C, and the same carboxy-terminal amino-acid (tyrosine) as N. Furthermore, almost all our preparations (see following) of the polypeptide hp 2α show both the peptides (F and S) characteristic of hp 1Fα and hp 1Sα. These finding are understandable if almost the whole of the amino-acid sequence of the hp 1α polypeptides occurs twice in a single molecule of hp 2α, and lead us to suggest the structures for the three haptoglobin α-polypeptides shown in outline in Fig. 2.

The general correctness of the proposed structures has now received confirmation from measurements of the molecular weights of the α-chains, and of the yields from them of N-terminal amino-acid (valine) and of C-terminal amino-acid (glutamine), as shown in Table 1.

Table 1

	Molecular weight*	N-valine† (per mg)	C-glutamine‡ (per mg)
hp 1Fα	8,860 ± 400	0.034 μM	0.063 μM
hp 1Sα		0.035 μM	
hp 2α	17,300 ± 1,400	0.021 μM	0.032 μM

* By Archibald method.
† By dinitrofluorobenzene method, uncorrected for losses during hydrolysis and chromatography.
‡ By carboxypeptidase method, uncorrected.

Evolution of haptoglobin genes. The substitution of one amino-acid residue in hp 1Fα by a different residue in hp 1Sα indicates that the genes Hp^{1F} and Hp^{1S} may differ only in a single base pair, as has been suggested for the genes controlling the hæmoglobin variants A, S, C, etc.[8]. On the other hand, the structure of hp 2α, in which most of the amino-acid sequence of the hp 1α polypeptides is repeated in a single molecule, indicates that the gene Hp^2 did not evolve from Hp^{1F} or Hp^{1S} by mutations involving small changes in base composition. The simplest hypothesis we can suggest (outlined in ref. 6) is that the gene Hp^2 was evolved from Hp^{1F} and Hp^{1S} by non-homologous crossing-over occurring in a heterozygous individual Hp^{1F}/Hp^{1S}. The suggested mechanism is illustrated diagrammatically in Fig. 3, and is similar to the type of process proposed by Serebrovsky[9] to account for chromosomal rearrangements in general. The net result of this non-homologous crossing-over is the production of one chromosome containing a small duplication and one with a corresponding deletion.

Chromosomal rearrangements and gene evolution. Duplication of existing genes as an important factor in evolution has been discussed previously by several workers (for example, Bridges[10], Lewis[11], and Ingram and Stretton[12]), but the consequences at the molecular level of *partial* gene duplications do not appear to have been foreseen. Our studies with the haptoglobin system indicate that partial gene duplications can lead to the production of a continuous polypeptide which 'bridges' the beginning of the duplication. We use the term 'gene' here, and in what follows, to mean that portion of the DNA coded for a single polypeptide chain, such as an α-chain of haptoglobin. No assumptions as to the length or nature of the DNA between such genes are

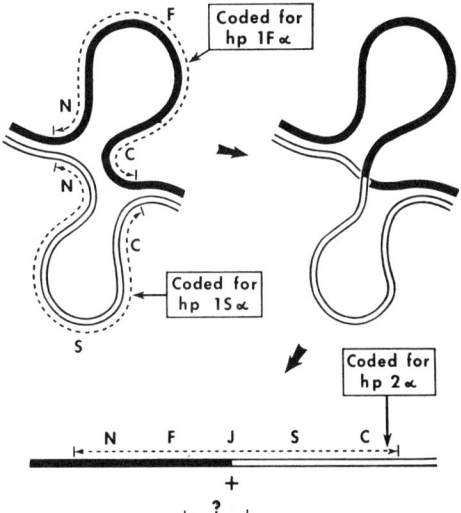

Fig. 3. Diagrammatic representation of the possible evolution of the gene Hp^2 from Hp^{1F} and Hp^{1S} by non-homologous crossing-over in a heterozygote. The parental chromatids involved are distinguished by double and heavy lines. The regions coded for the α-polypeptides are indicated by dotted lines, and the letters locate the approximate regions corresponding to the chymotryptic peptides N, F, S, C and J referred to in the text

made, although we assume that the beginnings and/or ends of genes are specified in some way, and we recognize that the relative frequency but not the possibility of occurrence of some of the events to be discussed may be influenced by the presence and the amount of any DNA not coded for amino-acid sequences.

Partial gene duplications may or may not be completely functional so far as protein synthesis is concerned, depending on the particular point in the chromosome where the duplication begins, and on the nature of the base pair code specifying amino-acids and the ends of polypeptide chains. On theoretical grounds we would expect coding restrictions to limit the number of duplications of random length leading to the production of a polypeptide bridging the beginning of the duplication. For example, less than one in three random duplications would make 'sense' at the beginning of the duplicated region if each amino-acid is coded by one particular triplet of base pairs. On the other hand, if an amino-acid can be specified in more than one way the requirements for continuity of the corresponding polypeptide may be met more frequently at the beginning of the duplication, but may fail to be met at some other point (see Crick et al.[13] for further discussion of this problem). If integral numbers of amino-acid coding units are involved, the requirements for continuity may be met still more frequently and the whole of the duplication may make 'sense'. Although partly functional and non-functional duplications are thus conceivable, our experiments indicate that completely functional partial gene duplications occur.

Once it is realized that a new gene able to code for an uninterrupted sequence of amino-acids can be formed from pre-existing genes by chromosomal rearrangements, the possible importance of this process for the evolution of new proteins becomes apparent. Three of the many conceivable examples of such evolutionary events, related to duplications which could have been formed by non-homologous crossing-over, are indicated diagrammatically in Fig. 4.

A duplication, I, within the boundaries of a single gene, leads to a repeated sequence of amino-acids in the corresponding polypeptide (subject to the involvement of an integral number of amino-acid coding units). Unexpected properties can result from such repetitions, as, for example, the polymerization tendencies of the polypeptide hp 2α. Although the repeated sequence in hp 2α is extensive, there appears to be no a priori reason why shorter ones could not exist. The consequences of the involvement of non-integral numbers of amino-acid coding units in a partial gene duplication will, as indicated here, depend on the nature of the code. For example, the new gene might give rise only to that part of the (pre-existing) polypeptide sequence which is coded by the DNA preceding the duplicated region, or the pre-existing sequence may be followed, as a result of the shift in reading of the DNA code (cf. Crick et al.[13]), by a completely new polypeptide sequence extending over part or almost all of the duplication. Any partial gene duplications which are non-functional in polypeptide synthesis, or give rise to phenotypically inactive products, may behave like amorphic mutations. However, unlike amorphic mutations due to deletions, they will revert to the normal gene occasionally, as a result of unequal but homologous crossing-over within the locus in homozygotes

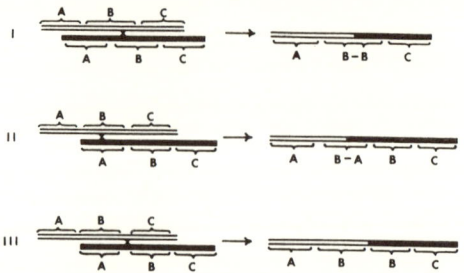

Fig. 4. Diagrammatic representation of the possible formation (by non-homologous crossing-over) and the consequences of duplications of several types. The brackets indicate DNA coded for polypeptide sequences. I, An intragenic duplication of the type suggested as involved in the evolution of Hp^2 from Hp^{1F} and Hp^{1S}; II, A duplication involving parts of two genes, leading to the formation of a new gene $B-A$ without loss of the pre-existing genes A and B; III, a duplication of the same extent as II, but formed by crossing-over between genes. The result is a classical gene duplication, such as may have been involved in the evolution of the hæmoglobin polypeptide chains

(see following), or as a result of unequal but homologous sister strand or intrachromatid crossing-over in heterozygotes[14].

A duplication involving parts of two genes, II, can lead to the formation of a new gene coded for a polypeptide sequence combining parts of the sequences of two pre-existing proteins if integral numbers of coding units are involved. Such a polypeptide might combine in one molecule properties formerly present in separate molecules. For example, in the case of enzymes it could have amino-acid sequences enabling it to combine with two substrate groups where its 'progenitors' could each combine with only one of them. (If non-integral numbers of coding units are involved the situation is similar to that discussed here.) Yet the evolution of such radically new proteins, by a single genetic event, need not be accompanied by loss of any genes for pre-existing proteins (Fig. 4). The alternate chromosome (not shown in the figure) formed during the non-homologous crossing-over leading to duplication II also has a new gene ($A-B$) which could code for a sequence of amino-acids with the portions corresponding to the genes A and B in the order A-B, in contrast to the order B-A for the duplication. This alternate chromosome, however, lacks the genes A and B, and so may have selective disadvantages.

Non-homologous crossing-over occurring between genes can lead to complete duplication of an existing gene (III).

Many types of chromosomal rearrangements, in addition to duplications, have been detected by genetic and cytological studies; for example, deletions, inversions, shifts, transpositions, etc. None of these would appear on a priori grounds to be confined to integral numbers of genes. We can therefore expect that protein variants will eventually be found corresponding to them, although until the base pair-amino-acid code is more completely known it is difficult to predict what proportion of these arrangements will fulfil the requirements for continuity in the corresponding polypeptides. The frequency of occurrence of cytologically detectable chromosomal rearrangements in naturally occurring populations of Drosophila and other species certainly indicates that re-arrangements are common evolutionary events.

Other proteins, in addition to the haptoglobins, have structures which suggest that chromosomal rearrangements were involved in their evolution. Thus, the porcine pituitary hormones α-melanotropin (αMSH) and adrenocorticotrophin (ACTH) both contain the same sequence of 13 amino-acids[15]. αMSH has an acyl group on the N-terminal amino-group and the terminal carboxyl-group is amidated. On the other hand, ACTH has a free N-terminal amino-group and 26 additional amino-acid residues on the C-terminal end. βMSH lacks the N- and C-terminal amino-acids of αMSH and has short, completely dissimilar sequences on both ends of the common sequence. Two amino-acid residues in the common sequence are interchanged in position. These structures are strongly suggestive of evolution involving chromosomal rearrangements at some stage. Many proteolytic enzymes (for example, trypsin and chymotrypsin) have portions of their sequences in common, but in general their structures differ greatly[16]. They could also have evolved by chromosomal rearrangements.

Consequences of unequal but homologous crossing-over at the Hp locus. Several consequences of the intragenic duplication probably involved in the gene Hp^2 can be predicted by considering other systems involving duplications. The 'Bar' allele in *Drosophila* is associated with a cytologically demonstrable duplication[17,18]. Homozygotes for 'Bar' are relatively frequently involved in unequal but homologous crossing-over[19], as a result of the displaced synapsis possible when there is a duplication in the genome. (Unequal but homologous crossing-over must be distinguished from non-homologous crossing-over, although both involve displaced synapsis. The latter, as indicated in Fig. 3, involves crossing-over between chromosomal regions which are not equivalent in their gene structure; it is consequently a rare event, and the subsequent action of the genes involved will be dependent on the requirements for continuity discussed here. Unequal but homologous crossing-over, on the other hand, involves crossing-over between chromosomal regions which are of equivalent structure but which are not in the same relative positions on the two chromosomes. Unequal but homologous crossing-over will not usually change the 'sense' of the DNA involved, and its frequency will be determined only by the frequency of displaced synapsis and by the map length of the corresponding duplicated or displaced chromosomal segment.) This unequal crossing-over in *Drosophila* gives rise to the formation of "double 'Bar'", associated with a cytological triplication, or to reversion to 'wild-type' which lacks the original 'Bar' duplication. Homozygous individuals of the genotype Hp^2/Hp^2 are in many ways analogous to homozygotes for the 'Bar' allele. Unequal crossing-over would therefore be expected to occur occasionally in such individuals, either with the production of a triplicated haptoglobin gene (instead of the usual duplicated Hp^2 gene) or with reversion to one of the original 'wild-type' genes Hp^{1F} or Hp^{1S} (see Fig. 5 I). In the absence of information about the nature of the gene Hp^2 these events would be interpreted as mutations and the Hp locus would be described as having a high mutation rate.

A rare and unusual type of haptoglobin (the Johnson type), discovered by Dr. E. R. Giblett in a Negro woman and her daughter in Seattle, is associated with a rare gene, Hp^{2J}, in heterozygous combination with the common Hp^{1S} allele. Preliminary

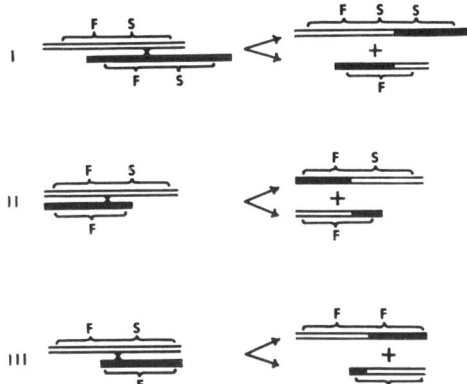

Fig. 5. Diagrammatic representation of the consequences of: I, unequal but homologous crossing-over occurring during displaced synapsis in a Hp^2/Hp^2 homozygote, leading to the possible evolution of a haptoglobin triplication or to reversion of Hp^2 to Hp^{1F} or Hp^{1S}; II and III, crossing-over within the locus in a heterozygous individual (Hp^2/Hp^{1F}) during the two possible synaptic configurations. This crossing-over will always be unequal and can lead to the production of four types of Hp^2 gene, and to the formation of a gamete containing the gene Hp^{1S} in an individual of genotype Hp^2/Hp^{1F}

tests, involving comparison of the mobilities of the α-chains in starch gels at different starch concentrations[20], indicate that the corresponding polypeptide $hp\ 2J\alpha$, is considerably larger than $hp\ 2\alpha$. This suggests that the gene Hp^{2J} could be the triplication formed by unequal but homologous crossing-over in a homozygote Hp^2/Hp^2 during displaced synapsis. This suggestion receives some confirmation from the recent discovery of the Johnson type in other widely separated racial groups (Dr. E. R. Giblett has found it in a Canadian Mennonite and two Hawaiian Chinese, Dr. K. Henningsen in a Dane, Dr. J. Hirschfeld in a Swede, Dr. R. L. Kirk in an Australian Dutch, Dr. J. Lundevall and Mrs. E. A. Fleischer in three Norwegians, Dr. B. Romat in a Kurdish Jew). It seems highly improbable that the gene Hp^{2J} in the Seattle Negroes and these other genes have a common ancestor; yet they appear to be identical as judged by the phenotypes. The cause of their identity could be a common mechanism of formation rather than descent from a common ancestor. If the gene Hp^{2J} does prove to be an intragenic triplication then the Johnson type (genotype Hp^{2J}/Hp^{1S}) provides an excellent example and a possible explanation at the molecular level of the stable position effect, since the Johnson phenotype is readily distinguished from Hp 2-2, in which the same genetic material is distributed equally between the two chromosomes.

Other unusual serum samples very similar to but not identical with the Johnson type have also been found and sent to us for testing. Unfortunately, all had been shipped considerable distances and were not freshly collected, so that the observed small differences are difficult to evaluate. However, differences would be expected in terms of the triplication hypothesis, since there should be eight different forms of a triplicated haptoglobin gene (see later) and consequently sixteen heterozygous combinations with the genes Hp^{1F} and Hp^{1S}.

Synapsis in heterozygotes Hp^2/Hp^{1F} and Hp^2/Hp^{1S} can occur in two ways (Fig. 5 II and III), as a result of the duplicate nature of the gene Hp^2. Crossing-over within the locus (which will always be unequal) should therefore lead eventually to the occurrence in the gene pool of all the four possible types of Hp^2. Thus, if we assume that the original Hp^2 gene was FS, then intragenic crossing-over in heterozygotes will lead to the formation of Hp^2 genes which are FF, SS, and SF. The relative chances of finding these types is difficult to predict, since we have insufficient knowledge about the frequency of the corresponding unequal crossing-over, and of the selective forces involved, to be able to estimate the equilibrium frequencies of the different genes. However, we may already have had some samples of hp 2α from individuals who are $Hp^{2(FF)}/Hp^{2(FF)}$ or $Hp^{2(FS)}/Hp^{2(FF)}$, as three out of fifteen single donor hp 2α preparations showed complete, or almost complete absence of the peptide S characteristic of hp 1Sα (Fig. 1), but clearly showed the corresponding F peptide. In any event, we predict that four types of Hp^2 gene are possible, and are undertaking a search for them.

In summary, we conclude that the haptoglobin system provides examples of the effect on the structure of a single polypeptide chain of both a 'point' mutation and chromosomal rearrangements. We suggest that proteins with radically changed properties can be formed as a consequence of the single genetic event of a chromosomal rearrangement involving non-integral numbers of genes. Chromosomal rearrangements of this type appear to provide a mechanism for achieving more rapid and extensive changes in protein structure during evolution than are possible by point mutations, even when preceded by gene duplication. New genes formed by chromosomal rearrangements need not be accompanied by loss to the organism of any pre-existing genes.

This work was supported in part by grants from the U.S. National Science Foundation, and the Medical Research Council, Canada. We thank Dr. W. E. Nance for his help during the preparation of the manuscript and acknowledge many helpful discussions of the work with him, Drs. J. F. Crow, N. E. Morton, and L. Sandler.

[1] Smithies, O., *Biochem. J.* **61**, 629 (1955).
[2] Smithies, O., and Walker, N. F., *Nature*, **176**, 1265 (1955).
[3] Smithies, O., and Walker, N. F., *Nature*, **178**, 694 (1956).
[4] Connell, G. E., Dixon, G. H., and Smithies, O., *Nature*, **193**, 505 (1962).
[5] Smithies, O., Connell, G. E., and Dixon, G. H., *Amer. J. Hum. Genet.*, **14**, 14 (1962).
[6] Smithies, O., Connell, G. E., and Dixon, G. H., *Proc. Sec. Intern. Cong. Hum. Genet.* (in the press).
[7] Ingram, V. M., *Biochem. Biophys. Acta.* **28**, 539 (1958).
[8] Ingram, V. M., in *Genetics, Macy Found. Symp.*, edit. by Sutton, H. E., 112 (1960).
[9] Serebrovsky, A. S., *Amer. Nat.*, **63**, 374 (1929).
[10] Bridges, C. B., *J. Hered.*, **26**, 60 (1935).
[11] Lewis, E. B., *Cold Spring Harb. Symp. Quant. Biol.*, **16**, 159 (1951).
[12] Ingram, V. M., and Stretton, A. O. W., *Nature*, **190**, 1079 (1961).
[13] Crick, F. H., Barnett, L., Brenner, S., and Watts-Tobin, R. J., *Nature*, **192**, 1227 (1961).
[14] Laughnan, J. R., *Proc. Symp. Mutation and Plant Breeding NAS-NRC 891*, 3, (1961).
[15] Li, C. H., *Vitamins and Hormones*, **19**, 313 (1961).
[16] Dixon, G. H., Kauffman, D., and Neurath, H., *J. Biol. Chem.*, **233**, 1373 (1958).
[17] Bridges, C. B., *Science*, **83**, 210 (1936).
[18] Muller, H. J., Prokofyeva-Belgovskaya, A. A., and Kossikov, K. V., *C.R. Acad. Sci. U.S.S.R.*, **2**, 78 (1936).
[19] Sturtevant, A. H., *Genetics*, **10**, 117 (1925).
[20] Smithies, O., *Arch. Biochem. Biophys.* (in the press).

THE FUSION OF TWO PEPTIDE CHAINS IN HEMOGLOBIN LEPORE AND ITS INTERPRETATION AS A GENETIC DELETION

CORRADO BAGLIONI*

DEPARTMENT OF BIOLOGY, MASSACHUSETTS INSTITUTE OF TECHNOLOGY

Communicated by S. E. Luria, August 22, 1962

Gerald and Diamond[1] have described an abnormal hemoglobin, called hemoglobin Lepore (Hb-Lepore$_{\text{Boston}}$). The present communication reports the results of a chemical investigation of Hb-Lepore$_{\text{Boston}}$ and an interpretation of the genetic events leading to the formation of a single peptide chain from the fusion of two different chains.

Abnormal hemoglobins electrophoretically identical to Hb-Lepore$_{\text{Boston}}$ have have been reported among Greeks,[2] Italians,[3] and Papuans[4] and designated Hb-Pylos, Hb-G, and Hb-Lepore$_{\text{Hollandia}}$ respectively. These hemoglobins are always found in concentrations of about 10–15 per cent of the total hemoglobin in the heterozygotes. One individual homozygous for the Hb-Pylos gene and two homozygous for the Hb-Lepore$_{\text{Hollandia}}$ gene have been reported.[2,4] They were characterized by the absence of normal adult hemoglobin (Hb-A) and of the minor component hemoglobin A$_2$(Hb-A$_2$).[2,4] Neeb et al.[4] have reported, however, the presence of traces of a hemoglobin component migrating like Hb-A$_2$ in Hb-Lepore$_{\text{Hollandia}}$ homozygotes; the identification of this trace component remains to be established.

The three normal human hemoglobins Hb-A, Hb-F (fetal hemoglobin), and Hb-A$_2$ consist of two α peptide chains which are under the control of a single structural gene and of two other peptide chains which differ in different hemoglobins: β chains in Hb-A, γ chains in Hb-F, and δ chains in Hb-A$_2$.[5] These are under the control of different structural genes.[5] The β gene is linked to the δ gene.[5] The β and δ chains have extremely similar amino acid compositions and sequences.[6-8] This similarity has suggested that the δ gene may have originated through a duplication of the β gene followed by independent evolution.[9]

The Hb-Pylos and Hb-Lepore$_{\text{Hollandia}}$ mutations seem to affect both the two linked genes β and δ, suppressing the synthesis of both Hb-A and Hb-A$_2$ in homozygotes.[2,4] The α and γ peptide chains are synthesized normally.[2,6]

Gerald et al.[10] have suggested that the Lepore abnormality is a "mutation possibly involving two cistrons." These authors fingerprinted Hb-Lepore$_{\text{Boston}}$ and Hb-Pylos and analyzed some of their tryptic peptides. The fingerprints were found to be indistinguishable from those of Hb-A$_2$. The composition of the tryptic peptides analyzed was the same as that of homologous peptides of Hb-A$_2$.[10] Hb-Lepore$_{Boston}$ and Hb-Pylos appear to be identical.[2, 10] Barnabas and Muller[6] reported that Hb-Lepore$_{Hollandia}$ shows some differences from Hb-A$_2$. One peptide, called A$_2$a,[7] is absent in Hb-Lepore$_{Hollandia}$. A peptide which in Hb-Lepore$_{Boston}$ has the composition of a δ chain peptide (δ5)[11] has in Hb-Lepore$_{Hollandia}$ the composition of a β chain peptide (β5).[6] This suggested that the non-α chain of Hb-Lepore$_{Hollandia}$ may have characteristics of both the β and δ peptide chains.

Materials and Methods.—*Hemoglobins:* Hb-A and Hb-A$_2$ were purified by column chromatography according to Schnek and Schroeder.[12] The hemoglobin solutions were concentrated as previously described[13] and were heat-denatured and digested with trypsin according to Ingram.[14] The tryptic digest of 20 mg of purified Hb-Lepore$_{Boston}$ was kindly donated by Park S. Gerald of the Children's Hospital (Boston, Mass.).

Fingerprinting: The original procedure of Ingram was followed for the ionophoresis,[14] while different chromatographic solvents were used: *#1*, pyridine/*i*soamyl/alcohol water (35/35/30);[15] *#2*, butanol/acetic acid/pyridine/water (75/15/50/60);[16] *#3*, butanol/acetic acid/pyridine/water (75/15/50/50). When elution of the peptides from the fingerprints was desired, the ionophoresis was prolonged for 4 hr, allowing the most basic peptides to run off the paper. The fingerprints were then developed in solvent #3.

Analysis of tryptic peptides: The fingerprints were stained with 0.01% ninhydrin and then washed with acetone. The peptides were eluted with 1 N acetic acid. The eluates were hydrolyzed and the amino acid composition was determined as previously indicated.[13]

Results.—Fingerprints of Hb-Lepore$_{Boston}$ were compared to fingerprints of Hb-A and of Hb-A$_2$. The fingerprints of Hb-Lepore$_{Boston}$ and Hb-A developed in solvent #3 are shown in Figure 1. The peptides of Hb-Lepore$_{Boston}$ indicated

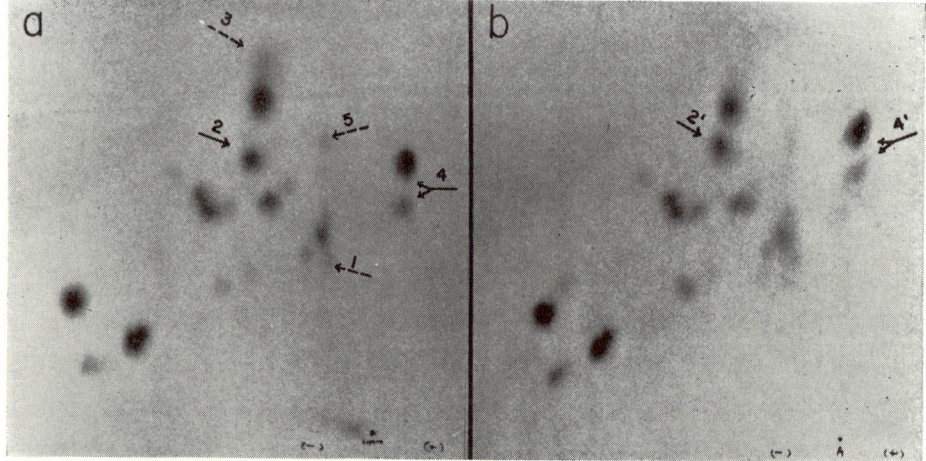

FIG. 1.—Photographs of fingerprints of Hb-Lepore$_{Boston}$ (*a*) and Hb-A (*b*) developed in solvent #3 (see *Methods*). The solid arrows in *a* and *b* indicate corresponding peptides that show different R$_f$. The arrows in *a* indicate the peptides of Hb-Lepore$_{Boston}$ which have been eluted from fingerprints and analyzed. The numbers in *a* refer to the peptides reported in Table 1: 1 = x_B1; 2 = x_B2; 3 = x_B4; 4 = x_B5 + x_B5 oxidized; 5 = x_B9. 2' in *b* indicate peptide β2 and 4' peptides β5 + β5 oxidized.[15]

by solid arrows in Figure 1a were found to have R_f's different from those of the corresponding peptides of Hb-A, indicated by solid arrows in Figure 1b. The fingerprints of Hb-Lepore$_{Boston}$ developed in solvent #1 or #3 were found to be indistinguishable from fingerprints of Hb-A$_2$, confirming the observation of Gerald et al.[10]

When, however, the fingerprints of Hb-A$_2$ and Hb-Lepore$_{Boston}$ were developed in solvent #2 (see Fig. 2) and stained for sulfur-containing amino acids and for

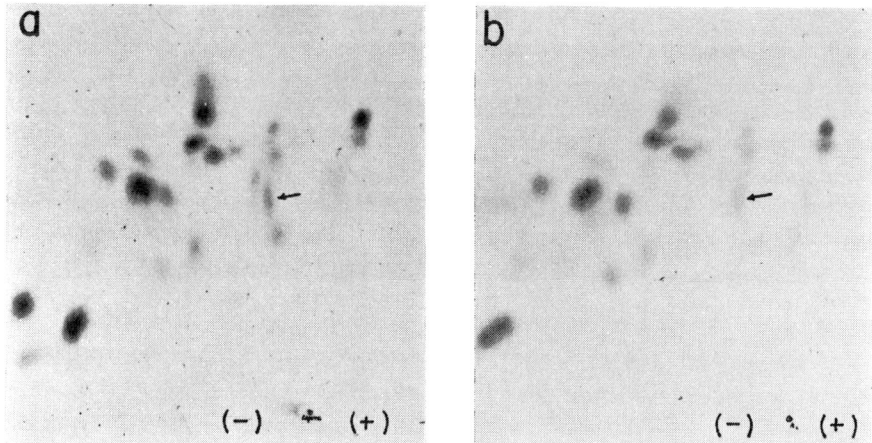

FIG. 2.—Photographs of fingerprints of Hb-Lepore$_{Boston}$ (a) and Hb-A$_2$ (b) developed in solvent #2 (see Methods). The arrow in b indicates peptide δ14; the arrow in a indicates the corresponding peptide of Hb-Lepore$_{Boston}$ (x_B13) that gives the same staining reaction as β13.

tyrosine, a difference between them became apparent. A Hb-A$_2$ peptide indicated by the arrow in Figure 2b gave a positive reaction for sulfur and tyrosine. A peptide which occupies an identical position in the fingerprints of Hb-Lepore$_{Boston}$ (arrow in Fig. 2a) showed a positive reaction for tyrosine but not for sulfur. This peptide has been identified as β13 in Hb-A.[15] In Hb-A$_2$, there is a peptide, δ14 (see later),[11] which occupies the same position in fingerprints as β13 in Hb-A but which gives a positive reaction for sulfur and tyrosine.[7] Analysis of peptide δ14 has shown the presence of a methionine residue.[8] No methionine is present in β13.[15] This difference between Hb-A and Hb-A$_2$ does not show in fingerprints developed in solvent #1 or #3 since the Hb-A$_2$ peptide δ14 overlaps peptide α9, which also contains methionine.[7, 15] The difference is noticed only in fingerprints run at different pH.[7]

A second difference between Hb-A$_2$ and Hb-Lepore$_{Boston}$ tryptic digests was observed by running a one-dimensional ionophoresis at pH 6.4.[18] A peptide, called A$_2$a, has been observed in ionograms of the tryptic digest of isolated δ chain.[18] This peptide was not present in Hb-Lepore$_{Boston}$[19] and in Hb-Lepore$_{Hollandia}$.[6]

A few peptides, which were separated in a sufficiently pure form by fingerprinting (indicated by arrows in Fig. 1a), were eluted and analyzed. The results of the amino acid analysis are shown in Table 1. The peptides analyzed have amino acid composition identical to homologous peptides of Hb-A$_2$.[7, 18]

Discussion.—In the following discussion, the position of the tryptic peptides along the amino acid sequence of the δ chain of Hb-A$_2$ and the non-α chain of Hb-

TABLE 1

Amino Acid Analysis of Hb-Lepore_Boston Peptides

Amino Acid†	x_B^1	x_B^2	Peptide* x_B^4	x_B^5	x_B^9
Lysine	1.02	1.01		1.13	1.11
Arginine			1.10		
Histidine	0.98				1.22
Glycine		0.87		1.97	2.07
Alanine		1.81		1.24	2.07
Valine	0.95	1.13	1.94	1.30	1.10
Serine				§3.57	§4.38
Leucine	1.25	1.25	2.33		
Threonine	0.82	0.74	0.91		
Methionine				0.76	
Glutamic acid	1.98		1.03	1.28	
Phenylalanine				2.02	1.12
Aspartic acid		1.19		2.64	2.93
Tyrosine			0.69		
Proline‡	++		++	++	

* The identification of the peptides is based on their positions in fingerprints, identical with those of β or δ chain peptides.[7, 15] The peptides are designated as indicated.[11] The x_B peptide chain is the non-α chain of Hb-Lepore_Boston.
† The molar ratio of the amino acids is given.
‡ Proline cannot be estimated quantitatively by this method of analysis; the presence of this amino acid, judged by the visual examination of the ionograms, is indicated by ++.
§ Serine and leucine, when present in high relative amounts, are incompletely resolved. The values reported represent the sum of the leucine plus serine molar ratios.

Lepore_Boston (here called x_B chain) is arbitrarily assigned according to the principle of homology.[20] This principle postulates that peptides having identical or very similar amino acid sequence or composition occupy corresponding positions along homologous peptide chains, which have differentiated from a common ancestor chain either in the course of evolution in different species or following gene duplications and independent evolution within the same species.[9, 20]

Following this principle of homology it has been possible to predict the amino acid sequence of animal hemoglobins from the composition of their tryptic peptides[21, 22] and the sequence of the human hemoglobin chains.[23, 24]

Thus, it may be justified to assume, until the complete amino acid sequence of the δ and x_B chains is established, that the order of the tryptic peptides in these chains corresponds to the order in the β chain. The similarities in the amino acid composition of the tryptic peptides are striking.[6, 7, 10, 18] Moreover, the amino acid sequences so far determined within δ chain peptides are identical to those of β chain peptides, except for a few amino acid substitutions.[8, 18]

The β chain differs from the δ chain in four positions: 9, β serine → δ threonine; 12, β threonine → δ asparagine; 22, β glutamic acid → δ alanine; and 50, β threonine → δ serine.[7, 18] Preliminary evidence has been obtained by Stretton for four more differences.[8] One of these has been localized in residue 126, β valine → δ methionine.[8] Additional differences are revealed by the presence of peptide A₂a in the tryptic digest of the δ chain.[18] This peptide has the sequence[25] $Asp(NH_2).Phe.Gly.Lys$.[8, 18] The only sequence homologous to this one in the β chain is the C-terminal ...$His.His.Phe.Gly.Lys$ sequence of peptide β12 (from residue 116 to 120).[23] This has suggested[8] two further amino acid differences between the β and the δ chains: 117, β histidine → δ asparagine and 116, β histidine → δ arginine or lysine, to account for the splitting of peptide A₂a (provisionally designated δ13) by trypsin. Stretton[8] has obtained an indication that residue 116 of the δ chain is arginine. The δ chain has, therefore, one more tryptic peptide than the β chain.

274

No difference in composition between other β and δ chain peptides has been observed.[8, 18] It so happens that $\beta 1$ is identical in composition to $\delta 1$, $\beta 4$ to $\delta 4$, $\beta 6$ to $\delta 6$, $\beta 7$ to $\delta 7$, $\beta 8$ to $\delta 8$, $\beta 9$ to $\delta 9$, $\beta 14$ to $\delta 15$, and $\beta 15$ to $\delta 16$ (because of the presence of the extra peptide $\delta 13$ in the δ chain).

The fingerprints of Hb-Lepore$_{Boston}$ and the amino acid analyses of the peptides isolated indicate that the x_B chain of this hemoglobin is made up of peptides characteristic of both the β and the δ chains. The N-terminal portion of the x_B chain is made up of δ-like peptides. $x_B 2$, $x_B 3$, and $x_B 5$ are identical in amino acid composition to the corresponding δ chain peptides[10] and $x_B 1$ and $x_B 4$ could not be distinguished from $\delta 1$ and $\delta 4$ (= $\beta 1$ and $\beta 4$ respectively). The C-terminal portion of the x_B chain is made up of β-like peptides (see Fig. 3). $x_B 13$ is identical in the fingerprinting analysis and in the specific staining reaction to $\beta 13$; $x_B 14$ and $x_B 15$ could not be distinguished from $\beta 14$ and $\beta 15$ (= $\delta 15$ and $\delta 16$ respectively). Moreover, the absence in Hb-Lepore$_{Boston}$ of peptide $\delta 13$ (A$_2$a), suggests that the x_B chain peptide corresponding to $\beta 12$ is similar if not identical to this β chain peptide.

Similarly, the non-α chain of Hb-Lepore$_{Hollandia}$ (here indicated as x_H) seems to have a δ-like N-terminal portion and a β-like C-terminal portion. The N-terminal δ-like portion of the x_H chain, however, seems to be shorter than the δ-like portion of the x_B chain. $x_H 5$ has indeed the composition of $\beta 5$.[6] Thus, the x_H chain may be δ-like from residue 1 to 40 approximately and β-like in the remaining portion of the chain (see Fig. 3).

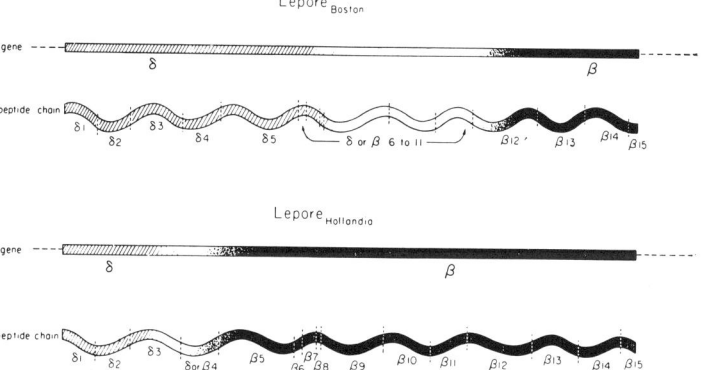

FIG. 3.—Schematic representation of the Lepore$_{Boston}$ gene (top), of the Lepore $_{Hollandia}$ gene (bottom), and of the corresponding peptide chains. The shaded areas indicate that part of the Lepore genes which appears to be derived from the δ gene and that part of the corresponding peptide chains which is δ-like. The area in solid color indicates that part of the Lepore genes which appears to be derived from the β gene and that part of the corresponding peptide chains which is β-like. The empty areas between shaded and solid color areas indicate those portions of the genes and of the peptide chains that may be derived either from β or from δ. The postulated joining of part of the β gene to part of the δ gene has occurred presumably in this area. The dotted lines along the peptide chains indicate schematically the peptide bonds which may be split by trypsin. The resulting peptides are indicated below and are numbered consecutively starting from the N-terminus. The symbols β or δ which precede the peptide numbers indicate whether the corresponding peptides are β-like or δ-like.

Conclusions.—There is now a large body of evidence which suggests that the structural gene for a protein determines its primary amino acid sequence.[26] The structural genes seem to be separated by regions, which establish boundaries between them and allow their function to be expressed individually.[27]

In the well-studied region r_{II} of phage $T4$ there are two adjacent cistrons, called A and B.[28] In the mutant $r1589$ of $T4$, there is a deletion of the region separating the A from the B cistron and the two cistrons appear to be joined, with loss of the A function but with preservation of the B function.[28] The insertion of certain deletions in the A cistron of $r1589$ prevents the B cistron from functioning.[27] This suggests that in $r1589$ a protein may be produced "which consists of part of the protein from the A cistron and part of the B cistron, joined together in the same polypeptide chain." [27]

The x_B peptide chain of Hb-Lepore$_{Boston}$ seems to consist of part of the β chain and part of the δ chain joined together. That part of the δ gene corresponding to the N-terminal portion of the δ chain is presumably joined to that part of the β gene which determines the C-terminal portion of the β chain. The fact that the β and δ genes are linked[5] suggests that the joining may be due to a deletion of parts of the β and δ genes and of the region separating them.

If this postulated joining of part of the β gene with part of the δ gene were due to a random deletion, it would be difficult to understand why the peptide chain made by the hybrid β-δ gene would have the same length as the β or δ chain. The evidence so far suggests that there is no amino acid sequence in the x_B chain that is not also present either in the β or in the δ chain. One would not expect to find such a situation if the corresponding genes were randomly joined by a deletion. A specific mechanism that accounts for the joining of the β to the δ gene in such an exactly complementary way is a nonhomologous crossing-over between corresponding points of the β and the δ genes, resulting in the formation of unequal products (see Fig. 4).

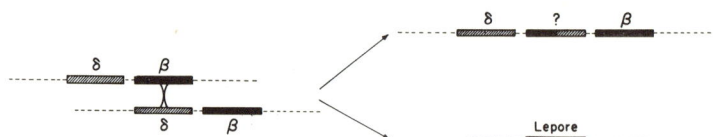

FIG. 4.—Schematic representation of the postulated nonhomologous crossing-over leading to the formation of the Lepore genes. The shaded area indicates the nucleotide sequence corresponding to the δ gene. The nonhomologous crossing-over is supposed to have occurred in the areas shown as empty in Figure 3.

Unequal crossing-over is known to involve frequently chromosomal duplications, such as the *bar* duplication of *Drosophila melanogaster*,[30] leading to recombinants with deletion of the duplication (*wild type*) or insertion of one extra *bar* duplication (*double-bar*). It has been suggested that the δ gene originated from a duplication of the β gene, because the primary sequence of the δ chain is extremely similar to the primary sequence of the β chain.[9] The similarity of the β and δ genes may increase the probability of nonhomologous crossing-overs. It is suggestive that Hb-Lepore$_{Boston}$ and Hb-Lepore$_{Hollandia}$ have been observed in different populations.[1-2] The corresponding genes may have been formed by different nonhomologous crossing-overs, which have involved different points of the β and δ genes (see Figs. 3 and 4).

Note added in proof: Unequal crossing-over has recently been invoked as the interpretation of variations in the length of the α chain of human haptoglobin.[31]

I wish to thank Vernon M. Ingram for help, advice, and criticism during this work and in the preparation of this manuscript, Park S. Gerald and Mary L. Efron for providing a sample of Hb-

Lepore$_{Boston}$ and for discussions and suggestions, and Antony O. W. Stretton for making available unpublished results. This work has been supported by research grants from the Medical Foundation, Inc., Boston, and from the National Science Foundation.

* Present address: International Laboratory for Genetics and Biophysics, Via Claudio 1, Naples, Italy.

[1] Gerald, P. S., and L. K. Diamond, *Blood*, **13**, 835 (1958).
[2] Fessas, P., G. Stamatoyannopoulos, and A. Karaklis, *Blood*, **9**, 1 (1962).
[3] Silvestroni, E., and I. Bianco, *Policlinico*, **65**, 203 (1958).
[4] Neeb, H., J. L. Beiboer, J. H. P. Jonxis, J. A. Kaars-Sijpesteijn, and C. J. Muller, *Trop. Geogr. Med.*, **13**, 207 (1961).
[5] Baglioni, C., in *Progresses in Molecular Genetics*, ed. H. Taylor (New York and London: Academic Press), vol. 1, in press.
[6] Barabas, J., and C. J. Muller, *Nature*, **194**, 931 (1962).
[7] Ingram, V. M., and A. O. W. Stretton, *Nature*, **190**, 1079 (1961).
[8] Stretton, A. O. W., unpublished.
[9] Ingram, V. M., *Nature*, **189**, 704 (1961).
[10] Gerald, P. S., M. L. Efron, and L. K. Diamond, *Am. J. Dis. Child.*, **102**, 514 (1961).
[11] The peptides obtainable by tryptic digestion are indicated as follows: the symbol of the peptide chain precedes arabic numerals indicating the order in which the peptides occur in the chain starting from the N-terminus.[15] This system of nomenclature is similar to the one recently recommended.[17]
[12] Schnek, A. G., and W. A. Schroeder, *J. Am. Chem. Soc.*, **83**, 1472 (1961).
[13] Baglioni, C., *J. Biol. Chem.*, **237**, 69 (1962).
[14] Ingram, V. M., *Biochim. Biophys. Acta*, **28**, 539 (1958).
[15] Baglioni, C., *Biochim. Biophys. Acta*, **48**, 392 (1961).
[16] Hill, R. L., R. T. Swenson, and H. C. Schwartz, *Blood*, **19**, 573 (1962).
[17] Gerald, P. S., and V. M. Ingram, *J. Biol. Chem.*, **236**, 2155 (1961).
[18] Ingram, V. M., and A. O. W. Stretton, *Biochim. Biophys. Acta*, **62**, 456 (1962) and in press.
[19] Gerald, P. S., and M. L. Efron, unpublished.
[20] Zuckerkandl, E., and L. Pauling, in *Horizons in Biochemistry*, ed. M. Kasha and B. Pullman (New York and London: Academic Press, 1962), p. 189.
[21] Diamond, J. M., and G. Braunitzer, *Nature*, **194**, 1287 (1962).
[22] Naughton, M. A., and H. M. Dintzis, these PROCEEDINGS, **48**, 1822 (1962).
[23] Braunitzer, G., R. Gehring-Muller, N. Hillschmann, K. Hilse, G. Hobman, V. Rudloff, and B. Wittman-Liebold, *Hoppe Seyler's Z. physiol. Chem.*, **325**, 283 (1961).
[24] Konigsberg, W., G. Guidotti, and R. G. Hill, *J. Biol. Chem.*, **236**, PC55 (1961).
[25] The amino acids are indicated by the first three letters of their names. Asparagine is indicated as Asp(NH$_2$).
[26] Yanofsky, C., D. R. Helinski, and B. D. Maling, in *Cellular Regulatory Mechanisms*, Cold Spring Harbor Symposia on Quantitative Biology, vol. 26, (1961), p. 11.
[27] Crick, F. H. C., L. Barnett, S. Brenner, and R. J. Watts-Tobin, *Nature*, **192**, 1227 (1961).
[28] Champe, S. P., and S. Benzer, *J. Mol. Biol.*, **4**, 288 (1962).
[29] Kunkel, H. G., R. Ceppellini, V. Muller-Eberhardt, and J. Wolf, *J. Clin. Invest.*, **36**, 1615 (1957).
[30] Bridges, C. B., *Science*, **83**, 210 (1936).
[31] Smithies, O., G. E. Connell, and G. D. Dixon, *Nature*, **196**, 232 (1962).

XI
Gene Organization and Protein Organization

Editor's Comments on Papers 27 Through 29

27 **Yanofsky, Helinski, and Maling:** *The Effects of Mutation on the Composition and Properties of the A Protein of* Escherichia Coli *Trypytophan Synthetase*

28 **Helinski and Yanofsky:** *Correspondence Between Genetic Data and the Position of Amino Acid Alteration in a Protein*

29 **Henning and Yanofsky:** *An Alteration in the Primary Strucutre of a Protein Predicted on the Basis of Genetic Recombination Data*

Charles Yanofsky and a number of coworkers accomplished a monumental task during the 1960s. They succeeded in putting to direct test many of the hypotheses developed during the 1950s after it became evident that DNA indeed was the genetic material. Phage experiments also contributed a great deal toward testing these hypotheses, but the Yanofsky group's experiments with *Escherichia coli* stand out in an exemplary way. With the tryptophan synthetase mutants of *E. coli,* they demonstrated a more direct relation between gene mutation and protein than had been done before.

Paper 27 reviews much of the background work and in fact outlines much of the work that was to be published in the next several years. Included is a section on suppressor mutations which makes the observation that these may be expected to be suppressor genes that are involved in transcribing transfer RNAs. This idea had earlier been suggested by Yanofsky and Lawrence. This in fact has turned out to be the case.

Once it became apparent that a genetic code existed which consisted of a series of codons presumably arranged in a linear array in a DNA strand, it was necessary to test for colinearity between the sequence of codons and the sequence of amino acids in the polypeptide theoretically arranged in that array by the codons. The experiments necessary to do this were reasonably simple to formulate, but a genetic system was needed that lent itself to recognizing rare recominants. Yanofsky early recognized the need and the adaptability of the tryptophan synthetase mutants of *Escherichia coli* to fill the need. In Paper 28 he and Helinski report the first of such experiments in which they show that several missense mutations affecting the tryptophan synthetase A gene, and causing amino acid substitutions in the A polypeptide, are indeed located in the chromosome regions expected from the determined location of the amino acid substitution in the altered polypeptide.

This was only the first of a series of papers on this subject. It is now well established both by subsequent work from Yanofsky's laboratory and from an English laboratory (Sarabhai et al., 1964) that there is colinearity between the genetic material in both bacteria and phage.

The original theory of the gene, now called the classical gene theory, pictured the gene as a unit of function, mutation, and recombination. The phage r_{II} mutant work of Benzer (1955) reduced the gene to the status of being only a unit of function. In Paper 29, Henning and Yanofsky present substantial evidence that recombination between different nucleotides in the same codon can occur. Apparently, the unit of recombination is a nucleotide, and crossing over can probably occur anywhere along the length of DNA strands whether within or between genes. In addition it was now recognized that a "mutation" could result by recombination within a codon.

Selected Bibliography

Benzer, S. 1955. Fine structure of a genetic region in bacteriophage. *Proc. Natl. Acad. Sci. 41:* 344.

Sarabhai, A. S., A. O. W. Strelton, S. Brenner, and A. Bolle, 1964. Co-linearity of the gene with the polypeptide chain. *Nature 201:* 13.

Yanofsky, C., and V. Horn. 1972. Tryptophan synthetase α-chain positions affected by mutations near the ends of the genetic map of trp A of *E. coli. J. Biol. Chem 247:* 4494–4498.

——, B. C. Carlton, J. R. Guest, D. R. Helinski, and V. Henning. 1964. On the colinearity of gene structure and protein structure. *Proc. Natl. Acad. Sci. U.S. 51:* 266–272.

——, G. R. Drapeau, J. R. Guest, and B. C. Carlton. 1967. The complete amino acid sequence of the tryptophan synthetase A protein (α subunit) and colinear relationship with the genetic map of the A gene. *Proc. Natl. Acad. Sci. U.S. 57:* 296–298.

27

Copyright © 1961 by the Biological Laboratory, Long Island Biological Association, Inc.

Reprinted from *Cold Spring Harbor Symp. Quant. Biol.*, **26**, 11–23 (1961)

The Effects of Mutation on the Composition and Properties of the A Protein of *Escherichia coli* Tryptophan Synthetase

CHARLES YANOFSKY, DONALD R. HELINSKI AND BARBARA D. MALING

INTRODUCTION

Only a little more than ten years ago the first case was described in which the mutationally induced loss of ability to catalyze a specific biochemical reaction was associated with the absence of the enzyme normally required for the reaction (Mitchell and Lein, 1948). More recently, it has been shown that mutational changes lead to many different types of enzyme alterations and that mutant proteins resembling the normal enzyme are often formed (Fincham, 1959; Yanofsky and St. Lawrence, 1960). With the development of techniques that can distinguish a mutant protein from the normal where the mutant protein differs by a single amino acid substitution (Ingram, 1958), it has become possible to study the effects of mutation in terms of alterations in the primary structure of specific proteins. Ingram and subsequent workers (Ingram, 1961) have, in fact, shown that several inherited abnormal forms of human hemoglobin differ from normal adult hemoglobin by single amino acid substitutions. Genetic methods have also been highly developed and now permit the mapping of mutational sites within genes and the ordering of these mutational sites with respect to one another. Furthermore, there is reason to believe that mutations often involve single nucleotide changes in DNA and that the smallest recombinable unit may be the nucleotide (Benzer, 1957; Benzer and Freese, 1958; Freese, 1959). These developments, plus advances in our understanding of the structure and composition of genetic material, have led to the hypothesis that the linear sequence of nucleotides in a DNA segment determines the linear sequence of amino acids in the corresponding protein. According to this concept the order of mutationally altered sites in a gene should correspond directly to the order of the positions at which amino acid substitutions occur in the corresponding mutant proteins. In view of the importance of this concept and the obvious way to examine it, several gene-protein systems are being studied at the present time as possible model test systems. Our recent work has been concerned with one of these systems, the A protein of the tryptophan synthetase of *Escherichia coli*, and the structural gene that controls its synthesis. The following report represents our present understanding of the effects of different types of mutations on this protein.

CHARACTERISTICS OF THE TRYPTOPHAN SYNTHETASE SYSTEM

The tryptophan synthetase system of *E. coli* consists of two protein components designated A and B (Crawford and Yanofsky, 1958; Yanofsky, 1959). These components may be separated by chromatography on ion exchange resins or by mild acidification, which precipitates only component B. Together these two proteins catalyze the following three reactions (Crawford and Yanofsky, 1958):

(1)   indole + L-serine → L-tryptophan

(2) indoleglycerol phosphate ⇌

indole + triose phosphate

(3) indoleglycerol phosphate + L-serine →

L-tryptophan + triose phosphate

Various lines of evidence suggest that reaction (3) is the physiologically important reaction and that free indole is not an intermediate in this reaction (Crawford and Yanofsky, 1958; Yanofsky and Rachmeler, 1958). Either component can be quantitatively assayed in any one of the three reactions in the presence of an excess of the second component. The three reactions have characteristic relative rates with either component limiting; these are given in Table 1. Component A in the absence of B has trace activity in reaction (2) and is inactive in the other two reactions. Component B is also slightly active by itself, but only in reaction (1). To obtain maximum reaction rates in any of the reactions, both components must combine with one another (Crawford and Yanofsky, 1958).

Wild-type strains of *E. coli* normally produce low levels of the A and B components of tryptophan synthetase. The A protein constitutes about 0.05 per cent of the extractable protein. Tryptophan auxotrophs, on the other hand, when grown on low levels of indole or tryptophan, produce approximately 20 to 50 times as much of these proteins (Yanofsky and Crawford, 1959). Presumably, repression is minimized when the endproduct level is low and, therefore, the mutants

TABLE 1. RELATIVE RATES OF THE TRYPTOPHAN SYNTHETASE REACTIONS

Preparation	Reaction			
	In → Tryp	InGP → Tryp	InGP → In	In → InGP
Component A (excess B)	100*	43	3.4	2.6
Component B (excess A)	100*	40	3.2	2.7
Component A alone	<0.2	<0.02	—	.02
Component B alone	3	<.1	—	<.001

* The indole → tryptophan reaction arbitrarily set at 100. In = indole; InGP = indoleglycerol phosphate; Tryp = tryptophan.

TABLE 2. AMINO ACID COMPOSITION OF THE A PROTEIN

	moles/ mole protein		moles/ mole protein
Lysine	14	alanine	41
Histidine	4	valine	19
Arginine	12	methionine	5
Aspartic (+ amide)	24	isoleucine	20
Threonine	10	leucine	28
Serine	12	tyrosine	7
Glutamic (+ amide)	31	phenylalanine	12
Proline	20	cysteic (as cystine or cysteine)	3
Glycine	20	tryptophan	0

form higher levels of enzyme. Most of the other enzymes in the tryptophan pathway are also formed in large amounts when auxotrophs are grown on limiting levels of indole or tryptophan. The A protein can be easily purified with good yields, the final steps involving chromatography on DEAE-cellulose and crystallization (Henning et al., in preparation). The purified material is homogeneous electrophoretically and in the ultracentrifuge. Sedimentation studies with highly purified A protein have led to an estimated molecular weight of 29,500 (Henning et al., in preparation). Amino acid analyses (Table 2) are consistent with this figure and have shown that all the common amino acids with the exception of tryptophan are present in the A protein. It is interesting that a protein concerned with the biosynthesis of tryptophan lacks this amino acid. The B protein of tryptophan synthetase is relatively labile when highly purified and has not been studied intensively.

Mutant strains with defects in the A or B protein are readily detected as tryptophan auxotrophs (Yanofsky and Crawford, 1959) and may be isolated by penicillin selection. Mutants with defective A proteins respond to either indole or tryptophan since the B protein they form can convert indole to tryptophan. B mutants, on the other hand, grow only on tryptophan. Enzyme activity measurements with extracts of the various A and B mutants permit the division of the A and B mutants into two sub-groups (Yanofsky and Crawford, 1959). Members of one of the sub-groups form a protein which is enzymatically active with the other normal component in one of the three reactions, while members of the other sub-group do not. All the altered B proteins (B mutants) are enzymatically inactive by themselves but will combine with component A and catalyze the conversion of indoleglycerol phosphate to indole. Similarly, all the mutationally altered A proteins, which are enzymatically inactive by themselves, combine with normal B; and the complex converts indole to tryptophan. Immunological studies are consistent with these findings —whenever an altered protein is detected enzymatically it can also be detected with the appropriate antiserum (Yanofsky and Crawford, 1959). On the other hand, mutants which lack any A or B activity also lack a protein that will react with anti-A or anti-B serum. Antibodies to the A or B protein are specific for the corresponding protein.

The fact that all the altered A and B proteins are enzymatically active with the normal second component is perhaps surprising, but is consistent with one interpretation of the nature of the A-B interaction. It is assumed that each of the proteins is primarily responsible for the catalysis of one of the reactions: A catalyzing reaction (2), and B catalyzing reaction (1). These are the same reactions which proceed at low rates with each protein alone. It is further assumed that each of the proteins activates the other with respect to the reaction the latter protein can perform by itself. Thus the combination of normal or altered B protein with normal A activates A in the indoleglycerol phosphate ⇌ indole reaction, while the combination of normal or altered A protein with B activates B in the indole → tryptophan reaction. This activation can be visualized simply as alterations of the surfaces of the two proteins as a result of association. Presumably, the new surface configuration created by the combination of the two normal proteins permits catalysis of reaction (3). Considering the types of mutants that have been obtained in the light of this interpretation, one would conclude that mutations which affect the catalytic activity of either of the proteins may not affect the ability to combine with the other protein. One might expect to recover some mutants in which one of the proteins is incapable of combining with the second component. However, if alterations affecting association do not affect activity, the inherent activity of the separate proteins would probably permit limited growth in the absence of tryptophan, assuming, as is likely, that such mutants would not be repressed.

Genetic studies using transducing phage Plkc have been performed with many ultraviolet-induced A and B mutants, and it appears that the A and B genes are located adjacent to one another on the *E. coli* chromosome (Yanofsky and Lennox, 1959; Yanofsky and Crawford, 1959). To date no mutant has been discovered which maps in one region and affects the other protein, but this possibility has not been rigorously examined. Several mutants with deletions including the A and B genes have been studied (Yanofsky and Lennox, 1959). As expected, these lack any enzymatically or immunologically detectable A or B protein (Yanofsky and Crawford, 1959).

Somewhat more extensive mapping has been performed with the A mutants and a cluster map of the strains that have been studied is shown in Fig. 1. The upper mutants form altered A proteins while the mutants on the lower half of the chart do not produce a detectable protein. The various mutants have been grouped into clusters, preliminary to final ordering of the mutationally altered sites. Each cluster includes those mutants that give less than 0.1 per cent recombination with one another in appropriate transduction tests. The total length of the A gene is about 2.5 map units (Maling and Yanofsky, 1961). Three-point tests are being performed to establish the order of the clusters and the order of the recombinable mutants within each cluster. Some of the A mutants give unexpectedly high recombination values, especially mutants A1 and A23 (and the others resembling A23).

To establish whether or not mutational changes which map in the same cluster represent repeat mutational events at the same site, transduction tests are carried out with *cys-tryp* double mutants as recipients (Maling and Yanofsky, 1961). The cysteine marker employed in these experiments is linked to the A region. The use of these double mutants as recipients eliminates reversion as a limiting factor in tests of genetic identity. This permits the detection of recombination frequencies down to a level where one might expect to distinguish between altered sites which are one or a few nucleotides apart. With this test it was shown that mutants A1, 3, 11, 26, 33, 37, 41, 45, and 48 do not recombine with one another and that the same is true for mutants A23, 24, 27, 28, 35, 36, and 53.

PROPERTIES OF THE MUTATIONALLY ALTERED A PROTEINS

Although quantitative enzymatic activity measurements and immunological tests failed to reveal significant differences between the altered A proteins (Maling and Yanofsky, 1961), differences were detected in other tests. The normal A protein is not precipitated when extracts are acidified to pH 4.0. The A proteins of six of the A mutants (A11, 26, 37, 41, 45, 48) were precipitated, but not inactivated, at this pH (Maling and Yanofsky, 1961). Apparently the mutational changes in these strains have affected both the solubility of the A proteins and their enzymatic activities. The A proteins of the other A mutants were not precipitated at pH 4.

Heat inactivation was also employed in an attempt to distinguish altered A proteins from one another (Maling and Yanofsky, 1961). The acid-precipitable A11 protein mentioned previously and the A protein from one other mutant, strain A34, were approximately as heat-sensitive as the normal A protein. On the other hand, the A proteins produced by seven other mutants (A23, 24, 27, 28, 35, 36, 53) were very labile to heating; only 5 to 15 per cent of the original A activity remained under conditions where 60 to 70 per cent of the wild-type A protein was stable to heating. In addition, there was one strain, A1, which formed a moderately heat-labile A protein. Three of the mutant proteins, those produced by strains A3, A33 and A46, were more stable to heating than the normal A protein. Various mixing experiments were performed with crude and partially purified preparations of the mutant A proteins, and in all cases the stability or lability seemed to be characteristic of the mutant protein itself and did not appear to be due to interactions with some other component of the preparation. Thus it is apparent that mutations in the A

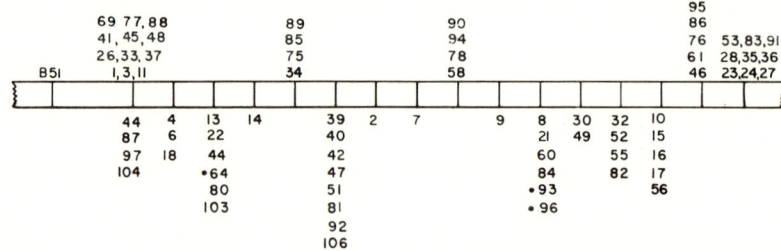

FIGURE 1. Cluster map of the A gene. The mutants listed on the upper half of the chart form altered A proteins while those on the lower half do not. The mutants marked • may form small amounts of A protein; they require further testing. Several of the clusters in the figure map close to one another, and three-point tests will be required to establish their relative order.

gene lead to the production of distinguishable altered A proteins.

Representatives of each of the different A protein types were examined in affinity tests to determine whether mutation affected the ability to combine with the B protein. In all cases the altered A protein had approximately the same affinity for the B protein as the normal A protein. This finding supports the interpretation mentioned previously that defective A proteins which are enzymatically active with the B protein are capable of combining with this protein.

One additional difference has been detected—of all the A mutants only strain A1 forms less A protein than B protein. Generally, the A-B ratio is between 1:1 and 2:1; in mutant A1 it is 1:2 (Maling and Yanofsky, 1961). Furthermore, this mutant forms much less A and B protein than any of the other mutants, although grown under the same conditions. It is apparent that the mutational change in strain A1 affects the rates of synthesis of the A and B proteins, as well as the structure of the A protein.

CORRELATION OF THE PROPERTIES OF ALTERED PROTEINS WITH THE GENETIC MAP

The map locations of the altered sites in the mutants discussed in the previous section are given in Fig. 1. All the mutants which form heat-labile proteins (A23 group) with approximately the same heat-lability map at the same site, at one end of the A region. Mutant A46 also maps at this end of the A region and forms an A protein which is slightly more heat-resistant than the normal A protein. At the other end of the A gene are located all the mutants (A11 group) which form acid-precipitable A proteins. In addition, two mutants, A3 and A33, which form heat-resistant proteins, and one mutant, A1, which forms a moderately heat-labile protein, map at the A11 site. Mutants A3 and A33 are clearly different from each other since A 3 reverts while A33 does not. Furthermore, a suppressor of A3 does not suppress A33.

The finding of four different mutant types, A1, A3, A11 and A33, which map at the same site is contrary to expectations. However, strain A33 does not revert; it may have mutational damage more extensive than a single nucleotide substitution. One mutant, strain A34, which maps near the A11 site, forms a protein which is indistinguishable from the normal protein in stability tests. Mutant A85 maps in the same cluster as A34 but grows slightly on minimal medium and thus must form a different type of altered A protein. Mutant A58 and other strains which map at or near the same site have not, as yet, been examined in stability tests; these mutants map somewhere near the center of the A gene. The mutational alterations which result in the absence of a detectable A protein are scattered throughout the A region.

It would appear from the limited sample analyzed to date that mutational alterations which lead to detectable altered proteins tend to be localized at the ends of the A gene. Whether this apparent localization has any meaning in terms of the function of the corresponding portions of the protein is, of course, not yet known. The genetic localization of mutant sites which lead to specific types of altered proteins has also been reported for the tryptophan synthetase of *Neurospora crassa* (Bonner, Suyama, and DeMoss, 1960).

Of the first 100 ultraviolet-induced A mutants which were isolated, only 32 formed an immunologically cross-reacting A protein. Most of the B mutants (38 out of 52), however, do form an altered B protein. Among the *Neurospora crassa* tryptophan synthetase mutants, 23 of the first 25 examined formed a detectable tryptophan synthetase-like protein (Suskind et al., 1955; Rachmeler, 1960). The estimate of altered protein producers is probably low, and many of the mutants now classified as non-protein formers probably produce a protein which will be detected when more sensitive detection techniques are developed. These figures have some bearing on the type of genetic code employed in DNA. Most of the reasonable genetic codes that have been proposed predict that a large percentage of mutations will lead to nonsense coding units and consequently the absence of a specific protein (Crick et al., 1957; Golomb et al., 1958). Although the data quoted above is not corrected for repeat events at the same site, it would appear that in some genes nonsense mutations are much more infrequent than would be expected on the basis of any of these codes. Perhaps we should consider more seriously the possibility that several different nucleotide sequences code for each amino acid.

DOUBLE MUTANTS

Strains carrying more than one mutational alteration in the A gene have been prepared by transduction using the various A mutants described previously (Maling and Yanofsky, 1961). The identity of each double mutant was established by transduction experiments using phage grown on each of the single parental mutants. Double mutants were prepared between various strains, both of which formed altered proteins; only one of which formed A protein; and neither of which formed a detectable A protein. These double mutants were then examined for A protein in enzymatic and immunological tests. Only when both parents were capable of forming A protein was A protein detected in the corresponding double mutant (Table 3) (Maling and Yanofsky, 1961). Double mutant proteins sometimes showed the properties of both parental proteins. For example, a double mutant derived from a parental strain which formed an acid-precipitable A protein and a parental strain which formed a heat-labile protein, formed an A protein which was both heat-labile and acid-precipitable. Furthermore, this protein was as heat-labile and acid-precipitable as the A proteins of the parental strains. On the other hand, when double mutants were prepared between parental strains which formed a heat-resistant protein and a heat-labile protein, respectively, the A protein produced was heat-resistant. In several cases the heat-stability of the double mutant protein was intermediate to that of the two parental proteins and in one case (double mutant A1–34) the A protein formed was considerably more heat-labile than the A protein of either of the parental strains. It is clear that various types of interactions occur in proteins bearing two mutational alterations, leading in most cases to a protein phenotype which is distinguishable from that of either of the parental strains (Maling and Yanofsky, 1961). Quantitative immunological tests were performed with many of the double mutant proteins, and in some cases it appeared that the double mutant protein did not combine as well with anti-A serum as the normal A protein (Maling and Yanofsky, 1961). Since the A:B ratio is generally very low in extracts of such double mutants, it is also possible that this result is due to the presence of some enzymatically inactive A protein which can combine with antibody.

FINGERPRINTING STUDIES WITH MUTANT AND NORMAL A PROTEINS

Fingerprinting studies have been performed with many of the mutant A proteins (Helinski and Yanofsky, unpublished). A portion of a proteolytic digest of an A protein is pipetted on the corner of a sheet

TABLE 3. A-PROTEIN PRODUCTION BY DOUBLE MUTANTS

Parental Strains	Form A-Protein	
	Parental Strains	Double Mutant
1; 23	+; +	+
3; 23	+; +	+
11; 23	+; +	+
33; 23	+; +	+
34; 23	+; +	+
3; 46	+; +	+
11; 46	+; +	+
33; 46	+; +	+
34; 46	+; +	+
11; 17	+; −	−
14; 17	−; −	−
23; 13	+; −	−
46; 2	+; −	−
46; 14	+; −	−

TABLE 4. FINGERPRINTING RESULTS WITH MUTANT PROTEINS*

Protein Examined	Digest	Observation (compared with wild type)
A23	T, T + C	change in position of one peptide
A27	T, T + C	change in position of one peptide
A35	T, T + C	change in position of one peptide
A36	T, T + C	change in position of one peptide
A46	T, T + C, C	no change detected
A3	T, T + C, C	a possible change, not clear
A11	T + C, C	one additional peptide
A26	T + C, C	" " "
A33	T + C, C	" " "
A34	T, T + C, C	no change detected
A75	T, T + C, C	" " "

* Unpublished data of Helinski and Yanofsky. T = trypsin; C = chymotrypsin; T + C = trypsin + chymotrypsin.

of filter paper, chromatographed in one direction, dried, and then subjected to high voltage electrophoresis in the other direction. Tryptic digests of the A protein yield approximately 25 intense and clearly separated peptide spots by this method. Since there are 14 lysine and 12 arginine residues in the A protein, the number of peptides obtained is only slightly lower than the expected theoretical number, 27. The discrepancy is in part due to the fact that several tryptic peptides remain at the origin. The peptide pattern obtained with the wild-type A protein is highly reproducible and does not vary appreciably with the time of digestion. In addition to the 25 major peptide spots there are usually several minor peptide spots. These can be partly attributed to trace chymotrypsin activity generally associated with highly purified trypsin. Chymotrypsin or trypsin + chymotrypsin digests of the A proteins similarly give reproducible peptide patterns. We have routinely examined each mutant protein following digestion with trypsin alone, chymotrypsin alone, and trypsin + chymotrypsin. The mutant proteins that have been examined to date are listed in Table 4 along with the peptide pattern change observed, if any. The altered A proteins from four mutants which form a heat-labile A protein and map at the same site have been examined in these studies. All of these mutants show the same single peptide difference from the wild-type pattern when trypsin or trypsin + chymotrypsin peptide patterns are examined. This finding, in conjunction with other similarities between these mutants, suggests that they represent repeat identical mutational events at the same site. The particular peptide involved (T + C peptide) has been isolated and analyzed from two of these mutants and from the wild-type strain. The results of these analyses showed that the mutant peptides contained phenylalanine in addition to the nine amino acids that were present in the wild-type peptide (Helinski and Yanofsky, unpublished). The sequence of the N-terminal three amino acids has been determined for both mutant and normal peptides (Carlton, Helinski, and Yanofsky, unpublished) and is identical, with the exception that in the mutant peptide the extra phenylalanine is in the N-terminal position. Our interpretation of the presence of the extra phenylalanine in the mutant peptide is based in part on the results of total amino acid analyses of the wild-type and mutant proteins. These analyses indicated that the mutant protein contained one extra arginine, but it was not possible to say which amino acid, if any, was present in reduced amounts. This extra arginine in the mutant protein would be expected to give rise to an additional trypsin-sensitive bond in the peptide chain. Our current interpretation of these findings is that in the mutant protein this arginine is adjacent to the phenylalanine in the mutant peptide (on the N-terminal side), while in the wild-type protein there is an amino acid other than arginine in this position (see Fig. 2). Thus, when both proteins are treated with trypsin, cleavage occurs on the C-terminal side of the arginine in the mutant peptide, but there is no hydrolysis of the bond on the N-terminal side of the phenylalanine in the wild-type peptide. When chymotrypsin is added it will not attack the bond adjacent to phenylalanine in the mutant peptide because this amino acid has a free α-amino group, and the phenylalanine will remain in the mutant peptide. When the corresponding wild-type trypsin peptide is attacked by chymotrypsin, phenylalanine is not N-terminal; and the bond between phenylalanine and the next amino acid will be split. Further evidence that is consistent with this interpretation is the finding that if the mutant protein is first treated with chymotrypsin and then with trypsin, the mutant peptide now appears at the same position as the wild-type peptide (Helinski and Yanofsky, unpublished). Studies are in progress to isolate the peptide on the N-terminal side of the mutant peptide and the corresponding peptide from the wild-type protein to determine the exact amino acid change. The other mutant which maps in this region, strain A46, forms a heat-resistant A protein. Digests of this protein do not show any peptide difference from the normal wild-type pattern. The peptide corresponding to the altered peptide in mutant A23 has been isolated from the mutant protein and found to have the same amino acid composition as the corresponding wild-type peptide.

The A proteins of two mutants which form acid-precipitable A proteins have been examined, and both mutant fingerprints contain one peptide in addition to those observed with the wild-type protein (Helinski and Yanofsky, unpublished). The heat-resistant proteins formed by mutants which map at the same site have also been examined, and the peptide pattern of one of these mutants has the identical change observed with the acid-precipitable proteins, that is, one extra peptide present in a chromatographically high position. Amino acid analyses have been performed on this extra peptide from both types of mutants, and the peptides have the same amino acid composition (Helinski and Yanofsky, unpublished). These results suggest that in both mutant types the amino acid change resulting in the appearance of this extra peptide probably was at the C-terminal position of the N-adjacent peptide. The other heat-resistant protein does not show any clear-cut peptide pattern difference from the wild-type pattern. Further studies with the acid-precipitable proteins and the heat-resistant protein which gives the extra peptide have shown that the same tryptic peptide fraction (obtained by paper electrophoresis) gives rise to the additional chymotrypsin peptide characteristic of these proteins. It appears likely, therefore, that amino acid changes have occurred at the same position in the A proteins of these two mutant types, giving different amino acids, each one of which adds a new chymotrypsin-sensitive bond in the protein (see Fig. 2). Total amino acid analyses of the A protein of one of these mutants (strain A33) showed the presence of an extra methionine residue. Since methionine bonds are often split by chymotrypsin, this may be the amino acid substituted or added in mutant A33. Two other mutants which were examined, strains A34 and A75, do not show any definite fingerprint differences.

It is clear from our fingerprinting studies, therefore, that although peptide pattern differences are often encountered, it is rare that a single amino acid substitution can be simply and unequivocally established.

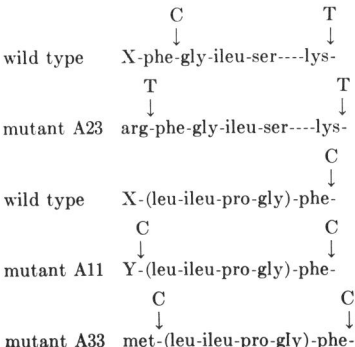

FIGURE 2. Interpretation of fingerprint differences. Trypsin is generally added first, then chymotrypsin. Bonds which should be split by trypsin or chymotrypsin are indicated by T and C, respectively.

Furthermore, in several of the mutants that have been analyzed, one would expect to find peptide differences in addition to those observed, if our interpretations of the nature of the amino acid changes are correct. It is also obvious that the fingerprinting technique will not detect all amino acid changes.

Fingerprinting analyses were carried out with double mutant proteins where the A protein of one or both of the parental types contributing to the double mutant strain showed a peptide pattern difference. These analyses were of particular importance since most of the double mutant proteins had different properties than the single mutant proteins. Thus it was possible to determine whether or not peptide pattern differences actually indicated the location of the primary structure alterations, or whether they reflected different stabilities of the altered proteins and corresponding differences in the degree or sites of digestion by proteolytic enzymes. The peptide patterns obtained with several of these double mutant proteins are shown in Fig. 3. In each case it was found that the peptide pattern differences characteristic of the proteins of both of the parental strains were present in the double mutant protein fingerprints, with no additional changes. Thus it is clear that peptide patterns reflect the corresponding primary structure changes in the proteins examined.

Incidentally, double mutants have been extremely useful as sources of mutant peptides. For example, the A protein of mutant A23 is labile and very difficult to purify. The A protein of the double mutant *A23-33* is very stable, purifies easily, and has both the A23 and A33 peptide alterations.

REVERSION STUDIES

Representatives of each of the various types of A mutants have been examined in reversion studies. Many of the mutants yield tryptophan-independent strains which grow more poorly than the wild-type strain in the absence of tryptophan, in addition to strains which are indistinguishable from the wild type. The slow-growing strains are either partial revertants, that is, strains in which the reversion occurred in the A gene, or suppressed mutants. Since partial revertants should be of considerable help in relating the genetic map to the protein map, an effort was made

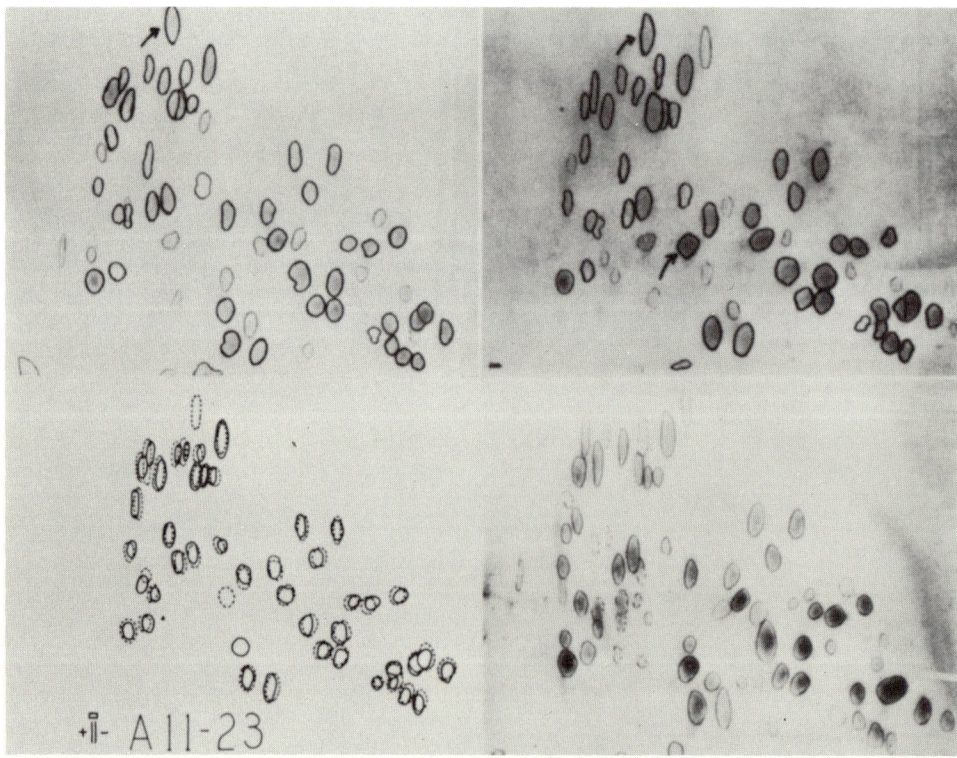

FIGURE 3. Trypsin + chymotrypsin fingerprints of the A protein of double mutants A*11-46* (upper left), A*23-33* (upper right), A*11-23* (lower left, a tracing) and the wild-type A protein. On the A*11-23* tracing the major wild-type peptides are outlined with a solid line and the A*11-23* peptides with a dashed line.

TABLE 5. Tests Used in Classifying Tryptophan-Independent Colonies

Tryp+ types	Colony size	InG accum.	5 me-Tryp sensitivity	Whole cell TSase†	Specific A	Activity B	Ratio A/B	Genetic test
A	=WT	0	=WT	=WT	=WT	=WT	1:1–2:1	FR
B	=WT	0	>WT	=WT	?	?	?	FR
C	=WT	0	>WT	>WT	?	?	?	FR
D	<WT	+	>WT	>WT	>WT	>WT	=WT	PR
E	≪WT	+	≫WT	>WT	>WT	>WT	<WT	PR
F	<WT	+	≫WT	<WT	<WT	>WT	<WT	PR
G	<WT	+	>WT	>WT	>WT	>WT	=WT	su
H	<WT	+	>WT	<WT	<WT	>WT	<WT	su

InG = indole-3-glycerol. 5 me-Tryp = 5-methyl tryptophan.
FR = full revertant (indistinguishable from wild type in growth rate and accumulation behavior).
PR = partial revertant.
† = indole to tryptophan reaction measurements with cell suspensions.

to develop a number of different tests which would detect differences between partial revertant types derived from the same mutant. The tests employed (Allen and Yanofsky, unpublished), and the revertant and partial revertant types encountered, are listed in Table 5. Also included are the two most frequently encountered suppressor types. Since many of the tests described can be performed quantitatively, it is possible to distinguish different partial revertant types within each of the groups listed.

Unfortunately, the analysis of reversion is rather complex, as can be seen from the possibilities considered in Fig. 4. Reversion may affect the nucleotide substituted in the original mutation, or different nucleotides in the same coding unit, or a nucleotide in a different coding unit. These changes would be expected to lead to amino acid substitutions at the same position in the first two cases, and at a different position if the reversion were in a separate coding unit. Genetic tests should distinguish the last case from the other two, but probably would not readily distinguish between the first two possibilities. Second-step reversion experiments with base analogs as mutagens may be of help here. The use of analogs in single-step reversion experiments presumably adds some specificity to the reversion process and may also indicate the nucleotide change in the original mutational event.

The reversion patterns of many of the A mutants and the effect of 2-aminopurine on the reversion of these mutants are shown in Fig. 5. These studies are not as yet completed, and the types listed represent the minimum number of distinguishable revertants obtained from each mutant. The data presented indicates that 2-aminopurine increases the reversion rate of about one-half of the ultraviolet-induced A mutants that have been examined. In general, mainly the number of wild-type-like colonies is increased, although in one case (mutant A7) 2-aminopurine also increased the frequency of suppressor mutations. As yet, 2-aminopurine has not led to an increase in any partial revertant type. It should also be pointed out that of the cluster of mutants which map at the same site, A1, A3, A11 and A33, the reversion frequency of only A1 is increased by aminopurine. Presumably only this mutant, of the several at this site, was formed by a GC → AT transition (Bautz and Freese, 1960).

Four of the partial revertants have been studied

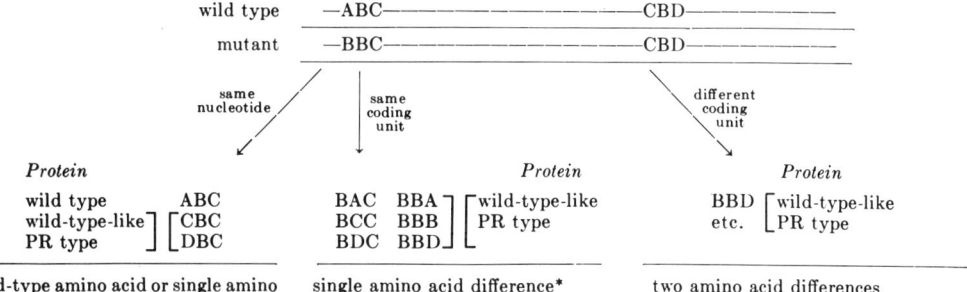

FIGURE 4. Reversion analysis. Possible events at the nucleotide and amino acid levels.

* Assuming that for each amino acid there is a unique nucleotide sequence.

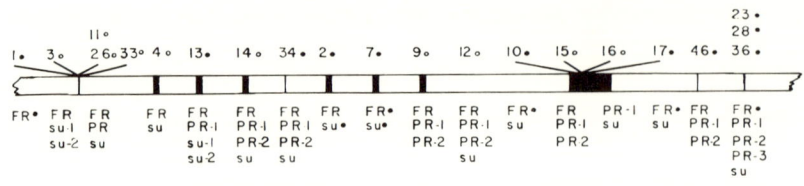

FIGURE 5. Reversion patterns of A mutants. Based on the work of Allen and Yanofsky, unpublished. FR = full revertant; PR = partial revertant.

more fully, and two of these have been examined by the fingerprinting method. Table 6 presents the results of enzymatic and immunological tests with the A proteins of these partial revertants. It is apparent that they are all abnormal. The A proteins in these strains are also distinguishable from the corresponding parental proteins in stability. This is shown in Fig. 6 where the results of heating experiments with the parental and partial revertant proteins are presented.

The A proteins from both A46 partial revertants were fingerprinted, and a single peptide difference was detected in tryptic + chymotryptic digests of the A protein of strain A46PR8 (Helinski and Yanofsky, unpublished). The mutant peptide contains one mole each of leucine, serine, and arginine, while the corresponding wild-type peptide contains these amino acids and, in addition, a second mole of leucine (Helinski and Yanofsky, unpublished). Whether there is one less amino acid in the partial revertant protein, or whether the amino acid replacing leucine is on the C-terminal end of the peptide adjacent to the mutant peptide, is not as yet known. The amino acid compositions of the corresponding peptides from mutant A46 and the other partial revertant have also been determined, and both are identical to the wild-type peptide (Helinski and Yanofsky, unpublished). Clearly, therefore, the amino acid change associated with the reversion in A46PR8 is not in the peptide presumably altered in mutant A46. This would suggest that A46PR8 is a second-site revertant. Genetic tests have been performed to examine this possibility and have led to the recovery of mutant A46 from this partial revertant, confirming this expectation. It is apparent, therefore, that in strain A46PR8, interactions between different portions of the A protein molecule result in a catalytically effective protein, although the protein is very inefficient (see Table 6). Second-site reversion has also been detected in bacteriophage (Feynman, personal communication; Jinks, 1961).

SUPPRESSOR MUTATIONS

Since suppressor mutations by definition include all genetic changes which reverse the effects of mutational alterations at other loci, it is not surprising that they have been found to act in a variety of ways (Fincham, 1960; Yanofsky and St. Lawrence, 1960). Relatively few instances of suppression have been examined at the enzyme level, and as a result we know little of the effects of suppression on specific proteins. In one of the best analyzed cases of suppression it has been concluded that the effect is indirect and that it does

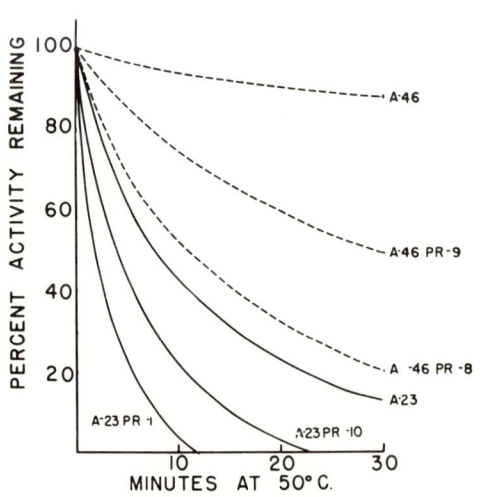

FIGURE 6. The heat stability of the A proteins from mutants A23 and A46 and their partial revertants.

TABLE 6. EXAMINATION OF THE A PROTEINS FROM PARTIAL REVERTANTS

Strain	A specific activity In → Tryp	Activity Ratio In → Tryp/ InGP → Tryp	Relative antibody inhibition
wild type	2.1	2.5	100%
A-23 PR-1	2.3	187	36%
A-23 PR-10	32	790	92%
A-46 PR-8	32	4200	92%
A-46 PR-9	95	4300	86%

not involve an alteration of the mutant protein (Suskind, 1957). Nevertheless, suppressor mutations remain a very interesting and important category since they represent the most obvious possible exception to the rule that the primary structure of a single polypeptide chain is determined by only one genic region.

In our recent studies on suppressor gene action (Brody and Yanofsky, in preparation) we have directed our attention to three questions. What are the normal functions of suppressor genes? Do suppressed mutants form specific proteins which differ from the corresponding mutant proteins? Does suppressor action occur before, during, or after the synthesis of specific proteins? In attempting to answer the first of these questions a large number of A mutants were examined (A1, 2, 3, 4, 7, 9, 10, 11, 12, 13, 14, 15, 17, 23, 33, 34, 46), only some of which were previously found to be suppressible. Suppressor mutations (and reversions) were induced in these strains by ultraviolet light, but instead of plating on the usual minimal medium, the irradiated suspensions were plated on a medium supplemented with all of the common amino acids except tryptophan, a mixture of B vitamins, and a nucleic acid hydrolysate. The $tryp^+$ colonies which appeared due to reversion or suppression were then replicated to minimal medium and the plates examined for colonies which could only grow on the supplemented medium. If suppressor genes are the genes controlling biosynthetic reactions involved in the synthesis of amino acids, vitamins, etc., one would expect to find that some of the suppressed mutants would be auxotrophs. No auxotrophs were found among the $tryp^+$ colonies although many of the colonies that appeared contained suppressed mutants (Brody and Yanofsky, in preparation). These findings suggest that most of the allele-specific suppressor genes that affect the different A mutants are concerned with cellular functions other than the biosynthesis of small molecules.

TABLE 7. THE ACTIVITY OF THE A PROTEINS FROM SUPPRESSED MUTANTS

Strain	A protein specific activity		InGP → Tryp / In → Tryp per cent
	In → Tryp	InGP → Tryp	
wild type	2.5	1	40
A36	22	0	—
A36su	6	0.25	4.2
A3	35	0	—
A3su	44	0.19	0.44
A11	31	0	—
A11su	55	0.48	0.87

The second question has been examined with suppressed mutants derived from A mutants that form characteristically altered A proteins. If suppression affects the primary structure of the A protein, one might expect to find a suppressed A protein with properties different than those of the mutant A protein. If suppression affects only a fraction of the A protein molecules, then one might expect to find two types of A protein in a suppressed mutant, the A protein of the original unsuppressed mutant and another A protein species. Fortunately, as can be seen from the data presented in Table 7, extracts of the suppressed A mutants have a label which can be used to identify the suppressed mutant protein. Unlike the A proteins of the unsuppressed mutants, the A protein fraction of suppressed mutants is capable of a limited conversion of indoleglycerol phosphate to tryptophan in the presence of the normal B component. This activity accounts for the growth of suppressed mutants in the absence of tryptophan. The A protein fractions from the suppressed mutants listed are fully active in the indole to tryptophan reaction.

These findings lead to the question—does the slight activity in the indoleglycerol phosphate to tryptophan reaction indicate that all the A protein molecules are considerably less active in this reaction (as in the partial revertants), or is there a mixture of A protein types in the suppressed mutants? The answer to this question was obtained by following the loss of enzymatic activity in the two reactions under conditions that were known to inactivate the A protein of the unsuppressed parents. If the indoleglycerol phosphate to tryptophan activity of the suppressed mutants was associated with a small amount of a second A protein species, this activity might be expected to have different stability properties than the majority of the A protein in these extracts. The mutants selected for this study form A proteins with distinctive properties —one forms a heat-labile A protein (A36), another a heat-stable A protein (A3), and a third an acid-precipitable A protein (A11). When the A proteins from the various suppressed mutants were examined in stability tests, it was found that the indole to tryptophan activity was associated with an A protein type that had the same stability as the A protein of the original unsuppressed mutants. In all cases, however, the A protein bearing the indoleglycerol phosphate to tryptophan activity was inactivated at a different rate—a rate characteristic of the wild-type protein (Brody and Yanofsky, in preparation). Furthermore, it was also shown that after prolonged heating of the A protein fraction from A36su the remaining activity had the wild-type ratio of the two enzymatic activities. Thus it appears that each of the three suppressed mutants examined forms two A proteins, one with the physical characteristics of the A protein of the parental unsuppressed mutant, and the other with the physical properties of the wild-type A protein. In other studies with two of these suppressed mutants it was possible to separate the two A proteins by chromatography on DEAE-cellulose columns (Crawford and Yanofsky, 1958; Brody and Yanofsky, in preparation). When enough of the wild-type-like component has been isolated, it will be examined in fingerprinting studies to determine if it actually has the wild-type primary structure.

Assuming that a small amount of wild-type protein is produced by the suppressed mutants, we are confronted with the problem of deciding whether this protein species is formed by a reaction between the mutant protein and some other cellular component or is synthesized with this configuration. Two approaches have been employed in attempting to answer this question. First, suppressor genes which affect altered sites at the ends of the A gene have been tested on other mutants that map close to these sites. If there is some exchange mechanism which operates at the ends of the A protein, one might expect these suppressors to be effective with many mutants altered at these ends. No cross-suppression was observed; the suppressors were allele-specific. Experiments were also performed in which the ratio of the activities in the two reactions, indole → tryptophan and indoleglycerol phosphate → tryptophan, was determined under conditions of tryptophan repression. It was reasoned that if suppression involves a bimolecular reaction between the mutant A protein and some cellular constituent, then lowering the concentration of the mutant A protein should increase the proportion of the wild-type-like A protein (especially since the data presented in Table 7 suggests that the mutant protein is in great excess). The results of these experiments showed that the characteristic ratio of the two activities in each of the three suppressed mutants examined was not appreciably affected by repression (Brody and Yanofsky, in preparation). Thus it would appear that these cases of suppression do not involve interactions between mutant proteins and other cellular constituents but probably reflect a limited synthesis of the wild-type-like protein.

A number of schemes have been considered as explanations for these observations, and those that we favor at this time are presented in Fig. 7. There is, of course, no direct experimental data as yet which supports any of the explanations given. They are offered within the framework of what seems reasonable on the basis of present knowledge. The possibilities under consideration assume that suppressor genes are either the genes that determine the primary structure of the amino acid-activating enzymes or the genes that determine the composition of sRNA's and thus their ability to combine with specific activating enzymes. In either case a suppressor mutation would cause a specificity change which would lead to a mistake in the attachment of a given amino acid to its sRNA. In the example cited the mutant protein would differ from the normal in having arginine at a position normally occupied by glycine. The suppressor mistake would involve the attachment of glycine to some of the sRNA molecules which are normally specific for arginine. This error could only occur infrequently, for it is almost certain that a high mistake level for any amino acid would be lethal. It is of interest in this regard that the suppressor genes we have studied

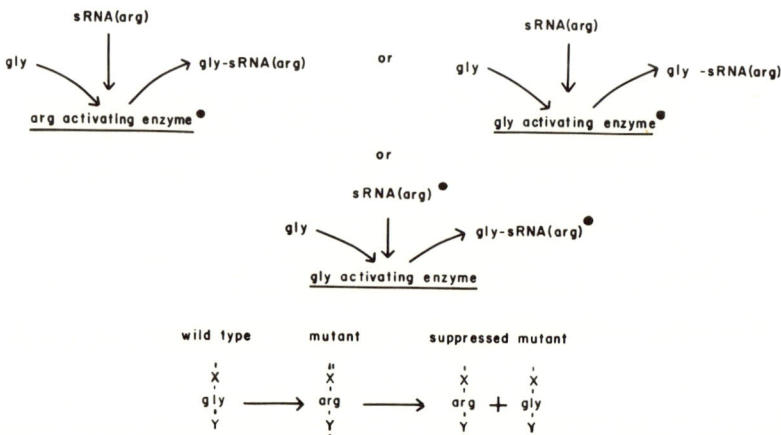

FIGURE 7. Hypothetical schemes to account for the results of the suppressor studies. ● = mutationally altered enzyme or sRNA. It is assumed that the altered enzyme or sRNA performs its normal function most of the time but can also make the above-mentioned mistakes.

never detectably alter more than 10% of the A protein molecules. Similar interpretations of suppression have been considered by other investigators (M. Lieb and L. Hertzenberg, personal communication; Benzer and Champe, 1961). Benzer and Champe have shown that different groups of T4rII mutants will grow in certain K-12(λ) strains of *E. coli* but not in others. This observation has been interpreted in much the same manner as we have interpreted suppressor mutations. Obviously, serious consideration of the possibilities offered depends on the demonstration that suppression does affect primary structure, and, hopefully, this will be decided in the near future.

CONCLUSIONS

The studies described in this paper demonstrate that mutational changes at several sites in the A gene lead to the formation of altered A proteins. Many of these altered A proteins are distinguishable from one another in addition to differing from the wild-type protein. Furthermore, distinguishable altered proteins are formed by several mutants which map at the same site. The altered A proteins produced by all of these strains were examined in peptide pattern studies, and a number of differences were detected. No more than a single peptide difference was observed with any of the mutant proteins, however, suggesting that in each case the mutational event led to a localized alteration in the primary structure of the A protein. Total amino acid analyses performed on most of the mutant proteins described here also indicate that the amino acid changes in the mutant proteins are slight indeed (Helinski and Yanofsky, unpublished). The peptide pattern studies performed with the double mutant proteins support the conclusion that primary structure alterations are localized and further show that the patterns obtained reflect the specific mutationally altered genetic sites which are combined in the double mutants. Together, the findings described leave no doubt that the A gene determines the primary structure of the A protein.

The number of mutants examined and the peptide alterations detected are insufficient to seriously test the concept of collinearity of amino acid sequence and nucleotide sequence. However, the fact that mutants A11 and A33 mapped at the same site and formed distinguishable A proteins which appeared to have amino acid changes in the same position certainly suggests that there is a correlation between the region on the genetic map at which a mutation occurs and the amino acid position that will be affected. The fact that several indistinguishable mutational changes at the same site lead to altered proteins with identical properties and with identical peptide pattern differences is also consistent with expectations.

In no case has it been possible as yet to demonstrate a single amino acid substitution in a mutant A protein. In most cases the amino acid changes in the mutant proteins examined result in a new point of cleavage by one of the proteolytic enzymes employed. This considerably complicates the analysis of the altered proteins. Our present data do not exclude the possibility that some of the altered A proteins differ from the normal by the addition or loss of a single amino acid rather than a substitution of an amino acid.

Reversion studies with selected A mutants have given a wealth of material for further study. The A proteins from two partial revertants derived from the same mutant were examined in peptide pattern studies, and one of them was found to have an amino acid change in a peptide which is unaltered in its composition in the original mutant. This finding suggests that the secondary mutational event in this strain occurred at a second site, a conclusion supported by genetic tests.

The suppressor studies demonstrate that suppression leads to the formation, in the suppressed mutants, of a small amount of a second A protein which resembles the wild-type A protein in several properties. It appears that this second protein is not formed by the interaction of the mutant protein with some other cell constituent but that its final structure is determined at the time of synthesis. Whether suppression does, in fact, restore the wild-type primary structure remains to be determined. If this is the case it will be interesting to test the explanations of suppressor gene action which were mentioned.

Acknowledgment

The work described in this paper was supported by grants from the National Science Foundation and the U. S. Public Health Service.

REFERENCES

BAUTZ, E., and E. FREESE. 1960. On the mutagenic effect of alkylating agents. Proc. Nat. Acad. Sci. U.S., *46:* 1585–1594.

BENZER, S. 1957. The elementary units of heredity. pp. 70–93. *A Symposium on the Chemical Basis of Heredity.* Baltimore: The Johns Hopkins Univ. Press.

BENZER, S., and S. P. CHAMPE. Modifications of *Escherichia coli* affecting the expression of specific phage mutations. Proc. Nat. Acad. Sci. U.S., in press.

BENZER, S., and E. FREESE. 1958. Induction of specific mutations with 5-bromouracil. Proc. Nat. Acad. Sci. U.S., *44:* 112–119.

BONNER, D. M., Y. SUYAMA, and J. A. DEMOSS. 1960. Genetic fine structure and enzyme formation. Federation Proc., *19:* 926–930.

CRAWFORD, I. P., and C. YANOFSKY. 1958. On the separation of the tryptophan synthetase of *Escherichia coli* into two protein components. Proc. Nat. Acad. Sci. U. S., *44:* 1161–1170.

———, ———. 1959. The formation of a new enzymatically active protein as a result of suppression. Proc. Nat. Acad. Sci. U.S., *45:* 1280–1288.

CRICK, F. H. C., J. S. GRIFFITH, and L. E. ORGEL. 1957.

Codes without commas. Proc. Nat. Acad. Sci. U.S., *43:* 416–421.

FINCHAM, J. R. S. 1959. The biochemistry of genetic factors. Ann. Rev. Biochem., *28:* 343–364.

———. 1960. Genetically controlled differences in enzyme activity. Advances in Enzymology, *22:* 1–43.

FREESE, E. 1959. The difference between spontaneous and base-analogue induced mutations of phage T4. Proc. Nat. Acad. Sci. U. S., *45:* 622–633.

GOLOMB, S. W., L. R. WELCH, and M. DELBRÜCK. 1958. Construction and properties of comma-free codes. Kgl. Danske Videnskarb. Selskab, Biol. Medd., *23:* 1–34.

HENNING, U., D. R. HELINSKI, F. C. CHAO, and C. YANOFSKY. Manuscript in preparation.

INGRAM, V. 1958. Abnormal human haemoglobins. I. The comparison of normal human and sickel-cell haemoglobins by "fingerprinting." Biochim. Biophys. Acta, *28:* 539–545.

———. 1961. *Hemoglobin and Its Abnormalities.* Springfield, Illinois: Charles C. Thomas.

JINKS, J. L. 1961. Heredity, in press.

MALING, B., and C. YANOFSKY. 1961. The properties of altered proteins from mutants bearing one or two lesions in the same gene. Proc. Nat. Acad. Sci. U.S., *47:* 551–566.

MITCHELL, H. K., and J. LEIN. 1948. A *Neurospora* mutant deficient in the enzymatic synthesis of tryptophan. J. Biol. Chem., *175:* 481–482.

RACHMELER, M. 1960. A study of the normal and mutationally altered forms of the tryptophan synthetase of *Neurospora*. Ph.D. Dissertation, Western Reserve University, Cleveland, Ohio.

SUSKIND, S. R. 1957. Gene function and enzyme formation. pp. 123–129. *The Chemical Basis of Heredity.* Baltimore: The Johns Hopkins Univ. Press.

SUSKIND, S. R., C. YANOFSKY, and D. M. BONNER. 1955. Allelic strains of *Neurospora* lacking tryptophan synthetase: A preliminary immunochemical characterization. Proc. Nat. Acad. Sci. U.S., *41:* 577–582.

YANOFSKY, C. 1959. A second reaction catalyzed by the tryptophan synthetase of *Escherichia coli*. Biochim. Biophys. Acta, *31:* 409–416.

YANOFSKY, C., and I. P. CRAWFORD. 1959. The effects of deletions, point mutations, reversions and suppressor mutations on the two components of the tryptophan synthetase of *Escherichia coli*. Proc. Nat. Acad. Sci. U.S., *45:* 1016–1026.

YANOFSKY, C., and E. S. LENNOX. 1959. Transduction and recombination study of linkage relationships among the genes controlling tryptophan synthesis in *Escherichia coli*. Virology, *8:* 425–447.

YANOFSKY, C., and M. RACHMELER. 1958. The exclusion of free indole as an intermediate in the biosynthesis of tryptophan in *Neurospora crassa*. Biochim. Biophys. Acta, *28:* 640–641.

YANOFSKY, C., and P. ST. LAWRENCE. 1960. Gene action. Ann. Rev. Microbiol., *14:* 311–340.

[*Editor's Note:* Material has been deleted at this point.]

Reprinted from *Proc. Natl. Acad. Sci.*, **48**(2), 173–183 (1962)

CORRESPONDENCE BETWEEN GENETIC DATA AND THE POSITION OF AMINO ACID ALTERATION IN A PROTEIN*

BY DONALD R. HELINSKI† AND CHARLES YANOFSKY

DEPARTMENT OF BIOLOGICAL SCIENCES, STANFORD UNIVERSITY

Communicated by Victor C. Twitty, December 22, 1961

The development of methods for the rapid detection of primary structure differences between proteins of similar structure has greatly facilitated studies on the effect of mutations on the primary structure of a protein.[1] Investigations with normal and abnormal forms of human hemoglobin have clearly established single amino acid differences between these proteins.[2] Similarly, some induced mutants of tobacco mosaic virus form proteins containing one to several amino acid changes.[3] Techniques for determining genetic fine structure have also been highly developed, and it is now possible to prepare detailed genetic maps of specific genes.[4] These advances in protein chemistry and genetic analysis permit an examination of the gene-protein relationship at the fine-structure levels of both gene and protein, and, accordingly, several microbial gene-protein systems are being studied at this level.[5–7] The present report is concerned with a study of one of these systems, the A gene-A protein system of the tryptophan synthetase of *Escherichia coli*. The experiments described in this paper deal with an examination of the primary structure of mutationally altered A proteins. The results obtained indicate that at least in two groups of mutants there is a correlation between the region of the gene altered by mutation and the location of the amino acid change in the corresponding altered A protein.

Materials and Methods.—The A proteins examined were isolated by the same or slight modifications of the procedure used for the purification of the wild-type A protein.[8] Proteolytic digestion and peptide pattern examination of the mutant proteins were performed as described elsewhere.[9] The trypsin and chymotrypsin preparations used were obtained from Worthington Biochemical Corp. In experiments with the A proteins of mutants A-23 and A-11, all of the protein preparations were heated for 5 min at 100°C prior to proteolytic digestion in order to inactivate trace amounts of other proteolytic enzyme(s) that occasionally were present as contaminants.[9] Amino acid analyses were carried out with a Spinco amino acid analyzer. The peptides were hydrolyzed in twice-distilled 5.7 N HCl in sealed, evacuated tubes at 105°C for 48 hr.

Characteristics of Mutants Examined.—The various mutant strains relevant to this study and the sites of their mutational alterations in the A gene are shown in Figure 1. The mutants listed were isolated in the K-12 strain of *E. coli* following ultraviolet irradiation.[10, 11] All of these mutants form altered A proteins that are fully active in the conversion of indole and serine to tryptophan in the presence of the B protein of tryptophan synthetase.[11] However, these altered A proteins are incapable of participating in the physiologically essential reaction, the conversion of indoleglycerol phosphate and serine to tryptophan. The A proteins of many of the mutants listed in Figure 1 have been shown to be distinguishable from each other and from the wild-type A protein on the basis of their physical properties.[11] Mutants A-1, A-23, A-28, A-35, and A-36 form heat-labile A proteins, while

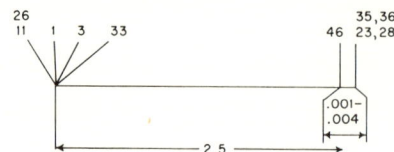

Fig. 1.—Sites of mutational alterations in the A gene in the mutants examined. Sites and distances determined by transduction with phage P1kc.[20] The relative order of A-46 and the A-23 group is not known.

mutants A-3, A-33, and A-46 form A proteins that are more heat-stable than the wild-type A protein. The A proteins of mutants A-11 and A-26 are precipitated at pH 4 while the A proteins of the other mutants and the wild-type A protein do not precipitate at this pH. Mutants A-3 and A-33 have been shown to be distinguishable in other tests;[11] A-3 reverts while A-33 does not, and A-3 is suppressible by a gene that does not affect A-33. Mutants A-11 and A-26 cannot be distinguished in reversion and suppression studies (as well as the tests already mentioned), and the same is true for mutants A-23, A-28, A-35, and A-36.[11, 12]

In deciding whether mutants that appear to map at the same site represent repeat or nonidentical changes at the same mutational site, the sensitivity of the test used to detect genetic separability is obviously of considerable importance. In studies with the A mutants, *cys-tryp* double mutants were used as recipients in transduction tests to increase the sensitivity of the recombination test.[11] It can be calculated that the use of *cys-tryp* double mutants as recipients in transduction recombination experiments should permit the detection of recombination frequencies of 0.0001 per cent before reversion in the donor or the recipient needs to be considered as an interfering factor. Previous studies using this method with many of the mutants listed at identical sites in Figure 1 have shown that there is no recombination at the 0.001 per cent level.[11] However, prototrophic recombinants were detected in these transduction experiments with A-46 and mutants of the A-23 type. The recombination frequencies detected were very low, ranging from 0.0003 to 0.002 per cent in different experiments. More extensive data have now been obtained, and the recombination frequency between mutants A-46 and A-23 has been estimated at 0.001 to 0.004 per cent. This estimate is based on the recovery of only 8 tryptophan-independent recombinants in transductions involving $> 4 \times 10^5$ possible events. The total recombinational length of the A gene is approximately 2.5 per cent. Although mutants A-23 and A-46 map extremely close to each other, A-23 behaves aberrantly in recombination experiments with many other A mutants. For example, A-46 gives 0.21 per cent and 0.45 per cent recombination with mutants A-17 and A-7, respectively, while A-23 gives 1.2 per cent and 1.4 per cent with the same two mutants.

Comparison of Proteolytic Digests of the A Proteins of Mutants Mapping at the A-11 Site.—The A proteins of four mutants that map at the same site (A-3, A-11, A-26, and A-33) have been examined for peptide pattern differences from the wild-type A protein. Trypsin (T), chymotrypsin (C), and trypsin plus chymotrypsin (T + C) peptide patterns of the A protein of mutant A-3 did not show a clear difference from the corresponding wild-type protein peptide patterns, while peptide patterns of the A proteins of mutants A-11, A-26, and A-33 did show differences (Fig. 2).[13] The peptide patterns of the A-11, A-26, and A-33 mutant proteins

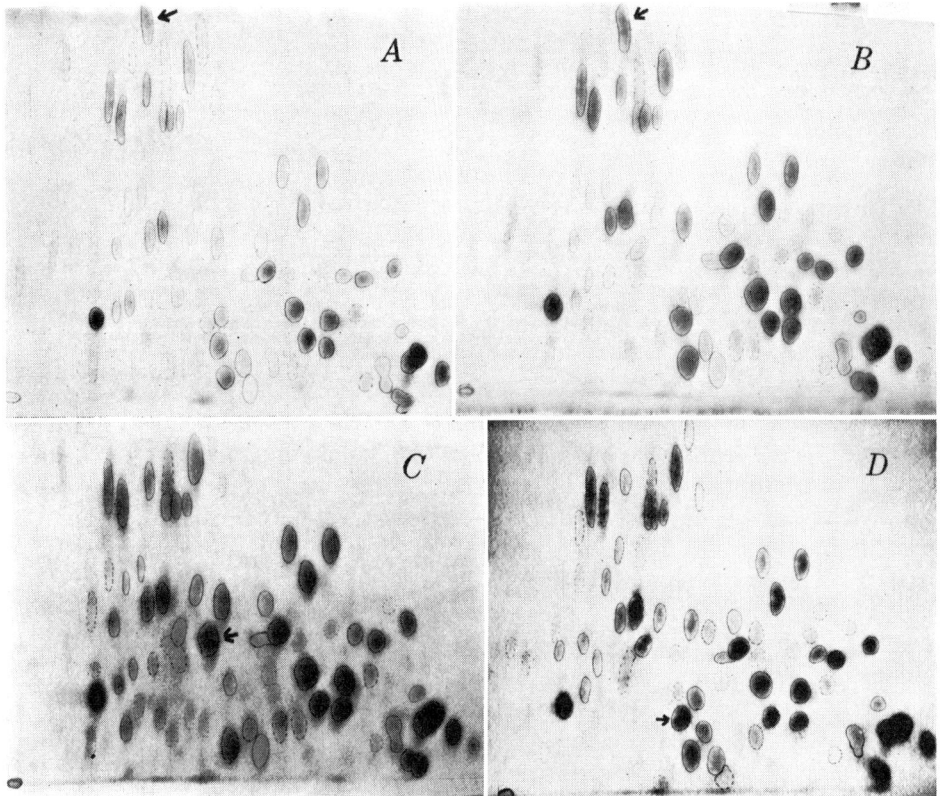

FIG. 2.—Trypsin plus chymotrypsin peptide patterns of the following A proteins: (a) A-11, (b) A-33, (c) A-23, and (d) wild-type. The arrows indicate the major peptides involved in the differences between the mutant and wild-type peptide patterns.

contained one peptide in addition to those observed in the corresponding wild-type pattern. This extra peptide was present in the same position in T + C or C peptide patterns of these mutant proteins. Amino acid analyses were carried out on the additional peptides in the T + C digests of the A proteins of mutants A-33 and A-11. The analyses have shown that these peptides have the same amino acid composition (Table 1). As illustrated in Figure 3, the presence of this additional peptide in the

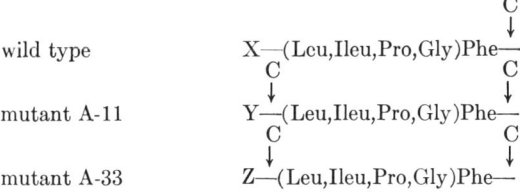

FIG. 3.—Interpretation of peptide alterations in A-33 and A-11. Chymotrypsin-sensitive bonds are designated by the letter C. X, Y, Z represent different amino acids.

A-11 and A-33 peptide patterns could be interpreted as being due to an amino acid change in the same position in the protein, resulting in a new chymotrypsin-sensitive bond. Since the A proteins of A-11 and A-33 have very different properties,

TABLE 1

COMPOSITION OF EXTRA PEPTIDES IN T + C DIGESTS OF THE A-11 AND A-33 PROTEINS

Amino acid	A-11 peptide	A-33 peptide
Proline	0.92	1.10
Glycine	1.21	1.28
Isoleucine	0.85	0.89
Leucine	1.10	0.80
Phenylalanine	0.92	0.94

The composition of each peptide is expressed as the molar ratios of the constituent amino acids. The peptides were isolated by a combination of paper chromatography and paper electrophoresis. In each case, approximately 15 mg of a T + C digest were applied as a 1 × 10 cm band on Whatman 3MM paper and chromatographed by the procedure described for obtaining peptide patterns.[9] The location of the particular peptide band was determined by comparison with a control peptide pattern of a T + C digest of the mutant protein. The area of the paper containing the peptide band was cut out and stitched to a new sheet of Whatman 3MM filter paper and the peptide purified further by electrophoresis at pH 3.7.[9] The peptide band was again located by comparison with a control peptide pattern and then eluted from the paper with 10 per cent acetic acid.

it is assumed that different amino acids are at this position in the two mutant proteins. Total amino acid analysis of the A-33 A protein clearly showed the presence of an additional methionine residue.[14] Since in some cases methionine bonds are hydrolyzed by chymotrypsin,[15] the substitution of methionine in the A-33 protein would account for the new chymotrypsin-sensitive bond in this protein.

If the interpretation considered for the position of the amino acid alterations in A-11 and A-33 is correct, then the extra peptide should be derived from the same region of both mutant proteins. Since the extra peptide is found in chymotrypsin digests of the A protein, trypsin digests of either mutant protein should contain a relatively larger tryptic peptide that contains the amino acid change and yields the extra peptide when treated with chymotrypsin. Several lines of evidence indicate that the extra peptide characteristic of digests of the A proteins of mutants A-11 and A-33 is derived from the same tryptic peptide. When trypsin digests of either of these two mutant proteins were fractionated on a Sephadex column, the tryptic peptide which gives rise to the extra peptide upon addition of chymotrypsin was found in the same fraction (Fig. 4). Furthermore, when these Sephadex fractions were further fractionated on DEAE-Sephadex columns, these tryptic peptides from both mutant digests were again found in approximately the same fractions.

Additional evidence for the similarity of the altered tryptic peptides from the A-11 and A-33 proteins was obtained in studies on the electrophoretic properties of these peptides. When trypsin digests of the A-11 and A-33 proteins were applied as a 1 × 2 cm band to Whatman 3MM filter paper and separated by high-voltage electrophoresis at pH 3.7,[9] the tryptic peptide that gives rise to the extra peptide upon chymotrypsin addition was found at the origin. The failure of this peptide to migrate electrophoretically and, as it was subsequently found, chromatographically, under the conditions used for fingerprinting accounts for the similarity of the T peptide patterns of A-11, A-33, and wild-type A proteins. Experiments are currently being carried out to further purify the relevant tryptic peptides from A-11, A-33, A-26, A-3, and wild-type A proteins.

The results of these studies with mutants A-11, A-26, and A-33 suggest that the amino acid changes in the A proteins of these strains are present at the same position. The fact that mutants A-11 and A-26 are indistinguishable in all tests performed to date further suggests that the A proteins of these particular mutants have the same amino acid change.

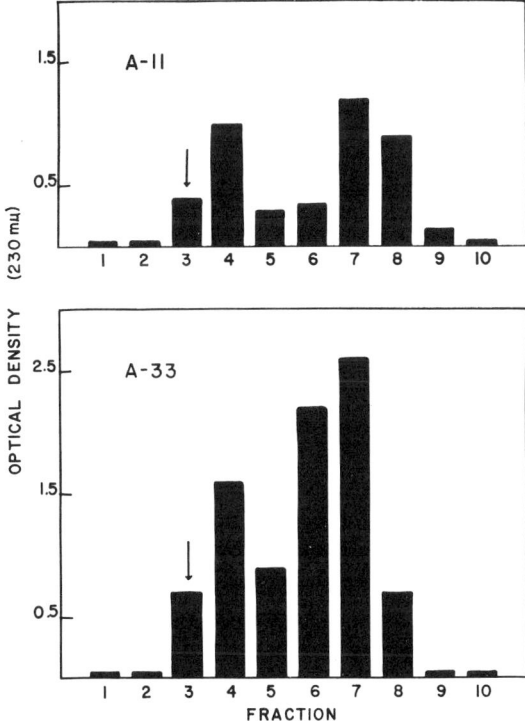

FIG. 4.—Elution patterns of trypsin digests of the A-11 and A-33 proteins. The arrows indicate the fractions containing the tryptic peptides altered in digests of the A-11 and A-33 proteins. The elution patterns were obtained by fractionation of 44 mg of a trypsin digest of A-33 protein and 30 mg of a trypsin digest of A-11 protein on a Sephadex G25 column (1.5 × 100 cm). The peptides were eluted with distilled water and located by measuring absorption at 230 mμ. The flow rate was approximately 10 ml per hr and fractions of 8–12 ml were collected. For the location of the altered tryptic peptides, an aliquot of each fraction was dried *in vacuo* and the peptides dissolved in 0.3 ml of 0.1 *M* ammonium carbonate, pH 8.3. Chymotrypsin was then added and the digestion mixture incubated at 25°C for 90 min. The digests were then dried *in vacuo* and peptide patterns obtained by the procedure described elsewhere.[9]

Studies on the Location of the Amino Acid Change in the A Proteins of Mutants Mapping at the A-23 Site.—The altered A proteins of four mutants, A-23, A-28, A-35, and A-36, all of which map at the same site and form heat-labile A proteins, have also been examined in peptide pattern studies. T + C peptide patterns of each of these mutant proteins showed an identical difference from the corresponding wild-type pattern. A T + C peptide pattern of the A protein of one of these mutants, A-23, is compared with a corresponding wild-type pattern in Figure 2. As was the case with mutants A-11, A-26, and A-33, the peptide pattern studies showed that the same peptide was altered in strains with mutational changes at the same site.

The corresponding mutant and wild-type peptides that differ in the A-23 and wild-type T + C peptide patterns were purified by paper electrophoresis and paper chromatography. The amino acid composition of these peptides is shown in Table 2. It is clear that the mutant peptide contains phenylalanine in addition to the nine amino acids in the wild-type peptide. The corresponding mutant

TABLE 2
Composition of A-23 Mutant Peptide and Corresponding Wild-Type Peptide from T + C Digests

Amino Acid	A-23 peptide	Wild-type peptide
Lysine	1.02	1.05
Aspartic (amide)	0.92	0.94
Serine	0.82	0.96
Glutamic (amide)	1.03	1.04
Proline	1.12	1.03
Glycine	1.32	1.06
Alanine	1.02	1.03
Valine	0.93	0.95
Isoleucine	1.04	0.95
Phenylalanine	0.84	—

The composition of each peptide is expressed as the molar ratios of the constituent amino acids. The peptides were isolated by the procedure described in Table 1.

peptide was also isolated from T + C digests of the A-35 protein and was found to have the same composition as the A-23 mutant peptide. The amino acids in the corresponding mutant and normal peptides have been partially sequenced, and the sequences are identical with the exception that the extra phenylalanine is present in the N-terminal position in the mutant peptide.[16]

Further information on the nature of the amino acid change in mutant A-23 was obtained from peptide pattern studies with trypsin digests of the A-23 protein. Comparison of a T peptide pattern of the A-23 protein with the corresponding wild-type pattern clearly showed the absence of one peptide and the presence of two new peptides (Fig. 5). One of the new peptides was located in exactly the

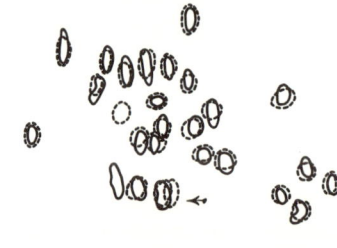

Fig. 5.—A tracing of the major peptides of the peptide pattern of a trypsin digest of the A-23 protein superimposed over a tracing of the corresponding peptide pattern of the wild-type A protein. The major peptides of the A-23 tracing are outlined with a dashed line and the wild-type peptides are outlined with a solid line.

same position as the altered peptide in the T + C peptide pattern of A-23. The other new peptide overlaps a normally occurring peptide and was located by staining A-23 and wild-type peptide patterns for tyrosine-containing peptides.[17]

Since two peptides in the A-23 T peptide pattern replace a single wild-type peptide, an amino acid alteration resulting in the formation of a new trypsin-sensitive bond was considered for the A-23 protein. Such a possibility was also suggested by total amino acid analyses of the A-23 protein, which showed the presence of an additional arginine.[14] The wild-type tryptic peptide which is absent in trypsin digests of the A-23 protein has been isolated,[18] and its composition is shown in Figure 6. This information and the peptide pattern studies suggested the following amino acid sequence for the altered region of the A-23 protein: basic amino acid-phenylalanine-glycine; and X-phenylalanine-glycine for the correspond-

wild type

T
↓
(Lys,Leu,Glu)Tyr—(Asp,Glu,Pro$_2$,Ala$_2$,Leu)**Gly**—Phe—Gly—
          ↑   ↑
          C   C
                                                        T
                                                        ↓
                        —(Asp,Glu,Ala,Pro,Val,Ileu,Ser)Lys

A-23

T
↓
(Lys,Leu,Glu)Tyr—(Asp,Glu,Pro$_2$,Ala$_2$,Leu)**Arg**—Phe—Gly—
          ↑   ↑
          C   C
                                                        T
                                                        ↓
                        —(Asp,Glu,Ala, Pro,Val,Ileu,Ser)Lys

FIG. 6.—Corresponding regions of wild-type and A-23 proteins. The trypsin and chymotrypsin-sensitive bonds are designated by T and C, respectively.

ing wild-type region. The arginine-phenylalanine-glycine sequence in the mutant protein would account for the peptide pattern differences, since the addition of trypsin first, and then chymotrypsin, would result in the hydrolysis of the peptide bond between the arginine and phenylalanine yielding a peptide with an N-terminal phenylalanine. An N-terminal phenylalanine bond has been shown to be resistant to hydrolysis by chymotrypsin.[19] The corresponding sequence in the wild-type protein, X-phenylalanine-glycine, would not be attacked by trypsin but would be hydrolyzed between phenylalanine and glycine by chymotrypsin. The net result of the combined action of trypsin and chymotrypsin on this region of the mutant and wild-type proteins would be one mutant peptide which differs from the corresponding wild-type peptide by the presence of an N-terminal phenylalanine. This peptide would be found in either a T or T + C peptide pattern. In addition, a second mutant peptide would be expected and this peptide should contain a C-terminal arginine in place of an X-phenylalanine C-terminal sequence in the corresponding wild-type peptide.

Such an arginine-containing peptide on the T + C peptide pattern was indicated when a comparison was made of the A-23 and wild-type T + C peptide patterns which had been stained for arginine-containing peptides.[9] The new arginine-containing mutant peptide gave a yellow-brown ninhydrin color reaction and was located slightly below the position of a corresponding wild-type peptide that gives a similar ninhydrin reaction but does not contain arginine.

Since analyses of trypsin peptides of the A-23 protein were complicated by the presence of an additional trypsin-sensitive peptide bond, attempts were made to isolate the arginine-containing mutant peptide from chymotrypsin digests of the A-23 protein. Chymotrypsin peptide patterns of A-23 and wild-type A proteins were stained for arginine-containing peptides, and an additional arginine-containing peptide was found in the A-23 peptide pattern at a slightly lower chromatographic position than the corresponding chymotryptic peptide of the wild-type protein. This arginine-containing peptide was isolated by fractionation of a chymotrypsin digest of A-23 protein on a DEAE-Sephadex column. Amino acid analyses were performed on this mutant peptide and the compositions of this peptide and the corresponding wild-type chymotryptic peptide are shown in Table 3.

TABLE 3

Composition of the Chymotryptic Mutant Peptide of A-23 and the Corresponding Wild-Type Peptide

Amino acid	Wild-type peptide*	A-23 peptide
Arginine	—	0.97
Aspartic (amide)	0.98	1.03
Glutamic (amide)	1.11	1.09
Proline	2.08	2.03
Glycine	1.05	—
Alanine	1.81	1.97
Leucine	1.07	0.99
Phenylalanine	0.92	0.90

The composition of each peptide is expressed as the molar ratios of the constituent amino acids. The A-23 peptide was isolated from a chymotrypsin digest of the A-23 protein by fractionation on a DEAE-Sephadex column. A solution of 38 mg of a chymotrypsin digest of the A-23 protein in 1 ml of pyridine-acetic buffer, pH 7.0 (0.02 M with respect to pyridine), was added to a 1.5 × 100 cm column of DEAE-Sephadex, A-50, which had been washed thoroughly with the pyridine-acetic buffer. A linear gradient was then applied to the column with 300 ml of the pyridine-acetic buffer in the mixing flask and 300 ml of 0.2 M acetic acid in the inlet flask. The flow rate was approximately 10 ml per hr and 8–12 ml fractions were collected. The fractionation was performed at 4°C. The fractions were subjected to paper electrophoresis at pH 3.7 and the peptide located by staining for arginine-containing peptides.[9] The mutant peptide was further purified by paper electrophoresis at pH 3.7.

* The analysis of this peptide was taken from the accompanying paper by Henning and Yanofsky.[18]

It is clear that the mutant peptide differs from the wild-type peptide by the substitution of an arginine for a glycine. The composition of this chymotryptic peptide and the peptide pattern information indicate that the sequence of amino acids in the altered region of the A-23 protein is as shown in Figure 6.

The A protein of mutant A-46 was also examined in peptide pattern studies. The mutational alteration in this strain maps very close to the alteration in strain A-23. No clear difference was observed between T or T + C peptide patterns of the A-46 and the wild-type A proteins.

Discussion.—Recombinational analyses and peptide pattern studies with a number of A protein mutants indicate that a correlation exists between the region of the genetic map at which a mutation occurs and the position of amino acid alteration in the A protein. Recombination studies with mutants A-11, A-33, and A-3 suggest that the mutational alterations in these strains are at the same site in the A gene. These mutants can be distinguished from one another in reversion[11, 12] and suppressor studies, and, in addition, the A protein of mutant A-11 has different properties from the A proteins of A-3 and A-33. Peptide pattern studies clearly indicate a difference between the A proteins of A-3 and A-11 or A-33. The peptide patterns of the A-11 and A-33 A proteins show the same peptide difference from the corresponding wild-type A protein patterns. The finding that the amino acid changes in the A proteins of A-11 and A-33 are probably in the same tryptic peptide and possibly at the same position in the A protein demonstrates a correlation between corresponding regions of the genetic map and the protein. Studies with the alkaline phosphatase system of *E. coli* have also suggested such a relationship.[5]

Mutant A-26 maps at the same site as A-11 and is indistinguishable from this mutant in reversion[12] and suppressor tests. The proteins of these two mutants have identical properties, and the peptide patterns of the two mutant proteins show the same difference from the corresponding wild-type patterns. These observations suggest that mutants A-11 and A-26 represent identical mutations. The same conclusion was derived from peptide pattern studies and other comparisons of four mutants, A-23, A-28, A-35, and A-36, that map at one site at the other end of the A gene. Final proof that these mutants represent repeat identical mutations will require amino acid analyses of the mutant peptides of each protein digest.

Mutant A-46 gives tryp$^+$ recombinants with low frequency when crossed with strain A-23, indicating that the genetic alterations in these strains are close to each other. Although the peptide patterns of the A-46 protein are similar to the corresponding patterns of the wild-type protein, evidence is presented in the accompanying paper[18] that the amino acid substitution in the A-46 protein involves the same amino acid in the wild-type protein that is replaced in the A-23 protein. This amino acid, glycine, is replaced by glutamic acid in A-46 and by arginine in A-23.

In the Watson and Crick model of DNA, a nucleotide can be considered to be the smallest possible recombinable unit. Since mutants A-23 and A-46 are recombinable and have mutational changes involving the same amino acid in their A proteins, it can be concluded that at least two nucleotides are required to specify this amino acid in the A protein. Furthermore, the fact that only single amino acids are changed as a result of mutational events rules out, at least in these cases, overlapping nucleotide codes in which each nucleotide is involved in the coding of adjacent amino acids. The same conclusion can be derived from similar observations with other proteins.[2, 3] In view of the fact that the mutational alterations in A-23 and A-46 affect the same amino acid in the A protein, the recombination distance between these sites is of considerable importance. The recombination values obtained suggest that the distance between the A-23 and A-46 mutational alterations is approximately $1/625$ to $1/2,500$ of the total recombinational length of the A gene. This estimate can be compared with the length of the A gene in the following way: if three nucleotide pairs code for each of the 280 amino acids in the A protein and if all of the nucleotides are directly involved in specifying the amino acids of the A protein, then the A gene would have a length of about 840 nucleotide pairs. Adjacent nucleotide pairs, therefore, would be expected to give recombination values of $1/840$ of the recombinational length of the A gene. This value agrees reasonably well with the observed recombination distance between A-23 and A-46.

In the recombination experiments with mutants A-23, A-46, and the other A mutants, it was noted that A-23 behaves aberrantly, giving very high recombination values with mutants with alterations near the same end of the A gene. Since the A-23 mutation involves a single amino acid change and presumably a change in a single nucleotide pair, this aberrant behavior would suggest that recombination frequencies are markedly influenced by the nature of the nucleotide differences in recombining genic regions.

Although the detected alterations in the A-23 and A-46 proteins involve single amino acid substitutions, our present information does not exclude the possibility that other undetected changes have occurred or that some of the other ultraviolet-induced mutants have more complex amino acid changes. Only single peptide alterations have been observed with any of the mutant proteins examined to date, however, indicating that in each case there is only a localized alteration in the primary structure of the A protein. Total amino acid analyses of the mutant proteins are consistent with this conclusion.[14]

The substitution of an arginine or a glutamic acid for glycine as a result of point mutations suggests a coding relationship between these three amino acids. If the mutational event in each of these two mutants involved a change of a single nucleotide, then it could be concluded that the composition of the coding units for arginine or glutamic acid does not differ from that of the coding unit for glycine by more than one nucleotide. A similar consideration can be applied to the A-11 region, where there are several different mutant types that map at or near the same site and, consequently, may have amino acid alterations in the same position in their A proteins. By examining many mutants at or near the same site, it should be possible to obtain useful information on the coding relationships between amino acids. If the code is degenerate, this approach will be particularly important.

It is apparent that it will be necessary to carry out similar studies on a number of mutants at different sites along the gene in order to conclusively test the concept of collinearity of amino acid sequence and nucleotide sequence. The correlation between the mutational site map and the positions of amino acid changes in the A proteins of the mutants examined, however, is consistent with this concept and establishes the fact that the genetic map does relate to the primary structure of the corresponding protein.

Summary.—Studies of the altered A proteins produced by several mutants of *E. coli* have indicated that at least in two groups of mutants there is a correlation between the region of the gene altered by mutation and the location of the amino acid change in the corresponding mutant protein. In addition, peptide pattern studies have supported genetic evidence indicating that several of the mutant strains in each group represent repeat identical mutations at the same site. The A protein of one mutant, A-23, has been shown to differ from the wild-type protein by the substitution of a glycine by an arginine. This substitution involves the same amino acid that is replaced in the protein of another A-protein mutant, A-46.[18] Recombinational analyses have shown that the mutationally altered site in the A gene of mutant A-46 is very near to but separable from the altered site in mutant A-23.

Note added in proof.—Preliminary electrophoretic studies on cellulose acetate by U. Henning indicate that the A proteins of mutants A3, A11, and A33 are less negatively charged (apparently to the same degree) than the wild type A protein. These findings are consistent with a change of the same amino acid in the A proteins of these mutant strains.

The authors are indebted to Virginia Horn and Janet Lind for their excellent technical assistance.

* This investigation was supported by grants from the National Science Foundation and the U.S. Public Health Service.

† Postdoctoral Fellow of the U.S. Public Health Service. Some of these studies have been reported in a preliminary communication: Helinski, D., and C. Yanofsky, *Federation Proc.*, **20**, 255 (1961).

[1] Ingram, V. M., *Biochim. et Biophys. Acta*, **28**, 539 (1958).

[2] Ingram, V. M., *Hemoglobin and Its Abnormalities* (Springfield, Illinois: Charles C Thomas Co., 1961).

[3] Tsugita, A., and H. Fraenkel-Conrat, *J. Molecular Biology* (in press).

[4] Benzer, S., in *The Chemical Basis of Heredity*, ed. W. D. McElroy and B. Glass (Baltimore: The Johns Hopkins Press, 1957), p. 70.

[5] Rothman, F., see discussion following the paper cited in reference 7.

[6] Streisinger, G., F. Mukai, W. J. Dreyer, B. Miller, and S. Horiuchi, in Cellular Regulatory

Mechanisms, Cold Spring Harbor Symposia on Quantitative Biology, vol. 26 (1961, in press).

[7] Yanofsky, C., D. R. Helinski, and B. Maling, in *Cellular Regulatory Mechanisms*, Cold Spring Harbor Symposia on Quantitative Biology, vol. 26 (1961, in press).

[8] Henning, U., D. R. Helinski, F. C. Chao, and C. Yanofsky, *J. Biol. Chem.* (in press).

[9] Helinski, D. R., and C. Yanofsky, manuscript in preparation.

[10] Yanofsky, C., and I. P. Crawford, these PROCEEDINGS, **45**, 1016 (1959).

[11] Maling, B. D., and C. Yanofsky, these PROCEEDINGS, **47**, 551 (1961).

[12] Allen, M., and C. Yanofsky, unpublished experiments.

[13] Initial studies involving the A-11 protein were performed on the protein isolated from the mutant A-11; however, since the A protein of the double mutant A-11-46 is more stable than the A-11 protein, the double mutant protein was used in later studies on the A-11 amino acid change. Similarly, in studies on the A-23 protein, the initial studies were performed with the protein from the mutant A-23. Later studies were carried out on the more stable A protein obtained from the double mutant A-3-23. In each case, the peptide pattern difference characteristic of the mutant in question was found to be present in the corresponding double mutant peptide pattern.[7]

[14] Helinski, D. R., and C. Yanofsky, unpublished experiments.

[15] Hirs, C. H. W., S. Moore, and W. H. Stein, *J. Biol. Chem.*, **235**, 633 (1960).

[16] Carlton, B. C., and C. Yanofsky, unpublished experiments.

[17] Acher, R., and C. Crocker, *Biochim. et Biophys. Acta*, **9**, 704 (1952).

[18] Henning, U., and C. Yanofsky, these PROCEEDINGS, **48**, 183 (1962).

[19] Neurath, H., and G. W. Schwert, *Chem. Revs.*, **46**, 69 (1950).

[20] Yanofsky, C., and E. S. Lennox, *Virology*, **8**, 425 (1959).

AN ALTERATION IN THE PRIMARY STRUCTURE OF A PROTEIN PREDICTED ON THE BASIS OF GENETIC RECOMBINATION DATA*

By Ulf Henning† and Charles Yanofsky

DEPARTMENT OF BIOLOGICAL SCIENCES, STANFORD UNIVERSITY

Communicated by Victor C. Twitty, December 22, 1961

Studies on the fine structure relationships between a gene and the corresponding protein are being carried out with the A protein of the tryptophan synthetase of *Escherichia coli*. The preceding paper[1] describes an investigation with altered A proteins from several mutants and demonstrates that mutational changes at or near the same site in the A gene lead to alterations in the same region of the A protein.

Studies[2,3] with a large number of A-protein mutants have shown that there are two closely linked sites at one end[4] of the A gene that have mutated frequently. All of the strains with mutational changes at one of these sites form a heat-labile A protein[2] that is distinguishable from the wild-type A protein in peptide pattern studies.[1] Strains with mutational changes at the second site form an altered A protein that is somewhat more heat-resistant than the wild-type A protein[2] but cannot be distinguished from the normal protein in peptide pattern studies.[1] The distance between the two mutational sites is approximately $1/625$–$1/2500$ of the total length of the map of the A gene.

Since the A protein appears to be a single polypeptide chain[5] containing approximately 280 amino acids,[6] one would predict that these two mutant types should have amino acid substitutions at or near the same position in the A protein if the number of nucleotide pairs in the A gene is a small multiple of the number of amino acids in the A protein and if distances on the genetic map bear some relation to the distances between corresponding amino acids in the A protein. This report is concerned with an examination of this prediction.

Materials and Methods.—Mutants: The mutant strains mentioned in this paper were obtained by penicillin selection[7] following ultraviolet irradiation of wild-type strain K-12 of *E. coli.*

Isolation of peptides and amino acid analyses: Tryptic digests of the isolated A proteins were prepared as described previously.[8] The trypsin-resistant core of the A protein was removed from the digests by isoelectric precipitation at pH 4.[10] The soluble peptides were separated by the method of Rudloff and Braunitzer[11] with minor modifications in order to improve the separation of the A protein tryptic peptides. In principle, the procedure consists of column chromatography of the acid-soluble supernatant on Dowex 1 × 2 (acetate form) starting with a collidine acetate buffer at pH 8.5. A pH gradient with 0.09 N acetic acid was applied and then a second pH gradient was used with 1.5 N acetic acid. The peptide discussed in this communication, TP3, was eluted near the end of the first pH gradient. The TP3 obtained by Dowex chromatography often contained minor contaminants and was further purified by paper chromatography or paper electrophoresis in the solvent system or buffer, respectively, used in peptide pattern studies.[1, 8] Complications resulting from purification by paper chromatography are mentioned in Table 1. Acid hydrolysis of peptides was performed in evacuated, sealed tubes in 5.7 N HCl at 105°C for 48 hr. Amino acid analyses were carried out with a Spinco amino acid analyzer.

Results.—Characteristics of mutant A 46 and the altered A protein it produces: Tryptophan auxotroph A 46 is a typical A mutant of *E. coli* K-12.[12-14] It accumulates indoleglycerol and can grow on a minimal medium supplemented with tryptophan or indole but it will not respond to anthranilic acid. It forms an altered A protein that is fully active in the conversion of indole plus serine to tryptophan in combination with normal B protein but is inactive in the other two tryptophan synthetase reactions, the reversible conversion of indoleglycerol phosphate to indole and the conversion of indoleglycerol phosphate plus serine to tryptophan. The A-46 protein has the same affinity for normal B protein as the wild-type A protein and is slightly more heat-resistant than wild-type A.[2] It protects wild-type A protein from neutralization by antibody to the wild-type A protein and appears to react with the antibody to the same extent as the normal A protein.[2]

The isolated A 46 protein was examined electrophoretically under various conditions to determine if there was a charge difference from the normal A protein. Different mobilities were observed on cellulose acetate (Fig. 1), at pH 9.2, indicating that the mutant protein has a higher negative net charge than the normal A protein. At pH 7.5, where the A protein is still negatively charged (the isoelectric point of the wild-type protein is at pH 5),[6] this charge difference could not be demonstrated.

The wild-type A protein can be crystallized under certain conditions from a 40–45 per cent ammonium sulfate solution.[6] Repeated attempts to crystallize the isolated A 46 protein, however, were unsuccessful. Instead of crystallizing, the originally slightly turbid protein solution congealed and after several days assumed a consistency similar to that of soft agar. The gel dissolved readily in water.

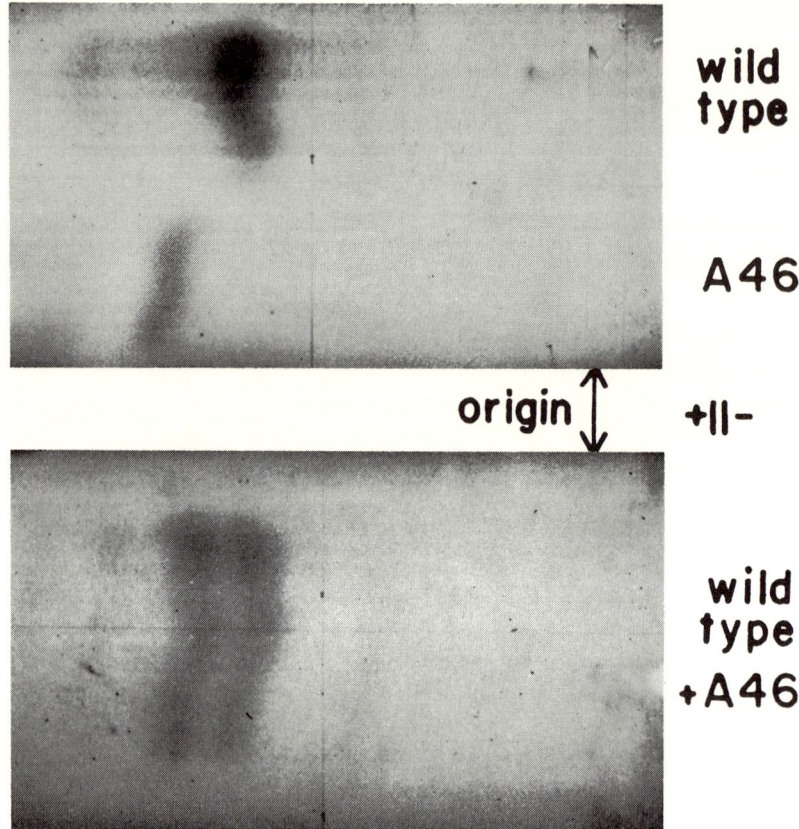

Fig. 1.—Electrophoretic comparison of the A proteins from wild type and mutant A 46. Approximately 30 μg of each purified protein was applied to cellulose acetate strips (5 × 10 cm.). Electrophoresis was performed in 0.04 M sodium diethylbarbiturate buffer, pH 9.2, at 4°C for 11 hr at 36 V/cm (3 mA). Ponceau S stain.[24]

Microscopical examination of the gel did not reveal any organized structure.

Examination of the wild-type and A 46 tryptic peptides corresponding to the tryptic peptide altered in mutant A 23: The amino acid substitution in mutant A 23 described in the preceding paper[1] occurs in a region of the A protein that corresponds to the wild-type tryptic peptide TP3. In spite of the absence of any differences in trypsin or trypsin plus chymotrypsin peptide patterns of the A-46 protein and the normal A, the fact that the genetic alterations in mutants A-23 and A-46 map close to one another prompted an examination of peptide TP3 from the A protein of wild type and A-46. The location of this peptide on fingerprints of tryptic digests of the A protein is shown in Figure 2. Peptide TP3 was isolated from the wild-type and A-46 proteins as described in the *Methods* section. The amino acid composition of this peptide from the two sources (Table 1) clearly showed that the mutant peptide contained a glutamic acid or glutamine in place of a glycine. Since wild-type TP3 contains two glycine residues, it was necessary to degrade the peptide further to determine which glycine was affected by the mutation. Chymotrypsin treatment of TP3 resulted in the liberation of three peptides designated TP3C1, TP3C2, and TP3C3 (Fig. 2). Fortunately, the glycine residues proved to be in different

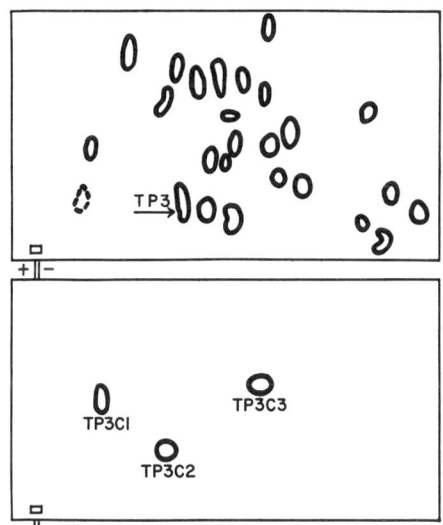

FIG. 2.—Tracings of peptide patterns. Upper figure: location of TP3 in the peptide pattern from a tryptic digest of wild-type A protein. Lower figure: chymotryptic peptides from isolated TP3. The same patterns were obtained with the corresponding peptides from mutant A-46. Origin: □; chromatography in vertical direction; electrophoresis, at pH 3.7, in horizontal direction; ninhydrin stain. The conditions employed in the peptide pattern studies have been described.[1,8]

TABLE 1

MOLAR RATIOS* OF CONSTITUENT AMINO ACIDS IN CORRESPONDING PEPTIDES

Amino acid	Wild type peptides				Mutant A 46 peptides				Mutant A 95 peptides			
	TP3†	TP3C1	TP3C2	TP3C3	TP3	TP3C1	TP3C2	TP3C3	TP3	TP3C1	TP3C2	TP3C3
Lysine	1.87		1.11	0.94	1.85		1.00	1.00	2.05		0.95	1.06
Aspartic	2.13	0.98	1.13		1.98	0.94	1.03		1.97	0.94	1.00	
Serine‡	1.09		1.10		1.16				0.97		1.07	
Glutamic	3.04	1.11	1.05	1.16	4.12	2.11	1.15	1.21	4.15	2.13	0.99	1.17
Proline	3.12	2.08	1.03		2.84	1.95	1.05		2.87	1.93	1.17	
Glycine	2.16	1.05	1.05		1.22		0.89		1.13		0.99	
Alanine	2.92	1.81	0.90		2.91	1.95	0.98		2.95	1.98	1.05	
Valine	1.07		1.05		0.94		0.93		1.05		0.92	
Isoleucine	1.00		0.83		1.01		0.92		0.96		0.90	
Leucine	1.89	1.07		1.04	1.88	1.08		0.84	2.08	1.08		0.98
Tyrosine	0.91			0.87	—**			0.95	0.83			0.88
Phenylalanine	0.96	0.92			0.93	0.96			1.02	0.92		
Ammonia§	4.24				4.19				4.27			

* Calculation of molar ratios according to Hirs, Moore, and Stein.[26]
† Hydrolysis for 72 hr did not change these molar ratios.
‡ Corrected for partial destruction upon acid hydrolysis.
§ Not extrapolated to zero time hydrolysis.
** Complete or partial loss of tyrosine upon acid hydrolysis was observed when the peptide was purified by paper chromatography. That the tyrosine is still unaltered after the chromatography can be seen from the analysis of TP3C3 which was derived from this TP3 and was purified by paper electrophoresis. The product of the degradation of tyrosine has been found but not yet characterized.

peptides (Table 1) and it was immediately evident that TP3C1 contained the alteration. TP3C1 isolated from mutant A-23 does not contain glycine but has one arginine residue;[1] thus, the same glycine residue is replaced in the two mutant proteins.

It was concluded from paper electrophoretic studies that the amino acid substitution in A-46 involved glutamic acid and not glutamine. Since the γ-carboxyl group of glutamic acid has a $pk'_2 = 4.25$,[15] it would be expected that at higher pH values peptide TP3C1 from mutant A-46 would have an extra negative charge when compared to the wild-type peptide if the change involved glutamic acid. Electrophoresis of the peptides at pH 6.7 (Fig. 3) clearly demonstrated a charge difference. Consistent with this result are the ammonia values in the amino acid analyses of

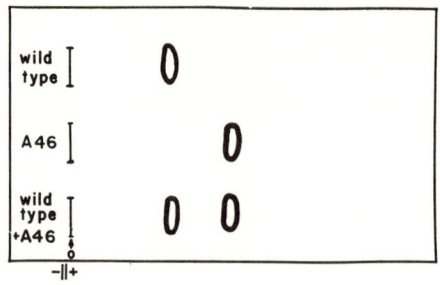

Fig. 3.—Electrophoresis of TP3C1 from the A proteins of wild type and mutant A-46 at pH 6.7. Approximately 0.06 μM of each peptide was applied to Whatman 3 MM paper, and electrophoresis was carried out in pyridine acetate buffer, pH 6.7,[25] at 55 V/cm (160 mA) for two hr at 20°C; ninhydrin stain.

TP3 from the wild-type and A-46 proteins which do not differ from each other (Table 1) and the greater electrophoretic mobility of the A-46 protein at pH 9.2 (Fig. 1). In the standard fingerprinting technique employed,[8, 9] electrophoresis is performed at pH 3.7. At this pH, the γ-carboxyl group of the glutamic acid residue evidently is not sufficiently dissociated to affect the migration of the peptide.

Examination of mutant A-95: Mutant A-95, an A mutant isolated recently, forms an altered A protein with properties identical to those of the A-46 protein. Genetic tests failed to detect recombination between mutants A-95 and A-46, while both gave identical recombination values with A mutants that map at other sites in the A gene. These findings suggest that A-46 and A-95 represent repeat mutations at the same genetic site. If, in addition, the mutational alteration was identical in both mutants, the same amino acid substitution should be present at the same position in their A proteins.

Peptide TP3 was isolated from a tryptic digest of the A protein of mutant A-95 and was digested with chymotrypsin to obtain the three peptides TP3C1, TP3C2, and TP3C3. The amino acid composition of all four peptides was determined and is given in Table 1. Since the same change observed in mutant A-46 was found in the A-95 protein, it is probable that the mutational alterations in these two mutants are identical. Three other A mutants, A-61, A-76, and A-86, which are also indistinguishable from A-46 in all tests performed to date, may also have the same amino acid substitution that was detected in A-46; the A proteins produced by these mutants have not yet been examined for the amino acid substitution.

Tentative structure of the tryptic peptide TP3: The result of chymotryptic treatment of TP3 agrees with expectations based on the usual specificity of chymotrypsin, since one tyrosine peptide and one phenylalanine peptide are liberated. TP3C2 does not have an aromatic amino acid and, therefore, is probably the C-terminal chymotryptic peptide of TP3. Furthermore, since TP3 is a tryptic peptide that contains two lysine residues, lysine is probably the N-terminal amino acid and TP3C2 is probably the N-terminal chymotryptic peptide of TP3. Several cases have been described[16, 17] in which an N-terminal basic amino acid is not liberated by trypsin from a peptide.[18] Other evidence suggesting the order TP3C3–TP3C1–TP3C2 for the chymotryptic peptides is the finding that the phenylalanine of TP3C1 is the N-terminal amino acid of TP3C2 obtained from trypsin plus chymotrypsin digests of the A protein from mutant A-23.[1] The phenylalanine is in the N-terminal position of TP3C2 from this mutant protein, since trypsin hydrolysis of the arginyl-phenylalanine bond leaves the phenylalanine N-terminal and thus resistant to release by chymotrypsin.[21] Consistent with this order are observations on the ninhydrin staining reactions of the various peptides. Peptides TP3, TP3C2, and TP3C3 stain blue while peptide TP3C1 stains yellow-brown; TP3 would be expected to stain yellow-brown if TP3C1 were its N-terminal chymotryptic peptide. These considerations plus the results of N-terminal amino acid sequence analyses of TP3C2[22] suggests the tentative structure of TP3 shown in Figure 4.

Peptide TP3 from

A 46   Lys-(Glu*, Leu)-Tyr-(Asp*,Glu*,Pro$_2$,Ala$_2$,Leu)-**Glu**-Phe-
         1   12
                             Gly-Ileu-Ser-Ala-Pro-(Asp*,Glu*,Val)-Lys
                                                                   22

         ←———TP3C3———→←———TP3C1———————→
Wild type Lys-(Glu*,Leu(-Tyr-(Asp*,Glu*,Pro$_2$,Ala$_2$,Leu)-**Gly**-Phe-
          1   2,3   4   5-11   12 13 ————TP3C2———————→
                             Gly-Ileu-Ser-Ala-Pro-(Asp*,Glu*,Val)-Lys
                             14 15 16 17 18   19-21   22

A 23   Lys-(Glu*,Leu)-Tyr-(Asp*,Glu*,Pro$_2$,Ala$_2$,Leu)-**Arg**-Phe-
         1   12
                             Gly-Ileu-Ser-Ala-Pro-(Asp*,Glu*,Val)-Lys
                                                                   22

FIG. 4.—The amino acid substitutions in the tryptic peptide TP3 from mutants A-46 and A-23. Evidence for, and possible objections to, this representation are given in the text. Glu*, Asp*,: unknown whether present as amides. Ammonia values (Table 1) suggest the presence of four amide nitrogens in the peptide.

Discussion.—The results presented in this paper demonstrate a single amino acid substitution of a glycine by a glutamic acid as the result of mutation in strain A-46. The electrophoretic behavior of the isolated A-46 protein is consistent with this substitution, since a higher net negative charge than that of the wild-type protein is observed. The site of this switch appears to be identical with the site of the replacement of a glycine by an arginine in the A protein of mutant A-23[1] (Fig. 4). Although this is the simplest interpretation of the data, two objections to the representation shown in Figure 4 remain to be removed by sequence analysis of the amino acids constituting TP3C1. The A-46 protein could have undergone two amino acid changes: glycine replaced by glutamine and, if the other glutamic acid of the peptide were present as the amide, replacement of this amide by glutamic acid. Another possibility, also consistent with the results obtained, is that the sequence 5-12 (Fig. 4) in TP3 is not identical in the A-23 protein and wild-type A. The position of the glycine residue in the wild-type peptide TP3C1 was deduced from the position of the arginine in this peptide from mutant A-23 and thus could be incorrect if more drastic changes than a single amino acid substitution had occurred in the A protein of mutant A-23. Regardless of the amino acid sequences in TP3C1 from the wild-type protein and the proteins from mutants A-23 and A-46, the prediction of the location of the amino acid change in the A-46 protein, based on recombination data with mutant A-23, was found to be true.

The replacement of glycine by glutamic acid and arginine in the two mutants suggests that different nucleotide changes in the coding unit specifying the glycine in question have occurred. If the deduced amino acid sequence is correct, it can be further assumed that different nucleotides in the same coding unit were changed by mutation, since the two mutant strains can recombine to give wild-type progeny.

Evidently, the substitution of glycine by glutamic acid or arginine in a certain position in the A protein causes the complete loss of enzymatic activity in the indoleglycerol phosphate–to–tryptophan and the indoleglycerol phosphate–to–indole reactions without affecting the immunological properties of the protein or the capacity to combine with normal B protein. The unaltered ability of these mutant A proteins to catalyze the indole-to-tryptophan reaction in combination with normal B supports the theory[14] that the A protein serves as an activator of the B protein in this reaction. Apparently, a region of the A protein unaffected by the amino acid changes in the mutant proteins is required for this function. One might also conclude that the glycine replaced in the A-46 and A-23 proteins is at, or affects, the active site of the protein. Replacement of this glycine by glutamic acid or arginine could alter the proper structure of this site so that it cannot be either maintained or formed in the sense of the "induced-fit" theory.[23] In view of the polarity of the residues introduced by mutation, this possibility is not difficult to visualize. It seems remarkable, in addition, that substitution of an uncharged residue by groups with opposite charges leads to opposite sensitivities to elevated temperatures. In this connection, it is also of interest that the presence of the extra negatively charged residue in the A protein of mutant A-46 leads to gel formation, which presumably is due to aggregation.

The studies with the A protein of mutant A-95 demonstrate that a mutational event classified as identical with that of mutant A-46 on the basis of genetic tests and comparisons of the properties of A proteins has, in fact, caused the identical amino acid substitution observed in mutant A 46. Since the A proteins of all five mutants that map at the A 46 site exhibit identical properties and since the A proteins of all nine mutants that map at the A-23 site also have identical properties, it would appear that the group of nucleotides coding for glycine in the normal A protein at the A-23–A-46 position is particularly susceptible to mutation following ultraviolet irradiation. Whether this susceptibility reflects a basic instability of this coding unit, a particular sensitivity to ultraviolet irradiation, or the possibility that only select classes of mutants forming certain types of altered A proteins can be recovered remains to be determined.

Summary.—The A protein of the tryptophan synthetase of mutant A-46 of *E. coli* K-12 was found to differ from the normal A protein by a single amino acid substitution, glutamic acid for glycine. The substitution results in alterations of several of the properties of the A protein. The glycine residue replaced by glutamic acid in the A-46 protein is the same glycine replaced by arginine in the A protein of the closely linked mutant, A-23.[1] The recombination observed between the two strains probably represents recombinational events between different nucleotides in the same coding unit.

Mutant A-95 and three other A mutants that map at the A-46 site form altered A proteins that exhibit the same properties as the A-46 protein. The A-95 protein was found to have the same amino acid substitution at the same position as the A-46 protein.

The authors are greatly indebted to Virginia Horn and Janet Lind for their excellent technical assistance.

* This investigation was supported by grants from the National Science Foundation and the U. S. Public Health Service.

† Fulbright grantee 1960–62.
[1] Helinski, D. R., and C. Yanofsky, these PROCEEDINGS, **48**, 173 (1962).
[2] Maling, B. D., and C. Yanofsky, these PROCEEDINGS, **47**, 551 (1961).
[3] Yanofsky, C., D. R. Helinski, and B. D. Maling, in *Cellular Regulatory Mechanisms*, Cold Spring Harbor Symposia on Quantitative Biology, vol. 26 (1961).
[4] The ends of the map of the A gene are assumed to be the most distal mutational alterations on this map.
[5] Carlton, B. C., and C. Yanofsky, *J. Biol. Chem.*, in press.
[6] Henning, U., D. R. Helinski, F. C. Chao, and C. Yanofsky, *J. Biol. Chem.*, in press.
[7] Davis, D. B., *J. Am. Chem. Soc.*, **70**, 4267 (1948); Lederberg, J., and N. Zinder, *J. Am. Chem. Soc.*, **70**, 4267 (1948); Adelberg, E. A., and J. W. Myers, *J. Bacteriol.*, **65**, 348 (1953).
[8] Helinski, D. R., and C. Yanofsky, in preparation.
[9] Katz, A. M., W. J. Dreyer, and C. B. Anfinsen, *J. Biol. Chem.*, **234**, 2897 (1959).
[10] Henning, U., and C. Yanofsky, unpublished results.
[11] Rudloff, V., and G. Braunitzer, *Hoppe-Seyler's Z. physiol. Chem.*, **323**, 129 (1961).
[12] Yanofsky, C., and E. S. Lennox, *Virology*, **8**, 425 (1959).
[13] Yanofsky, C., and I. P. Crawford, these PROCEEDINGS, **45**, 1016 (1959).
[14] Yanofsky, C., *Bacteriol. Rev.*, **24**, 221 (1960).
[15] Greenstein, J. P., and M. Winitz, *Chemistry of the Amino Acids* (New York and London: John Wiley & Sons, 1961), vol. 1, p. 487.
[16] Hirs, C. H. W., *J. Biol. Chem.*, **235**, 625 (1960).
[17] Wittmann, H. G., and G. Braunitzer, *Virology*, **9**, 726 (1959).
[18] Although TP3 is not the N-terminal peptide of the A protein, the presence of an N-terminal basic amino acid in such a tryptic peptide has precedence in a tryptic peptide from tobacco mosaic virus. In the amino acid sequence -thre-arg-arg-val-, the -arg-arg- bond is split preferentially by trypsin, leaving one arginyl residue N-terminal.[19], [20]
[19] Anderer, F. A., H. Uhlig, E. Weber, and G. Schramm, *Nature*, **186**, 922 (1960).
[20] Tsugita, A., D. T. Gish, J. Young, H. Fraenkel-Conrat, C. A. Knight, and W. M. Stanley, these PROCEEDINGS, **46**, 1463 (1960).
[21] Neurath, H., and G. W. Schwert, *Chem. Revs.*, **46**, 69 (1950).
[22] Carlton, B. C., and C. Yanofsky, unpublished results.
[23] Koshland, D. E., in "Symposium on Enzyme Reaction Mechanisms," *J. Cell. Comp. Physiol.*, **54**, suppl. 1, 245 (1959).
[24] Kohn, J., *Nature*, **180**, 986 (1957).
[25] White, F. H., *J. Biol. Chem.*, **236**, 1353 (1961).
[26] Hirs, C. H. W., S. Moore, and W. Stein, *J. Biol. Chem.*, **219**, 623 (1956).

XII
Suppressors and Proteins

Editor's Comments on Papers 30 Through 33

30 **Suskind and Kurek:** *On a Mechanism of Suppressor Gene Regulation of Tryptophan Synthetase Activity in* Neurospora crassa

31 **Garen and Siddiqi:** *Suppression of Mutations in the Alkaline Phosphatase Structural Cistron of* E. coli

32 **Capecchi and Gussin:** *Suppression* in Vitro: *Identification of a Serine-sRNA as a "Nonsense" Suppressor*

33 **Helinski and Yanofsky:** *A Genetic and Biochemical Analysis of Second Site Reversion*

In all the papers to this point we have been concerned primarily with the more direct relations between gene and protein. Now we turn our attention to another aspect of the role of the genetic material in the formation of the phenotype, the interactions that must occur within cells to produce a particular phenotype. These have generally been referred to as "gene interactions" with the usually implied understanding that this was not necessarily the direct interaction of genes themselves, but rather of their products. In Paper 30 by Suskind and Kurek we find an excellent example of a mutant of *Neurospora* that produces an altered protein which has enzyme activity as tryptophan synthetase, but only under certain conditions, for example, above 30°C or when it is partially purified from inactive extracts of the mutant. The mutant enzyme is obviously very similar to the wild-type enzyme, but differs in its sensitivity to certain compound(s) ordinarily expected to be found in living cells. A mutation at another locus relieves this inhibition, and it is suggested that this suppressor does so by lowering the concentration of inhibitor. This is the first good example of a possible molecular explanation for gene interaction involving suppressor genes.

The study of suppressor activity has led to important findings relating to gene action manifested at the protein level. Paper 31 deals with an interesting group of alkaline phosphatase mutants of *Escherichia coli* that produce neither alkaline phosphatase enzyme activity nor CRM. The evidence is all in favor of their carrying nonsense mutations, that is, mutant codons which do not code for any amino acid. These mutants are susceptible to suppression by a certain class of external suppressors (intergenic suppressors). In the presence of the suppressors, active enzyme and CRM are made. The explanation proposed is that the suppressor activity converts nonsense into sense. Subsequent work has shown that these suppressor genes are in reality tRNA genes.

Capecchi and Gussin present evidence showing that an amber nonsense codon is read by a serine sRNA (=tRNA) in the presence of a certain suppressor gene *(su-A)*. The logical conclusion is that the suppressor gene locus is a tRNA locus. Mutation has created a modified tRNA that is able to read the amber codon and introduce a serine into the polypeptide. This interpretation has since been verified for this and several other loci in *Escherichia coli* and a number of other organisms.

In Paper 32 it was shown that the suppression of a mutant gene which resulted in the appearance of a functional protein was the consequence of a second genetic change in the way "information" was interpreted by the protein-synthesizing machinery. In

Paper 33 it is made evident that a second mutation in the same gene can also result in suppression (second site reversion) in *Escherichia coli*. The original missense mutation results in a nonfunctional tryptophan synthetase, and a second missense mutation causing another amino acid substitution in the same polypeptide results mysteriously in a functional enzyme. This unexpected consequence of two amino acid substitutions returning a polypeptide to functional status can best be explained by assuming that the first mutation is corrected by the second mutation because the double mutant polypeptide has a tertiary structure more similar to the original wild-type enzyme.

Selected Bibliography

Brody, S., and C. Yanofsky. 1963. Suppressor gene alteration of protein primary structure. *Proc. Natl. Acad. Sci. U.S. 50:* 9–16.

Garen, A. 1968. Sense and nonsense in the genetic code. *Science 160:* 149–159.

Gilmore, R. A., J. W. Stewart, and F. Sherman. 1971. Amino acid replacements resulting from supersuppression of nonsense mutants of iso-*l*-cytochrome c from yeast. *J. Mol. Biol. 61:* 157–173.

Gorini, L. 1970. Informational suppression. *Ann. Rev. Genetics 4:* 107–134.

Hartman, P. E., and J. R. Roth. 1973. Mechanisms of suppression. *Adv. Genetics 17:* 1–105.

ON A MECHANISM OF SUPPRESSOR GENE REGULATION OF TRYPTOPHAN SYNTHETASE ACTIVITY IN NEUROSPORA CRASSA*

BY SIGMUND R. SUSKIND AND LORETTA I. KUREK

MCCOLLUM-PRATT INSTITUTE, JOHNS HOPKINS UNIVERSITY, BALTIMORE, MARYLAND

Communicated by B. H. Willier, December 22, 1958

There is abundant evidence that the potential of a cell to synthesize a specific enzyme is primarily under the control of a single genetic locus.[1, 2] However, repeated instances have been found in which modifier genes, called "suppressors," exert a profound effect on enzyme activity in certain mutant cells.[3-8] It is known, for example, that specific suppressor mutations are capable of partially restoring enzyme activity in mutants which lack this particular activity.[3, 5] These suppressor genes are distinct from the original enzyme locus, and, as is often the case, there is no linkage between the two loci.[3] Furthermore, the suppressor genes do not seem to be duplications of the original enzyme locus, for the suppressors often act in combination with certain mutant alleles only.[3] Consequently, it seems quite clear that a number of genetic regions may be capable of controlling the appearance of a single enzyme activity.

This paper is concerned with a discussion of the nature of suppressor gene control and, in particular, with whether it exists at the level of enzyme formation, either qualitative or quantitative, or of enzyme function. The basic problem would seem to hinge on the relative contributions of the original enzyme locus and the suppressor loci to the structure of the enzyme. If it were to be found in a given instance that the responsibility for the determination of enzyme structure residues entirely at a single locus, then an explanation of suppressor gene action would have to be sought in terms of those factors which may regulate the expression of enzyme activity.

An effort will be made here to review some of the results obtained by workers in several laboratories, who have been studying a particular mutant-suppressor system, the *td* system in *Neurospora crassa* and *Escherichia coli*. Some new evidence will be presented which suggests that, in at least one case in *Neurospora*, suppressor gene action is not concerned with providing genetic "information" essential for the synthesis of a structurally normal enzyme.

Experiments with *Neurospora* and *E. coli* have shown that certain tryptophan-requiring (*td*) mutants which are unable to utilize indole for growth and which lack tryptophan synthetase activity form, instead, a protein (CRM)[9] which is antigenically similar to the enzyme.[10-12] Although immunological experiments[13] have failed to reveal differences between the CRM proteins of different *td* mutants, recent enzymatic studies[14-16] suggest that characteristic differences may be present. It has been found that the CRM proteins of certain *td* mutants of *E. coli*[12,15] and *Neurospora*[16-18] are able to catalyze indole formation from indole glycerol phosphate as well as to synthesize indole glycerol phosphate from indole and triose phosphate. However, such CRM proteins are unable to catalyze a reaction with L-serine in the step necessary for tryptophan synthesis.[15, 16, 18] In some of the *td* mutants the CRM proteins have lost all their enzymatic activity, failing even to react with indole or indole glycerol phosphate.[16] It appears most likely, therefore, that the CRM proteins in the *td* mutants represent a spectrum of characteristic genetically altered tryptophan synthetase molecules in which one or more catalytic properties of the normal enzyme have been modified or lost.[11, 14-16]

The relationship between *td* mutant suppressibility and the presence of a CRM protein is of particular interest. One might reasonably ask whether the action of a suppressor gene which restores tryptophan synthetase activity in a certain *td* mutant is contingent on the presence of a CRM protein. While *td*-suppressor mutations have not been demonstrated in every CRM-forming mutant, all the suppressible mutants of *Neurospora* so far tested form a CRM protein.[16, 19] Those mutants which do not form CRM appear to be unsuppressible.[10, 11, 16, 19] Mutant strains which contain low levels of tryptophan synthetase activity following the introduction of a specific suppressor gene,[3] also continue to form large quantities of CRM.[10, 16] These results suggest that only a partial "repair" of damage to the enzyme system has occurred as a result of suppressor gene action. Some recent work with *E. coli* on the contrary, suggests that CRM may not be required for suppressibility.[20] However, it should be emphasized that the absence of CRM does not preclude the synthesis of other aberrant proteins lacking the necessary determinant groups for immunological cross-reaction with tryptophan synthetase.

It is clear from the work of Yanofsky and Bonner[3] that suppressor genes which are effective in partially restoring tryptophan synthetase activity in the *td* mutants of *Neurospora* exhibit a striking allele specificity, that is to say, a particular suppressor gene may affect one or more alleles of a series but not others of the same series. It may be inferred from such specificity relationships that certain suppressor loci contain "information" relevant to the structure of the enzyme. Indeed, these studies offer one of the strongest arguments for such an interpretation. Subsequent experiments[21] have disclosed a somewhat anomalous situation in which five "specific" suppressor genes were found to affect the same *td* allele. Oddly enough, no linkage has been observed between the *td* locus and any of these suppressor genes, and the five suppressor genes themselves appear to be randomly distributed in the genome. Contrary to expectation, tryptophan synthetase prepared from wild type or from any one of several suppressed mutant strains exhibited similar biochemical properties.[3] These results strongly suggest that several genes, located in widely different chromosomal regions, may be capable of "repairing" a damaged enzyme-forming system sufficiently to permit the cell to form enzyme which seems indistinguishable from the wild-type protein.

Our recent experiments with a temperature-sensitive tryptophan-requiring mutant of *N. crassa*, strain td_{24}, have provided some information on this important and perplexing aspect of gene-enzyme relationships. Strain td_{24} requires tryptophan for growth at 25° C., but it grows slowly without tryptophan above 30° C., and at this temperature also forms a slight amount of active tryptophan synthetase.[3] In all instances this mutant forms large quantities of a CRM protein.[11, 14] It has been possible to obtain highly *active* tryptophan synthetase from this mutant grown at 25° C., by suitable fractionation of crude *inactive* extracts.[14] A considerable quantity of CRM is present in both the crude inactive preparations and the active fraction. An inorganic inhibitor which completely inhibits the fractionated mutant enzyme but which has no effect on the wild-type enzyme at comparable concentrations can be isolated from the inactive td_{24} preparations. Whether this inhibitor is associated with an organic component in vivo is not known. The inhibitor can be found in other *td* mutants and also in wild-type strains. To date, the effects of the inhibitory material can best be duplicated by zinc.[19]

It appears from studies with strain td_{24} that gene mutation has in this instance resulted in the formation of an altered, metal-sensitive enzyme. The active enzyme which is obtained by the fractionation of crude inactive extracts appears to represent a conversion of CRM to active tryptophan synthetase by the dissociation of a metal-protein complex.[19]

The metal-sensitivity difference between the wild-type and td_{24} tryptophan synthetase preparations has provided one approach to the study of suppressor gene action. If a suppressor gene actually provides "information" essential for the synthesis of an enzyme possessing normal structure, then tryptophan synthetase formed in mutant td_{24} carrying the suppressor-24 (su_{24}) gene should be inhibitor-resistant, like the wild-type enzyme. If, on the other hand, the su_{24} gene in some manner converts an altered enzyme from an inactive to an active state without changing the basic structure of the protein, then the suppressed and the unsuppressed mutant enzymes might show similar inhibitor sensitivity. The results of a number of inhibitor experiments are summarized in Table 1. The data seem to favor the interpretation that suppressor action is based on control at the functional rather than the enzyme-forming level. Partially purified preparations of tryptophan synthetase from suppressed mutant strains td_{24} su_{24} and td_3 su_3 (a strain which is cross-suppressible with td_{24})[3] appear to be metal-sensitive, and in this manner resemble tryptophan synthetase from the unsuppressed mutant, td_{24}.

TABLE 1*

The Sensitivity of Wild-Type, *td* Mutant, and *td*-Suppressed Mutant Tryptophan Synthetases to Ashed td_{24}-Inhibitor and to Zinc

Partially Purified Tryptophan Synthetase	Concentration Range for 50 Per Cent Inhibition of Enzyme Activity	
	td_{24} Inhibitor (ml.)	Zinc Acetate (Molar Concentration of Zn^{++})
Strains:		
td_{24}	0.05 to 0.1	5×10^{-5} to 1×10^{-4}
td_{24} su_{24}	0.05 to 0.1	5×10^{-5} to 1×10^{-4}
td_3 su_3	0.05 to 0.1	5×10^{-5} to 1×10^{-4}
5256A (wild-type)	0.5 to 1.0	5×10^{-4} to 1×10^{-3}
5297a (wild-type)	0.5 to 1.0	5×10^{-4} to 1×10^{-3}
C-84 (histidine$^-$)†	0.5 to 1.0	5×10^{-4} to 1×10^{-3}

* Extracts were purified using protamine sulfate, ammonium sulfate, and alumina gel (see n. 11). The enzyme activity in all the preparations was adjusted to a comparable level. In all the procedures, triple-distilled water was employed. Ashed inhibitor was prepared from strain td_{24} by boiling an aliquot of a 25-80 per cent ammonium sulfate fraction of the td_{24} extract, centrifuging and discarding the precipitate, and ashing the supernatant solution. The ashed material was taken up in distilled water and brought to the volume of the original sample. Enzyme preparations from the different strains were treated with the ashed inhibitor or with zinc for 5 minutes in an ice bath. The substrate mixture, consisting of saturating concentrations of indole, L-serine, and pyridoxal phosphate (see n. 24), was added, and the tubes containing the complete system were incubated for 1 hour at 37° C. The reaction was then stopped with 0.2 ml. of 5 per cent sodium hydroxide, and the amount of indole that had disappeared in the control and the inhibitor or zinc-containing samples was measured by the p-dimethyl aminobenzaldehyde reaction (see n. 24).
† D. Hogness and H. K. Mitchell, *J. Gen. Microbiol.*, 11, 401, 1954.

These findings suggest that in some instances suppressor gene action in the *td* system may be concerned with the control of factors such as metal and coenzyme availability or the concentration of metal-binding agents. The results of current studies[19] clearly indicate that the synthesis, stability, and activity of tryptophan synthetase can be markedly influenced by the concentration of copper and zinc and of pyridoxal phosphate and by the presence of chelating agents. Certainly, the regulation of such factors in vivo may be influenced by many genes, and it would be reasonable to suppose that suppressor genes may be effective in such ways. In the case of mutant td_{24}, where the effect of a suppressor gene can be effectively mimicked in vivo by increasing the temperature and in vitro by fractionation, the suppressor gene does not appear to supply information concerning the structure of the enzyme. The latter role would seem to be reserved for the td locus itself.

Summary.—Evidence has been presented that suppressor gene action may in certain cases be concerned with the expression of enzyme activity rather than with enzyme formation. In the *Neurospora* tryptophan synthetase system, the suppressor gene does not appear to provide "information" necessary for the synthesis of a structurally normal enzyme.

* Contribution No. 245 of the McCollum-Pratt Institute. The investigation was supported by a research grant (RG-3080 M and G) from the National Institutes of Health, U.S. Public Health Service.

[1] N. H. Horowitz, *Federation Proc.*, 15, 818, 1956.
[2] D. M. Bonner, *Cold Spring Harbor Symposia Quant. Biol.*, 21, 163, 1956.
[3] C. Yanofsky and D. M. Bonner, *Genetics*, 40, 761, 1955.
[4] H. B. Glass, *Science*, 126, 683, 1957.
[5] J. Gots, in *Genetic Studies with Bacteria* ("Carnegie Institution of Washington Publications," No. 612 [1956]), p. 87.
[6] T. Yura, in *Genetic Studies with Bacteria* ("Carnegie Institution of Washington Publications," No. 612 [1956]), p. 63.
[7] B. Straus, *Am. Naturalist*, 89, 141, 1955.
[8] C. Yanofsky, these PROCEEDINGS, 38, 215, 1952.
[9] The quantitation of CRM in extracts of the td mutants lacking tryptophan synthetase activity is based on the fact that CRM and tryptophan synthetase exhibit equal affinity for tryptophan-synthetase-neutralizing antibody. One CRM unit is defined as the antigenic equivalent of one tryptophan synthetase unit and is determined under standard conditions by an enzyme neutralization method (see n. 11).
[10] S. R. Suskind, C. Yanofsky, and D. M. Bonner, these PROCEEDINGS, 41, 577, 1955.
[11] S. R. Suskind, *J. Bacteriol.*, 74, 308, 1957.
[12] P. Lerner and C. Yanofsky, *J. Bacteriol.*, 74, 494, 1957.
[13] S. R. Suskind, *Proc. VIIth Intern. Congress for Microbiology*, 1958.
[14] S. R. Suskind and L. I. Kurek, *Science*, 126, 1068, 1957.
[15] C. Yanofsky and J. Stadler, these PROCEEDINGS, 44, 245, 1958.
[16] J. A. DeMoss, A. Wust, A. Lacy, and D. M. Bonner, personal communication.
[17] J. A. DeMoss and D. M. Bonner, *Bacteriol. Proc.*, p. 112, 1908.
[18] S. R. Suskind and E. Jordan, submitted to *Science*.
[19] W. D. Mohler, M. Garrick, L. I. Kurek, and S. R. Suskind, unpublished results.
[20] C. Yanofsky, *Science*, 128, 843, 1958.
[21] C. Yanofsky and D. M. Bonner, *Genetics*, 40, 602, 1955.
[22] A. Nason, N. O. Kaplan, and S. P. Colowick, *J. Biol. Chem.*, 188, 397, 1951.
[23] S. R. Suskind, *Bacteriol. Proc.*, p. 104, 1958.
[24] C. Yanofsky, in *Methods in Enzymology*, ed. S. P. Colowick and N. O. Kaplan (New York: Academic Press, 1955), 2, 233.

Reprinted from *Proc. Natl. Acad. Sci.*, **48**(7), 1121–1127 (1962)

SUPPRESSION OF MUTATIONS IN THE ALKALINE PHOSPHATASE STRUCTURAL CISTRON OF E. COLI*

By A. Garen and O. Siddiqi

BIOLOGY DIVISION, UNIVERSITY OF PENNSYLVANIA

Communicated by Seymour Benzer, May 28, 1962

A suppressor mutation is defined as one that overcomes the defect caused by another mutation. Of particular interest are the "external" suppressors that are located outside the cistron in which the suppressed mutations occur, and are active only on some of the mutations within the cistron.[1,2] The present communication describes the properties of an external suppressor for a class of mutations in the structural cistron for alkaline phosphatase of *E. coli*.[3–5] The results suggest that this class of phosphatase-negative mutations produces a nucleotide configuration that is nonsense (unable to specify any amino acid) in the absence of the suppressor but which functions as a sense configuration when the suppressor is present.

Materials and Methods.—Bacterial strains: In all experiments, the Hfr was *E. coli* strain $K10$ and the F^- was *E. coli* strain $W1$. The characteristics of these strains have been described elsewhere.[3–6] Several of the P^- mutants were previously isolated in collaboration with C. Levinthal and E. Lin. The selective markers employed for the genetic crosses were T^+ and L^+ (ability to grow in a medium lacking threonine and leucine), S^r (ability to grow in the presence of 0.1 mg/ml of streptomycin), and M^+ (ability to grow in a medium lacking methionine).

Media: The composition of the low phosphate medium used to prepare cultures for alkaline phosphatase assays was as follows: 1.2×10^{-1} M tris buffer; 0.2% glucose; 6.4×10^{-5} M KH_2PO_4; 8×10^{-2} M $NaCl$; 2×10^{-2} M KCl; 2×10^{-2} M NH_4Cl; 3×10^{-3} M Na_2SO_4;

$1 \times 10^{-3}\ M$ $MgCl_2$; $2 \times 10^{-4}\ M$ $CaCl_2$; $1 \times 10^{-5}\ M$ $ZnCl_2$; $2 \times 10^{-6}\ M$ $FeCl_3$; 0.02% Difco casamino acids (phosphate free).

Genetic crossing and blendor experiments: The procedures have been described elsewhere.[6]

Preparation of cell extracts: Cultures were grown to saturation in low phosphate medium at 37°C with aeration. The cells were harvested and resuspended in a solution of 0.05 M tris buffer pH 7.5 + 0.02 M $MgCl_2$. This suspension was passed once through a French pressure cell to disintegrate the cells and afterwards was centrifuged at 10,000 g for 10 min. The supernatant was used for enzymatic activity and CRM assays.

Enzymatic activity assay: A sample of the cell extract was added to a solution containing 4 mg/ml p-nitrophenylphosphate in 1 M tris buffer pH 8.0 maintained at 32°C, and the rate of change in the optical density at 410 mμ was measured.

Immunological cross-reacting material (CRM) assay: A technique developed by Preer[16] was used. Rabbit antiserum was prepared by inoculations of purified normal enzyme as antigen. A constant concentration of the serum was assayed in microtubes against serial twofold dilutions of cell extract. The relative amount of CRM in an extract was estimated by the position of the precipitation band in the microtubes as compared with the position of the band formed by a standard solution of the normal enzyme. The CRM value is given as per cent of the normal enzyme, in categories of 100, 50, 25, 12, 6, and 3 per cent. Each extract preparation could be unambiguously classified in one of the categories.

Response to 5-fluorouracil:[7] Cultures of phosphatase-negative mutants were grown in low phosphate medium (without casamino acids) to a concentration of about 3×10^8 cells/ml. The cells were centrifuged, washed, and resuspended in an equal volume of low phosphate medium. To one portion of the culture, 20 μg/ml 5-fluorouracil was added; the untreated cells served as the control. After 6 hr of incubation at 37°C, the cultures were shaken with a few drops of toluene and subsequently assayed for enzymatic activity.

Results.—Effect of the suppressor on enzyme synthesis: The first indication of the existence of a suppressor in the phosphatase system came from the anomalous behavior of certain phosphatase-negative mutants in genetic crosses. Starting with a phosphatase-negative mutant of the *Hfr* strain, it was observed that enzymatic activity appeared when the P^- marker was transferred to an F^- strain (not isogenic with the *Hfr*).[3, 4] Of 220 independently isolated mutants examined, 15 responded in this way. The enhanced enzymatic activity in the genetic environment of the F^- strain suggested that the F^- harbored a suppressor for these P^- mutations. A comparison of the amounts of enzyme produced in nonsuppressed (*Hfr*) and suppressed (F^-) strains of the mutants is shown in Table 1. The main points to be noted are that essentially no enzyme is produced by any of the nonsuppressed strains, while in the suppressed strains the levels of both enzymatic activity and CRM range from 3 to 100 per cent of the values for the standard P^+ strain.

Genetic mapping of the suppressor: The map position of a suppressor for one of the P^- mutants (*U8*) was established by means of a Blendor experiment[8] which measured the time of transfer of a suppressor marker from an *Hfr* to an F^-. Two strains were required for this experiment, an F^- that did not contain a suppressor for *U8* (su^-) and an *Hfr* with the suppressor (su^+). The first strain was prepared by the following cross: *Hfr U8 su^- M^+ S^s* × *F^- U8 su^+ M^- S^r*. The purpose of this cross was to transfer the su^- marker to the F^-. A selection was made for $F^-M^+S^r$ recombinants, and it was found that about 0.5 per cent were su^-. This result furnished preliminary evidence that the *su* marker was transferred later than *M*, which enters 35 min after the start of mating. The second strain was prepared by isolating enzymatically active revertants from the *Hfr U8 su^-* strain. These revertants were of two types, one in which the reverse mutation was in the *P* cistron (P^+ revertant) and another in which the reverse mutation was a suppressor (*U8 su^+* revertant). The latter was used as the $Hfr\ U8\ su^+$ strain for the Blendor experiment. The amount of enzyme produced by the $Hfr\ U8\ su^+$ was nine per cent of the amount in the standard P^+ strain. This is approximately the same value previously obtained with the F^- strain of this mutant (see Table 1). It is possible that the same suppressor for $U8$ is involved in the Hfr and the F^-.

TABLE 1

ALKALINE PHOSPHATASE LEVELS IN NONSUPPRESSED AND SUPPRESSED MUTANT STRAINS

Suppressible P⁻ mutants	Nonsuppressed Strains		Suppressed Strains	
	Enzymatic activity	CRM	Enzymatic activity	CRM
U2	0.02	<1	13.	12
U5	0.02	<1	7.	6
U8	0.04	<1	12.	12
U20	0.03	<1	3.	3
U46	0.04	<1	24.	25
U56	0.07	<1	15.	12
E35	0.04	<1	3.	3
G5	0.03	<1	10.	12
G55	0.05	<1	18.	25
A7	0.02	<1	5.	6
H7	0.03	<1	5.	6
S16	0.04	<1	18.	25
S26	0.02	<1	100.	100
S42	0.05	<1	18.	25
S45	0.02	<1	18.	25

The P^- mutants were isolated after treatment of the standard P^+ strain with one of the following mutagenic agents: ultraviolet light (U), X ray (E), N-methyl-N'-nitro-N-nitrosoguanidine (G), 2-aminopurine (A), hydroxylamine (H), or ethylmethanesulfonate (S).

The nonsuppressed strains are the mutants as originally isolated in the Hfr. The suppressed strains were produced by crossing the Hfr mutants with an F^- strain that carried a nonsuppressible P^- mutation. The selective markers used in the crosses were T^+L^+ in the Hfr and S^r in the F^-. The recombinant class $T^+L^+S^r$ was tested for alkaline phosphatase activity, and it was found that about half the recombinants were enzymatically active. These were used as the suppressed strains.

For the enzymatic activity measurements, extracts of cultures grown in limiting phosphate medium were prepared as described under *Methods*. The CRM (immunological cross-reacting material) assays were done on the same extracts. All results in the table have been normalized to a standard cell concentration and are expressed as the per cent of the value obtained with the P^+ strain. Each experiment was done in duplicate on separate cultures. The enzymatic activities were reproducible to ±25%, and the CRM results to within the limits indicated in the section on *Materials and Methods*.

The blendor experiment was done by crossing $Hfr\ U8\ su^+\ S^s$ with $F^-\ U8\ su^-\ S^r$. At different times, samples were blended to stop mating and were then plated on a medium selective for $U8\ su^+\ S^r$ recombinants, to score for the entry of the su^+ marker into the F^-. The time of entry of su^+ was at 90 min, as compared to 9 min for the P marker. Thus, the suppressor for $U8$ is well separated from the P cistron, by a distance equivalent to about 75 per cent of the total length of the bacterial chromosome.[8]

The map position of a suppressor for another mutant, $S26$, was established by an analogous blendor experiment. The cross was between an $Hfr\ S26\ su^+S^s$ and an $F^-\ S26\ su^-\ S^r$, each strain being produced by the procedures described for the preceding cross. The $S26$ suppressor was transferred at about the same time as was the $U8$ suppressor, suggesting that the same suppressor may be involved with both mutants.

The two F^- strains prepared for the preceding blendor experiments, one containing an su^- marker for the $U8$ mutation and the other an su^- marker for the $S26$ mutation, were each crossed with all of the suppressible Hfr mutants to determine whether these su^- markers prevented suppression of the entire group of P^- mutations. None of the P^- mutations were suppressed when crossed into either F^-

strain. This is additional evidence that a single suppressor is responsible for suppression of all the P^- mutations listed in Table 1.

Genetic mapping of suppressible P^- mutations: The results in Table 1 show that the suppressible mutations have a wide range of responses to the suppressor. To determine whether they are located at different genetic sites, the following crosses were performed. (In all of these crosses, both the *Hfr* and F^- strains were su^-.) Each suppressible *Hfr* mutant was crossed against a P^- "deletion" mutant in which the genetic defect spans about two thirds of the P cistron. Five of the mutants ($U2$, $U5$, $U20$, $U56$, and $S45$) produced P^+ recombinants and therefore are located outside the span of the deletion. One of these ($U5$) and four of the mutants within the span ($U8$, $S26$, $E35$, and $U46$) were each crossed against the other suppressible mutants and against each other. In every case, P^+ recombinants were found. Thus, the number of suppressible sites must total at least seven, five within the span and two outside.

Properties of the enzyme produced by the suppressed mutants: A comparison of the enzymatic activity and CRM values for the suppressed mutants (Table 1) shows that the specific enzymatic activities per unit of CRM are approximately normal for all mutants. Two other properties, heat stability and electrophoretic mobility, were also examined to test for possible structural alterations in the enzymes. The standard P^+ enzyme is exceptionally heat-stable,[9] capable of surviving exposure of up to 90°C for 30 min without loss of activity. The enzymes from eight suppressed mutants ($U5$, $U8$, $U46$, $U56$, $S16$, $S26$, $S42$, $S45$) were as heat-stable at the standard enzyme. With regard to electrophoretic mobility, it is known that electrophoretically altered enzymes can result from mutations in the P cistron.[3, 4, 10] The electrophoretic behavior of the enzymes from six suppressed mutants ($U5$, $U8$, $U46$, $S45$, and $G5$) was found to be identical to that of the standard P^+ enzyme.

Complementation tests with suppressible mutants: Complementation has been observed between pairs of different P^- mutants.[11] The mechanism for complementation appears to be the formation of hybrid enzyme molecules containing one protein chain derived from each P^- cistron (the standard P^+ enzyme molecule is a dimer composed of two identical chains).[12] Of particular importance for the present work is the fact that complementation can occur with P^- mutants that by themselves produce no detectable enzymatic activity or CRM. Such mutants presumably are capable of producing the protein chains of the enzyme in a defective form. We have carried out complementation experiments with the suppressible mutants as a measure of their capacity to synthesize the protein chains. Five of the suppressible mutants ($U5$, $U8$, $U46$, $S26$, and $E35$) were tested for complementation against each of five nonsuppressible P^- mutants that are known to be capable of complementing.[11] There was no indication of complementation in any test involving a suppressible mutant.

Effect of 5-fluorouracil on suppressible P^- mutants: A physiologial method for reversing the phenotype of certain *rII* mutants of T4 phage (and also of phosphatase-negative mutants of *E. coli*) by the action of 5-fluorouracil (FU) has recently been reported.[7] FU appears to reverse specifically those mutational defects that result from the transition of a G-C to an A-T base pair in DNA. This finding makes it possible to test for such a base pair change in a P^- mutant: a positive response to

FU would be an indication that this change had occurred, although failure to respond would not necessarily mean the contrary. The 15 suppressible P^- mutants in Table 1 were tested (in an su^- strain) for their response to FU, the criterion being an increase in amount of enzymatic activity in cultures treated with FU. Positive results were obtained with all except two of the mutants (Table 2). The magnitude of the response to FU varied, the maximum value being 0.3 per cent of the activity of the standard P^+ strain. This is considerably below the level of suppression achieved with the su^+ strains (see Table 1) but nevertheless is significant.

TABLE 2

RESPONSE OF PHOSPHATASE-NEGATIVE MUTANTS TO 5-FLUOROURACIL

Suppressible P^- mutants	Enzymatic Activity		Response
	Without FU	With FU	
U2	0.03	0.2	+
U5	<0.03	<0.03	−
U8	0.03	0.13	+
U20	0.03	0.2	+
U46	0.04	0.14	+
U56	0.06	0.34	+
S16	0.03	0.16	+
S26	<0.03	0.48	+
S42	0.08	0.25	+
S45	<0.03	0.08	+
E35	<0.03	0.14	+
G5	<0.03	0.25	+
G55	0.03	0.4	+
A7	<0.03	0.04	(+)
H6	0.03	0.2	+
Nonsuppressible P^- mutants			
U3	<0.05	<0.05	−
U7	<0.05	<0.05	−
U18	<0.05	<0.05	−
U24	<0.05	<0.05	−
U38	0.13	0.13	−
U58	<0.05	<0.05	−
S10	<0.05	0.16	+
S36	<0.05	<0.05	−
S41	<0.05	<0.05	−

All of the mutants were Hfr strains that contained an su^- marker. The procedures involved in the FU tests are described in the section on *Materials and Methods*. The enzymatic activities are expressed as per cent of the value obtained with the P^+ strain without FU.

As a control, a group of P^- mutants not suppressible in the su^+ strain was also tested with FU. Only one of these responded to the FU treatment. Thus, the general responsiveness of the mutants in Table 1 to FU implies that they may have in common a G-C to A-T alteration in the P cistron.

Discussion.—All 15 of the suppressible P^- mutants are incapable of synthesizing any enzyme protein when the suppressor is not functioning, as indicated by the absence of CRM and enzymatic activity and by the failure to complement other P^- mutants. This behavior distinguishes these mutants from other P^- mutants that do not respond to this particular suppressor, since a majority of the latter do have the capacity to synthesize enzyme protein. These suppressible mutants may, therefore, all originate from nonsense mutations that block the formation of the protein molecule.

It is striking that suppression of one of the P^- mutants enables the fully normal amount of enzyme to be synthesized, thus completely restoring the function of the

P cistron despite the continued presence of a P^- mutation. This mutation would not have been detectable if the screening for P^- mutants had been carried out with a strain that contained the suppressor. Here, we have an extreme example of suppression, an absolute negative mutant being transformed into an apparently normal cell.

According to a recently proposed mechanism of suppression,[1, 2, 13] a suppressor strain might contain a new kind of transfer RNA-amino acid complex that generates a specific kind of mistake in proteins by substituting one particular amino acid for another. This could result in the restoration of a function in a mutant if the mistake happened to compensate for the amino acid defect produced by the mutation. Since such a mistake mechanism would affect all proteins, only a limited number of mistakes are likely to be tolerated by the cell.

In the special case of suppression of a nonsense mutation, this difficulty would not arise. While the same type of suppressor mechanism could operate, its effect would be limited to the nonsense site which would not recur elsewhere. Instead of producing mistakes, the suppressor would transform nonsense into sense. Since the only protein affected would be the one under the control of the cistron containing the nonsense mutation, there need not be any limitation to the efficiency of the process. The high efficiency of the alkaline phosphatase suppressor could be accounted for in this way.

Other evidence in support of the conclusion that the alkaline phosphatase suppressor acts on nonsense mutations is presented in an accompanying paper by Benzer and Champe,[14] in which they describe an ingenious test for nonsense mutations in the *rII* region of phage T4. They have found that the alkaline phosphatase suppressor can also suppress certain *rII* mutations[15] and that all of these concomitantly suppressible *rII* mutations appear to be of the nonsense type.

Summary.—A group of 15 phosphatase-negative mutants, involving at least seven separate sites in the structural cistron, are suppressible by an external suppressor. All of the mutants are incapable of synthesizing any enzyme protein when the suppressor is not functioning. In the presence of the functional suppressor, the amount of enzyme synthesized ranges from 3 to 100 per cent of normal for different mutants. The enzyme formed by the suppressed mutants is not detectably different from the normal enzyme in electrophoretic mobility, heat stability, or specific activity. It is proposed that the suppressor acts by converting nonsense mutations into sense.

We wish to express our gratitude to Suzanne Garen for her participation in these experiments and to Rebecca Greer for her technical assistance.

* This work was supported by grants from the National Science Foundation and National Institutes of Health (U.S. Public Health Service).

[1] Yanofsky, C., and P. St. Lawrence, *Ann. Rev. Microbiol.*, **14**, 311 (1960).
[2] Benzer, S., and S. P. Champe, these PROCEEDINGS, **47**, 1025 (1961).
[3] Levinthal, C., in *Structure and Function of Genetic Elements*, Brookhaven Symposia in Biology No. 12 (1959), p. 76.
[4] Garen, A., in *Microbial Genetics*, Society for General Microbiology (Cambridge University Press, 1960).
[5] Garen, A., C. Levinthal, and F. Rothman, *J. Chemie Physique*, **58**, 1068 (1961).
[6] Echols, H., A. Garen, S. Garen, and A. Torriani, *J. Mol. Biol.*, **3**, 425 (1961).
[7] Champe, S., and S. Benzer, these PROCEEDINGS, **48**, 532 (1962).
[8] Jacob, F., and E. L. Wollman, in *The Biological Replication of Macromolecules*, ed. F. K. San-

ders, Symposia of the Society for Experimental Biology, No. 12 (New York: Academic Press 1958), p. 75.

[9] Garen, A., and C. Levinthal, *Biochim. Biophys. Acta*, **38**, 470 (1960).

[10] Bach, M. L., E. R. Signer, C. Levinthal, and I. W. Sizer, *Fed. Proc.*, **20**, 255 (1961).

[11] Garen, A., in preparation.

[12] Rothman, F., in preparation.

[13] Yanofsky, C., D. Helinski, and B. Maling, in *Cellular Regulatory Mechanisms*, Cold Spring Harbor Symposia on Quantitative Biology, vol. 26 (1961), p. 11.

[14] Benzer, S., and S. Champe, these PROCEEDINGS.

[15] Benzer, S., S. Champe, A. Garen, and O. Siddiqi, in preparation.

[16] Preer, J. R., Jr., and L. B. Preer, *J. Protozool.*, **6**, 88 (1959).

Suppression *in Vitro*: Identification of a Serine-sRNA as a "Nonsense" Suppressor

MARIO R. CAPECCHI and GARY N. GUSSIN

Abstract. *In a cell-free system, with RNA from a suppressible mutant of bacteriophage R17 as messenger, no functional coat protein of the bacteriophage is synthesized unless serine-accepting soluble RNA from the suppressor strain, Escherichia coli S26R1E, is present.*

The molecular mechanism by which the amber suppressor genes in *Escherichia coli* reverse the effects of mutations in other genes recently has been shown to be a misreading of the genetic code. The amber suppressor genes constitute one class of suppressors; there is evidence that other suppressors in *E. coli* (1, 2) and in yeast (3) act in a similar manner.

Mutations suppressible by amber suppressor genes can occur in many different cistrons, regardless of the cistron's function. These amber mutations—mutations suppressible by amber suppressor genes—have been isolated in the rII region of bacteriophage T4 (4), in the entire T4 genome (5), in the *E. coli* structural gene specifying alkaline phosphatase (6), in the bacteriophage lambda (7), and in the RNA-containing (8) bacteriophage f2 (9).

Benzer and Champe (10) and Garen and Siddiqi (6) first suggested that the amber mutation produces a nonsense codon—a nucleotide triplet which does not code for any amino acid and thus interrupts polypeptide synthesis in a nonpermissive cell (one lacking an amber suppressor). They further postulated that this codon would be read as an amino acid code word in a permissive (suppressor-containing) host, thereby permitting protein synthesis to continue. This hypothesis received strong support from the work of Brenner and his colleagues, who showed that amber mutants in the head protein of bacteriophage T4D produce a normal yield of released chain fragments in a nonpermissive *E. coli* host, but make a mixture of about 60 percent complete protein and 40 percent fragments in the permissive host (11).

Four genetically distinguishable amber suppressor genes in *E. coli* have been identified (12). The action of one of these, the *su-A* suppressor gene contained in strain S26R1E of Garen and Siddiqi, has been investigated extensively. In the T4D head protein (13), in alkaline phosphatase (12), and in f2 coat protein (14), a glutamine or tryptophan residue in the wild-type protein is always replaced by serine in the mutant protein produced in the *su-A*–containing host. Thus, the amber nonsense codon has arisen through mutations from triplets coding for glutamine and tryptophan. An alteration of the protein-synthesizing machinery of the permissive host has changed its ability to recognize the nonsense codon, so that part of the time it is recognized as a code word for serine.

Since amber suppression is based on the misreading of the code in the permissive cell, an alteration of any component that guarantees the accuracy of reading might result in suppression. Four possibilities are: (i) an altered aminoacyl synthetase which transfers an amino acid to a class of sRNA molecules which normally recognizes the nonsense triplet, but does not bind any amino acid; (ii) a new species of sRNA specific for a particular amino acid, yet capable of recognizing the nonsense codon; (iii) mutant ribosomes which distort the natural mRNA-sRNA complex, thereby permitting the recognition of a nonsense codon by a normally occurring sRNA species (2); and (iv) the presence of base analogs or an inaccurate RNA polymerase, which would produce an altered messenger (for example, fluoracil inserted in place of uracil) (4).

We show here that the component active in suppression in strain S26R1E is a serine-accepting sRNA species not present in the nonpermissive isogenic strain S26. This result was made possible by the development of an assay, in vitro, for suppression by S26R1E of an amber mutation in the coat-protein cistron of the RNA-containing bacteriophage R17. The RNA isolated from this mutant directs the synthesis of functional coat protein in a cell-free protein-synthesizing system derived entirely from the permissive host S26R1E. No functional protein is produced in a system derived from S26. However, a mixed system, consisting of serine-accepting sRNA from S26R1E added to an otherwise nonpermissive system, produces functional coat protein.

Preparation of the assay system. *Escherichia coli* bacterial strains S26 and S26R1E, an isogenic pair obtained from A. Garen, are nonpermissive (Su⁻) and permissive (Su⁺), re-

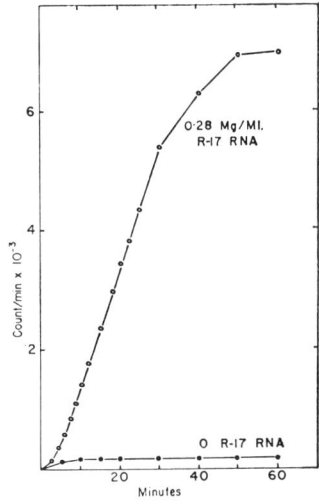

Fig. 1. Kinetics of lysine incorporation in vitro in the presence and absence of R17 RNA. The 1-ml reaction mixture contained 0.25 ml of "preincubated" *E. coli* B S-30, 0.8 mg of stripped sRNA from *E. coli* B, and 0.5 μc of C^{14}-lysine (1.93 × 10^7 count min^{-1} μmole^{-1}). Each point represents the radioactive material in 50 μl of reaction mixture, which was precipitable by hot trichloroacetic acid (TCA).

Table 1. Capacity of sRNA's to accept amino acids: results in mμmole of amino acids per milligram of sRNA. The conditions used for these assays were those of Berg et al. (26).

Amino acid	sRNA		
	Su⁺	Serine-accepting	Leucine-accepting
Alanine	2.21	0.024	0.025
Arginine	2.14	.026	.022
Isoleucine	0.895	.015	.083
Leucine	3.51	.028	3.26
Serine	1.81	1.64	0.027

spectively. *Escherichia coli* strain C600G, a permissive host used as an indicator strain by Campbell (7), was obtained from M. Meselson and converted to F⁺ so that it would support the growth of R17. *Escherichia coli* B and *E. coli* Hfr₁ are nonpermissive strains. Strain Hfr was obtained with a stock of R17 from A. Graham.

For the preparation of *E. coli* extracts (S-30) (8), cells were grown in a glycerol-casamino acid medium (15). Cells were harvested during exponential growth, washed twice in tris-magnesium buffer (0.01M magnesium acetate and 0.01M tris, pH 7.5), and frozen until used. The S-30 extracts were prepared according to the method of Nirenberg and Matthaei (16) except for two modifications: (i) NH₄Cl was used in place of KCl and (ii) 1.5 ml of tris-magnesium buffer per gram of cells was used instead of 3.0 ml per gram.

The S-30 extracts were 'preincubated' (8) with an ATP-generating system, chilled, and immediately used in incorporation experiments. For cell-free protein synthesis, the reaction mixtures contained per milliliter: one-fourth volume of preincubated S-30, 0.003 mmole ATP, 0.002 mmole GTP, 0.005 mmole phosphoenolpyruvate, 20 μg pyruvate kinase, 4×10^{-5} mmole of each amino acid, 0.010 mmole magnesium acetate, 0.075 mmole of NH₄Cl, 0.05 mmole tris (pH 7.8), 0.01 mmole glutathione, stripped sRNA (8), and R17 RNA. The temperature of incubation was 36°C.

Total sRNA was prepared by the procedure of Zubay (17). The biological activity of all sRNA species except those protected by the attachment of a chosen amino acid to the sRNA terminal adenylic acid residue was destroyed by oxidation with periodate. Stripped Su⁺ sRNA was charged with (esterified to) either serine or leucine, saturating amounts of Su⁻ activating enzymes being used. The reaction was stopped after 20 minues at 35°C by precipitation with ethanol. The precipitate was suspended in distilled water and dialyzed for 2 hours at 4°C. The RNA solution was then centrifuged at 15,000 rev/min for 20 minutes in a Serval SS-2 centrifuge to remove denatured protein. The supernatant was made 0.1M with sodium acetate, pH 4.7. A 12-fold excess of sodium periodate was added to the sRNA solution, and the oxidation was allowed to proceed for 30 minutes at 30°C in the dark. The reaction was stopped by the addition of excess glucose, and the mixture was precipitated with ethanol. The precipitate was dissolved in distilled water, and the solution was dialyzed for 24 hours. To test the capacity of the oxidized sRNA preparations to accept amino acid, the sRNA's were first stripped by incubating with 0.1M tris, pH 8.8, at 37°C for 3 hours. The oxidation procedure was adopted to assure protection of sRNA charged with serine. P. C. Zamecnik pointed out to us that the normal procedure for periodate oxidation is not successful because of the oxidation of the amino and hydroxyl groups of serine and the subsequent lability of the oxidation product.

Bacteriophage R17 was grown and purified according to the procedure of Gesteland and Boedtker (15). Titers of 1 to 3×10^{12} pfu/ml (8) were routinely obtained. The R17 RNA was extracted with phenol and then precipitated from the extract with ethanol. The ratio of the absorbancy at 260 mμ to that at 280 mμ of the R17 RNA so prepared is between 1.9 and 2.0.

For the isolation of R17 amber mutants, a concentrated phage suspension was incubated in 0.2M NaNO₂, pH 4.6, at 37°C for 8 to 9 hours. After dilution to stop the reaction, the suspension of phage was plated on S26R1E; individual plaques were picked, and each was tested for growth on S26. After nitrous acid treatment of a phage suspension which originally contained 10¹³ viable particles per milliliter, 10⁵ particles per milliliter survived. Among the survivors, about 1 to 3 in 500 were mutants.

For the growth of the R17 mutant am11B, a mutant stock containing 2×10^{11} pfu/ml, 0.8 percent revertants, was used to infect 16-liter cultures at a phage–bacteria ratio of about 3 to 1, and an optical density of 0.6 to 0.8. Stocks thus obtained all contain 5 to 7×10^{11} pfu/ml, with 1 to 2 percent revertants. The phage particles were purified in the same manner as wild-

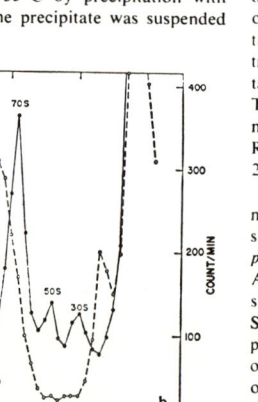

Fig. 2. Sucrose-gradient analysis of the total amino acid–incorporating system after 20-minute incubation with R17 RNA and either C¹⁴-lysine and C¹⁴-arginine or H³-histidine. The 200-μl reaction mixture contained 50 μl of preincubated *E. coli* B S-30, 104 μg of R17 RNA, 160 μg of stripped *E. coli* B sRNA. (*a*) The reaction mixture contained 0.2 μc of C¹⁴-lysine (7.25 × 10⁵ count min⁻¹ μmole⁻¹) and 0.2 μc of C¹⁴-arginine (7.75 × 10⁵ count min⁻¹ μmole⁻¹); (*b*) the reaction mixture contained 5 μc of H³-histidine (5.25 × 10⁶ count min⁻¹ μmole⁻¹). A 100-μl portion of the reaction mixture was layered on a 5-ml linear sucrose gradient (5 to 20 percent) and centrifuged at 38,000 rev/min for 2.3 hours at −12°C. Each fraction was first assayed for optical density at 260 mμ and then treated with 6 percent TCA at 90°C for 20 minutes to prepare the samples for radioactivity analysis. Of the acid-insoluble (hot TCA) radioactive material originally put on the gradient, 85 percent was recovered. The optical density of the 30S peaks in this and succeeding experiments is higher than might be expected because of the absorption of phage RNA.

type phage and the percentage of revertants remained constant throughout this procedure.

Amino acids labeled with C^{14} or H^3 were obtained from New England Nuclear Corporation, Boston, Mass. The specific activities of the C^{14} amino acids in $\mu c/\mu mole$ were: alanine, 100; arginine, 239; isoleucine, 221; lysine, 210; serine, 120; H^3-histidine had a specific activity of 1.1 mc/μmole. Pyruvate kinase, ribonuclease, and deoxyribonuclease were obtained from Worthington Biochemical Corporation.

Development of the assay for amber suppression. The kinetics of C^{14}-lysine incorporation in vitro, with a preincubated S-30 extract in the presence and absence of added R17 RNA, are shown in Fig. 1. Under optimum conditions, there is a 30- to 50-fold stimulation of amino acid incorporation with the addition of R17 RNA.

If we assume that the product synthesized in vitro has an amino acid composition similar to that of R17 coat protein (*18*), then we can calculate that approximately 260 amino acids per ribosome are being incorporated into protein. This corresponds to two molecules of phage coat protein per ribosome in the reaction mixture. From Fig. 1, one can see that preincubation virtually eliminates protein synthesis mediated by endogenous *E. coli* messenger. All experiments reported here were performed with a preincubated S-30 homogenate.

After 20 minutes of incubation with R17 RNA as messenger, newly synthesized polypeptide sedimented in six discrete regions, with sedimentation constants of approximately 112S, 85S, 70S, 30S, 20S, and less than 4S (Fig. 2a).

The radioactivity which is associated with fractions sedimenting at 70S, 85S, and 112S represents newly made chains attached to ribosomes. The 85S and 112S peaks arise from the combination of an R17 RNA chain with one and two ribosomes, respectively; the 70S peak represents ribosomes from which the RNA messenger has been detached or degraded, but which have not released their polypeptide product. All the more slowly sedimenting material consists of protein products released from ribosomes. Particularly relevant is the newly synthesized material which sediments at 30S.

We believe that this material consists of complete coat protein for the following reasons: (i) autoradio-

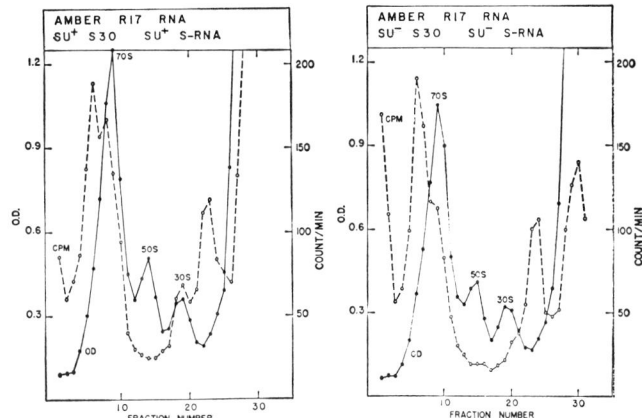

Fig. 3. Sucrose-gradient analyses of amino acid incorporation directed by am11B RNA. Each reaction mixture contained per 200 μl: 100 μg of am11B RNA containing less than 2 percent contamination from wild type, 0.2 μc of C^{14}-isoleucine (8.4 × 10^7 count min^{-1} μmole^{-1}), and either 50 μl of preincubated Su$^+$ S-30 and 160 μg of stripped Su$^+$ sRNA (left) or 50 μl of preincubated Su$^-$ S-30 and 160 μg of Su$^-$ stripped sRNA (right). A 100 μl sample of the reaction mixture was layered on a 5-ml sucrose gradient after a 20-minute period of incubation at 36°C. The gradients were centrifuged at 38,000 rev/min for 2.4 hours at -12°C. Approximately 85 percent of the acid-insoluble (hot TCA) radioactive material that had been put on the gradients was recovered.

grams of peptides resulting from digestion of the 30S material by trypsin reveal a peptide pattern (fingerprint) identical to that obtained from coat protein (*19*); (ii) sucrose-gradient analysis of cell-free reaction mixture in which the radioactive label was exclusively in histidine reveals (Fig. 2b) that no radioactivity is present in the 30S peak. This is in agreement with the fact that the coat protein of R17, like that of f2 and MS2, does not contain histidine (*18*).

The rapid rate of sedimentation of this coat protein synthesized in vitro suggests that it consists either of a coat protein aggregate or of protein subunits bound to the R17 RNA which, in the reaction mixture, normally sediments at about 30S. Mild

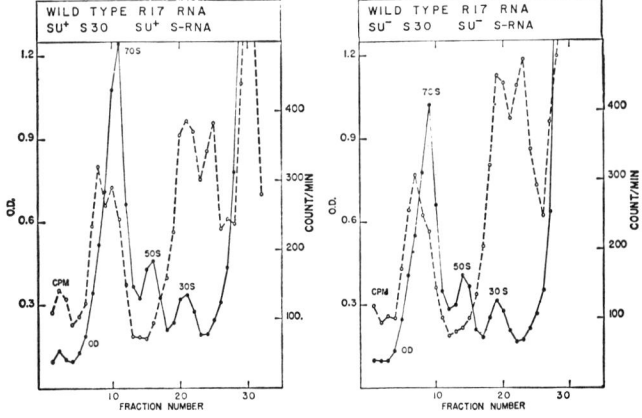

Fig. 4. Sucrose-gradient analyses of amino acid incorporation directed by wild-type RNA in systems derived from Su$^+$ and Su$^-$ bacteria. These experiments were run in parallel with, and under the same conditions as, those described in Fig. 3. In each 200-μl reaction mixture there were 104 μg of wild-type R17 RNA. On the left is the result of incorporation in the Su$^+$ system, on the right, that in the Su$^-$ system.

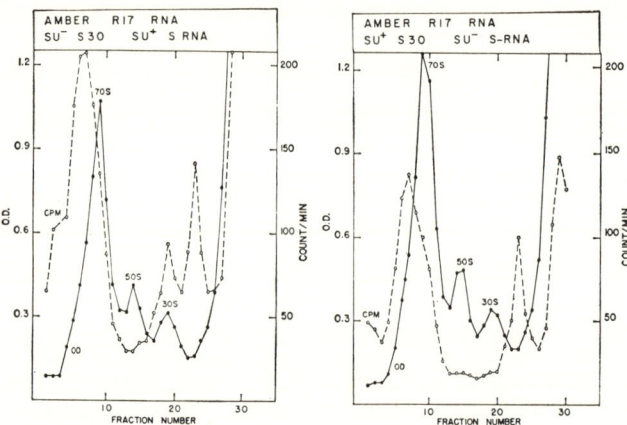

Fig. 5. sRNA mixing experiment. The sucrose-gradient analysis of protein synthesis in vitro with am11B RNA as messenger was run in parallel with, and under the same conditions as, the experiment shown in Fig. 3. The only difference was that to the reaction mixture containing 50 μl of preincubated Su^+ S-30, was added 160 μg of stripped Su^- sRNA (right) and to the reaction mixture containing Su^- S-30 was added stripped Su^+ sRNA (left).

treatment with ribonuclease (0.2 μg/ml for 3 minutes at 0°C) of an incubation mixture prior to its being layered on a sucrose gradient shifts the newly synthesized 30S material to the top of the gradient, thereby demonstrating that it was previously bound to free R17 RNA molecules.

The nature of the 20S peak is not yet clear. It definitely is not coat protein, since (i) it contains histidine and (ii) its autoradiogram is different from, and far more complex than, that of coat protein (19). Thus it must arise either from a mixture of different proteins or from a protein of high molecular weight relative to that of coat protein. It is most tempting to believe that the 20S material might be R17-specific RNA synthetase.

Notani, Engelhardt, Konigsberg, and Zinder (14) have unambiguously demonstrated that a number of their amber mutations in f2 are in the coat-protein cistron. All of the coat-protein mutants (su-3 class) grow on strain S26R1E, but not on strain S26 or on strain C600. In contrast, a number of other mutants (su-1 class), identified by complementation tests as belonging to another cistron, grow on C600 as well as on S26R1E. A search was undertaken for R17 amber mutants (20) which had the same plating and complementation properties as the su-3 group of f2. Among those found, one, am11B, was chosen for studies in vitro. Parallel with the experiments in vitro, we have undertaken chemical experiments to demonstrate directly a difference in the amino acid composition of its coat protein. Fingerprinting of tryptic digests, a separation of the tryptic peptides by electrophoresis in one dimension followed by chromatography in the second dimension, reveals the disappearance of one normal peptide and the appearance of a new one compared to wild-type R17 coat protein. This result gives us reason to believe that am11B is a coat-protein mutant. The results of the studies in vitro that are described next confirm this identification.

When am11B RNA is used as messenger in the cell-free system, there are no gross differences either in the kinetics or in the final amount of protein synthesis between the Su^+ (S26R1E) and Su^- (S26) systems. Under conditions in which synthesis mediated by wild-type RNA is identical in the two systems, there is a 15 to 20 percent greater incorporation of amino acids in the Su^+ than in the Su^- system when RNA from mutant phage is used.

The reaction mixtures are supplemented with stripped sRNA (about 1 mg/ml) isolated from Su^+ and Su^- bacteria. This addition stimulates protein synthesis threefold, regardless of whether the mRNA used is from wild-type or am11B bacteriophage. The influence of added sRNA on the degree of amino acid incorporation is an important factor in the success of experiments described below.

Sucrose gradient analyses of Su^+ and Su^- reaction mixtures, incubated with am11B RNA, are shown in Fig. 3. Each reaction mixture contained

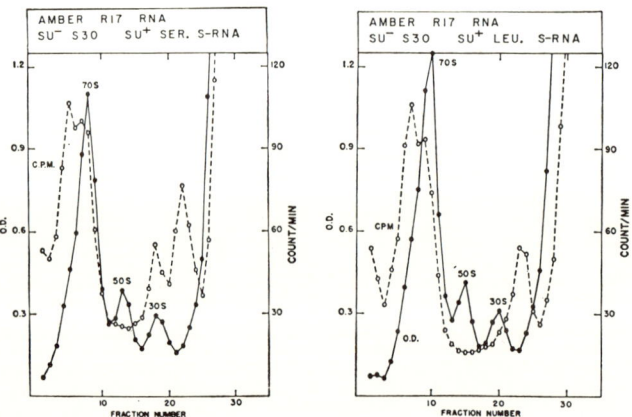

Fig. 6. Addition of Su^+ serine- or Su^+ leucine-accepting sRNA's to an otherwise nonpermissive protein-synthesizing system directed by am11B RNA. In each 200-μl reaction mixture there were 95 μg of am11B RNA, 50 μl of preincubated $Su-$ S-30, 120 μg of stripped Su^- sRNA, 0.2 μc of C^{14}-isoleucine (8 × 10^7 count min^{-1} μmole^{-1}), and approximately 5 μg of Su^+ sRNA having only serine-accepting activity (left) or leucine-accepting activity (right). A 100-μl sample of the reaction mixture was put on a 5-ml sucrose gradient after a 20-minute period of incubation at 36°C. Approximately 85 percent of the acid-insoluble (hot TCA) radioactive material, which had been put on the gradients, was recovered.

added sRNA which was homologous to the S-30 used. A difference between the two gradients is apparent. The 30S peak, previously identified as R17 coat protein synthesized in vitro and bound to R17 RNA, is absent in the gradient of the Su⁻ reaction mixture (right side of Fig. 3). This observation is consistent with the hypothesis that am11B is a mutation in the R17 cistron specifying coat protein. As a control, the experiment was performed in parallel tubes with wild-type R17 RNA as messenger (Fig. 4). Since both bacterial strains permit the growth of wild-type phage, both the 30S and 20S proteins should be synthesized in vitro in either system when wild-type RNA is used as messenger. This is, in fact, observed.

We have repeated all of these experiments, with identical results, using different preparations of S-30's, sRNA's, R17 wild-type and am11B RNA's, and also using amino acids other than isoleucine as label. These results confirm prior indications that suppression occurs during messenger RNA translation; and most important, the presence or absence of the 30S peak constitutes as assay for suppression in vitro.

The amount of 30S protein synthesized in the Su⁺ system is reduced relative to that of the 20S protein when am11B RNA is used as messenger. This reduction can be used as a measure of the efficiency with which the Su⁺ system can suppress the amber mutation in vitro. The level observed is consistent with the observation that suppression of amber mutations in the T4 head protein by the *su-A* suppressor gene is approximately 60 percent efficient in vivo (*11*).

With this assay for suppression in vitro, we are in the position, by means of mixing experiments, to discover which component or components of the protein-synthesizing system are responsible for the suppression of the amber mutation. Experiments in which Su⁺ sRNA was added to the incubation mixture containing Su⁻ S-30 (Fig. 5, left), and vice versa (Fig. 5, right), were run in parallel with those shown in Figs. 3 and 4. Since addition of Su⁺ sRNA to the S-30 prepared from Su⁻ cells permits synthesis of functional coat protein (protein capable of binding to the added R17 RNA), and since, conversely, the Su⁺ S-30 cannot produce 30S protein in the presence of Su⁻ sRNA, suppression must be caused by a difference in the respective sRNA pools in Su⁺ and Su⁻ bacteria.

Weigert and Garen (*12*) and Notani et al. (*14*) have shown that suppression by S26R1E results in the insertion of serine into mutant protein, usually in place of a glutamine residue in the wild type. This fact, coupled with our finding that suppression in this strain is due to a difference between its sRNA pools and those of the isogenic strain S26, leads us to predict that the specific sRNA acting as the suppressor is a new or altered serine-accepting sRNA capable of inserting serine at the nonsense site. To test this hypothesis, we prepared Su⁺ sRNA which contained only biologically active serine-accepting sRNA's. This we achieved by charging stripped Su⁺ sRNA with serine, using an unfractionated, nucleic acid–free mixture of amino acid activating enzymes (*21*) prepared from S26 (Su⁻) bacteria. As a control, a second portion of Su⁺ sRNA was charged with leucine. The two sRNA preparations were then treated with sodium periodate to destroy any uncharged sRNA by oxidation of the 2'- and 3'-hydroxyl groups on the ribose of the terminal adenylic acid residue of sRNA (*22*). To test the effectiveness of this method in protecting only a specific sRNA, the oxidized sRNA was stripped, and the amino acid–accepting activity for each of five different amino acids was examined and compared to that of the original Su⁺ sRNA (Table 1). Protection of the serine- and leucine-accepting sRNA's is better than 90 percent, whereas other sRNA species were destroyed to the extent that their residual accepting activity was only 1 percent. The anomaly with isoleucine could be due either to contamination by C^{14}-leucine of the C^{14}-isoleucine used to test the isoleucine-accepting activity, or to contamination by C^{12}-isoleucine of the C^{12}-leucine employed for protection. The addition of sRNA's which have the 2'- and 3'-hydroxy groups of ribose oxidized to aldehydes does not inhibit protein synthesis in vitro.

The effect of adding either Su⁺ serine sRNA or Su⁺ leucine sRNA to the protein-synthesizing system mediated by am11B RNA is shown in Fig. 6. The reaction mixtures contained Su⁻ S-30, 0.9 mg per milliliter of Su⁻ sRNA, and 0.5 mg per milliliter of either Su⁺ serine-accepting or Su⁺ leucine-accepting sRNA. The results clearly implicate Su⁺ serine-sRNA as the requirement for suppression of amber mutations by S26R1E. The experiment was repeated with wild-type RNA instead of am11B RNA as messenger, and no appreciable difference between the serine and leucine gradients was observed.

Molecular models for suppression. The C^{14}-labeled polypeptide sedimenting in the 30S region of the sucrose gradient has been identified by fingerprinting as phage coat protein. Its high sedimentation value is due to the binding of subunits of phage coat protein, synthesized in vitro, to the phage RNA, which was added as messenger. This formation of a 30S complex between input phage RNA and protein synthesized in vitro allows us to examine selectively the synthesis of intact coat-protein molecules and thus to develop an assay for suppression in a cell-free system.

The assay consists of testing the capacity of the amino acid incorporating system to read the nonsense codon in the coat-protein cistron of the R17 amber mutant am11B as a serine codon, and thereby to permit the synthesis of intact phage coat protein. We have shown that the only requirement for this translation of the amber codon is the addition of an Su⁺ serine-accepting sRNA to the otherwise totally Su⁻ system.

Based on studies of the pattern of mutation to, and reversion from, nonsense, Brenner et al. and Garen (*23*) have proposed that the amber triplet is UAG (*8*). This triplet could, in fact, be unreadable, or it could be a specific signal to terminate a completed polypeptide chain. In the first case, there would be no sRNA species able to recognize the amber codon, and chain termination might be merely a means of freeing ribosomes for the synthesis of new protein. In the second case, one might imagine an sRNA species capable of recognizing the UAG triplet, but unable to bind any amino acid.

With these possibilities in mind, two types of models can be proposed for the reading of the amber codon by a specific serine-accepting sRNA. These can be designated as (i) precise-reading models and (ii) ambiguous-reading models.

One precise-reading model could be based on a mutation in a serine-accepting sRNA anticodon, enabling it to recognize the amber codon. Since there is a UCG serine codon (*24*), the change of a single base in the corresponding serine-accepting sRNA would permit it to pair exactly with the amber codon UAG. However, that species of sRNA capable of recognizing the UCG serine codon would now be eliminated unless there existed more than one cistron for the UCG-specific sRNA. Furthermore, this model does not explain the release of a large proportion of polypeptide fragments in the Su+ cell, unless we include the additional hypothesis that an amber-specific chain-terminating sRNA exists.

A second precise-reading model is based on the assumption that an sRNA species exists that normally recognizes the amber codon as chain-terminating. A modification of this sRNA could enable it to accept serine. The production of a mixture of complete and incomplete protein would be a consequence of this model if the charging of the altered amber sRNA with serine were inefficient—that is, if not all such sRNA molecules became charged.

In the case of the ambiguous-reading model, no assumption about the nature of chain termination is necessary. Suppression could be caused by a modification of a serine-accepting sRNA, which destroys the specificity of its attachment to a codon without altering its anticodon. For example, modification of the overall structure of the serine-accepting sRNA species which recognizes the serine codon UCG might permit this sRNA species to recognize either the nonsense (UAG) or the serine (UCG) codon. This model requires no additional hypotheses to explain the production of a mixture of fragments and completed chains in the Su+ cell.

The amber suppressor mutation, in the second and third models, could be a mutation either in the sRNA cistron or in a gene coding for an enzyme which could modify the serine-accepting or amber-specific sRNA species assumed to act as the suppressor. Likely candidates for such a task are enzymes which produce the unusual bases in sRNA (for example, methylated bases, pseudouracil, or dihydrouracil). Our results do not allow us to distinguish between enzymatic and genetic modification of the specific sRNA involved. It is possible, therefore, that the new serine-accepting sRNA is only the indirect product of the su-A suppressor gene.

Finally, the conditions selected for the mixing experiment (Fig. 6) enable us to determine whether the suppressor mutation is dominant in vitro. Since, in these experiments, the incubation mixtures contained a twofold excess of Su− serine-accepting sRNA over Su+ serine-accepting sRNA, we conclude that the su-A suppressor is dominant. This is consistent with the findings of Signer, Beckwith, and Brenner (25) that a cell heterozygous for amber suppressor gene is permissive.

References and Notes

1. C. Yanofsky, D. R. Helinski, B. C. Maling, in *Cold Spring Harbor Symp. Quant. Biol.* **28**, 11 (1961); L. Gorini and E. Kataja, *Proc. Nat. Acad. Sci. U.S.* **51**, 487 (1964); J. R. Beckwith, *Biochim. Biophys. Acta* **76**, 162 (1963).
2. J. Davies, W. Gilbert, L. Gorini, *Proc. Nat. Acad. Sci. U.S.* **51**, 883 (1964).
3. D. C. Hawthorne and R. K. Mortimer, *Genetics* **47**, 1085 (1960); T. R. Manney, *ibid.* **50**, 109 (1964).
4. S. Benzer and S. P. Champe, *Proc. Nat. Acad. Sci. U.S.* **47**, 1025 (1961).
5. R. H. Epstein, A. Bolle. C. M. Steinberg, E. Kellenberger, E. Boy de la Tour, R. Chevalley, R. S. Edgar, M. Susman, G. H. Denhardt, A. Lielausis, in *Cold Spring Harbor Symp. Quant. Biol.* **28**, 375 (1963).
6. A. Garen and O. Siddiqi, *Proc. Nat. Acad. Sci. U.S.* **48**, 1121 (1962).
7. A. Campbell, *Virology* **14**, 22 (1961).
8. Abbreviations and definitions: RNA, ribonucleic acid; mRNA, messenger RNA; sRNA, soluble RNA; ATP, adenosine 5′-triphosphate; GTP, guanosine 5′-triphosphate; pfu, plaque forming units; S-30, a bacterial homogenate from which whole cells, cell-wall fragments, and cell-membrane fragments have been removed by centrifugation at 30,000g for 30 minutes; "preincubation," refers to incubation of the S-30 in the absence of labeled amino acids to eliminate amino acid incorporation directed by endogenous *E. coli* messenger RNA; "charged sRNA," is sRNA to which an amino acid is attached by a phosphate-ester linkage between the terminal adenylic acid of sRNA and the carboxyl group of the amino acid; "stripped sRNA," is sRNA from which the amino acid has been removed by hydrolysis of the phosphate-ester linkage; UAG and UCG are symbols representing the sequence of nucleotides in specific codons, the letters represent the nucleotides uridine, adenosine, guanosine, and cytidine 5′-monophosphate; TCA, trichloroacetic acid.
9. N. D. Zinder and S. Cooper, *Virology* **23**, 152 (1964).
10. S. Benzer and S. P. Champe, *Proc. Nat. Acad. Sci. U.S.* **48**, 1114 (1962).
11. S. Brenner and A. O. W. Stretton, *J. Cellular Comp. Physiol.* **64**, Suppl. 1, 43 (1964).
12. M. G. Weigert and A. Garen, *J. Mol. Biol.*, in press.
13. A. O. W. Stretton and S. Brenner, *ibid.*, in press.
14. G. W. Notani, D. L. Engelhardt, W. Konigsberg, N. D. Zinder, *ibid.*, in press.
15. R. F. Gesteland and H. Boedtker, *J. Mol. Biol.* **8**, 496 (1964).
16. M. W. Nirenberg and J. H. Matthaei, *Proc. Nat. Acad. Sci. U.S.* **47**, 1588 (1961).
17. G. Zubay, *J. Mol. Biol.* **4**, 347 (1962).
18. M. D. Enger and P. Kaesberg, *Proc. Nat. Acad. Sci. U.S.*, in press.
19. M. R. Capecchi, manuscript in preparation.
20. G. N. Gussin, in preparation.
21. E. M. Martin, C. Yegian, G. Stent, *Z. Vererbungslehre* **94**, 303 (1963).
22. P. C. Zamecnik, M. L. Stephenson, J. P. Scott, *Proc. Nat. Acad. Sci. U.S.* **46**, 811 (1960).
23. S. Brenner, A. O. W. Stretton, S. Kaplan, *Nature*, in press; A. Garen, personal communication.
24. M. W. Nirenberg, P. Leder, M. Bernfield, R. Brimacombe, J. Trupin, F. Rottman, C. O'Neal, *Proc. Nat. Acad. Sci. U.S.*, in press.
25. E. R. Signer, J. R. Beckwith, S. Brenner, in preparation.
26. P. Berg, F. H. Bergmann, E. J. Ofengand, M. Dieckmann, *J. Biol. Chem.* **236**, 1726 (1961).
27. We thank Professors J. D. Watson and W. Gilbert for their encouragement and helpful discussion. Supported in part by grants GM-09541-03 and -04 from the U.S. Public Health Service.

A Genetic and Biochemical Analysis of Second Site Reversion*

Donald R. Helinski† and Charles Yanofsky

From the Department of Biological Sciences, Stanford University, Stanford, California

(Received for publication, October 26, 1962)

The A gene-A protein system of the tryptophan synthetase of *Escherichia coli* is being examined in studies on various aspects of the gene-protein relationship. The A proteins of a number of A mutants have been examined and several amino acid substitutions due to mutations in the A gene have been described (1–4). Reversion studies with several A mutants have shown that reversion can result from either a mutational change at the same site as the original mutation or at a second site within the A gene (4, 5). Second site reversions have also been described in the alkaline phosphatase gene (6), the h_{III} region of phage T4D (7) and the r_{II} region of phage T2 (8). Peptide pattern studies of the A protein of one second site revertant strain, A46PR8, have detected a peptide difference from the corresponding wild-type A protein peptide patterns (9). This report is concerned with the determination of the amino acid substitution associated with the second site mutational change in this strain and the effect of this substitution on the enzymatic activity of the A protein.

EXPERIMENTAL PROCEDURE

Mutant A46 was obtained by penicillin selection after ultraviolet irradiation of wild-type strain K-12 of *E. coli* (10, 11). Crosses were performed by transduction with the phage P1kc. The procedures used have been described by Lennox (12).

The A proteins examined in this study were prepared by the procedure described for the purification of the wild-type A protein (13). Performic acid oxidation of the A proteins was carried out for 3 hours at 0° by the method of Hirs (14). The procedure used for proteolytic digestion of the A proteins and the method employed for obtaining peptide patterns are described elsewhere (15). Trypsin (five times recrystallized, Lot No. TR706) and chymotrypsin (Lot No. CD576-81) were obtained from Worthington Biochemical Corporation. The peptides TP8C1 and TP8C2 from the wild-type and mutant protein digests were purified by a combination of paper chromatography and paper electrophoresis as previously described (1). The peptides TP8 and TP3C1 were purified by a column chromatographic procedure with Dowex 1-X2 which is a modification (2) of the method of Rudloff and Braunitzer (16). A final purification of the peptides was carried out by paper electrophoresis as previously described (1). The peptides or proteins were hydrolyzed in

twice distilled 5.7 N HCl in sealed, evacuated tubes at 105° for 48 hours. Amino acid analyses were performed with a Spinco amino acid analyzer. The composition of each peptide is expressed as the molar ratios of the constituent amino acids.

RESULTS

Genetic Analysis of Revertants Derived from Mutant A46—When a population of A46 cells is plated on a medium lacking tryptophan, spontaneous revertant colonies of various sizes appear. After these colonies are purified by single colony isolations, three distinguishable classes are easily recognized; one resembles the wild-type, the second grows somewhat slower than the wild-type and accumulates indole glycerol, while the third type grows considerably more slowly than members of either of the other two classes and also accumulates indole glycerol. Genetic analyses with an isolated colony of the second type have shown that this partial revertant resulted from a reversion at the A46 site (4).

Genetic analyses carried out with partial revertants of the third type, typified by the strains designated A46PR7 and A46PR8, demonstrated that this reversion event did not occur at the A46 genetic site. This conclusion is based on the recovery of mutant A46 from these partial revertants in appropriate crosses performed by transduction. In addition to recovering mutants that were indistinguishable from A46, a second mutant type, designated PR8, appeared in these crosses. This second mutant type, when crossed back to A46, gave two classes of tryptophan-independent colonies: one class was indistinguishable from A46PR7 or A46PR8, whereas the second class was indistinguishable from wild-type. These findings indicate that the reversion events leading to strains A46PR7 and A46PR8 occurred at a genetic site distinct from the A46 site and, in addition, that this second alteration, when present in an otherwise wild-type A gene, leads to the formation of an inactive A protein. This second alteration has been designated the PR8 alteration.

Mapping experiments performed to locate the PR8 mutational site on the genetic map are summarized in Fig. 1. A three point cross performed to locate the PR8 site relative to the A46 site is diagrammed in Fig. 2. Of 103 recombinants obtained from this cross that were + at the PR8 and A46 sites, 81 were T15⁻ and 22 were T15⁺. Of 85 recombinants examined that carried the A46 and PR8 alterations, 20 were T15⁻ and 65 were T15⁺. These results favor order A in Fig. 2.

Since the PR8 genetic alteration partially reverses the effect of the A46 mutation and leads to a strain that is capable of growing in a medium lacking tryptophan, it was of interest to examine the effect of this alteration on several other A mutants. The PR8 genetic alteration was transduced into mutants A23, A2,

---

* This investigation was supported by grants from the United States Public Health Service and the National Science Foundation.

† Postdoctoral fellow of the United States Public Health Service. Present address, Department of Biology, Princeton University, Princeton, New Jersey.

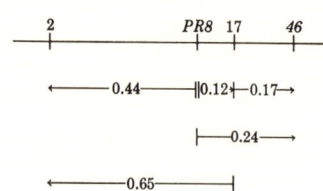

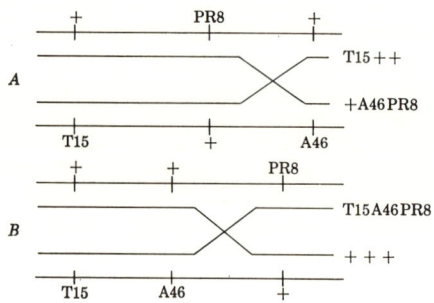

FIG. 1. Sites of mutational alterations in the A gene in the mutants examined. Sites and distances determined by traduction with phage P1kc (11).

FIG. 2. The two possible arrangements of PR8 and A46 relative to T15 and the respective predominant recombinant types expected. The T15 marker corresponds to a block in anthranilic acid formation.

A17, and A58 and in no case did the combination lead to a strain with partial or complete independence of the requirement for indole or tryptophan for growth.

Location of Second Amino Acid Change in Partial Revertant Protein—Peptide patterns of trypsin plus chymotrypsin digests of the A proteins of strains A46PR8 and A46PR7 are identical and differ from the corresponding wild-type pattern by the position of one peptide, designated TP8C2 (Fig. 3). The TP8C2 peptides were isolated from trypsin plus chymotrypsin digests of the A46PR7 and wild-type proteins and their amino acid compositions determined (Table I). It is clear that the two peptides differ by the presence of an additional leucine in the wild-type peptide. This peptide difference is not found with similar digests of the A46 protein and, thus, this change appears to be associated with the reversion event in this strain. Further information on the amino acid change associated with this second mutational difference between the two proteins was obtained from a comparison of peptide patterns of tryptic digests. The A46PR7 and A46PR8 tryptic peptide patterns were identical and differed from the corresponding wild-type pattern by the absence of one major peptide, TP8. No additional peptide was evident in the partial revertant peptide pattern. The composition of the wild-type TP8 peptide is shown in Table I.

The tryptic peptide TP8 and the peptide altered in the trypsin plus chymotrypsin digest of the partial revertant protein, TP8C2, are related in that TP8C2 can be derived from TP8 by chymotrypsin action. The composition of the other product of chymotryptic digestion of TP8, peptide TP8C1, is also shown in Table I. The peptide TP8C1 is normally found in a trypsin plus chymotrypsin peptide pattern, but it occupies a position at which at least one other peptide appears. Consequently, the effect of the second mutation on TP8C1 could not be determined from an examination of the trypsin plus chymotrypsin peptide pattern alone. In addition peptide TP8C1 is further hydrolyzed to two smaller peptides, TP8C1A and TP8C1B, upon prolonged chymotrypsin digestion. The relationship between these peptides is illustrated in Fig. 4.

Determination of Second Amino Acid Change in Partial Revertant Protein—The first clue to the identity of the mutant amino acid was obtained from total amino acid analyses of performic acid-oxidized partial revertant and wild-type A proteins. As shown in Table II, these amino acid analyses clearly indicated that the partial revertant proteins contain one more half-cystine than either the wild-type or A46 A protein. The wild-type amino acid substituted by the cysteine was not obvious from the total amino acid analyses.

Since the peptides, containing the mutant amino acid, were apparently missing in peptide patterns of trypsin or trypsin plus chymotrypsin digests of the partial revertant proteins, the possibility was considered that their absence was due to interpeptide disulfide bond formation as a result of an additional sulfhydryl group in the partial revertant protein. A large peptide containing a disulfide bond might remain at the origin of the peptide patterns or be lost during the preparation of the digest because of its insolubility. In order to avoid such a possibility, digests were prepared from performic acid-oxidized partial revertant and wild-type A proteins. When a peptide pattern of a trypsin plus chymotrypsin digest of the performic acid-oxidized A46PR7 protein was compared with the corresponding wild-type pattern, a new peptide was observed in the partial revertant peptide pattern. This peptide was purified and its amino acid composition determined. The amino acid composition of this new peptide, corresponding to the TP8C1 peptide, is shown in Table I and is similar to that of the wild-type TP8C1 peptide except that it contained one less tyrosine and an additional leucine and cysteine (as cysteic acid). This difference suggested a substitution of cysteine for tyrosine in the partial revertant protein. This substitution would account for the presence of the Leu–Ser–Arg and Gly–Tyr–Thr–Cys–Leu peptides in the digests of the partial revertant protein. It would also account for the absence of the wild-type peptides TP8, TP8C2, and, probably, TP8C1 from the partial revertant peptide patterns.

This conclusion was reinforced by a comparison of tryptic digests of performic acid-oxidized A46PR7 and wild-type A proteins. A peptide pattern of a tryptic digest of the performic acid-oxidized partial revertant protein also contained a new peptide. The composition of this peptide, corresponding to the TP8 peptide, is also shown in Table I. This partial revertant peptide differed from the wild-type TP8 peptide by the presence of cysteine (as cysteic acid) and the absence of one tyrosine.

Tyrosine and Sulfhydryl Group Analyses of A Proteins—The loss of a tyrosine from the partial revertant protein was substantiated by a spectrophotometric estimation of the tyrosine content of the A46PR7 and A46PR8 proteins. With the method of Goodwin and Morton (17), the partial revertant proteins were found to contain one less tyrosine than is present in the A46 or wild-type protein (Table III). Since the peptide patterns of the A46PR7 and A46PR8 A proteins show identical differences from the corresponding wild-type peptide patterns, and since the A46PR8 and A46PR7 proteins each contain one more half-

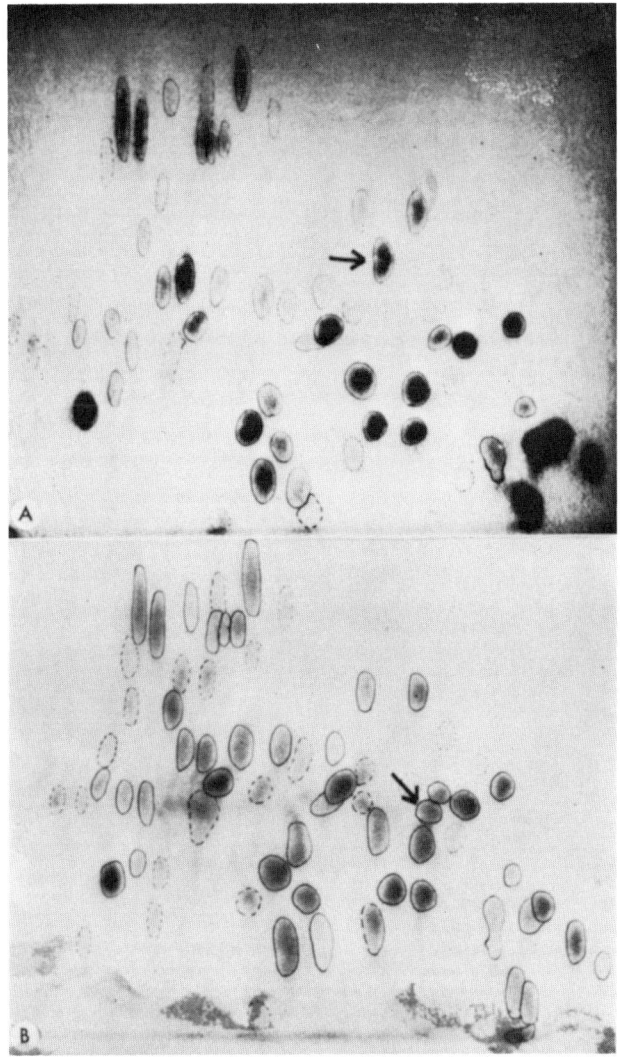

FIG. 3. Trypsin plus chymotrypsin peptide patterns of the wild-type A protein (*upper*) and A46PR7 A protein (*lower*). The *arrows* indicate peptide TP8C2 in the wild-type pattern and the corresponding peptide in the A46PR7 pattern.

cystine and one less tyrosine than the wild-type protein, it is clear that the proteins possess the same amino acid substitution associated with the secondary mutational event. This suggests that A46PR7 and A46PR8 represent identical mutational events.

The presence of the additional cysteine residue in the partial revertant protein raises the question of the oxidation state of this residue in the A protein. It has been found that the wild-type A protein exhibits one free sulfhydryl group and, thus, the additional cysteine in the partial revertant protein could be present in disulfide linkage with the cysteine normally found free in the wild-type A protein. To obtain some information on this question, the partial revertant, A46, and wild-type A proteins were examined for their content of reacting sulfhydryl groups by the p-hydroxymercuribenzoate spectrophotometric method of Boyer (18). As indicated in Table IV, considerable variation was found between different preparations of the partial revertant proteins. Several preparations of the partial revertant proteins exhibited approximately two free sulfhydryl groups while others exhibited considerably less than two. It is clear, however, that certain preparations of the partial revertant proteins do exhibit two free sulfhydryl groups.

Examination of A46PR7 A Protein for A46 Amino Acid Change —The peptides altered in the partial revertant protein digests are different from the peptide containing the amino acid change in the A protein of the original mutant, A46. The tryptic peptide, TP3, previously shown to contain the A46 amino acid change, glutamic acid in place of glycine, is in the same position

TABLE I
Composition of peptides from digests of wild-type and partial revertant proteins

Amino acid	TP8 peptide		TP8C1 peptide		TP8C2 peptide	
	Wild-type	A46PR7	Wild-type	A46PR7	Wild-type	A46PR7
Arginine	1.02	*			1.01	1.07
Threonine	0.98	0.97	1.00	1.05		
Serine	1.02	1.01			1.08	1.02
Glycine	1.22	0.97	1.00	0.98		
Leucine	2.04	1.97			1.24	
Tyrosine	1.74	0.95	1.77	0.73	1.91	0.91
Cysteic acid†		1.15		1.12		

* The arginine in this peptide was lost during the analysis. However, the presence of arginine in this peptide was clearly indicated by a strong positive reaction to a specific arginine-staining reagent (15) given by this peptide in the peptide pattern.

† Analyses for cysteic acid were performed on peptides obtained from performic acid-oxidized proteins.

```
                         TP8
              Gly-Tyr-Thr-Tyr-Leu-Leu-Ser-Arg
                          |
                          | Chymotrypsin
                          | (or some trypsin preparations)
                          ↓
       TP8C1                              TP8C2
   Gly-Tyr-Thr-Tyr        +           Leu-Leu-Ser-Arg
        |
        | Chymotrypsin
        ↓
   TP8C1A    TP8C1B
   Gly-Tyr + Thr-Tyr
```

FIG. 4. Relationship of various peptides derived from wild-type tryptic peptide TP8. The sequence of amino acids in the TP8 peptide was taken from unpublished studies of B. C. Carlton and C. Yanofsky.

TABLE II
Cysteic acid content of performic acid-oxidized A proteins

Protein preparation	Moles of cysteic acid per mole of A protein*
Wild-type	3.1
A46	3.2
A46PR7	4.1
A46PR8	4.0
PR8	4.1

* These values were obtained from analyses of 48-hour acid hydrolysates of the performic acid-oxidized A proteins. The procedure used for performic acid oxidation is given in the text.

TABLE III
Spectrophotometric determination of number of tyrosine residues in A proteins

Protein preparation	Moles of tyrosine per mole of A protein*
Wild-type	6.7
A46	6.7
A46PR7	5.7
PR8	5.9

* The absorption measurements were performed with solutions of 1.50 mg of each A protein in 3.0 ml of 0.1 M sodium hydroxide. The values were calculated with the formula of Goodwin and Morton (17) with the absorption of 280 mμ and the intersection wave length of 294.4 mμ.

TABLE IV
Spectrophotometric determination of protein sulfhydryl groups

Free sulfhydryl groups were determined by the addition of a 4-fold excess of *p*-hydroxymercuribenzoate to 2 mg of each A protein in 2.5 ml of 0.1 M potassium phosphate buffer, pH 7.0. The increase in absorption was measured at 255 mμ as described by Boyer (18). The concentrations of free sulfhydryl groups were calculated from the increase in absorption and the molar extinction coefficient of the monomercaptide of the A protein (255 mμ = 8.45 × 10^3) as determined by Henning *et al.* (13). The values obtained by this procedure were in good agreement with the results with the titration method (18) in the case of several of the A protein preparations that were analyzed by both methods. Each value shown in the table represents a different A protein preparation.

Protein preparation	Moles reactive —SH groups per mole of protein
Wild-type	1.03
Wild-type	1.00
A46	0.86
A46	0.58
A46PR7	1.82
A46PR7	0.74
A46PR8	1.37
A46PR8	1.03
PR8	1.80
PR8	1.35

in peptide patterns of the wild-type and A46 proteins (2). This peptide was also in the same position on the tryptic peptide patterns of the partial revertant proteins. Amino acid analysis of the peptide TP3 from tryptic digests of the A46PR7 protein clearly demonstrated that this peptide contained the same glutamic acid for glycine substitution as is present in the A46 protein.[1] Since the partial revertant protein contains the A46 amino acid change in addition to the change of cysteine in place of tyrosine, it is evident that the partial revertant protein differs from the wild-type protein by two amino acid substitutions and from the A46 protein by a single amino acid substitution.

Examination of PR8 A Protein—As mentioned previously, two mutant types were recovered from strain A46PR8. One was genetically indistinguishable from A46, whereas the second, PR8, gave wild-type recombinants and A46PR8 recombinants in transduction crosses with A46. The A protein of one isolated

[1] U. Henning and C. Yanofsky, unpublished observations.

TABLE V

Composition of TP3C1 peptide from chymotryptic digests of PR8, A46, and wild-type A proteins

Amino acid	PR8	A46*	Wild-type*
Aspartic†	0.98	0.94	0.98
Glutamic†	0.94	2.11	1.11
Proline	2.07	1.95	2.08
Glycine	1.08		1.05
Alanine	2.04	1.95	1.81
Leucine	1.00	1.08	1.07
Phenylalanine	0.88	0.96	0.92

* These analyses were taken from the data of Henning and Yanofsky (2).

† Not determined whether present in the amide or free acid form.

```
Wild-type   ---Thr-Tyr-Leu------//-------Gly-Phe-Gly---
A46         ---Thr-Tyr-Leu------//-------Glu-Phe-Gly---
A46PR8      ---Thr-Cys-Leu------//-------Glu-Phe-Gly---
PR8         ---Thr-Cys-Leu------//-------Gly-Phe-Gly---
```

FIG. 5. Summary of amino acid changes in A46, A46PR8, and PR8 A proteins.

PR8 strain was purified and peptide patterns were obtained from trypsin and trypsin plus chymotrypsin digests of the protein. These peptide patterns showed the same peptide pattern differences from the corresponding wild-type patterns that were found with the partial revertant peptide patterns. Furthermore, analyses of the tyrosine (Table III) and cysteine (Tables II and IV) contents of the A protein of the PR8 strain clearly indicated that one less tyrosine and one more half-cystine was present. These results suggest that the PR8 A protein probably has the same amino acid substitution in peptide TP8, cysteine in place of tyrosine, that is characteristic of the partial revertant protein. The other amino acid change which is present in the partial revertant protein, glutamic acid for glycine, is, however, absent from the PR8 A protein. This was evident upon isolation and analysis of the TP3C1 peptide from a chymotryptic digest of the PR8 A protein. The amino acid composition of this peptide, which is a peptide containing the amino acid change characteristic of the A46 protein, is identical to that of the corresponding wild-type peptide (Table V). Thus, these results indicate that the PR8 A protein possesses the amino acid substitution characteristic of the second mutation but not the primary mutational change found in the partial revertant protein.

DISCUSSION

The presence of a mutational change at a second site in the A46PR8 A gene is clearly indicated by both genetic and protein structure studies. The second mutational change, which has resulted in a substitution of a cysteine for a tyrosine in the A protein, can be separated from the primary mutational change of the partial revertant strain. This second mutational change, when present alone in the A gene, leads to a functionally ineffective A protein. The amino acid changes in the partial revertant protein and the change in the protein of the PR8 mutant recombinant are summarized in Fig. 5. Studies on the location of this second mutational site by recombinational analysis indicate that the primary and secondary mutational sites in the partial revertant strain are separated by a distance approximately equivalent to $\frac{1}{10}$ of the known length of the A gene. This would correspond approximately to 30 amino acids if there is a correspondence between the known recombinational length of the A gene and the size of the A protein.

The finding that the A gene containing the second mutational change alone directs the formation of a mutant A protein gives the rather ironic situation whereby a protein with two amino acid changes is functional, whereas the proteins with either change alone are inactive. A similar situation has been observed for two alkaline phosphatase mutants of *E. coli* (6). It is also possible that the interaction of two amino acid changes resulting in a restoration of enzymatic activity may account for several of the cases of mutational site interaction reported in phage T2 by Crick *et al.* (8). The finding of a second site partial revertant, as described in this study, also emphasizes the importance of careful interpretation of reversion data. Other cases of second site reversion have been observed in the A gene-A protein system[2] and in other systems. These observations indicate the possibility that changes at different locations in a gene (protein) are capable of restoring catalytic activity to an enzymatically ineffective protein, thereby permitting growth on a minimal medium.

The tyrosine to cysteine change found in the partial revertant protein is of particular interest in that it is the first instance that has been reported of an amino acid substitution involving cysteine. Since the amino acid substitutions are presumably associated with a single nucleotide change, the coding units AUU and GUU that have been assigned to tyrosine and cysteine, respectively, by the Nirenberg (19) and Ochoa (20) groups, are in agreement with the tyrosine to cysteine change observed in these studies.

The significance of the compensation of the glycine to glutamic acid change by a tyrosine to cysteine change probably will only become apparent when information has been obtained on the tertiary structure and the active site of the A protein. In addition, it is important to know whether the additional cysteine is functionally important in the disulfide or free sulfhydryl state in the A46PR8 protein. In order to clearly establish this point it will be necessary to use several of the methods available for sulfhydryl group analysis and to examine the relevant proteins under various conditions of enzymatic activity. Stabilization of the secondary and tertiary structure of proteins by disulfide bonds has been amply demonstrated in the case of several proteins. However, the sulfhydryl group of a cysteine residue has no well established role in determining the tertiary structure of proteins.

In several proteins tyrosine has been shown to play a relatively important role in determining tertiary structure. The tyrosine side chain may contribute significantly to the hydrophobic bonding which appears to play an important role in maintaining tertiary structure (21). Alternatively tyrosine has the ability under certain conditions to form a hydrogen bond between its phenolic hydroxyl group and a carboxyl group and this bond may be of importance in the A protein. Evidence for such a bond has been obtained with insulin and ribonuclease (22).

When considering the mechanism by which the tyrosine to cysteine substitution partially compensates for the substitution of glycine by glutamic·acid, it is, of course, necessary to know

[2] M. Allen and C. Yanofsky, manuscript in preparation.

the restrictions on the number of different amino acid changes that can reverse the A46 change. To date only two second site partial revertants of A46 have been examined thoroughly, A46PR7 and A46PR8. Both of these appear to possess the same compensating amino acid change, tyrosine to cysteine. It would be of interest to determine the effect of a change of tyrosine to some other amino acid on the activity of the A46 protein. This might indicate whether it is the loss of tyrosine from that position in the A protein, or the addition of cysteine, or both, that is responsible for the partial restoration of enzymatic activity.

The ability of the tyrosine to cysteine substitution to reverse the effects of amino acid changes other than the glycine to glutamic acid substitution was examined by transducing the PR8 genetic alteration into several other A protein mutants. Combination of the PR8 mutational alteration with any of these mutant sites apparently did not result in any detectable restoration of the activity of any of these mutant proteins. One of the mutants examined in this way, A23, has the amino acid change glycine to arginine at the same position as the glycine to glutamic change. Thus, the tyrosine to cysteine substitution cannot compensate for the inactivating effect of an arginine at the same position occupied by the glutamic acid in the corresponding A46 mutant protein.

Several other substitutions at the A46 site have also been reported (4). It has been shown that serine, alanine, or valine at this site instead of glycine results in an enzymatically active A protein. Since the combination of glycine at the A46 site with cysteine at the second mutational site results in an inactive protein, it would be of interest to determine the effect of introducing the tyrosine to cysteine change in the protein with alanine, serine, or valine at the A46 site. Such a study could yield valuable information on the types of interactions between amino acids present in different regions of the protein which affect catalytic activity.

SUMMARY

Genetic analysis of a partial revertant obtained from an A protein mutant strain, A46, indicated that reversion resulted from a mutation at a site distinct from the A46 site. Examination of the A protein produced by this partial revertant strain revealed that the protein had two amino acid differences from the wild-type protein. One of the amino acid changes, glutamic acid in place of glycine, was identical to the amino acid change characteristic of the A46 mutation (2). The other amino acid change, resulting from the reversion event, involved the substitution of a cysteine for a tyrosine. A strain carrying the second mutational change alone was obtained by appropriate crosses. This strain produced an enzymatically inactive A protein which was found to have the tyrosine to cysteine change but not the glycine to glutamic acid change.

Acknowledgment—The authors are indebted to Virginia Horn, Sharon Ward, and Linda Blumenthal for their excellent technical assistance.

REFERENCES

1. HELINSKI, D. R., AND YANOFSKY, C., *Proc. Natl. Acad. Sci. U. S.*, **48**, 173 (1962).
2. HENNING, U., AND YANOFSKY, C., *Proc. Natl. Acad. Sci. U. S.*, **48**, 183 (1962).
3. YANOFSKY, C., HENNING, U., HELINSKI, D. R., AND CARLTON, B. C., *Federation Proc.*, in press.
4. HENNING, U., AND YANOFSKY, C., *Proc. Natl. Acad. Sci. U. S.*, **48**, 1497 (1962).
5. YANOFSKY, C., HELINSKI, D. R., AND MALING, B., in *Cellular regulating mechanisms, Cold Spring Harbor symposia on quantitative biology, Vol. 26*, Long Island Biological Association, Cold Spring Harbor, Long Island, New York, 1961, p. 11.
6. GAREN, A., LEVINTHAL, C., AND ROTHMAN, F., *J. chim. phys.*, **58**, 1068 (1961).
7. JINKS, J. L., *Heredity*, **16**, 241 (1961).
8. CRICK, F. H. C., BARNETT, L., BRENNER, S., AND WATTS-TOBIN, R. J., *Nature*, **192**, 1227 (1961).
9. HELINSKI, D., AND YANOFSKY, C., *Federation Proc.*, **20**, 255 (1961).
10. YANOFSKY, C., AND CRAWFORD, I. P., *Proc. Natl. Acad. Sci. U. S.*, **45**, 1016 (1959).
11. MALING, B. D., AND YANOFSKY, C., *Proc. Natl. Acad. Sci. U. S.*, **47**, 551 (1961).
12. LENNOX, E. S., *Virology*, **1**, 190 (1955).
13. HENNING, U., HELINSKI, D. R., CHAO, F. C., AND YANOFSKY, C., *J. Biol. Chem.*, **237**, 1523 (1962).
14. HIRS, C. H. W., *J. Biol. Chem.*, **219**, 611 (1956).
15. HELINSKI, D. R., AND YANOFSKY, C., *Biochim. et Biophys. Acta*, **63**, 10 (1962).
16. RUDLOFF, V., AND BRAUNITZER, G., *Hoppe-Seyler's Z. physiol. Chem.*, **323**, 129 (1961).
17. GOODWIN, T. W., AND MORTON, R. A., *Biochem. J.*, **40**, 628 (1946).
18. BOYER, P. D., *J. Am. Chem. Soc.*, **76**, 4331 (1954).
19. MARTIN, R. G., MATTHEI, J. H., JONES, O. W., AND NIRENBERG, M. W., *Biochem. and Biophys. Research Communs.*, **6**, 410 (1962).
20. SPEYER, J. F., LENGYEL, P., BASILO, C., AND OCHOA, S., *Proc. Natl. Acad. Sci. U. S.*, **48**, 441 (1962).
21. TANFORD, C., *Physical chemistry of macromolecules*, John Wiley and Sons, Inc., New York, 1961, p. 515.
22. SCHERAGA, H. A., *Protein structure*, Academic Press, Inc., New York, 1961, p. 241.

XIII
Supramolecular Organization

Editor's Comments on Paper 34

34 Giles, Case, Partridge, and Ahmed: *A Gene Cluster in* Neurospora crassa *Coding for an Aggregate of Five Aromatic Synthetic Enzymes*

Complementation studies were one of the factors that led to the finding that many proteins exist in their physiologically active state as multimers, or aggregates of polypeptides. Second site reversions, as described in Paper 33, made clear how important the tertiary and quaternary structural states are in determining a protein's functioning. Giles and coworkers introduced a third facet. Genetic analysis combined with biochemical analysis shows that some enzymes may aggregate into a supramolecular particle capable of carrying out several steps in a biosynthetic pathway.

Selected Bibliography

Case, M. E., and N. H. Giles. 1971. Partial enzyme aggregates formed by pleiotropic mutants in the arom gene cluster of *Neurospora crassa. Proc. Natl. Acad. Sci. U.S. 68:* 58–62.

De Moss, J. A., and J. Wegman. 1965. An enzyme aggregate in the tryptophan pathway of *Neurospora crassa. Proc. Natl. Acad. Sci. U.S. 54:* 241.

Harding, R. W., D. F. Caroline, and R. P. Wagner. 1970. The pyruvate dehydrogenase complex from the mitochondrial fraction of *Neurospora crassa. Arch Biochim. Biophys. 138:* 653–661.

A GENE CLUSTER IN NEUROSPORA CRASSA CODING FOR AN AGGREGATE OF FIVE AROMATIC SYNTHETIC ENZYMES*

By Norman H. Giles, Mary E. Case, C. W. H. Partridge, and S. I. Ahmed

DEPARTMENT OF BIOLOGY, KLINE BIOLOGY TOWER, YALE UNIVERSITY

Communicated July 21, 1967

The widespread occurrence of functionally related clusters of genes, the operon of Jacob and Monod,[1] in various bacteria (procaryotes) is now well established,[2] and several distinctive characteristics of these genetic systems have been defined.[3] By contrast, comparative genetic evidence has appeared to indicate that such systems do not occur in eucaryotes. Recently, however, studies with histidine-requiring mutants in *Neurospora*, yeast, and *Aspergillus* have established the existence of a cluster of three genes having a number of properties characteristic of bacterial operons.[4] These results prompted the present studies of a gene cluster, first detected in *Neurospora* by Gross and Fein,[5] in which mutation produces polyaromatic auxotrophs. Comparative investigations of almost 500 newly induced mutants, employing combined complementation, recombination, and biochemical techniques, indicate that this cluster contains five structural genes coding for the enzymes controlling steps two through six in the polyaromatic pathway prior to chorismic acid.

Evidence that this gene cluster exhibits properties characteristic of bacterial operons has been obtained, such as biochemical pleiotropy, polarity effects, and the asymmetrical genetic localization of completely noncomplementing mutants, some of which are suppressible by an apparent nonsense suppressor.[6] However, certain mutants in the cluster have properties which distinguish them from mutants in typical bacterial operons. These differences appear to result primarily from the fact that the five enzymes coded for by this gene cluster remain associated as a multienzyme aggregate. Since this genetic system has a number of features not characteristic of typical bacterial operons, and since there is, as yet, no evidence for operator or regulator genes affecting these activities, it seems best, at the present stage in these investigations, to refer to this system by the more noncommittal designation of *arom* gene cluster (or *arom* region), rather than as the *arom* operon.[7] Indeed, the present studies suggest that the primary function of the cluster may not be related to gene regulation, but may well be to code for an enzyme aggregate which provides a channeling mechanism effecting an intracellular separation of two potentially competing pathways in aromatic metabolism.

Materials and Methods.—(1) *Origin and initial classification of mutants:* Approximately 500 aromatic amino acid-requiring mutants were isolated, principally by filtration-concentration in strains closely related to wild-type 74A, following treatments with ultraviolet or with the chemical mutagens ethyl methanesulfonate or N-methyl-N-nitro-N-nitrosoguanidine. The mutants studied are polyaromatic auxotrophs which typically require for growth tryptophan, phenylalanine, tyrosine, and para-aminobenzoic acid. The following levels of supplements (amounts per ml) were used throughout, unless otherwise noted: 40 µg/ml each of L-tryptophan, L-phenylalanine, and L-tyrosine, plus 0.25 µg/ml para-aminobenzoic acid. Complementation tests to classify mutants employed mixed conidial suspensions on Fries minimal agar plates without sorbose.

(2) *Crossing procedures:* Sterility in crosses has proved to be a considerable problem with many of the *arom* mutants. However, adequate data have been obtained from two factor crosses,

sometimes employing either of two adjacent proximal markers, *pe* (peach) or *arg-12*, indicating the relative positions of mutants within the *arom* cluster. Subsequent use in three-point crosses of double *arom* mutants induced in an *arom-1* strain has served to check the two-point cross data.

(3) *Biochemical methods:* (a) *Enzyme preparations:* Mycelium was routinely grown for 72 hr at 25° in standing culture. Fries minimal was supplemented with limiting tryptophan (16 μg/ml) and double the usual concentration of other required supplements. Harvested mycelium was freeze-dried, powdered, extracted in 0.1 M KPO$_4$ buffer, pH 7.4, containing 0.1 mM α-thioglycerol, treated with excess protamine sulfate, and supernate used to obtain a precipitate in the 30–50% (NH$_4$)$_2$ SO$_4$ saturation range. Protein determinations were made using biuret reagent.[8]

(b) *Enzyme assays:* Spectrophotometric assays were carried out with a total volume of 0.6 ml, fluorometric assays with 0.4 ml. All assays were performed at 37°C.

DHQ synthetase was measured by the production of DHS in the presence of excess dehydroquinase. The reaction mixture contained 0.20 mM DAHP, 50 mM pH 7.4, KPO$_4$ buffer, 0.05 mM α-thioglycerol, 0.13 mM NAD, 0.17 mM CoCl$_2$, 1/60 vol of an *arom-2* enzyme preparation (source of excess dehydroquinase), and 1/30 vol of the enzyme preparation being assayed. The change in absorption at 240 mμ was followed over the initial linear reaction period.

Dehydroquinase was measured by a modification of the method of Gross and Fein.[5] The reaction mixture contained approximately 0.3 mM crude DHQ, 50 mM pH 7.4 KPO$_4$ buffer, 0.2 mM EDTA, and 1/30 vol of enzyme preparation.

DHS reductase was measured by a modification of the method of Gross and Fein.[5] The reaction mixture contained 4.2 mM shikimic acid, 1.67 mM NADP, 42 mM pH 10.6 glycine buffer, and 1/12 vol of dialyzed enzyme preparation.

Shikimic acid kinase was measured either by (*a*) loss of shikimic acid or (*b*) production of anthranilic acid in the presence of excess *arom-5* extract. In method (*a*), a modification of the method of Fewster,[9] the reaction mixture contained 2.5 mM shikimic acid, 5 mM ATP, 5 mM MgCl$_2$, 50 mM pH 8.0 Tris HCl buffer, and 0.03 ml of enzyme preparation in a total of 0.4 ml. In method (*b*), the reaction mixture contained 2.5 mM shikimic acid, 5 mM ATP, 5 mM MgCl$_2$, 5 mM L-glutamine, 1 mM NADP, 1 mM PEP, 37 mM pH 8.0 Tris HCl, 8 mM β-mercaptoethanol, 3/40 vol of an *arom-5* enzyme preparation, and 1/20 vol of the enzyme preparation being assayed. The reaction was stopped with 1/4 vol of 1 M HCl, and the anthranilic acid extracted for fluorescence measurement.

EPSP synthetase was measured either by (*a*) loss of SAP or (*b*) by the above fluorometric method with the substitution of *arom-4* mutants (or of *E. coli* mutant, 83-24). In (*a*), a modification of the method of Levin and Sprinson,[10] *E. coli* strain 83-3 was used. The incubation mixture contained 1 mM PEP, 1 mM SAP, 50 mM citrate buffer, pH 5.6, 0.1 mM DTT, and enzyme fraction in a total of 1.5 ml. In (*b*), with mutant *E. coli* extract, the reaction mixture was supplemented with 1 mM NADH and 7.5 mM L-phenylalanine, and SAP (1.8 mM) was substituted for shikimic acid.

(c) *Abbreviations used:* EDTA, ethylenediaminetetraacetic acid; ATP, adenosine 5′-triphosphate; NAD, nicotinamide-adenine dinucleotide; NADH, reduced nicotinamide-adenine dinucleotide; NADP, nicotinamide-adenine dinucleotide phosphate; PEP, phosphoenolpyruvic acid; DTT, dithiothreitol; Tris HCl, tris(hydroxymethyl)aminomethane hydrochloride; E-4-P, D-erythrose-4-PO$_4$; DAHP, 3-deoxy-D-arabinoheptulosonic acid-7-PO$_4$; DHQ, 5-dehydroquinic acid; DHS, 5-dehydroshikimic acid; SA, shikimic acid; SAP, shikimic acid-5-PO$_4$; EPSP, 3-enolpyruvylshikimic acid-5-PO$_4$; and CA, chorismic acid.

Results.—(1) *Classification of mutants by complementation and biochemical analyses:* The polyaromatic auxotrophic mutants were first classified on the basis of complementation analysis and then examined for their biochemical defects by *in vitro* complementation with bacterial mutants carrying known defects and by direct enzymatic assays. The mutants have all been shown to be defective for one or more of the last six enzymatic reactions in aromatic biosynthesis prior to CA (Fig. 1). Mutants in one group were found to be equivalent to the *arom-3* mutants of Gross and Fein[5] and lack CA synthetase activity.

The second group of mutants (the *arom* gene cluster or region) is complex both

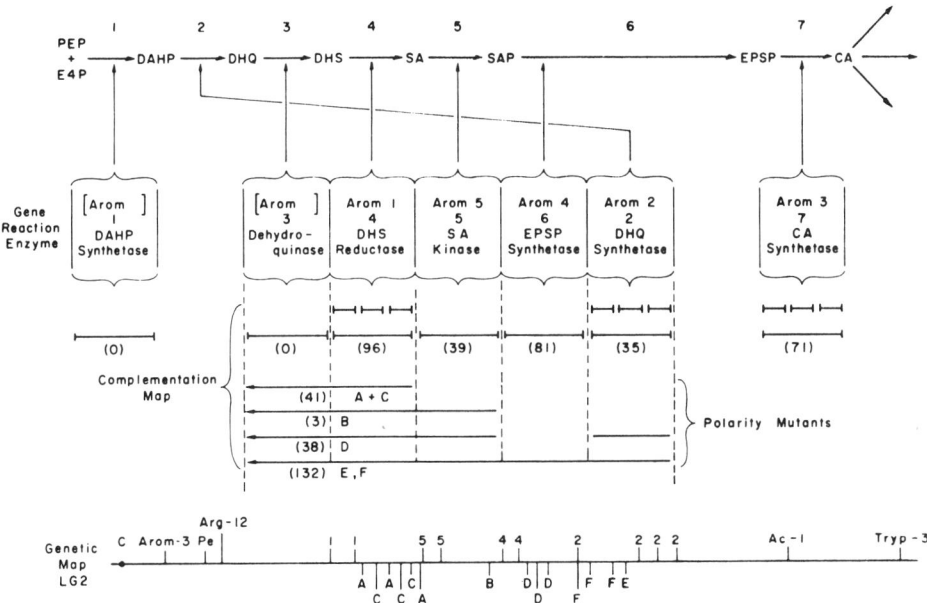

FIG. 1.—The organization of the *arom* gene cluster in *Neurospora crassa* on the basis of complementation, biochemical, and genetic recombination data. Reactions in the polyaromatic pathway prior to chorismic acid, indicated in detail by Giles,[7] are shown at the top of the figure (abbreviations are explained in the text). Designated below are the structural genes which code for the indicated enzymes catalyzing specific reactions. (The presumptive gene(s) controlling the first reaction and the identified (*arom-3*) gene controlling the seventh reaction, neither of which is within the *arom* cluster, are also included in the figure.) On the complementation map, numerals in parentheses indicate the number of mutants of a particular type. The short bars on the complementation maps of the *arom-1*, *-2*, and *-3* genes serve only to indicate the occurrence, but not the detailed pattern, of allelic complementation for these mutants. Categories of polarity mutants are indicated by the letters *A* through *F*. The symbols on the genetic map are as follows: *L.G. II* (linkage group II), *Cen* (centromere), *Pe* (peach), *Arg-12* (arginine-12), *Ac-1* (acetate-1), *Tryp-3* (tryptophan-3). On the genetic map, the approximate positions of single gene mutants within the *arom* clusters are indicated above the line, those of polarity mutants, below.

genetically and biochemically and consists of two major categories. The first major category, designated single gene mutants, consists of four clear-cut classes of complementing mutants each defective for a single activity but possessing, in general, high levels of the other four activities, either equivalent to or higher than the wild-type level (Table 1). These include classes equivalent to the *arom-1* (deficient for DHS reductase) and *arom-4* (deficient for EPSP synthetase) groups previously identified,[5] plus two newly identified groups, *arom-2* (deficient for DHQ synthetase) and *arom-5* (deficient for shikimic acid kinase).

The second major category of mutants within the *arom* region, designated polarity mutants, differs in certain notable respects from the first. These mutants either fail to complement with mutants in two, three, or all four of the single gene groups, or exhibit marked quantitative reductions in complementation responses (based on growth tests) with two or more of the mutant groups (e.g., the distinction between A and C types and *arom-1* mutants is based on this latter characteristic). In addition, assay data (Table 1) indicate that these mutants are biochemically pleiotropic, either lacking or having markedly reduced activities for all five enzymes.

TABLE 1
Specific Activities as a Per Cent of Wild Type in Various *arom* Mutants for the Five Enzymes Coded for by the *arom* Gene Cluster in *Neurospora crassa*, plus the Specific Activity for an Independently Controlled Related Enzyme*

Mutant category	Dehydroquinase Induced*	Dehydroquinase Constitutive	DHS Reductase	SA Kinase	EPSP Synthetase	DHQ Synthetase
Single gene mutants						
arom-1 (1183)	5.5	165	4	107	107	153
(80)	14.6	148	1	110	81	133
arom-2 (81)	0	118	176	115	104	0
(82)	0	163	168	130	99	0
arom-4 (49)	16.5	251	225	260	0	243
(1050)	19.0	305	251	270	0–2	325
arom-5 (56)	16.0	128	105	0	82	119
(1146)	14.0	156	93	2	84	126
arom-3 (47)	10.6	208	248	297	252	300
(87)	15.0	297	285	313	360	300
Polarity mutants						
A (58)	30.0	0	5	13	25	0
A (1131)	16.5	0–1	2	4	7	0
C (63)	8.8	0	2	0	7	0
C (1148)	30.8	0	3	0	4	0
B (1136)	8.1	0–1	4	8	20	0
B (14)	29.0	0	7	1	35	0
D (75)	17.7	1	2	1	2	0
D (1199)	18.3	0	2	0	9	0
E (54)	4.7	0	3	3	1–2	0
E (34)	0.4	0	3	0	1	0
F (25)	0	0	3	0–1	0–1	0

All cultures were grown in the absence of quinic acid. Assays were performed on material precipitated by ammonium sulfate solution between 30 and 50% saturated, except in the case of dehydroquinase where the material precipitated below 30% saturation was also assayed. The wild-type specific activities expressed as mμmoles/min/mg protein were as follows: DHS reductase, 25; dehydroquinase, 13; DHQ synthetase, 5; SA kinase, 21; EPSP synthetase, 46. SA kinase was assayed by method (a), and EPSP synthetase by method (a) (see *Methods* section). E_{240} for DHS was taken as 12,000. Assays of crude extracts indicated no significant differences in relative activities from the values in the above table.

* Since no induced dehydroquinase (see text) was detected in wild type grown without exogenous inducer, the specific activities of this enzyme are given directly based on protein values of 0–50% saturated $(NH_4)_2SO_4$ samples compared with constitutive dehydroquinase for which specific activities are based on protein values from 30–50% saturated samples. The induced form of the activity is defined as that portion of the total which is stable at 71° for 10 min in 100 mM pH 8.0 Tris HCl plus 1 mM EDTA. The constitutive form is defined as that portion lost under these conditions.

Extreme mutants of the F type fail to complement with all other *arom* region mutants and essentially lack all five activities. E mutants are also noncomplementing, but differ in having dehydroquinase activity. The complementation responses of other mutants, when represented on a complementation map (Fig. 1), tend, in general, to exhibit a marked polarity, and for this and other reasons these mutants have been designated polarity mutants. It should be noted that polarity mutants, even those classified within the same category, exhibit considerable diversity in their complementation responses.

One initially puzzling result of these studies was the absence of an anticipated class of single gene mutants, i.e., one lacking dehydroquinase activity alone. In all mutants studied to date a loss of, or marked reduction in, dehydroquinase activity is accompanied by a loss of, or marked reduction in, the other four enzyme activities coded for by the *arom* region. Recent biochemical evidence indicates the probable explanation, i.e., that *Neurospora crassa* has two distinct dehydroquinase activities, one constitutive and one inducible, the latter apparently coded for by a gene not in the *arom* cluster. Detailed evidence for this distinction, based in part on thermolability (Table 1) and zone centrifugation studies, will be presented in a later paper.[11]

(2) *Genetic mapping of mutants:* Crossing data confirm previous evidence[5] that the two groups of polyaromatic mutants occur in the right arm of linkage group II (Fig. 1). These data indicate the order of the four identified genes—each of which maps in a distinct, localized part of the region—and establish the relative positions of certain polarity mutants (Fig. 1). The polarity mutants map at various positions in the *arom* region, their locations corresponding, in general, to those predicted on the basis of their polarized complementation patterns. The completely noncomplementing mutants map asymmetrically at one end of the region among the *arom-2* mutants. The precise localization of other types of polarity mutants relative to the other three structural genes is not yet clear.

(3) *Density gradient centrifugation studies:* Representative comparative results from zone centrifugation[12] (Fig. 2 and Table 2) indicate that in wild type and all single gene mutants examined to date, the activities coded for by the *arom* region are associated in an enzyme aggregate with a molecular weight of *ca.* 200,000. By contrast, at least three of the five activities in wild-type *Salmonella* and *E. coli* are easily separable by this same procedure, and all three have much lower molecular weights than those determined for similar activities in *Neurospora*.

The results of comparable studies with polarity mutants are quite different from those with wild-type and single gene mutants. The only constitutive activities so far examined in gradients, EPSP synthetase, and SA kinase have been found to have much lower molecular weights (less than 100,000).

Discussion.—On the basis of the previously summarized characteristics of the polyaromatic auxotrophic mutants, the organization and functioning of the *arom* region in wild-type *Neurospora crassa* is interpreted as follows: This region is considered to be a functionally integrated cluster of five contiguous structural genes

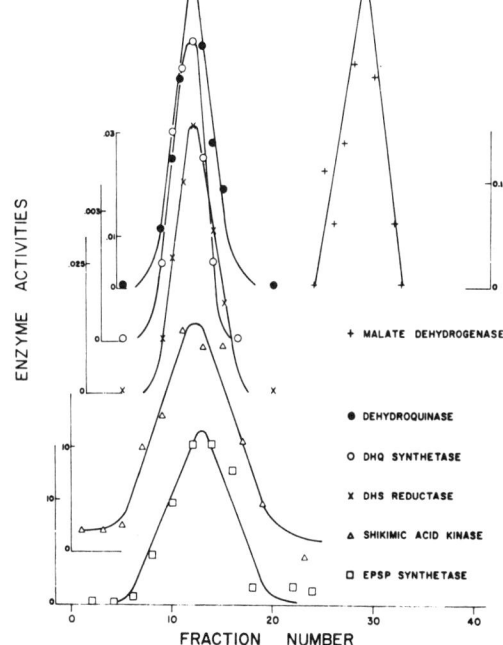

Fig. 2.—Distribution, after centrifugation in a sucrose density gradient, of activities of the five aromatic synthetic enzymes coded for by the *arom* gene cluster in *Neurospora crassa* (and of the reference activity—malate dehydrogenase) from an *arom-3* mutant (lacking CA synthetase). Conditions as described in Table 2. Assays of shikimate kinase and EPSP synthetase were performed by method (*b*) (see *Methods* section). Centrifugation of a crude extract and of an alcohol-precipitated subfraction from the ammonium sulfate fraction indicated no significant shift in distribution.

+ MALATE DEHYDROGENASE
● DEHYDROQUINASE
○ DHQ SYNTHETASE
× DHS REDUCTASE
△ SHIKIMIC ACID KINASE
□ EPSP SYNTHETASE

TABLE 2

SEDIMENTATION CONSTANTS OF ENZYME ACTIVITIES OF WILD TYPE (74A), *arom* GENE CLUSTER MUTANTS LACKING SINGLE ACTIVITIES ONLY, AND AN *arom-3* MUTANT (LACKING CA SYNTHETASE)

Mutant category	No. of strains	Enzymic Activity				
		DHQase	DHS Reductase	EPSP Synthetase	SA Kinase	DHQ Synthetase
arom-1	3	11.5	(none)	10.9	10.9	11.6
arom-2	2	11.3	11.4	11.1	11.1	(none)
arom-4	2	11.1	11.1	(none)	11.1	11.1
arom-5	4	11.3	11.2	11.2	(none)	11.3
arom-3	1	11.3	11.3	11.4	11.3	11.3
Wild type	1	11.1	11.3	11.4	11.1	11.6

Samples were precipitated by ammonium sulfate solution between 32 and 48%, saturated, and applied in a volume of 0.4- to 4.6-ml gradients of 5–20% sucrose in 100 mM, pH 8.0 Tris HCl, and in pH 7.4 KPO$_4$, each containing 0.1 mM EDTA and 0.1 mM DTT. Centrifugations were carried out for 16–17 hr at 35,000 rpm in a Spinco model L2 ultracentrifuge using an SW50 rotor, at 3.5°.
Procedures and computations followed Martin and Ames.[12] Endogenous *Neurospora* malate dehydrogenase was used as the standard (S 4.77)[13] after comparison with several standard purified enzymes ranging throughout the gradient. Assay methods are described in the *Methods* section. EPSP synthetase and shikimate kinase were assayed by method (*b*). Average values are given.

coding for five different polypeptides, each carrying the active site for one of the five enzymes catalyzing steps two through six in the prechorismic acid part of the polyaromatic pathway. The cluster is transcribed in a polarized fashion via a single messenger RNA. The initiation of transcription occurs in, or adjacent to, the *arom-2* gene, which is the most proximal gene in the cluster (the most distal in the chromosome arm) and codes for the polypeptide carrying the active site for DHQ synthetase. Transcription proceeds distally within the cluster, involving, in order, the four other structural genes, as shown in Figure 1. Normally, after translation of the messenger RNA (which presumably starts with the codons specifying the DHQ synthetase polypeptide), all five different polypeptides remain associated (at least two apparently as multimers on the basis of intragroup complementation evidence) in a single aggregate.

The four known classes of single gene mutants are interpreted as resulting from missense mutations in four of the five structural genes in the *arom* cluster, the resulting single amino acid substitutions giving polypeptides each lacking a normal specific active site. However, these polypeptides are retained as part of an intact aggregate in which the other four activities are, in general, comparable to those in the wild type.

Polarity mutants are clearly very different in a number of important characteristics from single gene mutants. The most significant polar mutants are the noncomplementing ones essentially lacking all five activities coded for by the *arom* gene cluster, since these mutants provide the most cogent evidence that the cluster constitutes a supragenic functional unit. The evidence that these mutants map at one end of the cluster (within the *arom-2* gene), that some are suppressible by a presumptive nonsense suppressor,[6] and that one mutant yields both complete revertants indistinguishable from wild type and other revertants equivalent to certain *arom-2* mutants,[6] leads to the interpretation of these mutants as nonsense (polarity) mutants in the first (proximal) structural gene of the cluster.

The A, B, and C polarity mutants are interpreted as nonsense mutants occurring within any one of the structural genes located within the cluster, distal to the *arom-2* gene. Whether D mutants are located within or distal to the *arom-2* gene is not yet clear. The biochemical evidence that all five enzyme activities are either absent or markedly reduced, and that any remaining detectable activities are associated with proteins having molecular weights much less than that of the aggregate present in wild-type and in missense mutants is taken to indicate that the nonsense (chain-terminating) mutations in these polarity mutants prevent the formation of a normal aggregate. The marked "reverse polarity" effect characteristic of these mutants (e.g., in A and C mutants, the strong reduction or essential absence, on the basis of direct assay, of activities coded for by genes located proximal to A and C mutant sites) presumably results from the fact that the *arom* region normally codes for an enzyme aggregate. Thus, the normal expression of any of the five enzyme activities apparently depends upon the presence in a complete aggregate of a specific number and arrangement of the postulated five kinds of complete polypeptide chains. However, the possibility also exists that some of these mutants result from frame shifts, or from missense mutations which in some way affect aggregation. In particular, additional evidence will be required to intrepret the D mutants, which constitute an exceptional class in having an interrupted rather than a continuous complementation pattern when ordered in accord with the genetic map.

The organization of the *arom* gene cluster as discussed here differs somewhat from that inferred from initial studies,[7] since at that time no definitive recombination data were available, nor had the evidence been obtained for two different dehydroquinase activities, with polarity mutants possessing only the inducible activity. Further evidence concerning the physical organization of the enzyme aggregate controlled by the *arom* gene cluster must await current attempts at purification and direct disruption by physical and/or chemical treatment of the wild-type aggregate.

There remains the question as to the functional significance of this supragenic unit. It is evident that this gene cluster has many features characteristic of bacterial operons. However, certain other features, in particular the presence of an enzyme aggregate, serve to distinguish this cluster from most, although not all,[14] bacterial operons. In addition, there is at present no genetic evidence for regulatory genes or for an operator gene in this system. Thus, a possible regulatory significance of the cluster remains to be elucidated. However, the present studies do suggest at least one functional role for this gene cluster based on the evidence that *Neurospora crassa* possesses two distinct dehydroquinases: a constitutive form, which is part of the aromatic enzyme aggregate in the biosynthetic pathway, and an inducible form, which is part of an aromatic degradative pathway.[11,15] It appears that the *arom* enzyme aggregate in *Neurospora* may provide a channeling mechanism for segregating these two potentially competing pathways in aromatic metabolism—one synthetic and one degradative—which have one step in common, namely, the conversion of DHQ to DHS. Other authors have recently proposed a role of enzyme aggregates in providing channeling mechanisms.[16] A gene cluster transcribed via a single messenger would appear to provide a potentially efficient method of coding for an enzyme aggregate, since such a system should facilitate the juxtaposed synthesis, both in time and space, of the several different polypeptide chains forming an aggregate.

Summary.—Genetical and biochemical evidence is presented for the existence in *Neurospora crassa* of a functionally integrated cluster of five contiguous structural genes coding, apparently via a single polycistronic mRNA, for an enzyme aggregate catalyzing reactions two through six in the aromatic biosynthetic pathway prior to chorismic acid. Although this gene cluster has many features characteristic of bacterial operons, no clear evidence has yet been obtained that it plays a role in gene regulation. However, one function of the cluster may be to facilitate the formation of the enzyme aggregate which apparently acts as a channeling mechanism to effect the separation of two potentially competing pathways in aromatic metabolism, one synthetic and one degradative. Evidence is presented for the occurrence in *Neurospora* of two dehydroquinases, one constitutive and one inducible, catalyzing the one step common to both pathways.

The authors would like to thank Drs. D. B. Sprinson and Carl M. Stevens for supplies of SAP and DAHP, and of DHQ, respectively; also Drs. B. D. Davis and F. Gibson for bacterial mutant strains, and Mr. Thomas Seale for the *Neurospora* nonsense suppressor. Excellent technical help was provided by Mrs. Delilah Gomes, Mrs. Clara Sampson, and Miss Rosemary Brett. Portions of this manuscript were written while one of the authors (N. H. G.) was a Guggenheim Fellow in the Department of Genetics of the John Curtin School for Medical Research at the Australian National University. He wishes to express his appreciation to Prof. David Catcheside for his hospitality and to Dr. Colin Doy for stimulating discussions.

* These studies were aided by Atomic Energy Commission contract AT (30-1)-3098 and by National Science Foundation grants GB 2176 and GB 5687. One of the authors (S. I. A.) is a postdoctoral trainee on a National Institutes of Health Genetics Training grant, GM 397.

[1] Jacob, F., and J. Monod, *J. Mol. Biol.*, **3**, 318 (1961).
[2] Demerec, M., these PROCEEDINGS, **51**, 1057 (1964).
[3] Ames, B. N., and R. G. Martin, *Ann. Rev. Biochem.*, **33**, 235 (1964).
[4] Fink, G. R., *Genetics*, **53**, 445 (1966).
[5] Gross, S. R., and A. Fein, *Genetics*, **45**, 885 (1960).
[6] Seale, T., *Genetics*, in press; Case, M. E., N. H. Giles, and C. W. H. Partridge (Abstr.) *Genetics*, **56**, 548 (1967).
[7] Giles, N. H., *Natl. Cancer Inst. Monograph*, **18**, 341 (1965).
[8] Munkres, K. D., and F. M. Richards, *Arch. Biochem. Biophys.*, **109**, 466 (1965).
[9] Fewster, J. A., *Biochem. J.*, **85**, 388 (1962).
[10] Levin, J. G., and D. M. Sprinson, *J. Biol. Chem.*, **239**, 1142 (1964).
[11] Giles, N. H., C. W. H. Partridge, S. I. Ahmed, and M. E. Case, these PROCEEDINGS, in press.
[12] Martin, R. G., and B. N. Ames, *J. Biol. Chem.*, **236**, 1372 (1961).
[13] Munkres, K. D., *Arch. Biochem. Biophys.*, **112**, 340 (1965).
[14] Henning, U., G. Dennert, R. Hertel, and W. S. Shipp, in *Cold Spring Harbor Symposia on Quantitative Biology*, vol. 31 (1966), p. 227; Bauerle, R. H., and P. Margolin, in *Cold Spring Harbor Symposia on Quantitative Biology*, vol. 31 (1966), p. 203.
[15] Tatum, E. L., S. R. Gross, G. Ehrensväld, and L. Garnjobst, these PROCEEDINGS, **40**, 271 (1954).
[16] Lynen, F., *Angew. Chem.*, **77**, 929 (1965); Davis, R. H., in *Organizational Biosynthesis*, ed. H. J. Vogel, J. O. Lampen, and V. Bryson (New York: Academic Press, in press).

XIV
One Gene–One Chromomere

Editor's Comments on Paper 35

35 Judd, Shen, and Kaufman: *The Anatomy and Function of a Segment of the X Chromosome of* Drosophila melanogaster

One aspect of the gene–protein or gene–function relationships that is only now beginning to be actively explored is to identify what the gene really is in eukaryotes. The concept of the gene obtained from studying prokaryotes is that of a stretch of DNA which contains the necessary nucleotide sequences in the form of codons to transcribe messenger RNA with the necessary information to translate a specific amino acid sequence in a polypeptide. The problem is somewhat clouded in the eukaryotes, however, because they have more DNA per haploid set of chromosomes than would conceivably be necessary to code for the number of different kinds of polypeptides. Eukaryotes have 10 to 100,000 times more DNA per nucleus (haploid) than prokaryotes, but it is certain that they do not have literally thousands more different polypeptides than prokaryotes. Metabolism in the two types is not that different. This paper is an example of one way to approach the problem, which can certainly be considered one of the more important in current biological research.

Selected Bibliography

Daneholt, B., J.-E. Edström, E. Egyhazi, B. Lambert, and U. Ringborg. 1969. Physicochemical properties of chromosomal RNA in *Chironomus tentans* polytene chromosomes. *Chromosoma* 28: 379–398, 399–417, 418–429.

Hochman, B. 1974. Analysis of a whole chromosome in *Drosophila. Cold Spring Harbor Symp. Quant. Biol. 38:* 581–590.

Judd, B. H., and M. W. Young. 1974. An examination of the one cistron: one chromomere concept. *Cold Spring Harbor Symp. Quant. Biol. 38:* 573–580.

Lefevre, G., Jr. 1974. The one band–one gene hypothesis: evidence from cytogenetic analysis of mutant and non-mutant rearrangement breakpoints in *Drosophila melanogaster. Cold Spring Harbor Symp. Quant. Biol. 38:* 591–600.

Sorsa, Veikko. 1974. Organization of chromomeres. *Cold Spring Harbor Symp. Quant. Biol. 38:* 601–608.

THE ANATOMY AND FUNCTION OF A SEGMENT OF THE X CHROMOSOME OF DROSOPHILA MELANOGASTER[1]

B. H. JUDD, M. W. SHEN AND T. C. KAUFMAN[2]

Department of Zoology, University of Texas at Austin
Austin, Texas 78712

Manuscript received September 27, 1971
Revised copy received December 20, 1971

ABSTRACT

An average size chromomere of the polytene X chromosome of *Drosophila melanogaster* contains enough DNA in each haploid equivalent strand to code for 30 genes, each 1,000 nucleotides long. We have attempted to learn about the organization of chromosomes by asking how many functional units can be localized within a chromomere. This was done by 1) recovery of mutants representative of every cistron in the 3A2–3C2 region; 2) the characterization of the function of each mutant type and grouping by complementation tests; 3) the determination of the genetic and cytological position of each cistron by recombination and deletion mapping. The data clearly show one functional group per chromomere. It is postulated that a chromomere is one cistron within which much of the DNA is regulatory in function.

THE chromomeres of polytene chromosomes are DNA-rich regions which alternate with DNA-poor interchromomeric spaces. Chromomeres range in width from less than 0.1 μ to about 0.5 μ and are so constant in type and sequence that they can be used to construct cytological maps of polytene chromosomes which correlate precisely with gene order in genetic linkage maps (see LINDSLEY and GRELL 1968). There is no doubt that the chromomeric pattern reflects in some fashion the genetic organization of the chromosome.

Both PAINTER (1934) and MULLER (MULLER and PROKOFYEVA 1935) by cytogenetic analysis of chromosome rearrangements concluded that one or a few genes are represented by one chromomere. Such results were used by MULLER and PROKOFYEVA to estimate that the number of genes in *Drosophila melanogaster* is between five and ten thousand. BRIDGES (1935) inferred that each of the faint chromomeres corresponds to one locus and that some of the heavy-walled capsules, consisting of two heavy bands and a lighter one between, contain three loci. BRIDGES counted 3,540 distinct bands in *D. melanogaster* chromosomes. This number was subsequently increased to 5,072 by the more detailed studies of C. B. BRIDGES (1938) and P. N. BRIDGES (1942).

Work on the puffing patterns in the giant polytene chromosomes of Chironomous (BEERMANN 1962) led to the concept that each puff originates from a

---

[1] This investigation was supported in part by U.S.P.H.S. Research Grants GM 12334 and HD 03803 and by U.S.P.H.S. Training Grant GM 00337.
[2] Present address: Department of Zoology, University of British Columbia, Vancouver, B. C., Canada.

single band of the chromosome. BEERMANN's (1961) demonstration that the production of a particular secretion granule by a specialized lobe of the salivary gland is correlated with the puffing of a specific band in the 4th chromosome supported the concept that the information necessary for the production of that protein resides within that single band. GROSSBACH's (1969) analysis of the salivary gland secretory proteins showed a strong correlation between certain puffs and the synthesis of cell-specific polypeptide units. BEERMANN (1967) postulated that the transcriptional unit is the chromomere. The rapidly labelled, chromosomal RNA of Chironomus is a heterogeneous, high molecular weight species (DANEHOLT et al. 1969a), as is that produced by a single Balbiani ring or puff (DANEHOLT et al. 1969b). However, the question of whether there is only one function or several specified by the information in a single chromomere remains to be answered.

RUDKIN (1965) calculated from his measurements of the amount of DNA per haploid X-chromosome strand that an average chromomere contains about 3×10^4 nucleotide pairs, with the range extending over several orders of magnitude. This means that an average chromomere contains enough DNA for about 30 genes, each 1,000 nucleotide pairs long. By RUDKIN's measurement, each haploid X-chromosome strand is made up of about 3×10^7 nucleotide pairs. This figure compares rather favorably with measurements of LAIRD and MCCARTHY (1969), who found the total DNA in the haploid complement of *Drosophila melanogaster* to be 1×10^8 nucleotide pairs. The X chromosome represents slightly less than 20% of the genome or just under 2×10^7 nucleotide pairs. These measurements show that there is sufficient DNA in the X chromosome to code for 20,000 to 30,000 genes of 1,000 nucleotide pairs in average length.

LAIRD and coworkers (DICKSON, BOYD and LAIRD 1971) have concluded from their data on renaturation kinetics of salivary gland DNA fragments that about 95% is single-copy or unique-sequence DNA. Thus most of the DNA sequences in euchromatin must be replicated uniformly during the formation of polytene chromosomes. On the face of it, these data would indicate that there is sufficient nucleotide sequence information in each chromomere to specify many functions. The data we present here, however, show that associated with 15 adjacent bands and interbands in the X chromosome are 16 complementation groups on essentially a one band:one cistron basis. We will attempt to reconcile these apparently divergent sets of data about chromosome organization and fit them to a model of the functional subunits of chromosomes.

MATERIALS AND METHODS

The experimental plan had three components: 1) the recovery of mutants representative of every locus within a small, cytologically and genetically defined segment of the X chromosome, 2) the characterization of the function of the recovered mutant types and their grouping by complementation tests, 3) the determination of the genetic and cytological position of each locus.

Mutant Recognition and Recovery: The chromosome region chosen for analysis extends from band 3A2 to 3C2 of BRIDGES' (1938) chromosome map. Genetically, the region is about 0.75 map units long, with the zeste and white loci located near the left and right margins, respectively.

The screening procedure for obtaining mutants in this region consisted of crossing males

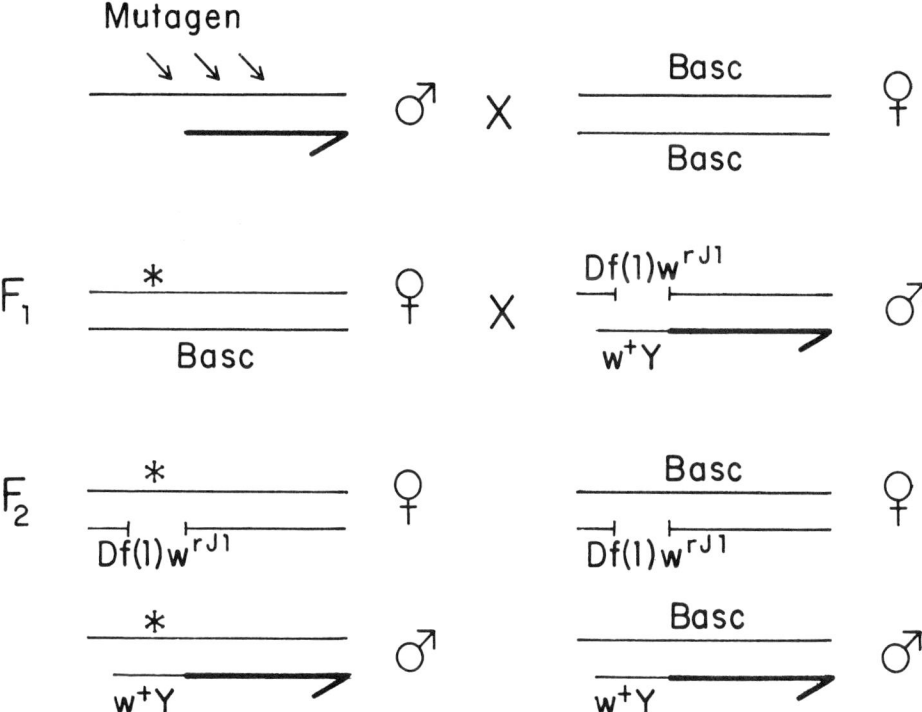

FIGURE 1.—The mating scheme employed for recovering mutants in the X chromosome regions 3A2–3C2. The X chromosome is indicated by the thinner line; the thicker line depicts the Y chromosome. A mutant induced in the region of interest is shown by *.

which had been treated with various mutagens to females homozygous for an X-chromosome balancer, either *Basc* or *FM7* (MERRIAM 1968). The structure of the *Basc* balancer chromosome and description of other mutants employed in this study can be found in LINDSLEY and GRELL (1968). The F_1 females produced by this cross were mated individually to males carrying an X chromosome deficient for the 3A2–3C2 region. This deletion, designated as $Df(1)w^{rJ1}$, is balanced by the Y chromosome, $w^+ \cdot Y$, in which a section of the X chromosome extending from 2D1–2 to 3D3–4 has been inserted into the long arm of the Y chromosome. The presence of this duplicated X-chromosome segment allows males with $Df(1)w^{rJ1}$ to survive. Offspring from this cross were classified for any mutations induced in the 3A2–3C2 region. The mating scheme is diagrammed in Figure 1.

It should be noted that each F_1 female was heterozygous for an X chromosome exposed to the mutagen (X^*); thus half of the offspring produced carried copies of that chromosome. Daughters which received the X^* also carried $Df(1)w^{rJ1}$ which "uncovers" the zeste–white region of the mutagenized chromosome. A mutant induced in the 3A2–3C2 interval was easily classified by examination of this female class. Should an induced mutant be a recessive lethal, as proved most often to be the case, this class of females was absent. The lethal could be recovered, however, because it was also present in $X^*/w^+ \cdot Y$ male sibs. Such males, when mated to females carrying compound-X chromosomes and the $w^+ \cdot Y$, $(C(1)DX/w^+ \cdot Y)$, established a balanced stock. Each mutant was also carried in a female line balanced against $In(1)dl$-49, $y\ Hw\ m^2\ g^4$.

Mutagens: The mutagenic agents employed were 250Kv X rays, ethyl methanesulfonate (EMS), and N-methyl-N^1-nitro-N-nitrosoguanidine (NNG). Seven separate X-ray series were

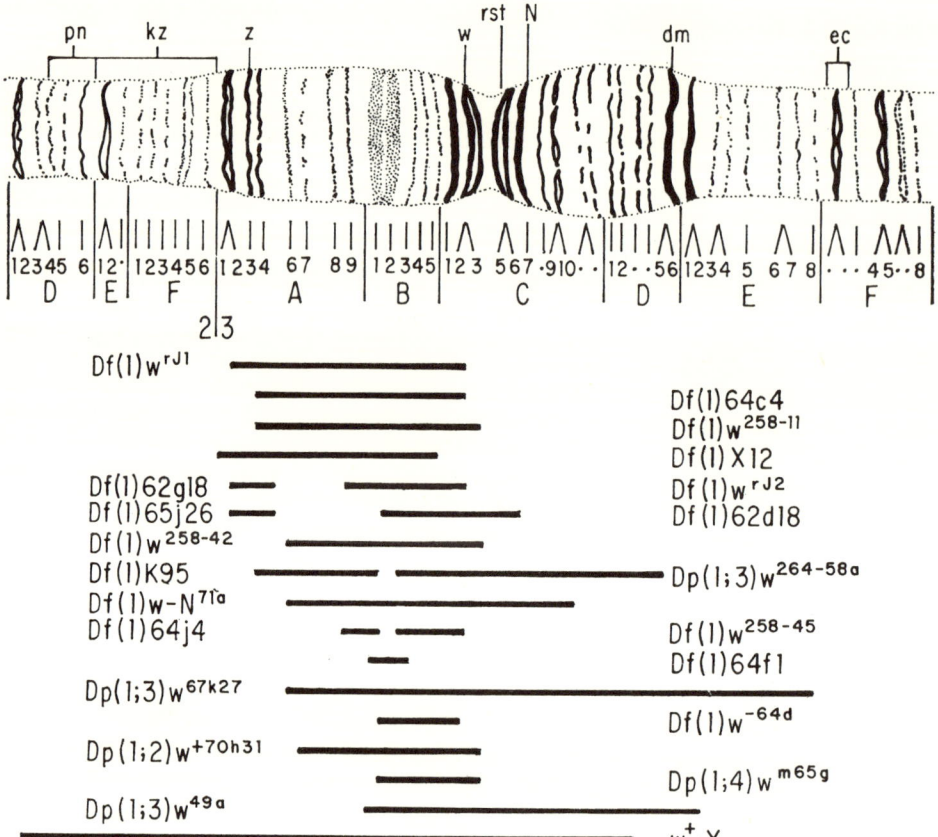

FIGURE 2.—A drawing of a segment of the X chromosome showing the cytological extent of the deletions and duplications used for determining the location of each complementation group. Each solid line shows the segment which is deleted or duplicated in the chromosome indicated.

run with doses ranging from 1,000 r to 7,000 r. Males treated were from Oregon-R or Amherst wild-type strains; in some experiments the males carried the mutants $z\ spl\ sn^3$ as X-chromosome markers. In all experiments, males were 2 to 4 days of adult age at the time of treatment. Each male was mated with five *Basc* or *FM7* females for 48 hr, after which he was discarded. The females were transferred to fresh culture medium every 2 or 3 days depending on the mutagen dose administered to the males.

EMS was used as a 0.025 M solution in 1% sucrose after the method of LEWIS and BACKER (1968). Males were fed the solution for 18–24 hr and then mated to virgin females in vials containing a standard brewer's yeast-cornmeal-Karo syrup-agar medium.

NNG was used as concentrations of 0.0025 M, 0.005 M, and 0.01 M in 2% sucrose (KAUFMAN 1969). Feeding was carried out in the same manner as with EMS.

Sex-linked lethals induced by other investigators were also screened to determine whether any of them were located in the 3A2–3C2 region. The lethals obtained from other workers and the agents used are given in the RESULTS section.

Complementation mapping: After mutants were produced and established in stock cultures, each was mated once again to the deletion for the 3A2–3C2 region, $Df(1)w^{rJ1}$, and to other, shorter deletions that were available. This test confirmed the 3A2–3C2 location of the mutant. Its interaction with the shorter deletions permitted a rough placement of the mutation within the region. Mutants were also mated *inter se* to determine if complementation occurred in transconfiguration heterozygotes. That is, mutant m_1 was crossed to mutant m_2 and the F_1 heterozygote, m_1/m_2, was examined to determine whether its phenotype was standard or mutant. In this manner, complementation groups were identified, each group consisting of all those mutants which showed a mutant phenotype in heterozygous combinations. In the case of noncomplementing lethals, this heterozygous class would be absent.

A linear array of complementation units was established by crossing members of each group to the several deletions or duplications that have at least one break point within the 3A2-3C2 region. There are areas where linear order could not be established with certainty by deletion mapping because rearrangements with the necessary break points were not available. In such cases, recombination tests described below were necessary to establish an unambiguous order.

As the number of mutants increased, all *inter se* crosses could not be performed; thus each new mutant was placed in one of three divisions by deletion mapping against $Df(1)w^{rJ2}$ and $Df(1)w^{258-45}$ and then crossed with all members of all the complementation groups that fell within its division. In this manner, any multigroup deletions could be detected and intralocus complementation could also be observed. The cytological extent of the deletions and duplications used in this study are given in Table 1, and Figure 2.

Recombination mapping: A genetic map of the region was constructed for comparison with the complementation and cytological maps. This was accomplished using F_1 females which were heterozygous for two complementing mutations and the flanking marker mutants y (yellow body) or z (zeste eye) located on the left at map positions 0.0 and 1.0, respectively, and either w (white eye) or spl (split bristles) on the right at 1.5 and 3.0, respectively. Such females were then mated to either $y\ w\ spl\ sn^3$ or $z\ spl\ sn^3$ males, depending on which marker mutants the females carried. The offspring were scored for recombinants. Since most of the crosses involved

TABLE 1

Cytological break points of deletion and duplication chromosomes used in this study

DELETIONS	Cytological break points	Source
$Df(1)w^{rJ1}$	3A1–2; 3C2–3	JUDD
$Df(1)64c4$	3A3–4; 3C2–3	JUDD
$Df(1)w^{258-11}$	3A3–4; 3C3–5	DEMEREC
$Df(1)X12$	2F6–3A1; 3B5–3C1	FALK
$Df(1)62g18$	3A1–2; 3A4–6	JUDD
$Df(1)65j26$	3A1–2; 3A4–6	JUDD
$Df(1)K95$	3A3–4; 3B1–2	FALK
$Df(1)w\text{-}N^{71a}$	3A4–6; 3C10–11	LEFEVRE
$Df(1)w^{258-42}$	3A4–6; 3C3–5	DEMEREC
$Df(1)w^{rJ2}$	3A8–9; 3C2–3	JUDD
$Df(1)64j4$	3A8–9; 3B1–2	JUDD
$Df(1)64f1$	3A9–B1; 3B3–4	JUDD
$Df(1)62d18$	3B1–2; 3C6–7	JUDD
$Df(1)w^{-64d}$	3B1–2; 3C2–3	LEFEVRE
$Df(1)w^{258-45}$	3B2–3; 3C2–3	DEMEREC
DUPLICATIONS		
$Dp(1;3)w^{67k27}$	3A4–6; 3E8–3F2	LEFEVRE
$Dp(1;2)w^{+70h31}$	3A6–8; 3C2–3;31	LEFEVRE
$Dp(1;4)w^{m65g}$	3B1–2; 3C3–5;101	LEFEVRE
$Dp(1;3)w^{49a}$	3A9–B2; 3E2–3;81	LEFEVRE
$Dp(1;3)N^{264-58a}$	3B2–4; 3D5–6;80	LEFEVRE
$w^+ \cdot Y(kz^+\text{-}spl^+KL\cdot bb^+KS)$	2D1–2; 3D3–4	LINDSLEY

lethals, the only males which survived were recombinants carrying the normal alleles of the two lethal loci being examined. The reciprocal double mutant class, being lethal, was not recovered, but it was assumed to occur with a frequency equal to that of the nonlethal crossover class observed. For recombinational analysis, care was taken to select only those lethals which were not leaky, i.e., those approaching 100% lethality. Occasional escapers were encountered, however, and progeny testing was required to determine whether or not surviving males were recombinant for the lethal loci.

Cytological mapping: All mutations which mapped as multisite lesions by complementation tests were examined cytologically, as were the majority of those that behaved as point mutations. Several duplications with one or both break points in or near the 3A–3C region were also examined.

Male larvae proved to be most useful for determining break points in the deletion and duplication chromosomes. Females heterozygous for the rearranged chromosomes were also examined.

RESULTS

Complementation tests of the mutants recovered by the screening procedure outlined above showed that 121 of them behaved as single-site mutations and grouped into fourteen complementation units. Including the z and w loci, there are sixteen complementation groups in this segment of the 3A2–3C2 region. Twelve of the complementation groups map in the region between z and w, whereas the remaining two are positioned to the left of z. Two of the mutants which act as single-site changes are associated with X-chromosome inversions. Each is inseparable from its inversion break point which falls in the 3A–3C

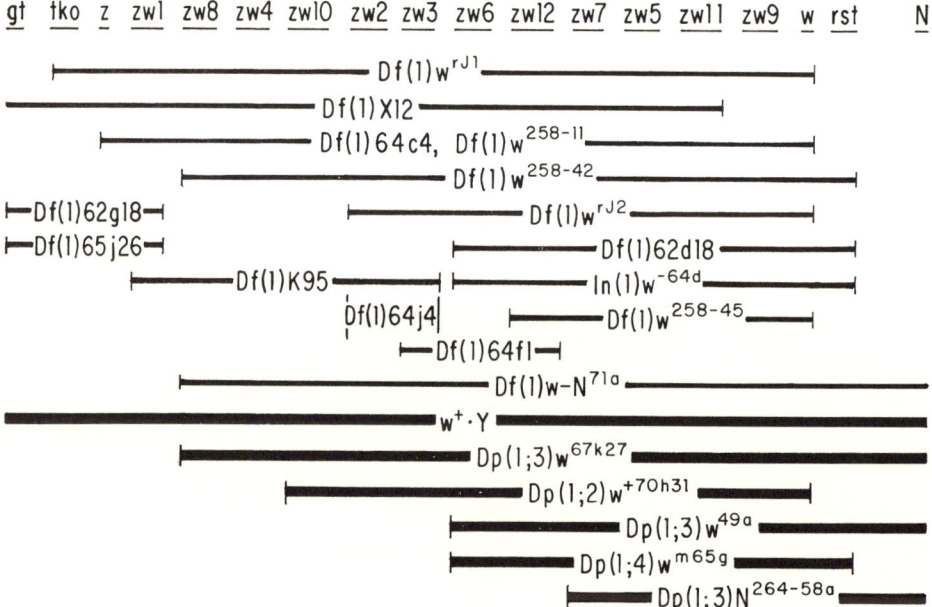

FIGURE 3.—A complementation map of the 3A–3C region of the X chromosome. The thin lines indicate the complementation groups recognized by *cis-trans* tests. The lines of intermediate thickness show the extent of each deletion. The thicker lines show which complementation groups are covered by each duplication.

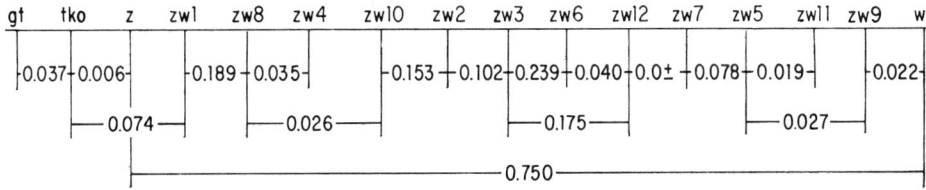

FIGURE 4.—A recombination map of the region from the *gt* locus to the *w* locus. The distances are expressed in map units.

region. One of them, *g31*, with break points at 3A3–4 and 6B2–C1 has a vesiculated (*vs*) phenotype apparently associated with the proximal break point.

Five of the induced mutations mapped as multisite changes. These, along with previously described deletions and duplications (LINDSLEY and GRELL 1968), were employed for the genetic and cytological analysis. The map constructed by deletion and recombination mapping shows a non-ambiguous linear order for the 16 complementation groups (Figure 3). Recombination values are summarized in Figure 4.

The twelve complementation units which are positioned between the *z* and *w* loci were designated as *zw1* through *zw12*, in the order in which they were discovered. Two loci, *gt* and *tko*, are positioned to the left of *z*.

In our experiments, slightly more than 2% (112/5256) of induced recessive lethals in the X chromosome fall in the 3A2–3C2 interval. This section represents about 1% of the X chromosome DNA (RUDKIN 1965) and also about 1% of the genetic map (0.75 map units/70 ± map units). The distribution of the mutations recovered in our tests is shown in Figure 5. Complementation groups *zw1* and *zw2* contain relatively large numbers of lethals while *gt* and *zw9* are represented by only two alleles each. It is not possible, however, to draw conclusions about relative mutability because, despite the high efficiency of the genetic screen, some types of mutants, such as non-leaky lethals and those with obvious morphological changes, are easily recognized; whereas semi-lethals or subtle morphological variants can be overlooked. There is no obvious correlation between chromomere size and the number of recovered mutant alleles in our tests.

Rather precise positioning of the complementation units in the chromomeres of the 3A2–3C2 region can be made by comparing the cytological length of each deletion and duplication with its genetic length. A remarkably good correlation between complementation groups and chromomeres emerges; essentially there is one function per chromomere. This relationship is diagrammed in Figure 5.

Our cytological map differs somewhat from that presented by BRIDGES (1938). We observe 8 bands in section 3A, 5 in section 3B. BRIDGES' map shows 10 bands in 3A, but only 4 in section 3B. The assignments of the *z* locus to band 3A3 (GANS 1953) and of the *w* locus to 3C2 (LEFEVRE and WILKINS 1966) are confirmed by our results.

A comparison of our map with that of BERENDES (1970), who employed the electron microscope in constructing his map, shows some interesting similarities

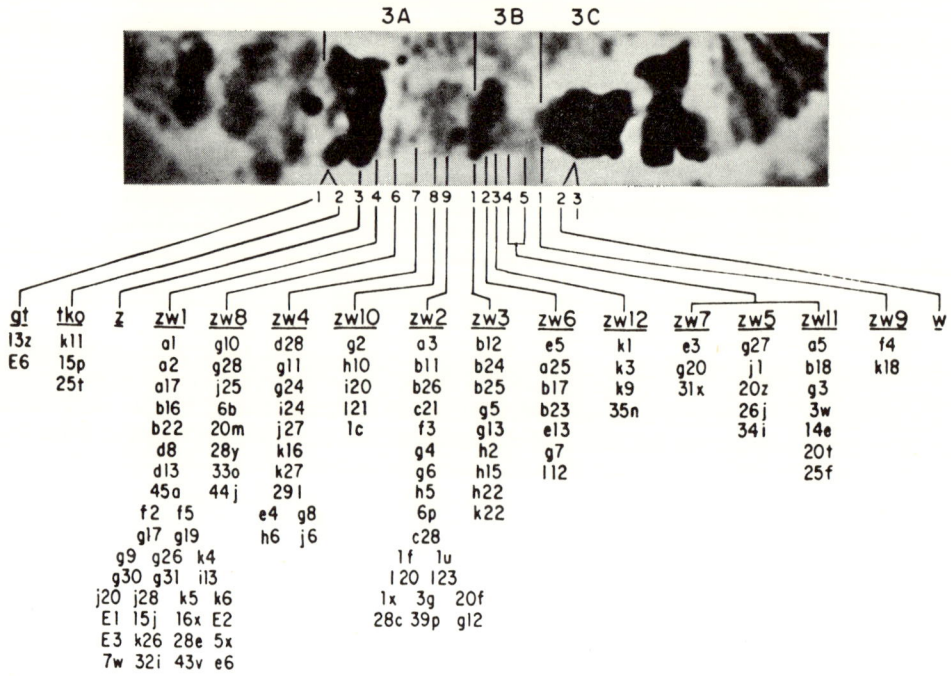

FIGURE 5.—A photograph of the 3A-3C region of the X chromosome of a *Drosophila melanogaster* female of genotype $gt\,w^a$. The complementation groups with the point mutations found in each are shown below. The cytological position of each complementation group as determined by deletion mapping is indicated.

and some differences. First, we believe that BERENDES misjudged the boundary between sections 2 and 3 when reconciling his map with that of BRIDGES (1938). In our opinion it would be more accurate to homologize BRIDGES' chromomere 3A1-2 with 3A3 on BERENDES' map. Of particular interest is that BERENDES found four very small chromomeres between the rather prominent bands 3B1-2 and 3C1, whereas BRIDGES identified only two. Although we have been unable to detect such a sixth band in 3B, using light microscopy, the relationship between the number of functional groups and chromomeres is precisely one for one it we accept BERENDES' map for this segment of the chromosome. It should be pointed out that the total number of bands seen by BRIDGES also matches the number of functional groups we have discovered.

A brief description of each of the complementation groups is presented here to establish the relationships between position and function. [For a more detailed examination of the physiology and development of selected mutants in each group see SHANNON *et al.* 1972]. The phenotype and complementation patterns for each functional group are as follows:

gt: giant. The giant locus has been known for a long time (BRIDGES and GABRITSCHEVSKY 1928). Larval development is extended 4-5 days longer than normal with an increase in larval, pupal, and imago size. The polytene chromosomes apparently undergo an additional replication resulting in some cells with chromosomes of double thickness. Penetrance of the giant phenotype is rather low for most alleles.

Allele *13z* (synonymous with *1(1)13z*; see KAUFMAN 1970) is a recessive lethal induced by NNG. In combination with *gt* about one-third are giant in phenotype. *13z/Df(1)w^{rJ1}* heterozygotes have greatly reduced viability but survivors are normal in phenotype.

Allele *E6* was induced by EMS treatment. It was recognized from an experiment where *Df(1)62g18* was used as the screen. *E6/Df(1)w^{rJ1}* heterozygotes are reduced in viability (about 50%) and survivors are normal in phenotype.

Our interpretation is that the *gt* locus, which maps 0.04 units to the left of *tko*, lies in 3A1. This chromomere is not deleted from *Df(1)w^{rJ1}* but is reduced or missing in *Df(1)65j26*, *Df(1)62g18* and *Df(1)X12*. This accounts for the complementation in *gt/Df(1)w^{rJ1}* and the noncomplementation with the three deletions which remove this band. We are unable to explain the reduced viability in *13z/Df(1)w^{rJ1}* and *E6/Df(1)w^{rJ1}* heterozygotes.

tko. Three alleles are known for this locus: *k11* was induced by M. L. ALEXANDER in an experiment where males were injected with dimethyl sulfoxide and treated with 2,000 r of X ray; both *25t* and *15p* are NNG-induced. All alleles are lethal in combination with *Df(1)w^{rJ1}*, *Df(1)X12*, *Df(1)65j26*, and *Df(1)62g18*. Allele *25t* acts as a semi-lethal in homozygous females or hemizygous males. Survivors exhibit a fine bristle phenotype and are very sensitive to shock such as striking the culture bottle sharply on a pad. Such a blow causes the adults to fall to the bottom of the culture tube and remain immobile for a few seconds. After recovery, there generally is a refractory period lasting about an hour during which sensitivity to shock is greatly reduced.

Cytological data indicate that *tko* is most likely to be in chromomere 3A2. This is the leftmost band removed by *Df(1)w^{rJ1}*. Genetically, *tko* maps about 0.006 units to the left of *z*.

z: zeste. For a detailed description of this locus, see GANS (1953). The *z* mutant produces a lemon-yellow eye color in homozygous females. A deletion of the white locus acts as a dominant suppressor of this mutant as do point mutations located in the rightmost part of white locus. Hemizygous *z w$^+$* males have wild-type eye color. Because of this interaction between the *z* and *w* loci, no point mutations at *z* were detectable in the experiments reported here. This resulted from using a deletion that included the white locus in the mutant recognition scheme. GANS placed the *z* locus in band 3A3 of BRIDGES' (1938) map. Our cytological observations confirm this position.

l(1)zw1. Almost 25% of the mutants recovered in our experiments were located in *zw1*. A total of 34 independent recessive lethal *zw1* alleles were found. Of these, 20 were X-ray induced; 2 were produced from an experiment by ALEXANDER in which males were injected with dimethyl sulfoxide and then X-rayed; 1 was recovered by ALEXANDER from a male injected with ethylenimine and treated with X rays. There were also 8 NNG-induced alleles and 3 produced by EMS in an experiment where *Df(1)62g18* was used as the screening chromosome.

All alleles fail to complement with *Df(1)w^{rJ1}*, *Df(1)X12*, *Df(1)62g18*,

$Df(1)65j26$, $Df(1)64c4$, $Df(1)K95$, and $Df(1)w^{258-11}$; the locus is not covered by $Dp(1;3)w^{67k27}$. Examination of the break points of these rearrangements (Table 1) showed that the chromomere missing in all of these deletions is 3A4. Allele $g31$ is a lethal associated with an inversion in the X chromosome. One break point appears to be between 3A3 and 3A4 and the other following 6B1–2. On the basis of these observations, we place the locus of $l(1)zw1$ in band 3A4.

$l(1)zw8$. There are 8 alleles representing this group. Three were X-ray induced while 5 were recovered from nitrosoguanidine-treated males. All alleles are recessive lethals; two ($6b$ and $44j$) are temperature sensitive, producing both males and homozygous females at 18°C. Survivors show reduced bristle length and thickness, and often the posterior verticals and posterior scutellars are missing. The wing shows incisions on the inner margin and often the wing veins are netted at the distal ends. Females reared at 18°C have a reduced fertility and become completely sterile when placed at 25°C for 48 hr.

The $zw8$ locus is the leftmost group covered by $Dp(1;3)w^{67k27}$. It fails to complement with $Df(1)X12$, $Df(1)64c4$, $Df(1)K95$, $Df(1)w^{258-11}$ and $Df(1)w^{258-42}$ in addition to $Df(1)w^{rJ1}$. These results place $zw8$ in band 3A6. As pointed out above, we have observed only 8 chromomeres in section 3A. To make our map correspond as closely as possible with that of BRIDGES (1938), we have deleted bands shown as 3A5 and 3A10 by BRIDGES (1938), and have added 1 band in section 3B (3B5).

$l(1)zw4$. Twelve lethal alleles are known for this group. Four are X-ray induced; 3 are from ethylenimine plus X ray (ALEXANDER); 3 were recovered from ethylenimine treatment alone (ALEXANDER); one resulted from nitrosoguanidine exposure, and one was spontaneous (G. LEFEVRE).

Mutants in this group map 0.04 units to the right of $zw8$ and they fail to complement with $Df(1)X12$, $Df(1)64c4$, $Df(1)K95$, $Df(1)w^{258-11}$, $Df(1)w^{258-42}$, $Df(1)w$-N^{71a} and $Df(1)w^{rJ1}$. The locus is covered by $Dp(1;3)w^{67k27}$ but not by $Dp(1;2)w^{+70h31}$. From the break points of these rearrangements (Table 1), we conclude that $zw4$, most logically, is in 3A7.

$l(1)zw10$. The 5 alleles in this group are all semilethal as males or homozygous females. Surviving flies have reduced roughened eyes. Bristles on the head and thorax are sometimes missing. The wings are abnormal with thin texture and thickened veins. Both sexes are sterile. Based on phenotype and map position, this locus may be synonymous with abe, abnormal eye, reported by FAHMY and FAHMY (1959).

Viability of females heterozygous for a $zw10$ allele and $Df(1)X12$, $Df(1)64c4$, $Df(1)K95$, $Df(1)w^{258-11}$, $Df(1)w^{258-42}$, $Df(1)w$-N^{71a} or $Df(1)w^{rJ1}$ is extremely low, with very few survivors. The $zw10$ locus is the leftmost one covered by $Dp(1;2)w^{+70h31}$. It is most logically placed in chromomere 3A8.

$l(1)zw2$. This is the second largest group with twenty independent mutations. Nine are X-ray induced (two from G. LEFEVRE; two from S. ABRAHAMSON and one from M. L. ALEXANDER); eight were induced by NNG; two have their origin from males injected with ethylenimine and then X-rayed (ALEXANDER); one is from EMS treatment. Most alleles are recessive lethals; mutant $a3$ is semilethal with survivors showing a rough eye. Heterozygotes of *a3* and a *zw2* deletion or any other *zw2* mutant except *g6* are lethal. *a3/g6* heterozygotes are viable but sterile.

All alleles fail to complement with $Df(1)64j4$, $Df(1)w^{rJ2}$, $Df(1)w^{258-42}$, $Df(1)w$-N^{71a}, $Df(1)w^{258-11}$, $Df(1)K95$, $Df(1)64c4$, $Df(1)X12$, or $Df(1)w^{rJ1}$. The group is covered by $Dp(1;2)w^{+70h31}$ and $Dp(1;3)w^{67k27}$. The cytological examination of chromosomes with these rearrangements leads to the conclusion that $l(1)zw2$ is in band 3A9.

l(1)zw3. Nine alleles of independent origin are known for this locus. Five are X-ray induced; one is from an ethylenimine plus X-ray experiment (ALEXANDER) and two are from ethylenimine treatment only (ALEXANDER); one was recovered from ethyl methanesulfonate-treated males. All have recessive lethal phenotype and fail to complement with $Df(1)64f1$, $Df(1)64j4$, $Df(1)w^{rJ2}$, $Df(1)w^{258-42}$, $Df(1)w^{258-11}$, $Df(1)K95$, $Df(1)w$-N^{71a}, $Df(1)X12$, $Df(1)64c4$ or $Df(1)^{rJ1}$. $Dp(1;2)w^{+70h31}$ and $Dp(1;3)w^{67k27}$ cover the locus. One allele, *b12*, is inseparable from an X-chromosome inversion with break points at 3B1 and following 12F2. The cytological position of $l(1)zw3$ is in band 3B1.

l(1)zw6. Seven alleles are known; 6 are X-ray induced and one was supplied by B. HOCHMAN from an experiment using ICR 170 as the mutagen. All have a recessive lethal phenotype.

$l(1)zw6$ is placed in chromomere 3B2 on the basis of its failure to complement with deletions $In(1)w^{-64d}$, $Df(1)64f1$, $Df(1)62d18$, $Df(1)w^{rJ2}$, $Df(1)X12$, $Df(1)w^{258-42}$, $Df(1)w^{258-11}$, $Df(1)64c4$, $Df(1)w$-N^{71a}, or $Df(1)w^{rJ1}$. The locus is covered by $Dp(1;2)w^{+70h31}$, $Dp(1;3)w^{49a7}$, $Dp(1;4)w^{m65g}$, or $Dp(1;3)w^{67k27}$.

l(1)zw12. Three of the four alleles in this group were produced in an experiment by ALEXANDER where males were injected with dimethyl sulfoxide and then treated with X rays. The fourth allele was induced by NNG. All have a recessive lethal phenotype.

Complementation tests show that all alleles fail to complement with $Df(1)64f1$, $Df(1)w^{258-45}$, $In(1)w^{-64d}$, $Df(1)62d18$, $Df(1)w^{rJ2}$, $Df(1)w^{258-42}$, $Df(1)X12$, $Df(1)w$-N^{71a}, $Df(1)w^{258-11}$, $Df(1)64c4$, or $Df(1)w^{rJ1}$. The group is covered by $Dp(1;2)w^{+70h31}$, $Dp(1;3)w^{49a7}$, $Dp(1;4)w^{m65g}$, $Dp(1;3)w^{67k27}$. It is placed in chromomere 3B3 on the basis of these observations.

l(1)zw7. Three alleles are known in this group; two are X-ray induced and one had its origin from an experiment using NNG as the mutagen. All are recessive lethals.

All alleles fail to complement with $Df(1)w^{258-45}$, $In(1)w^{-64d}$, $Df(1)62d18$, $Df(1)w^{rJ2}$, $Df(1)w^{258-42}$, $Df(1)X12$, $Df(1)w^{258-11}$, $Df(1)w$-N^{71a}, $Df(1)64c4$ or $Df(1)w^{rJ1}$. They are covered by $Dp(1;2)w^{+70h31}$, $Dp(1;3)N^{264-58a}$, $Dp(1;3)w^{49a7}$, $Dp(1;4)w^{m65g}$ or $Dp(1;3)w^{67k27}$. These data place the locus most probably in 3B4. Difficulty in placement is due to the lack of a rearrangement with a break point to the immediate right of the group and because the bands are quite faint in this region.

l(1)zw5. Five alleles are known in this complementation group. Two are X-ray induced and three have their origin from NNG-treated males. Four of the mutants are recessive lethals while one, *g27*, is a semilethal. Survivors of *g27* show a roughened, reduced, dark color eye and lack a variable number of orbital, ocellar, and vertical bristles. The vibrissae are sparse, and the wings have thickened veins and incisions along inner margins. Surviving males are sterile. It is our opinion that *l(1)zw5* is synonymous with previously described mutants *ves*, vestigium, and *dwg*, deformed wing (see LINDSLEY and GRELL 1968).

The pattern of complementation with deletions and duplications is the same as that for *l(1)zw7*. *l(1)zw5* complements with *l(1)zw7* and maps 0.078 map units to the right of it. We do not have chromosome rearrangements with break points immediately right or left of this group; therefore, it is difficult to establish a cytological position. We tentatively place this group in 3B5 which we believe we have observed in particularly favorable cytological preparations.

l(1)zw11. Seven alleles comprise this group. Two were derived from X-ray experiments, four from NNG treatments, and one produced by ICR 170 was supplied to us by B. HOCHMAN. All are recessive lethals.

The locus fails to complement with the same array of deletions as do *l(1)zw7* and *l(1)zw5*. It is covered by the same duplications that cover those two groups as well. *l(1)zw11* maps 0.02 units to the right of *l(1)zw5*. Available rearrangements do not have break points in positions favorable for localizing *l(1)zw11*. We tentatively place it in the rightmost section of 3B. It is in this region that BERENDES describes six bands; 3B1 and 2 are rather wide and diffuse while 3B3, 4, 5, and 6 are faint and narrow. We believe we have seen three of these faint and narrow chromomeres in addition to the prominent 3B1, 2. However, we have not been able to verify the presence of the sixth band. If we accept BERENDES' map constructed from his electron microscope study of the region, we then have a possible site for *l(1)zw11* in band 3B6.

l(1)zw9. Both alleles recovered here are X-ray induced semi-lethals that exhibit reduced aristae with the third antennal segment blunt and rounded. The prothoracic leg of males has a second sex comb on the fourth tarsal segment. Males are greatly reduced in fertility and females are sterile. Both alleles in heterozygous combination with a deletion for the locus cause a marked reduction in viability. This locus is almost certainly synonymous with spare arista, *sa*, described by RAYLE and GREEN (1968).

The complementation pattern with available deletions and duplications is the same as with loci *l(1)zw7*, *l(1)zw5* and *l(1)zw11*, except *l(1)zw9* complements with *Df(1)X12*. It maps 0.027 units to the right of *l(1)zw5* and 0.022 units to the left of *w*. We place it in the band 3C1.

w: white. A rather large number of white locus mutations were recognized from the series of experiments we carried out. Only those which included mutants of loci in addition to white were analyzed here. The locus is associated with band 3C2 (LEFEVRE and WILKINS 1966). (For a more complete description see LINDSLEY and GRELL 1968.)

DISCUSSION

The results of the cytogenetic analysis of the 3A–3C region of the X chromosome lead to the conclusion that there is one essential function expressed by the genetic information contained in each chromomere. This correlation is supported by the findings of workers in other laboratories. HOCHMAN (1971) has found 33 lethal complementation groups in the fourth chromosome of *D. melanogaster* and estimates that 36 to 37 vital loci should exist. This number when added to the five loci at which visible mutants are known is in good agreement with the band number of about 50. In another study LIFSCHYTZ and FALK (1968, 1969) have isolated 105 lethals in the proximal portion of the X chromosome. These mutants form a complementation map of 34 functional units. By using the cytology and gene localization of SCHALET, LEFEVRE and SINGER (1970) in this region, it can be seen that here again there is reasonably good correlation between the number of functional units and the number of chromomeres. A probable reason for the lower number of functional units than chromomeres in these examples is that the regions are larger than the zeste to white interval and have not yet reached saturation. The point of note, however, is that in neither case have the investigators found more functional units than chromomeres.

The one chromomere: one function relationship holds regardless of the amount of DNA the chromomere contains. Within the region analyzed, the chromomeres vary rather widely in DNA content. The faintest bands are among the smallest observed in the X chromosome while some such as 3A3–4 and 3C2 are among the largest. RUDKIN (1965) states that the smallest measurable band contains about 5,000 nucleotide pairs, whereas the larger ones are at least one order of magnitude greater. RUDKIN estimates that 0.9% of the DNA of the X chromosome resides in the z–w interval, which means that there are approximately 3×10^5 nucleotide pairs in the 13 chromomeres. This averages about 25,000 nucleotide pairs per chromomere, or enough for about 25 genes.

In order to reconcile single functions with such lengths of DNA, two alternatives may be considered. The first is that some fraction of the DNA of a chromomere is not transcribed and may have no function other than to act as a structural part of the chromosome. Evidence bearing directly on this possibility is not available. The alternative is that essentially all of the DNA is transcribed and if this be the case, two possibilities are immediately obvious: 1) there are additional functions, either regulatory or structural in nature, which have gone undetected by our tests; 2) multiple copies of a single gene comprise a chromomere. This last alternative is essentially the master-slave concept postulated by CALLAN (1967), extended by EDSTRÖM (1968), and strongly supported by THOMAS (1970).

The conclusion that there is one function per chromomere depends directly on whether the array of detected mutations really has saturated the functional elements of the region. Theoretically, almost all types of mutants can be recovered by the screen we employed. In practice, however, it is not possible to examine all aspects of the individuals in each F_2 culture. Obvious morphological changes and lethals will be detected most easily. An important question which remains open is what proportion of genes are capable of mutating to recessive lethal form or to a form which causes an easily detectable morphological change.

The majority of the loci recognized are capable of mutating to lethal or semi-lethal states. Lethal alleles of zeste were not found nor were there any at the white locus. It has been known for a number of years that a deletion of the entire white locus has little effect on viability (LEFEVRE and WILKINS 1966), and the very close functional tie between *z* and *w* is made even stronger by our discovery that the zeste locus also performs a dispensable function. Nonlethal alleles of zeste were not recovered with the screen we employed because the white locus deletion acts as a dominant suppressor of *z* (see GANS 1953). Had the zeste locus not been previously described, we may have failed to identify it in these experiments. Of the 16 loci we identified, only two, *z* and *w*, are dispensable for development of viable individuals. However, mutants at both of these loci produce striking morphological changes. Locus *zw10* is characterized by semi-lethal alleles only; *zw9* is also semi-lethal and the escapers have the sparse arista phenotype described by RAYLE and GREEN (1968). However, both *zw9* and *zw10* mutants have greatly reduced viability in females heterozygous for deletion of the loci. Clearly our sample of mutants is biased toward those that interrupt functions essential for development. The remarkable aspect of our results is that those loci which are capable of changing in this fashion are distributed so regularly—one per chromomere.

The sample of mutants recovered is not distributed in a Poisson fashion among the groups we have characterized. Statistical test shows that *zw1*, with more than 30 representatives, and *zw2*, with 20, are represented by significantly more mutants than the remaining loci. It is not possible to design a rigorous test for determining the number of mutationally silent units in the 3A–3C region. However, if all units fall within the mutation frequency range that we observed, the probability that there are one or more functional units that have escaped our notice is very low. Therefore, we can be reasonably confident that the relationship between complementation units and chromomeres would not change by increasing the sample size.

Regarding the mutability of each locus and the types of mutants induced by the various mutagens, several points can be made. The seven mutations associated with chromosome rearrangement were recovered from experiments using X rays as the mutagen. In contrast to the results of LIFSCHYTZ and FALK (1968), the large majority of X-ray induced changes are single site mutations not associated with visible chromosome rearrangement. These were recovered at all of the recognized complementation groups with the exception of *gt* and *l(1)zw10*.

The other mutagen which was used extensively was NNG. A total of 34 mutants were obtained from this series (KAUFMAN 1970), and they were grouped into twelve of the complementation units. No representatives were found at *zw3*, *zw6*, or *zw9*. The distribution of mutants induced by NNG is not significantly different from that seen from X-ray experiments. None of the chemically induced changes were multisite lesions, nor where any associated with chromosome breakage.

A few mutants were induced with other mutagens including EMS and ICR 170, but the sample sizes are too small to allow any evaluation of distribution or predominant types. A number of mutants were induced in the 3A2–3C2 region by EMS treatment, but a majority of them were lost because one or more independent lethals were induced in the X chromosome in addition to the change in the screened region. These of course will die as F_2 males. In those instances where the change in region 3A–3C was also a lethal, all F_2 flies carrying the treated X chromosome died. EMS is such an efficient mutagen that about 56% of treated X chromosomes sampled carried at least one lethal.

It might be argued that the reason we see one complementation group per chromomeres is that the majority of the mutants induced are deletions which extend over several adjacent functional units and that the sample of single base change mutations is so small that an accurate picture of complementation pattern has not been achieved. The mutants induced by X rays may well represent deletions of the sizes necessary to give the pattern observed. We would expect then that the chemical mutagens, which predominantly elicit point mutations, might, at least occasionally, cause changes which complement with each other but fail to do so with those of X-ray origin. The complementation tests within and between groups of mutants induced by different mutagens expose complementation units identical to those found in the X-ray series. The best approach to this problem, of course, is to demonstrate that selected mutations within each complementation group are revertable.

An important point to consider is the nature of the alleles which our complementation tests identify as changes within a single functional unit. Do all alleles cause the same misfunction at the same point in developmental sequence or is there heterogeneity which might indicate that multiple functions are somehow packaged into what appears to be a single cistron? A detailed study of several aspects of the developmental pattern seen in sample of alleles within each complementation group (SHANNON et al., 1972) leads to the conclusion that all mutants within a complementation unit act in apparently similar fashion. If there are several different functions involved, their control must be such that a mutation in any one of them acts in a *cis*-dominant way to shut off the entire array so that complementation does not occur. If this is the case, the unit our tests have identified may be the equivalent of the operon of procaryotes.

The solution to this problem may be approached through the gene products that are produced from a single chromomere. DANEHOLT et al. (1969b) have characterized the rapidly labelled RNA in Balbiani ring 2 of chromosome IV of Chironomous. They found the RNA to be as heterogeneous in size as that produced by the rest of the genome but more of the label was distributed in high molecular weight species from the Balbiani ring than from the remainder of the chromosomes.

DANEHOLT (1970) also analyzed the base ratios of newly synthesized RNA from this Balbiani ring and showed them to be quite different from those found in the nucleolus or in chromosome I–III. On the other hand, no differences were found between base ratios of rapidly labelled, high molecular weight RNA molecules of different sizes within Balbiani ring 2 and within chromosome I–III. The similarity between the RNA molecules of different sizes produced in a single chromosome puff and the evidence that Balbiani ring 2 is most likely producing an RNA for only one or a few polypeptides (BEERMANN 1966; GROSSBACH

1969) led DANEHOLT et al. to the view that the RNA is constructed of repeated message sequences. There is evidence from the work of KEYL (1965) on polytene chromosomes, and from that of CALLAN (1967) and GALL and CALLAN (1962) on lampbrush chromosomes, that the chromomere might represent a large number of repeated units. This idea is supported by data of THOMAS et al. (1970) who found that DNA fragments from some eucaryotes will form circles with a surprisingly high frequency, whereas procaryotic DNA does not. They conclude that a rather sizeable fraction of the genome in eucaryotes is composed of regions containing tandemly repeated sequences. On the other hand, DICKSON, BOYD, and LAIRD (1971) conclude from their study of the renaturation kinetics of *Drosophila hydei* DNA that most sequences are present once per haploid cell and that DNAs from polytene cells, though reduced in the rapidly annealing fraction, have essentially the same nucleotide diversity as those from nonpolytene nuclei.

The paradox that these data create can be resolved when the biological function of the DNA is determined. For the moment, however, it would appear that the chromosome is a complex unit devoted to a single essential function. Though the concept that a chromomere consists of a family of repeated sequences is very attractive, we prefer the view that a major fraction of the DNA in a chromomere performs regulatory functions and that there are one or, at most, a few structural genes which are translated into polypeptides. We propose that the chromomere is a single complementation unit consisting of a relatively short length of structural DNA together with a variable length of regulatory DNA. Within this unit, mutations act in a *cis*-dominant fashion. This means that a mutation in any part of the chromomere would interfere with the function of the entire complex and would fail to complement with mutations in any of the other parts.

Our data do not preclude the possibility that the complementation groups correspond to the interbands. Cytologically it is not generally possible to determine whether a chromosome is broken in a band or an interband space. Some of the deletions appear to have breaks that are within a chromomere such that part of the band is still present. In such instances, the function that is associated with the diminished chromomere is impaired. Unfortunately we do not have a firm case where one deletion removes the rightmost part of a band and another the leftmost. Such a case may exist at the vermilion locus, however. LEFEVRE (1971) analyzed deficiencies in the 10A region and showed that those which remove a part of band 10A1-2 either from the left or from the right cause a vermilion mutant phenotype, as does the heterozygote for two such left- and right-encroaching deletions. From this he concluded that the *v* locus is physically within the chromomere. This implicates the chromomere as an integral part of the vermilion cistron but does not rule out the interband as having a role as well. It must be pointed out that our model of the cistron would predict an impairment of the vermilion function even in a heterozygote for two partial non-overlapping deletions because of the *cis*-dominant nature of the functional unit. Thus, if both band and interband contain the complementation unit, we would expect that cases of non-complementation between cytologically non-overlapping deletions of a band-interband unit will be discovered.

The possibility exists that the condensed chromomeric DNA is entirely regulatory in function whereas the structural gene or genes are located in the interband region. A chromosome model using this as one of the basic postulates has been proposed by CRICK (1971). It should be possible by careful analysis of various types of chromosome rearrangements to test this postulate as well as other predictions that may be generated from his model.

We are indebted to Drs. GEORGE LEFEVRE, JR., CHARLES D. LAIRD, LEONARD ROBBINS and HERMANN BULTMANN for their critical review of this manuscript and for their many useful comments. We are also indebted to Drs. GEORGE LEFEVRE, JR., M. L. ALEXANDER, B. HOCHMAN, SEYMOUR ABRAHAMSON, W. J. WELSHONS, and R. FALK for supplying some of the mutants and chromosome rearrangements used in this study.

LITERATURE CITED

BEERMANN, W., 1961 Ein Balbiani-Ring als Locus einer Speicheldrüsenmutation. Chromosoma (Berl.) **12**: 1–25. ——, 1962 Riesenchromosomen. Protoplasmatologia VI D. Wien: Springer. ——, 1966 Gen-Regulation in Chromosomen höher Organismen. In: *Jahrbuch 1966 der Max-Planck-Gesellschaft zur Förderung der Wissenschaften e. V.*, S 69–87. Göttingen: Hubert & Co. ——, 1967 Gene action at the level of the chromosome. In: *Heritage from Mendel*, p. 179–201. Edited by R. A. BRINK. University of Wisconsin Press, Madison.

BERENDES, H. D., 1970 Polytene chromosome structure at the submicroscopic level. I. A map of region X, 1–4E of *Drosophila melanogaster*. Chromosoma (Berl.) **29**: 118–130.

BRIDGES, C. B., 1935 Salivary chromosome maps. With a key to the banding of the chromosomes of *Drosophila melanogaster*. J. Heredity **26**: 60–64. ——, 1938 A revised map of the salivary gland X-chromosome of *Drosophila melanogaster*. J. Heredity **29**: 11–13.

BRIDGES, C. B. and E. GABRITSCHEVSKY, 1928 The giant mutation in *Drosophila melanogaster*. Part I. The heredity of giant. Z. indukt. Abstamm.-U. Vererbungslehre **46**: 231–247.

BRIDGES, P. N., 1942 A new map of the salivary gland 2L-chromosome of *Drosophila melanogaster*. J. Heredity **33**: 403–408.

CALLAN, H. G., 1967 The organization of genetic units in chromosomes. J. Cellular Sci. **2**: 1–7.

CRICK, F. H. C., 1971 General model for the chromosomes of higher organisms. Nature **234**: 25–27.

DANEHOLT, B., 1970 Base ratios in RNA molecules of different sizes from a Balbiani ring. J. Mol. Biol. **49**: 381–391.

DANEHOLT, B., J. E. EDSTRÖM, E. EGYHÁZI, B. LAMBERT and U. RINGBORG, 1969a Chromosomal RNA synthesis in polytene chromosomes of *Chironomus tentans*. Chromosoma (Berl.) **28**: 399–417. ——, 1969b RNA synthesis in a Balbiani ring in *Chironomus tentans* salivary gland cells. Chromosoma (Berl.) **28**: 418–429.

DICKSON, E., J. B. BOYD and C. D. LAIRD, 1971 Sequence diversity of polytene chromosome DNA from *Drosophila hydei*. J. Mol. Biol. (in press.)

EDSTRÖM, J. E., 1968 Masters, slaves and evolution. Nature **20**: 1196–1198.

FAHMY, O. G. and M. J. FAHMY, 1959 *Drosophila melanogaster*; New mutants. Report of O. G. and M. J. FAHMY. Drosophila Inform. Serv. **33**: 82–94.

GALL, J. G. and H. G. CALLAN, 1962 (H^3) uridine incorporation in lampbrush chromosomes. Proc. Natl. Acad. Sci. U.S. **48**: 562–570.

GANS, M., 1953 Étude génétique et physiologique du mutant z de *Drosophila melanogaster*. Bull. Biol. France et Belg. (Suppl.) **38**: 1–90.

GROSSBACH, U., 1969 Chromosomen-Aktivität und biochemische Zelldifferenzierung in den Speicheldrüsen von *Camptochironomus*. Chromosoma (Berl.) **28**: 136–187.

HOCHMAN, B., 1971 Analysis of chromosome 4 in *Drosophila melanogaster*. II: Ethyl methanesulfonate induced lethals. Genetics **67**: 235–252.

KAUFMAN, T. C., 1969 N-methyl-N'-nitro-N-nitrosoguanidine mutagenesis in the *zeste white* region of the X chromosome of *Drosophila melanogaster*. M.A. thesis. Univ. of Texas at Austin. ——, 1970 The genetic anatomy and function of the 3A1–3C7 region of the X chromosome of *Drosophila melanogaster*. Ph.D. dissertation. Univ. of Texas at Austin.

KEYL, H. G., 1965 Duplikationen von Untereinheiten der chromosomalen DNS während der Evolution von *Chironomus thummi*. Chromosoma (Berl.) **17**: 139–180.

LAIRD, C. D. and B. J. MCCARTHY, 1969 Molecular characterization of the Drosophila genome. Genetics **63**: 865–882.

LEFEVRE, G., JR., 1971 Salivary chromosome bands and the frequency of crossing over in *Drosophila melanogaster*. Genetics **67**: 497–513.

LEFEVRE, G., JR. and M. D. WILKINS, 1966 Cytogenetic studies on the white locus in *Drosophila melanogaster*. Genetics **53**: 175–187.

LEWIS, E. B. and F. BACHER, 1968 Method for feeding ethyl methanesulfonate (EMS) to Drosophila males. Drosophila Inform. Serv. **43**: 193.

LIFSCHYTZ, E. and R. FALK, 1968 Fine structure analysis of a chromosome segment in *Drosophila melanogaster*. Analysis of X-ray-induced lethals. Mutation Res. **6**: 235–244. ——, 1969 Analysis of ethyl methanesulfonate-induced lethals. Mutation Res. **8**: 147–155.

LINDSLEY, D. L. and E. H. GRELL, 1968 Genetic variations of *Drosophila melanogaster*. Carnegie Inst. Washington Publ. **627**.

MERRIAM, J. R., 1968 *Drosophila melanogaster*; New mutants. Report of John R. Merriam. Drosophila Inform. Serv. **43**: 64.

MULLER, H. J. and A. A. PROKOFYEVA, 1935 The individual gene in relation to the chromomere and the chromosome. Proc. Natl. Acad. Sci. U.S. **21**: 16–26.

PAINTER, T. S., 1934 Salivary chromosomes and the attack on the gene. J. Heredity **25**: 465–476.

RAYLE, R. E. and M. M. GREEN, 1968 A contribution to the genetic fine structure of the region adjacent to *white* in *Drosophila melanogaster*. Genetica **39**: 497–507.

RUDKIN, G. T., 1965 The relative mutabilities of DNA in regions of the X chromosome of *Drosophila melanogaster*. Genetics **52**: 665–681.

SCHALET, A., G. LEFEVRE and K. SINGER, 1970 Preliminary cytogenetic observations on the proximal euchromatic region of the X chromosome of *D. melanogaster*. Drosophila Inform. Serv. **45**: 165.

SHANNON, M. P., T. C. KAUFMAN, M. W. SHEN and B. H. JUDD, 1972 Lethality patterns of selected lethal and semi-lethal mutations in the zeste-white region of *Drosophila melanogaster*. (manuscript submitted to Genetics).

THOMAS, C. A., JR., 1970 The theory of the master gene. In: *The Neurosciences: Second Study Program*. Edited by F. O. SCHMITT. Rockefeller Univ. Press. New York.

THOMAS, C. A., JR., B. A. HAMKALO, D. N. MISRA and C. S. LEE, 1970 The cyclization of eucaryotic DNA fragments. J. Mol. Biol. **51**: 621–632.

XV
Regulation of Protein Synthesis in Eukaryotes

Editor's Comments on Paper 36

36 Thompson and Horning: *Regulatory Interactions Involving Two Hemoglobin Loci of Chironomus*

Closely related to the problem considered in Paper 35 is the question of regulation in the eukaryotes. This is a particularly difficult problem to attack because no one has yet come up with a good experimental system. Thompson and Horning have suggested one approach that shows promise. This is a phase of work that will be prominent in future studies of the gene–protein relationship.

Selected Bibliography

Beerman, W., ed. 1972. *Developmental Studies on Giant Chromosomes*. Springer-Verlag, New York.
Britten, R. J., and E. H. Davidson. 1969. Gene regulation for higher cells: a theory. *Science 165:* 349–357.
Brown, D. D., and I. B. David. 1969. Developmental genetics. *Ann. Rev. Genetics. 3:* 127–154.
Davidson, E. H. 1968. *Gene Activity in Early Development*. Academic Press, New York.
Georgiev, G. P. 1969. Histones and the control of gene action. *Ann. Rev. Genetics 3:* 155–180.
Gross, S. R. 1969. Genetic regulatory mechanisms in the fungi. *Ann. Rev. Genetics 3:* 395–424.
Kenney, F. T., et al., eds. 1973. *Gene Expression and Its Regulation*. Plenum Press, New York.
Metzenberg, R. L. 1972. Genetic regulatory systems in *Neurospora*. *Ann. Rev. Genetics 6:* 111–132.

Regulatory Interactions Involving Two Hemoglobin Loci of *Chironomus*

Peter Thompson[1] and M. J. Horning[1]

Received 17 April 1972—Final 14 Aug. 1972

Loci encoding two very similar monomeric hemoglobins among the 10–12 larval hemoglobins of Chironomus tentans *show opposite, apparently compensatory, changes in activity in the presence of a spontaneous regulatory mutant. The regulatory mutant maps near the structural loci for both hemoglobins in the right arm of chromosome 3. The locus showing a regulatory increase in rate of synthesis does so in both* cis *and* trans *alleles. The locus showing decreased activity, however, undergoes nearly complete inactivation in* cis. *The regulatory mutant may represent a lesion of a promoter serving the latter.*

INTRODUCTION

Individual larvae of *Chironomus tentans* have 10–12 monomeric hemoglobins which are combined in solution in their hemolymph (Thompson and English, 1966; Thompson *et al.*, 1968), constituting about 40% of the total hemolymph protein. These hemoglobins are encoded by structural genes located in at least three regions of chromosome 3 (Tichy, 1970) and show a wide range of structural similarity and difference (Thompson, unpublished). Two of the most similar forms show interdependent regulation, as evidenced by compensatory increase and decrease in the presence of a regulatory mutant (Thompson and Patel, 1972). Close linkage of the regulatory factor to the structural loci of both hemoglobins has been demonstrated.

Although the magnitude of reciprocal increase is much more striking, this case in many respects is similar to the compensatory increase in δ chains

---

This study was supported by National Science Foundation Grant GB-12901. Dedicated to Prof. Curt Stern in honor of his 70th birthday.

[1] Department of Zoology, University of Georgia, Athens, Georgia.

which accompanies β-chain deficiency in human β thalassemia. The human hemoglobinopathy also involves a regulatory mutant which is closely linked to the structural loci in question (Moore and Pearson, 1964). Its heterozygous expression is intermediate, as in the present case, and various lines of evidence from reticulocyte systems (Clegg *et al.*, 1968; Nienhuis *et al.*, 1971; Nathan *et al.*, 1971) indicate that the reduced β chain actually reflects reduced transcription.

Other aspects of the expression of β thalassemia have been cited in interpreting that condition as a mutation of the promotor serving the β-globin structural gene in binding a specific RNA polymerase for the onset of transcription (Thompson and Patel, 1972). In particular, it has been found that the β-chain decrease applies only to an allele linked to the regulatory mutation in *cis* (Schwartz *et al.*, 1957). The compensatory increase in δ chains applies equally to both *cis* and *trans* alleles (Ceppellini, 1959; Huisman *et al.*, 1961), as if a diffusible factor had stimulated it.

The present study was undertaken to examine more fully the interdependent regulation of *Chironomus* hemoglobins, particularly the *cis* and *trans* effects of the regulatory mutant on the two loci. Parallels to the human case have been found in *cis–trans* expression.

MATERIALS AND METHODS
Crosses

A strain of *Ch. tentans* homozygous for the regulatory mutant, Regulator-Jemmerson (R^{Jem}), was mated to a standard strain of *Ch. pallidivittatus* supplied by Ulrich Clever. These sibling species have comparable numbers of hemoglobins with similar electrophoretic patterns (Tichy, 1970), but only two components appear to be identical in primary structure, and several of the corresponding hemoglobins have somewhat different electrophoretic mobilities (Thompson, unpublished). It is of especial relevance that the *Ch. tentans* hemoglobin at R_f 0.61 has a *Ch. pallidivittatus* counterpart at R_f 0.58 (reference dye bromphenol blue). Larvae of the F_1 were examined in the last instar to establish percentage quantities of these two components in the presence of the regulatory factor.

The most abundant normal component of both *Ch. tentans* and *Ch. pallidivittatus* has an R_f of 0.50. It is this component which is decreased in R^{Jem} lines, while $Hb^{0.61}$ of *Ch. tentans* increases. The primary structures of the $Hb^{0.50}$ counterparts of both species and the F_1 were compared by peptide mapping.

A control mating of normal *Ch. tentans* and *Ch. pallidivittatus* was made in order to test the quantitative and qualitative intermediacy of the F_1 in the absence of the regulator mutant.

Polyacrylamide Gel Electrophoresis

Methods of electrophoresis and scanning of gels for quantitation have been described previously (Thompson and Patel, 1972).

Peptide Mapping

The isolation, digestion with trypsin, and fingerprinting of individual hemoglobins have been described previously (Thompson and Patel, 1972).

RESULTS

Decreased $Hb^{0.50}$ in *Cis*

Peptide mapping of the $Hb^{0.50}$ counterparts from *Ch. tentans* and *Ch. pallidivittatus* showed that they differ in at least two peptides in spite of their similar electrophoretic mobilities (Fig. 1). The pattern from *Ch. tentans* includes two distinct spots which are not seen in the fingerprint from *Ch. pallidivittatus*. The latter must have peptides from which the *Ch. tentans* peptides have derived by amino acid substitution, but their position is not evident. They could easily be among the several spots in the vertical axis above the origin. Hybrid larvae from the control (Fig. 1) produced a pattern combining all the peptides of both species.

Although peptide mapping is not well suited to estimating relative quantities of mixed components, the pattern from isolated $Hb^{0.50}$ of F_1 hybrid larvae with the regulatory factor is qualitatively identical to that of *Ch. pallidivittatus* (Fig. 2), lacking the two *tentans* peptides. This finding indicates that the R^{Jem} mutant does not act equally on both $Hb^{0.50}$ alleles in reducing the quantity of that hemoglobin. Its relative quantity is nearly halved in the R^{Jem} species hybrids, and the fraction remaining is entirely or almost entirely of the *Ch. pallidivittatus* type. Apparently, then, the mutant reduces $Hb^{0.50}$ only in *cis*.

Compensatory Increases in *Cis* and *Trans*

The F_1 from a control hybridization of *Ch. tentans* and *Ch. pallidivittatus*, involving a strain of the former species which has no regulatory mutant, is qualitatively and quantitatively intermediate between the parental lines in its hemoglobin pattern. The $Hb^{0.50}$ component, which is by far the most abundant hemoglobin in normal strains of both species, constitutes about the same fraction (44%) of the total hemoglobin of the F_1 as well. The species-specific allelic forms, $Hb^{0.58}$ and $Hb^{0.61}$, are approximately halved in quantity. Presumably, each is represented in the F_1 by one allele rather than two.

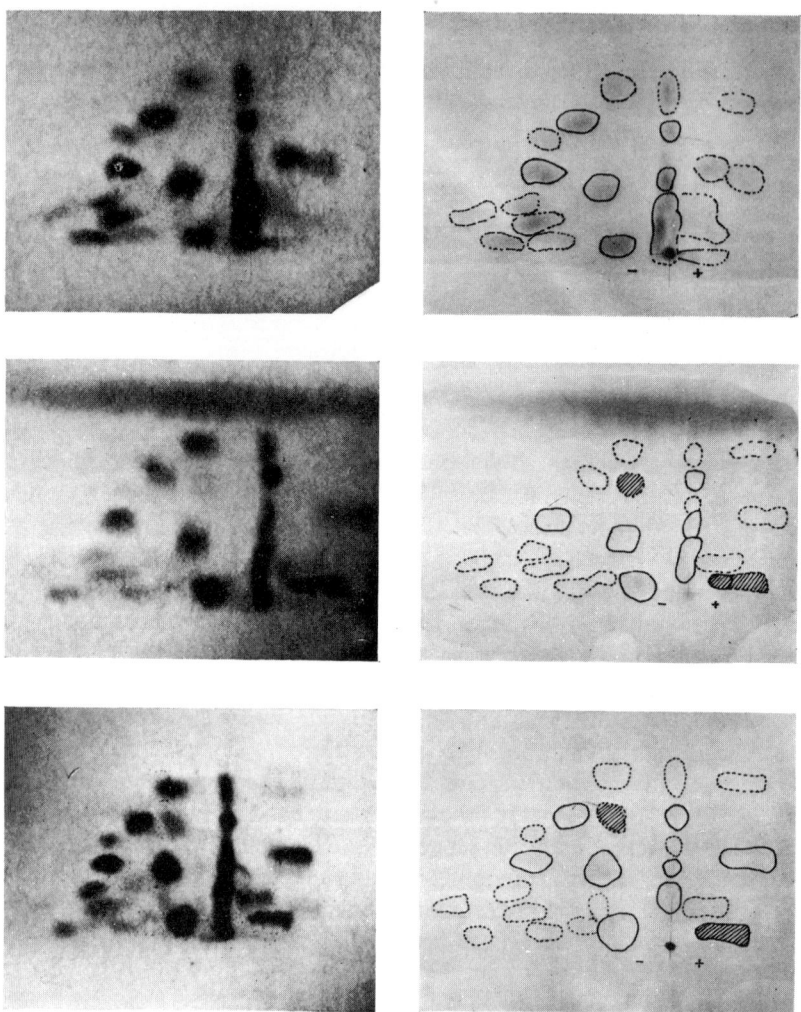

Fig. 1. Peptide maps of the tryptic digests of $Hb^{0.50}$ from *Ch. pallidivittatus* (above), normal *Ch. tentans* (below), and normal *tentans* × *pallidivittatus* hybrids (middle). Species-specific *tentans* spots are shaded in the tracing.

$Hb^{0.58}$ is somewhat more abundant, as it is in the parental species (Fig. 3 and Table IA).

This intermediacy of the hybrid to parental lines is not found, however, in the cross of *Ch. pallidivittatus* to *Ch. tentans* with the regulator R^{Jem} (Fig. 4 and Table IB). The major hemoglobin $Hb^{0.50}$ is substantially reduced by inactivation of the *cis* allele, as described above. $Hb^{0.61}$ is produced at a

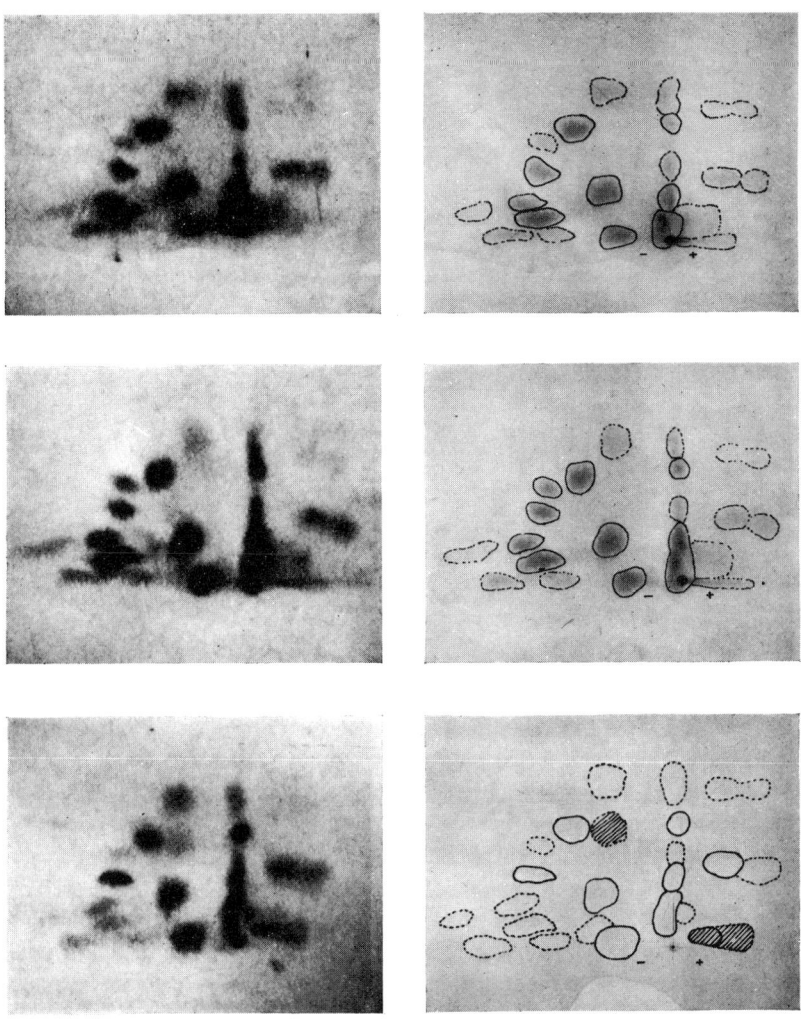

Fig. 2. Peptide maps of the tryptic digests of $Hb^{0.50}$ from *Ch. pallidivittatus* (above), *Ch. tentans* homozygous for the R^{Jem} mutant (below), and *tentans* × *pallidivittatus* hybrids heterozygous for the R^{Jem} mutant (middle).

level intermediate to the abnormally high level (42%) of pure Jemmerson mutant strains and the absence of that component which typifies *Ch. pallidivittatus*. The essential point is that the *trans*-allelic $Hb^{0.58}$, as well as *cis*-allelic $Hb^{0.61}$, is increased under the influence of the regulator to a level substantially higher than in the normal species hybrids (Fig. 3). Hence, while the regulatory decrease of $Hb^{0.50}$ occurs only in *cis*, the increase in closely related forms occurs both in *cis* and in *trans*.

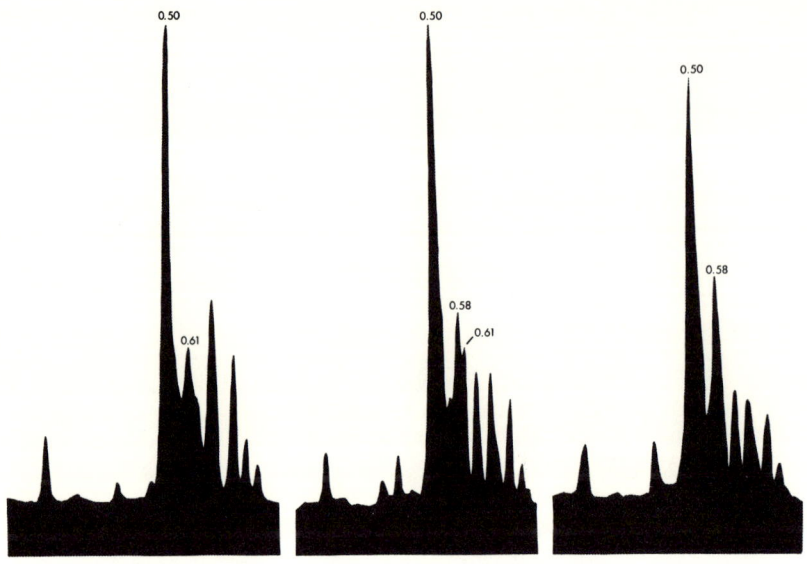

Fig. 3. Scans of polyacrylamide gel separations of multiple hemoglobins from normal *Ch. tentans* (left), *Ch. pallidivittatus* (right), and hybrids (middle).

DISCUSSION

The *cis–trans* effects of the Jemmerson regulatory mutant parallel closely the situation encountered in human β thalassemia. If the regulatory effect occurs at the transcriptional level, as has been demonstrated in thalassemia, these regulatory interactions support a model based on competition between two closely related structural loci (in the sense of recent evolutionary duplication and limited structural divergence) for a common RNA polymerase or some factor affecting polymerase specificity. This model assumes the duplication of promotor sites as part of the ancestral globin locus, whether by assymmetrical exchange or by transposition. Identical functions of the two globin loci would remove any selective pressure for both to transcribe efficiently, and, while initially both loci should compete equally well for the same RNA polymerase, it is expected that one should have accumulated promotor mutations more freely. This would result in two structurally similar forms ($Hb^{0.50}$ and $Hb^{0.61}$), of which one ($Hb^{0.50}$) is normally much more abundant.

Under such a model, the R^{Jem} mutant would involve a lesion of the promotor site serving for the initiation of transcription of the $Hb^{0.50}$ structural gene. Close linkage of the regulatory mutant to the structural locus encoding $Hb^{0.61}$ near the right tip of chromosome 3 has been reported previously (Thompson and Patel, 1972), and new data from deletions and other rearrangements (Thompson and McMichael, unpublished) place the $Hb^{0.50}$

Regulatory Interactions Involving Two Hemoglobin Loci of Chironomus

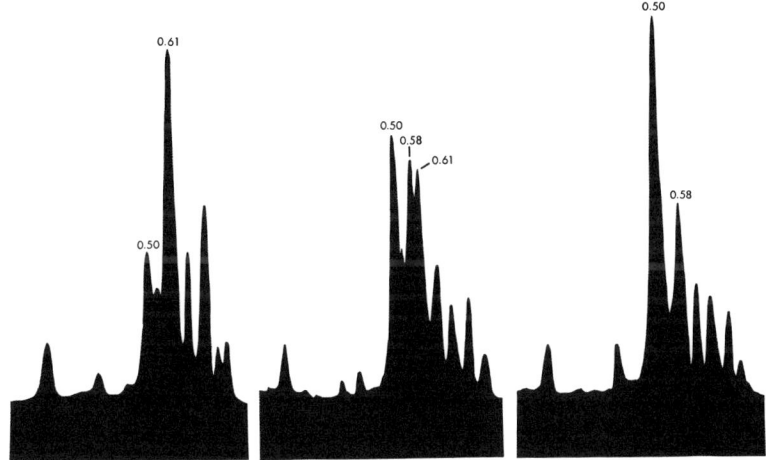

Fig. 4. Scans of polyacrylamide gel separations of multiple hemoglobins from *Ch. tentans* homozygous for R^{Jem} (left), *Ch. pallidivittatus* (right), and hybrids (middle).

locus nearby (in section 15), as predicted by the promotor-lesion hypothesis. On this basis, transcription of the adjacent structural locus for $Hb^{0.50}$ is reduced, while its *trans* allele with its promotor in no way impaired shows normal transcription. If competition for a more or less specific and rate-limiting transcriptional factor existed between alleles, the *trans* allele might actually show a detectable *increase* in absolute quantity of its globin product—this is seen in humans when the β-thalassemia factor is placed in *trans* with the HbS mutant of the β-chain locus (Schwartz *et al.*, 1957).

If initiation at the intact $Hb^{0.50}$ site were already as rapid as allowed at a given level of initiating nucleoside triphosphate, however, the most striking effect would be an increased transcription at the related $Hb^{0.61}$ locus through greater availability of the common transcriptional factor. This availability of factor should extend to both *cis* and *trans* alleles. On the other hand, in a situation of nonspecific transcription by a generalized RNA polymerase, the reduced binding of enzyme at one locus would have no substantial effect (certainly no preferential one) on any other locus.

This evidence of interdependent expression of evolutionarily similar hemoglobin loci may provide a preliminary indication that highly specific elements of transcription function in eukaryotic systems. The extent of this specificity may be gauged by the fact that only two hemoglobins (the two most similar) of the total array of about ten show interdependent change with the regulatory mutant.

Travers and Burgess *et al.* (1969; Travers and Burgess, 1969; Travers, 1971)

Table I. Relative Qualities (%) of Hemoglobin Components[a]

	A. In normal species hybrids		
Component (R_f)	Ch. pallidivittatus	Normal Ch. tentans	tentans × pallidivittatus
0.10	3.8	3.5	3.5
0.32	b	1.1	1.5
0.42	3.8	c	3.4
0.47	0.0	0.8	0.0
0.50	43.0	45.9	44.4
0.55	b	0.6	1.2
0.58	22.8	c	12.4
0.61	b	15.1	5.8
0.67	8.0	16.9	8.5
0.73	9.9	10.1	9.7
0.79	6.1	3.9	5.8
0.83	2.4	2.1	3.0
	B. In hybrids heterozygous for the regulatory factor		
Component (R_f)	Ch. pallidivittatus	Ch. tentans R^{Jem}/R^{Jem})	tentans (R^{Jem})/ pallidivittatus
0.10	3.5	5.2	3.8
0.32	b	2.7	0.7
0.42	3.6	c	1.6
0.47	0.0	0.6	0.0
0.50	42.9	11.9	22.2
0.55	b	7.2	4.2
0.58	23.1	c	20.1
0.61	b	42.0	15.6
0.67	8.6	10.3	12.6
0.73	10.0	12.2	7.1
0.79	5.9	3.0	8.3
0.83	2.4	4.2	3.8

[a] Mean values for 20 larvae per series.
[b] *tentans* specific.
[c] *pallidivittatus* specific.

have demonstrated that the σ subunit of *Escherichia coli* RNA polymerase plays an important role in transcription from specific DNA templates of phage T4. This specificity involves initiation after binding of the enzyme to DNA, presumably by recognition of promotor sequences at the 3' end of the functional strand of several structural genes. In the context of the present findings, which have suggested the competitive binding of a limiting factor for specific transcription at two promotor sites, it is relevant that σ is released from the promotor complex during or shortly after initiation. Consequently, a transcriptional site in the *E. coli* system has the capacity to bind only one σ subunit at a time. In this system, the efficiency of transcription at one locus would have no substantial effect on the availability of σ at another locus. The role of σ in phage T4 seems to be the specific increase of transcription among a number of phase-specific genes, and quantities of σ may not in that case be limiting.

In the transition from vegetative to sporulating phase in *Bacillus subtilis*, Losick (1970) identified a σ factor in vegetative polymerase which is no longer present in the polymerase of sporulating cells. Furthermore, transcription by the latter enzyme is not stimulated by vegetative σ. The core enzyme also differs in one of its β subunits from vegetative polymerase. This raises the possibility that site recognition for specific initiation may also reside to some extent in a β subunit, a possibility suggested by Burgess (1971) on theoretical grounds. More recently, however, Losick (cited by Travers, 1971) has identified a σ factor in sporulating cells which apparently is not easily isolated under the usual experimental conditions.

The situation with eukaryotic polymerases is outwardly similar to that of *B. subtilis*. Both calf thymus and rat liver yield two forms of polymerase, one from nucleoplasm and the other from nucleoli. These enzymes have subunits comparable in size to the polymerase of *E. coli*. The nuclear polymerase, in turn, consists of two types: one has a subunit structure of [(190,000), (150,000), (35,000), (25,000)], and the other is [(170,000), (150,000), (35,000), (25,000),]. Apart from this duality of nuclear forms, however, there is remarkably little heterogeneity of polymerases as a possible basis of functional specificity (Weaver *et al.*, 1971). Furthermore, these isolated polymerases appear to have RNA products which compete well with whole nuclear RNA fractions in hybridization-competition experiments (Blatti *et al.*, 1970), indicating a nonspecific transcriptional activity.

The failure to find σ-like functions in the isolation of these eukaryotic RNA polymerases may relate to the difficulty of assaying activity (Sugden and Sambrook, 1970). On the other hand, the isolation in quantity of an individual σ would not be expected if eukaryotic transcription were controlled by a large array of diverse σ chains providing positive control at restricted chromosomal sites, and with limited (or limiting) numbers of each species of subunit.

The present findings implicate a σ-like factor with high specificity in site recognition as the agent of compensatory control. Another category of positive control element in both bacterial and eukaryotic systems is comprised of the catabolite gene-activator proteins (CAP), which are receptor molecules for cyclic AMP (Zubay *et al.*, 1970*a*). While there is still uncertainty whether CAP stimulates transcription by the modification of RNA polymerase or by a direct change at the promotor site, Zubay *et al.* (1970*b*) have shown that CAP may bind to DNA in the presence of cyclic AMP. Regarding the transcriptional competition model proposed here, however, it should be mentioned that CAP does not form a strong and stable complex with DNA

(Zubay et al., 1970a). As a candidate for the factor competed for and stably bound by hemoglobin loci, it has the same drawback that bacterial σ factor has—large numbers of molecules would probably not be bound to a single transcriptional site at any stage. It is quite possible, of course, that the unstable association of σ or CAP in *in vitro* studies of bacterial transcription is very unlike the behavior of positive control factors in eukaryotes.

The most striking implication of such a model for interdependent regulation of two hemoglobin loci is that a competitive interaction exists between only two loci among the ten or so of the entire hemoglobin system. Since these two hemoglobins are structurally the most similar of the array, it would be assumed that promotor–polymerase relationships have evolved to the point where other loci are transcribed independently or by nonspecific recognition (as with core enzyme in *E. coli*). Obviously, the transcription of the two major loci would involve an extraordinary degree of specificity. In view of the prevalence of these hemoglobins among the proteins of *Chironomus*, however, such specificity is not out of the question. Their compensatory synthesis, their linkage relations, and the *cis* and *trans* aspects of regulation strongly support this model.

REFERENCES

Blatti, S. P., Ingles, C. J., Lindell, T. J., Morris, P. W., Weaver, R. F., Weinberg, F., and Rutter, W. J. (1970). Structure and regulatory properties of eucaryotic RNA polymerase. *Cold Spring Harbor Symp. Quant. Biol.* **35**:649.

Burgess, R. R. (1971). RNA polymerase. *Ann. Rev. Biochem.* **40**:711.

Burgess, R. R., Travers, A. A., Dunn, J. J., and Bautz, E. K. F. (1969). Factor stimulating transcription by RNA polymerase. *Nature* **221**:43.

Cepellini, R. (1959). The genetical control of protein structure: The abnormal human hemoglobins (Discussion). In Wolstenholme, G. E. W., and O'Conner, C. M. (eds.), *Biochemistry of Human Genetics*, Little, Brown, Boston, p. 135.

Clegg, J. D., Weatherall, D. J., Na-Nakorn, S., and Wasi, P. (1968). Haemoglobin synthesis in β-thalassemia. *Nature* **220**:664.

Huisman, T. H., Punt, K., and Schaad, J. D. G. (1961). Thalassemia minor associated with hemoglobin B_2 heterozygosity. *Blood* **17**:747.

Losick, R., Sonenshein, A. L., Shorenstein, R. G., and Hussey, C. (1970). Role of RNA polymerase in sporulation. *Cold Spring Harbor Symp. Quant. Biol.* **35**:443.

Moore, M. M., and Pearson, H. A. (1964). Human hemoglobin gene linkage with an apparent crossing over between thalassemia and delta loci. *Clin. Res.* **12**:34.

Nathan, D. G., Lodish, H. F., Kan, Y. W., and Houseman, D. (1971). Beta thalassemia and translation of globin messenger RNA. *Proc. Natl. Acad. Sci.* **68**:2514.

Nienhuis, A. W., Laycock, D. G., and Anderson, W. F. (1971). Translation of rabbit haemoglobin messenger RNA by thalassaemic and non-thalassaemic ribosomes. *Nature New Biol.* **231**:205.

Schwartz, H. C., Spaet, T. H., Zuelzer, W. W., Neel, J. V., Robinson, A. R., and Kaufman, S. F. (1957). Combinations of hemoglobin G, hemoglobin S and thalassemia occurring in one family. *Blood* **12**:238.

Sugden, B., and Sambrook, J. (1970). RNA polymerase from Hela cells. *Cold Spring Harbor Symp. Quant. Biol.* **35**:663.

Thompson, P. E., and English, D. S. (1966). Multiplicity of hemoglobins in the genus *Chironomus*. *Science* **125**:75.

Thompson, P. E., and Patel, G. (1972). Compensatory regulation of two closely related hemoglobin loci in *Chironomus tentans*. *Genetics* **70**:275.

Thompson, P. E., Bleecker, W., and English, D. S. (1968). Molecular size and subunit structure of the hemoglobins of *Chironomus tentans*. *J. Biol. Chem.* **243**:4463.

Tichy, H. (1970). Biochemische und cytogenetische Untersuchungen zur Natur des hamoglobin-Polymorphismus bei *Chironomus tentans* und *Chironomus pallidivittatus*. *Chromosoma* **29**:131.

Travers, A. A. (1971). Control of transcription in bacteria. *Nature New Biol.* **229**:69.

Travers, A. A., and Burgess, R. R. (1969). Cyclic re-use of the RNA polymerase sigma factor. *Nature* **222**:537.

Weaver, R. F., Blatti, S. P., and Rutter, W. J. (1971). Molecular structures of DNA-dependent RNA polymerases (II) from calf thymus and rat liver. *Proc. Natl. Acad. Sci.* **68**:2994.

Zubay, G., Schwartz, D., and Beckwith, J. (1970a). Mechanisms of activation of catabolite-sensitive genes: A positive control system. *Proc. Natl. Acad. Sci.* **66**:104.

Zubay, G., Schwartz, D., and Beckwith, J. (1970b). The mechanism of activation of catabolite-sensitive genes. *Cold Spring Harbor Symp. Quant. Biol.* **35**:433.

Author Citation Index

Archer, R., 305
Adelberg, E. A., 313
Adinolfi, N., 265
Agol, I. J., 50
Ahmed, S. I., 350
Aksoy, M., 257
Allen, D. W., 257
Allen, E., 82
Allen, F. H., Jr., 248, 257
Allen, M., 305, 339
Allison, A. C., 265
Altenburg, E., 52
Ames, B. N., 350
Anderer, F. A., 313
Anderson, E. P., 191
Anderson, R. L., 138
Anderson, W. F., 382
Anfinsen, C. B., 313
Armstrong, E. F., 120
Atwater, J., 257
Axelrod, J., 191

Bach, M. L., 328
Bacher, F., 370
Baglioni, C., 277
Baier, J. C., Jr., 80
Balog, J., 257
Bancroft, W. D., 119
Barabas, J., 277
Barnett, L., 270, 277, 340
Barnicot, N. A., 257
Barratt, R. W., 226, 245
Bartlett, H. H., 121
Basilo, C., 340
Bassett, H. L., 120
Bateson, W., 3, 12, 20, 119
Bauerle, R. H., 350
Bautz, E. K. F., 293, 382
Beadle, G. W., 119, 125, 138, 139, 151, 161, 166, 174, 188, 201, 206, 226
Beale, G. II., 120
Bearn, A. G., 235, 257
Becker, 125

Becker, E., 138
Beckwith, J. R., 334, 383
Beermann, W., 369, 372
Beiboer, J. L., 277
Benzer, S., 209, 226, 230, 277, 281, 293, 304, 327, 328, 334
Berendes, H. D., 369
Berg, P., 334
Bergmann, F. H., 334
Bernfield, M., 334
Beveridge, J. M. R., 156
Bianco, I., 277
Blakeslee, A. F., 120
Blatti, S. P., 382, 383
Bleecker, W., 383
Boedtker, H., 334
Bolen, H. R., 50
Bolle, A., 281, 334
Bond, P. A., 246
Bonner, D. M., 161, 166, 174, 201, 206, 293, 294, 321
Boyd, J. B., 369
Boy de la Tour, E., 334
Boyden, A., 80
Boyer, P. D., 340
Boyer, S. H., 248, 257
Bradley, T. B., 248, 257
Braganca, B., 191
Braunitzer, G., 265, 277, 313, 340
Brenner, S., 245, 270, 277, 281, 334, 340
Bridges, C. B., 50, 51, 52, 138, 139, 265, 270, 277, 369
Bridges, P. N., 369
Brimacombe, R., 334
Britten, R. J., 372
Brody, S., 317
Brown, D. D., 372
Buchanan, J. M., 226
Buchbinder, L., 80
Buettner-Janusch, J., 260
Burgess, R. R., 382, 383
Burkart, A., 50
Burnett, F. M., 80

Burton, R., 191
Butenandt, A., 138
Butler, E. T., 151
Buxton, B. H., 120

Cahn, R. D., 246
Callan, H. G., 369
Campbell, A., 334
Capecchi, M. R., 334
Carlton, B. C., 281, 305, 313, 340
Caroline, D. F., 342
Carter, C. E., 226
Case, M. E., 230, 342, 350
Caspari, E., 10, 138, 139
Castle, W. E., 20, 156
Catcheside, D. G., 206, 230, 245
Cavallini, D., 265
Cepellini, R., 265, 277, 382
Champe, S. P., 277, 293, 327, 328, 334
Chao, F. C., 294, 305, 313, 340
Chase, H. B., 98
Chevais, S., 138
Chevalley, R., 334
Chieffi, G., 265
Clancy, C. W., 138
Clegg, J. D., 382
Coddington, A., 246
Cohen, G., 206
Cohen, L. H., 226
Cohen, M., 201
Cole, L. J., 80
Cole, R. D., 265
Colowick, S. P., 321
Connell, G. E., 270, 277
Coonradt, V. L., 226
Cooper, G., 81
Cooper, S., 334
Coutagne, 20
Craig, L. C., 265
Crampe, 20
Crawford, I. P., 206, 235, 239, 305, 313, 340
Creighton, T. E., 213
Crick, F. H. C., 239, 270, 277, 340, 369
Crocker, C., 305
Cuénot, L., 20
Cullis, A. F., 265
Cutolo, E., 191

Dagley, S., 188
Daneholt, B., 352, 369
Danneel, R., 98
Darbishire, 20
David, I. B., 372

Davidson, E. H., 372
Davies, D. R., 265
Davies, J., 334
Davis, B. D., 174
Davis, B. J., 246
Davis, D. B., 313
Davis, R. H., 143, 350
Dawes, E. A., 188
Delage, Y., 10
Delbrück, M., 10, 294
de Marco, C., 265
Demerec, M., 350
DeMoss, J. A., 206, 293, 321, 342
Denhardt, G. H., 334
Dennert, G., 350
De Vries, H., 2, 21
de Winton, D., 120
deZeeuw, J. R., 226, 230
Diamond, J. M., 277
Diamond, L. K., 277
Dickerson, R. E., 265
Dickson, E., 369
Dieckmann, M., 334
Dintzis, H. M., 195, 277
Dixon, G. H., 270, 277
Dobzhansky, T., 50, 53
Dodge, B. O., 151
Doermann, A. H., 174
Doy, C. H., 213
Drapeau, G. R., 281
Dreyer, W. J., 304, 313
Dubinin, N. P., 51, 53
Dunn, J. J., 382
Dunn, L. C., 12, 98

Eaton, O. N., 98
Echols, H., 327
Edelman, G. M., 143
Edgar, R. S., 334
Edman, P., 209
Edsall, J. T., 265
Edström, J.-E., 352, 369
Efroimson, W. P., 51
Efron, M. L., 277
Egyházi, E., 352, 369
Ehrensväld, G., 350
Einsele, W., 98
Eirich, F., 209
Ellinger, A., 166
Emerson, R. A., 51
Emerson, S., 174, 188
Engelhardt, D. L., 334
Enger, M. D., 334
English, D. S., 382, 383

Ephrussi, B., 119, 125, 138, 139, 161
Epstein, R. H., 334
Everest, A. E., 121
Eyster, W. H., 120

Fahmy, M. J., 369
Fahmy, O. G., 369
Falk, R., 370
Farabee, 20
Farr, A. L., 201
Fein, A., 350
Ferguson, L. C., 83
Ferry, L., 51
Fessas, P., 277
Fewster, J. A., 350
Fincham, J. R. S., 174, 206, 213, 230, 245, 246, 294
Fink, G. R., 350
Fisher, R. A., 51
Fleischmann, P., 156
Fölling, A., 124
Ford, E. B., 51
Foster, J. W., 188
Fraenkel-Conrat, H., 209, 304, 313
Franklin, E. C., 235
Franklin, N. C., 209
Fraps, G. S., 120
Freese, E., 293, 294
Fries, N., 151
Frisch, K. von, 120
Fudenberg, H. H., 57
Fürth, von, 20

Gabritschevsky, E., 369
Gall, J. G., 369
Gans, M., 369
Garen, A., 246, 317, 327, 328, 334, 340
Garen, S., 327
Garlick, J. P., 257
Garnjobst, L., 226, 350
Garrick, M., 321
Garrod, A. E., 10, 12, 161
Gehring-Muller, R., 277
Geissman, J. A., 87
Geissman, T. A., 87
Georgiev, G. P., 372
Gerald, P. S., 277
Gershenson, S. M., 51
Gessard, 20
Gesteland, R. F., 334
Gilbert, W., 334
Giles, N. H., 188, 206, 226, 230, 235, 342, 350

Gilmore, R. A., 317
Gish, D. T., 313
Gisvold, O., 120
Glass, H. B., 51, 321
Glaubach, S., 156
Glick, D., 156
Goldberg, L., 191
Goldschmidt, R., 51, 120, 151
Golly, J. A., 143
Golomb, S. W., 294
Goodenough, U. W., 143
Goodwin, T. W., 340
Gorini, L., 317, 334
Gots, J., 321
Gratzer, W. B., 265
Graubard, M. A., 51
Green, M. M., 370
Greenstein, J. P., 313
Grell, E. H., 370
Griffith, J. S., 293
Gross, S. R., 350, 372
Grossbach, U., 370
Grossman, W. I., 166
Guaita, von, 21
Guest, J. R., 281
Guidotti, G., 265, 277
Guirard, B. M., 188
Gunsalus, C. I., 201
Gussin, G. N., 334
Gutter, F. J., 235

Haacke, von, J. W., 21
Haagen Smit, A. J., 139, 166
Haddox, C. H., 188, 226
Hagiwara, T., 120
Haldane, J. B. S., 10, 12, 28, 51, 120, 151, 209, 265
Hamkalo, B. A., 370
Hamlett, G. W. D., 51
Hanson, F. B., 51
Harder, R., 122
Harding, R. W., 342
Harnly, M. H., 138, 139
Harris, H., 235
Harris, J. I., 209
Harris, R. S., 209
Hart, R. G., 265
Hartmann, A., 191
Hartman, P. E., 317
Hartwell, L. H., 143
Haschek, L., 121
Havinga, E., 209
Hawthorne, D. C., 334
Heidenthal, G., 98

Helinski, D. R., 277, 281, 294, 304, 305, 313, 328, 334, 340
Henning, U., 281, 294, 305, 313, 338, 340, 350
Hertel, R., 350
Heyes, F., 51
Hibbert, 122
Hicks, R. A., 80
Hildemann, W. H., 57
Hill, R. G., 277
Hill, R. J., 265
Hill, R. L., 257, 277
Hillschmann, N., 277
Hilse, K., 277
Hirs, C. H. W., 305, 313, 340
Hirschfeld, L., 57
Hobman, G., 277
Hochman, B., 352, 370
Hogden, C. G., 201
Holzel, A., 191
Horiuchi, S., 304
Horn, V., 281
Horowitz, N. H., 161, 174, 188, 226, 321
Houlahan, M. B., 188, 226
Houseman, D., 382
Huisman, J. H. J., 260
Huisman, T. H. J., 265, 382
Hunt, J. A., 258, 265
Hussey, C., 382
Hutchinson, W. D., 248, 257

Ingles, C. J., 382
Ingram, V. M., 206, 209, 257, 258, 265, 270, 277, 294, 304
Inman, O. L., 120
Irwin, M. R., 80, 83
Israels, A. L. M., 265
Itano, H. A., 209, 248, 253, 257
Ives, P. T., 53

Jacob, F., 206, 327, 350
Jervis, G. A., 124
Jinks, J. L., 294, 340
Johnson, I. J., 120
Jollos, V., 51
Jones, O. W., 340
Jones, R. T., 258, 265
Jonxis, J. H. P., 277
Jordan, E., 206, 321
Jordan, H. E., 84
Jorgensen, E. C., 87
Judd, B. H., 352, 370

Kaars-Sijpesteijn, J. A., 277

Kaesberg, P., 334
Kalckar, H. M., 191
Kan, Y. W., 382
Kaplan, N. O., 246, 321
Kaplan, S., 334
Karaklis, A., 277
Karrer, 100
Karstens, W. K., 120
Kataja, E., 334
Katz, A. M., 313
Kauffman, D., 270
Kaufman, S. F., 257, 382
Kaufman, T. C., 370
Keeble, F., 120
Keeler, C. E., 156
Kellenberger, E., 334
Keller, O., 166
Kendrew, J. C., 265
Kenney, F. T., 372
Keyl, H. G., 370
Khouvine, Y., 125, 138, 139
Kikkawa, H., 139
Klebs, G., 120
Knight, C. A., 313
Knox, W. E., 166
Kohn, J., 313
Komrower, G. M., 191
Konigsberg, W., 277, 334
Koshland, D. E., 313
Kossikov, K. V., 270
Kotake, 125
Kröning, F., 98
Krumwiede, C., 81
Kühn, A., 139
Kühn, F., 125
Kuilman, L. W., 120
Kunkel, H. G., 257, 265, 277
Kurek, L. I., 206, 321

Lacy, A. M., 206, 321
Laird, C. D., 369, 370
Lambert, B., 352, 369
Landsteiner, K., 57, 81
Lata, M., 246
Laughnan, J. R., 270
Laustsen, O., 151
Law, L. W., 138
Lawrence, W. J. C., 120, 151, 161
Laycock, D. G., 382
League, B. B., 52
Leder, P., 334
Lederberg, E. Z., 188
Lederberg, J., 174, 313
Lee, C. S., 370

Lehmann, H., 265
Lefevre, G., Jr., 352, 370
Lein, J., 174, 188, 201, 294
Leloir, L. F., 191
Lengyel, P., 340
Lennox, E. S., 294, 305, 313, 340
Lepkovsky, S., 166
Lerner, P., 321
Levin, J. G., 350
Levine, P., 81
Levine, R. P., 143
Levinthal, C., 246, 327, 328, 340
Levy, A. L., 209
Levy, J., 156
Lewis, E. B., 265, 270, 370
Li, C. H., 270
Liebold, B., 265
Lielausis, A., 334
Lifschytz, E., 370
Lillie, F. R., 82
Lindegren, C. C., 151
Lindell, T. J., 382
Lindsley, D. L., 370
Lipmann, F., 161
Liquori, A. M., 209
Little, C. C., 80
Lodish, H. F., 382
Loper, J. C., 246
Losick, R., 382
Lowry, C. H., 201
Lucas, C. C., 156
Lynen, F., 350

McCarthy, B. J., 370
McClintock, B., 51
McElroy, W. D., 188
Mäkelä, P. H., 143
Maling, B. D., 277, 294, 305, 313, 328, 334, 340
Mallison, H., 121
Mamoli, 125
Mangelsdorf, P. C., 120
Manney, T. R., 334
Margolin, P., 350
Markert, C. L., 201, 246
Marquart, L., 120
Marrack, J. R., 81
Marryat, D. C. E., 120
Martin, E. M., 334
Martin, N., 257
Martin, R. G., 340, 350
Matsuda, G., 258
Matsumura, S., 156
Matthaei, J. H., 334, 340

Maxwell, E., 191
Maxwell, J., 138
Mehlquist, G. A. L., 87
Mendel, G., 156
Merriam, J. R., 370
Metzenberg, R. L., 372
Meyering, C. A., 265
Meyers, J. W., 313
Michel, E., 156
Michl, H., 209
Miller, B., 304
Miller, E. S., 120
Miller, P., Jr., 81
Milner, R. T., 121
Misra, D. N., 370
Mitchell, H. K., 143, 156, 161, 166, 174, 188, 201, 226, 294
Mitchell, M. B., 188, 226
Möbius, K., 120
Mohler, W., 206, 321
Mohr, O. L., 51
Møller, F., 246
Monod, J., 350
Moore, A. R., 120
Moore, M. M., 382
Moore, S., 265, 305, 313
Morgan, L. V., 139
Morgan, T. H., 51, 52, 139
Morris, P. W., 382
Morrison, G. A., 188
Mortimer, R. K., 334
Morton R. A., 340
Mott-Smith, L. M., 52
Mueller, J. H., 174
Muirhead, H., 265
Mukai, F., 304
Muller, C. J., 277
Muller, H. J., 28, 52, 53, 98, 120, 270, 370
Muller, R., 265
Muller-Eberhard, U., 265, 277
Munch-Petersen, A., 191
Munkres, K. D., 350
Murayama, M., 257

Na-Nakorn, S., 382
Nasarenko, I. I., 52
Nason, A., 321
Nathan, D. G., 382
Naughton, M. A., 195, 209, 277
Neeb, H., 277
Neel, J. V., 195, 209, 257, 382
Nelson, N. J., 209, 226, 230, 235
Neurath, H., 270, 305, 313
Newmeyer, D., 201, 226

Nienhuis, A. W., 382
Nirenberg, M. W., 334, 340
North, A. C. T., 265
Notani, G. W., 334
Nuttall, G. H. F., 81
Nye, J. F., 166

Ochoa, S., 340
Ofengand, E. J., 334
Offermann, C. A., 52
Ogura, S., 53
Ohno, S., 260
Oliver, C. P., 52
O'Neal, C., 334
Onslow, M. W., 120, 151, 156
Oparin, A. I., 161
Orgel, L. E., 293
Ornstein, L., 246
Overton, A., 245
Owen, R. D., 201

Painter, T. S., 52, 370
Partridge, C. W. H., 206, 209, 226, 230, 235, 246, 350
Patel, G., 383
Pateman, J. A., 230, 246
Patterson, J. T., 52, 53
Pauling, L., 209, 248, 257, 265, 277
Pearson, H. A., 382
Pease, M. S., 156
Perkins, D. D., 226
Perutz, M. F., 209, 265
Peterson, E. A., 235
Phillips, D. C., 265
Philip-Smith, E., 120
Phipps, I. F., 120
Pittenger, T. H., 226
Plagge, E., 125, 139
Plough, H. H., 53
Plunkett, C. R., 28, 53
Pommerehne, F., 166
Preer, J. R., Jr., 328
Preer, L. B., 328
Price, J. R., 120, 122, 151, 161
Pritchard, R. H., 209
Prokofyeva-Belgovskaya, A. A., 270, 370
Provost, D. J., 81
Punnett, R. C., 119
Punt, K., 382

Rachmeler, M., 206, 294
Randall, R. J., 201
Ranney, H. M., 257
Raspail, 21

Rayle, R. E., 370
Reichert, E. T., 81
Rhinesmith, H. S., 209, 248, 257
Richards, F. M., 235, 350
Riggs, A., 265
Ringborg, U., 352, 369
Robbins, W. J., 151
Roberts, D. B., 246
Robertson, A., 121
Robinson, A. R., 257, 382
Robinson, E., 248, 257
Robinson, G. M., 100, 120, 121, 122
Robinson, H. W., 201
Robinson, R., 120, 121, 122
Roboz, E., 166
Robson, E. B., 235
Rogers, C. H., 120
Rokitzky, P. T., 53
Rörig, 21
Rosebrough, N. J., 201
Rossi-Fanelli, A., 265
Rossmann, M. G., 265
Roth, J. R., 317
Rothman, F., 304, 327, 328, 340
Rottman, F., 334
Rudkin, G. T., 370
Rudloff, V., 265, 277, 313, 340
Russell, E. S., 98
Russell, W. L., 98
Rutter, W. J., 382, 383
Rutzler, J. E., 119

Safir, S. R., 53
St. Lawrence, P., 294, 327
Sambrook, J., 382
Sanders, D., 82
Sando, C. E., 121
Sanger, F., 209
Sanwal, B. D., 246
Sarabhi, A. S., 10, 281
Sasaki, K., 81
Saunders, E. R., 119, 121
Schaad, J. D. G., 382
Schalet, A., 370
Schaumann, K., 98
Scheinberg, I. H., 209
Scheraga, H. A., 340
Schlesinger, M. J., 246
Schmid, L., 121
Schneider, H., 20
Schnek, A. G., 277
Schoenheimer, R., 161
Schramm, G., 313

Schroeder, W. A., 209, 248, 257, 258, 260, 265, 277
Schultz, J., 52, 53
Schultz, W., 98
Schwartz, D., 383
Schwartz, H. C., 257, 277, 382
Schwartz, I. R., 257
Schwartz, V., 191
Schwert, G. W., 305, 313
Scott, J. P., 334
Scott-Moncrieff, R., 120, 121, 151
Seale, T., 350
Searle, G. O., 121
Sebens, T., 265
Seebald, A., 121
Seevers, C. H., 98
Serebrovsky, A. S., 53, 270
Shannon, M. P., 370
Shapiro, N. T., 51
Shelton, E. F., 260
Shelton, J. B., 260
Shelton, J. R., 260
Shen, M. W., 370
Sherman, F., 143, 317
Sherman, M. S., 121
Shibata, K., 121
Shipp, W. S., 350
Shooter, E. M., 257
Shore, V. C., 265
Shorenstein, R. G., 382
Siddiqi, O., 246, 328, 334
Sidoroff, B. N., 51
Signer, E. R., 328, 334
Silvestroni, E., 277
Singer, K., 370
Singer, S. J., 209, 235, 257
Siniscalco, M., 235, 265
Sizer, I. W., 328
Skinner, E. R., 257
Smith, E. E. B., 191
Smith, E. W., 257
Smithies, O., 270, 277
Sober, H. A., 235
Sonenshein, A. L., 382
Sorsa, V., 352
Spackman, D. H., 265
Spaet, T. H., 257, 382
Spencer, D. A., 98
Speyer, J. F., 340
Spitzer, J. L., 209
Sprinson, D. M., 350
Srb, A. M., 161, 226, 230
Stadler, J., 321
Stadler, L. J., 53

Stamatoyannopoulos, G., 195, 277
Stanley, W. M., 313
Stanton, E., 51
Stein, W. H., 265, 305, 313
Steinberg, C. M., 334
Stent, G., 334
Stephens, S. G., 265
Stephenson, M. L., 334
Stern, C., 53
Stewart, J. W., 143, 317
Stocker, B. A. D., 143
Stockes, J. L., 188
Stone, W. S., 52
Stormont, C., 83
Strandberg, B. E., 265
Strauss, B. S., 188, 321
Streisinger, G., 209, 304
Stretton, A. O. W., 258, 265, 270, 277, 334
Strominger, J. L., 191
Sturgess, V. C., 120
Sturtevant, A. H., 10, 50, 51, 53, 139, 151, 270
Sugden, B., 382
Suskind, S. R., 206, 294, 321
Susman, M., 334
Suyama, Y., 293
Swenson, R. T., 257, 277

Tammes, T., 121
Tanford, C., 340
Tatum, E. L., 125, 138, 139, 151, 161, 166, 174, 201, 206, 350
Thimann, K., 139
Thomas, C. A., Jr., 370
Thompson, D. H., 53
Thompson, E. O. P., 209
Thompson, P. E., 382, 383
Tichy, H., 383
Timoféeff-Ressovsky, N. W., 53
Tiselius, A., 201
Tocantins, L. M., 257
Torbert, J. V., 257
Torriani, A. M., 201, 327
Travers, A. A., 382, 383
Trimple, H. C., 156
Troland, L. T., 151
Trupin, J., 334
Tschermak, E. v., 121
Tsugita, A., 304, 313
Tuppy, H., 209

Uhlig, H., 313
Umbarger, H. E., 174
Umbreit, W. W., 201

Author Citation Index

Van Atta, E. W., 53
Van der Scheer, J., 81
Vinograd, J. R., 248, 257, 258

Wagner, R. P., 143, 188, 226, 342
Walker, N. F., 270
Wallenius, G., 265
Warner, N., 57
Waschkau, A., 121
Wasi, P., 382
Watts-Tobin, R. J., 270, 277, 340
Weatherall, D. J., 382
Weaver, R. F., 382, 383
Weber, E., 313
Wegman, J., 342
Weidel, W., 138
Weigert, M. G., 334
Weinberg, F., 382
Welch, L. R., 294
Wells, H. G., 81
Wells, I. C., 209
Wheldale, M., 120, 121
White, F. H., 313
Whitfeld, P., 226
Wiener, A. S., 57, 81
Wilkins, M. D., 370
Will, G., 265
Williams, R. D., 121
Willstätter, R., 121
Wilson, A. C., 246
Wilson, E. B., 2, 4, 10
Winge, O., 151

Winitz, M., 313
Wit, F., 121
Wittman-Liebold, B., 277
Wittmann, H. G., 313
Wolf, J., 265, 277
Wollman, E. L., 327
Wood, W. A., 201
Woodward, C. R., 188
Woodward, D. O., 206, 235, 246
Woodward, V. W., 226, 230
Wright, C. I., 156
Wright, S., 10, 53, 87, 98, 151, 156, 161
Wust, C., 321
Wyckoff, M. M., 235

Yanofsky, C., 9, 201, 206, 213, 235, 277, 281, 293, 294, 304, 305, 313, 317, 321, 327, 328, 334, 338, 339, 340
Yegian, C., 334
Young, J., 313
Young, M. W., 352
Yura, T., 235, 321

Zamecnik, P. C., 334
Zechmeister, L., 121
Zeleny, C., 53
Zinder, N., 174, 313, 334
Zollinger, E. H., 121
Zubay, G., 334, 383
Zuckerkandl, E., 265, 277
Zuelzer, W. W., 257, 382

Subject Index

Acetylcholine, 158
Adenine biosynthesis, 214
Adenosine monophosphate (AMP), 232
Adenosine monophosphate succinate (AMP-S), 232
Adenylosuccinase, 214, 227, 231
β-Alanine, 177
Alkaline phosphatase, 322, 335
Allelic differences, 29, 53, 90, 107
p-Aminobenzoic acid, 148
Amylase, 152
Anthocyanidins, 100
Anthocyanins, 99, 110
Anthoxanthins, 99, 110
    apigenin, 102
    kaempferol, 102
    luteolin, 102
    quercetin, 102
Anthranilic acid, 164
Antimorphic mutations, 41, 90
Antirrhinum majus, 104
"A" polypeptide, of *E. coli* tryptophan synthetase, 282, 295, 306, 335
Arginine, 158, 168
Atropinesterase, 152

Bacterial synthesis, of v^+ substance, 134
Biogenesis, of anthocyanins and anthoxanthins, 110
Biotin, 147
Bridge's forked deficiency, 31, 128
Bridge's plexate deficiency, 31
Burkart's Blond translocation, 31

Calliphora erythrocephala, 126, 132
Callistephus hortensis, 107
Campanula trachelium, 106
Carbomoxyhemoglobin, 250
Carotinoids, 99
Cell sap, 101
Chalkone, butein, 121
Cheiranthus cheiri, 105
Chironomus, 373
Chromosomal duplications, 276

Chromosomal rearrangements, 266
Chymotrypsin, 298, 306, 336
Collinearity, between genetic material and amino acid sequence, 280ff.
Complementation, 214, 227, 231, 236, 356
Conidia, 227
Copigmentation, 101
Cornyene—*Bacterium mediolanum*, 126
Creatine, 158
Crystallography, 58
Cyanidin, 100
Cyanidin 3-glycoside, 107
Cytological mapping, 358

Dahlia variabilis, 107
Delphinidin, 100
Deoxyribonucleic acid (DNA), 353
Diazo reaction, 136
Diethylaminoethylcellulose (DEAE), 238
3, 5-Dimonosides, 108
Dominance relationships, 114
Dopa oxidase, 89
Dosage differences of genes, effects of, 35
Double mutants, 285
Dove, 59
Drosophila melanogaster, 353
    eosin-eyed, 30
    mutant, 28, 126, 128
Drosophila simulans, 36

Enzyme complexes, 343ff.
Ephestia kuhniella, 126, 129
Erythrocytes, 83, 189, 250
Escherichia coli, 167, 282, 295, 318, 322, 329, 335, 380
Evolution
    of biochemical synthesis, 157
    of hemoglobins, 261
Eye color "hormones," in insects, 126, 128
Eye pigment, in insects, 126, 128

Fagopyrum esculentum, 106, 113
Flavones, 102

Subject Index

Flavonols, 102
Flower color variation, 99
Flower pigments, 99
5-Fluorouracil, 325

Galactosemia, 189
Galactowaldenase, 189
Gene-protein relationship, 1, 174, 194, 196, 202, 207
Genes
  clusters, 343
  complementation, 212ff.
  duplications, 261, 266
  evolution, 261, 266
  frequencies, 158
  functions, 1, 29
  mutations, 29, 207
  pigment production, 17, 25, 104
  rearrangements, 29
Genetic deletions, 271
Geranium psilostemon, 112
Glutamate dehydrogenase (GDH), 227, 236
Glutamic acid, 207
Glycine, 158
Guinea pig, 23, 88
Gynandromorphs, 128

Haptoglobin genes, 266
Hemoglobin, 207, 249, 261, 271, 373
Heterocaryons, 214, 227, 231
Heterophile antigen, 78
Hydrangea hortensis, 113
Hydroxyanthranilic acid, 162
Hydroxyflavylium salts, 100
Hyperploidy, 29
Hypomorphic mutations, 40, 90
  apricot, 32
  eosin, 31, 128
  forked, 33, 128
  scute, 33, 128
Hypoploidy, 29
Hypoxanthine, 214

Immunogenetics, 58
  bovine twins, 82
  F_1 male pearlnecks x ringdove females, 59
  pearlneck *(Spilopelia chinensis)*, 59
  ringdove *(Sheptopelia risoria)*, 59
Indispensable function, of genes, 169
Indole, 164, 202
Indole glycerol phosphate (InGP), 202, 306, 318
Inheritance, of color variations, 104

Intracellular physiology, 89
Iso-quercitrin, 109

Kynurenine, 126, 135, 162

Lamprey hemoglobin, 261
Lathyrus odoratus, 105, 108
Linum usitatissimum, 105
Lymph transfusion, 132

Macroconidia, 214
Malvidin, 100
Malvidin 3-galactoside, 108
Matthiola incana, 105
Melanic differentiation, 92
Melanic pigmentation, 92
Melanins, 88
Mendel's law, examples from mice, 22
Methionine, 158, 298
Mirabilis jalapa, 106
Mohr's notch-8 deficiency, 30
Monoacetylmorphinease, 154
3-Monoglucoside, 109
Multiple allelomorphs, 47

Neomorphic mutations, 43
  Bar eye, 44
  Blond, 44
  Hairy wing, 43
Neurospora crassa, 144, 157, 162, 167, 177, 196, 202, 214, 227, 231, 236, 285, 318, 343
Neurospora sitophila, 146
Nicotinic acid, 162
Nigella, 163
Nonsense suppressor, 329
Nucleoprotein, 160
Nucleotide transferase, 189

One gene–one band hypothesis, 353
One gene–one enzyme hypothesis, 12, 17, 25. 167
One gene–one polypeptide hypothesis, 249
a-Oxytryptophane, 135

Pantothenate, 174
Pantothenicless mutants, 177
Pantoyl lactone, 177
Papaver rhoeas, 105
Partial revertant protein, 336
Pelargonidin, 100
3-Pentose glycoside, 108
Peonidin, 100

Pelargonidin 3-glycoside, 107
Petunidin, 101
Pharbitis nil, 105
Phaseolus multiflorus, 104
Phenylalanine, 310
Pigmentation
 flowers, 99
 guinea pig, 88
 mice, 13, 22
Polarity mutants, 343
Polytene chromosomes, 353
Precipitin reaction, 58
Primula acaulis, 104
Primula sinensis, 105
Protein alterations, by mutation, 282ff.
Protein complexes, 231
Pyridoxal phosphate, 202
Pyridoxine, 146

Rabbit, 89, 152
Regulation, of gene action, 373
Ribonucleic acid (RNA), 1, 329

Second site reversions, 335
Serine, 202
Serine-sRNA, 329
Sickle cell hemoglobin S, 207, 250
Species-specific substances, 58
Streptocarpus, 108
Suppression, of mutations, 290, 316ff.

Thalassemia, 373
Thiamine, 147
Thiazole, 147
Transcription, 378
Translocation, 263
Transplantation, in *Drosophila*, 129
Trypsin, 375
Tryptic peptides, 298, 306, 336
Tryptophan, 126, 133, 162
Tryptophan auxotrophs, 282, 306
Tryptophan synthetase, 196, 202, 282, 295, 318, 335
Twins, bovine, 82
Tyrosine, 177, 273, 335

Uricase, 152

Vascular anastomoses, 82
Verbena hybrida, 108
Vermilion, in mosaics, 128
Vertebrate hemoglobins, 261
Vitis heterophylla, 113

Xanthic pigment, 92
Xanthophyllase, 152
Xanthophylls, 103
Xanthurenic acid, 162

Zea mays, 105